NATO ASI Series

Advanced Science Institutes Series

A series presenting the results of activities sponsored by the NATO Science Committee, which aims at the dissemination of advanced scientific and technological knowledge, with a view to strengthening links between scientific communities.

The Series is published by an international board of publishers in conjunction with the NATO Scientific Affairs Division

A Life Sciences	Plenum Publishing Corporation
B Physics	London and New York
C Mathematical and Physical Sciences	Kluwer Academic Publishers Dordrecht, Boston and London
D Behavioural and Social Sciences	
E Applied Sciences	
F Computer and Systems Sciences	Springer-Verlag Berlin Heidelberg New York
G Ecological Sciences	London Paris Tokyo
H Cell Biology	

The ASI Series Books Published as a Result of
Activities of the Special Programme on
SENSORY SYSTEMS FOR ROBOTIC CONTROL

This book contains the proceedings of a NATO Advanced Research Workshop held within the activities of the NATO Special Programme on Sensory Systems for Robotic Control, running from 1983 to 1988 under the auspices of the NATO Science Committee.

The books published so far as a result of the activities of the Special Programme are:

Vol. F25: Pyramidal Systems for Computer Vision. Edited by V. Cantoni and S. Levialdi. 1986.

Vol. F29: Languages for Sensor-Based Control in Robotics. Edited by U. Rembold and K. Hörmann. 1987.

Vol. F33: Machine Intelligence and Knowledge Engineering for Robotic Applications. Edited by A.K.C. Wong and A. Pugh. 1987.

Vol. F42: Real-Time Object Measurement and Classification. Edited by A.K. Jain. 1988.

Vol. F43: Sensors and Sensory Systems for Advanced Robots. Edited by P. Dario. 1988.

Series F: Computer and Systems Sciences Vol. 43

Sensors and Sensory Systems for Advanced Robots

Edited by

Paolo Dario

Scuola Superiore di Studi Universitari
e di Perfezionamento S. Anna
Via Carducci 40
I-56100 Pisa, Italy

and

Centro "E. Piaggio"
University of Pisa
Via Diotisalvi, 2
I-56100 Pisa, Italy

Springer-Verlag
Berlin Heidelberg GmbH

Proceedings of the NATO Advanced Research Workshop on Sensors and Sensory Systems for Advanced Robots, held in Maratea, Italy, April 28 – May 3, 1986.

ISBN 978-3-642-83412-7   ISBN 978-3-642-83410-3 (eBook)
DOI 10.1007/978-3-642-83410-3

Library of Congress Cataloging-in-Publication Data. NATO Advanced Research Workshop on Sensors and Sensory Systems for Advanced Robots (1986: Maratea, Italy) Sensors and sensory systems for advanced robots/edited by Paolo Dario. p. cm.—(NATO ASI series. Series F. Computer and systems sciences; vol. 43) "Proceedings of the NATO Advanced Research Workshop on Sensors and Sensory Systems for Advanced Robots, held in Maratea, Italy, April 28 – May 3, 1986"—"Published in cooperation with NATO Scientific Affairs Division."
1. Robotics—Congresses. 2. Detectors—Congresses. I. Dario, Paolo, 1951-. II. North Atlantic Treaty Organization. Scientific Affairs Division. III. Title. IV. Series: NATO ASI series. Series F, Computer and system sciences; vol. 43. TJ210.3.N376 1986 629.8'92—dc 19

© Springer-Verlag Berlin Heidelberg 1988

PREFACE

This volume contains papers presented at the NATO Advanced Research Workshop (ARW) on "Sensors and Sensory Systems for Advanced Robots", which was held in Maratea, Italy, during the week April 28 - May 3, 1986. Participants in the ARW, who came from eleven NATO and two non-NATO countries, represented an international assortment of distinguished research centers in industry, government and academia.

Purpose of the Workshop was to review the state of the art of sensing for advanced robots, to discuss basic concepts and new ideas on the use of sensors for robot control and to provide recommendations for future research in this area.

There is an almost unanimous consensus among investigators in the field of robotics that the addition of sensory capabilities represents the "natural" evolution of present industrial robots, as well as the necessary premise to the development of advanced robots for nonindustrial applications. However, a number of conceptual and technical problems still challenge the practical implementation and widespread application of sensor-based robot control techniques. Crucial among those problems is the availability of adequate sensors.

For these reasons, the analysis of transducer technologies potentially usable for robotics applications constituted the core of the ARW. Problems related to the development of new sensors for robots, especially those intended for operation in unstructured environments, were addressed with particular emphasis.

As the development of artificial sensory systems requires multidisciplinary efforts, the ARW gathered together scientists belonging to different cultural areas, including neurophysiology, psychology, biophysics, materials science, mechanical, chemical, electronics and biomedical engineering, computer science and automatic control. The often difficult interaction between sensor scientists and robotics scientists was pointed out. Efforts aimed at facilitating a closer collaboration between sensor designers and users were recommended.

Other aspects of artificial sensory systems design were discussed also during lectures, working groups and round tables within the ARW. In particular, the merit of a bionic versus a purely "mechanistic" approach to the design of sensory systems and sensor-based robot control architectures was debated. In this context, the different cultural background of participants stimulated many observations of both practical and speculative value.

The book is organized in five sections. The four papers contained in **Section 1** are concerned with the biological model, and illustrate the solutions adopted by the human sensory systems in order to manage the interactions with the environment.

The three papers included in Section 2 introduce the general problem of providing a robot with sensory capabilities. In particular, these papers are concerned with vision, proximity and tactile sensing.

The four papers of **Section 3** illustrate the principles of operation of some classes of sensors for robotics applications

Section 4 includes eight papers which describe a number of different sensors already used or potentially usable for nonconventional visual, proximity or environment sensing.

The six papers contained in **Section 5** deal with the high level integration of different sensory systems in robots intended for operating predominantly in unstructured environments.

The **Appendix** includes a few short papers on specific applications, which were presented within the working groups, and the list of participants.

Although it is almost impossible to render in a written text the "flavour" of the many formal and informal discussions that animated the Workshop, the variety of experiences and viewpoints expressed by the participants is reflected, in its essence, by the papers included in this book.

In closing, I wish to acknowledge the support of all the speakers in contributing excellent presentations, and of all the participants in ensuring stimulating discussions as well as creating a very friendly atmosphere.

I would also like to express my appreciation of the NATO Scientific Affairs Division and of the Panel on "Sensory Systems for Robotic Control" who recommended and provided most of the financial support to the Workshop. In particular I wish to acknowledge the late Mario Di Lullo and Jean Vertut, who were invaluable in providing advice and encouragement.

The other members of the Scientific Committee of the ARW, Professors Ruzena Bajcsy, Michael Brady, Bernard Espiau and Marc Raibert assisted me with very useful comments and suggestions.

Furthermore, the grants from IBM Italia S.p.A. and from Siemens AG, which provided substantial financial support for some participants, are gratefully acknowledged. The assistance of the "Azienda Autonoma Soggiorno e Turismo" of Maratea and of the "Banco di Lucania" allowed the participants to find some (short) breaks in the busy schedule of the Workshop and to enjoy the splendid natural environment of Maratea.

I am also very indebted to Massimo Bergamasco, Claudio Domenici and Antonino Fiorillo, who collaborated with me in the organization of the ARW, as well as to Daniela Fantozzi and Iolanda Giusti for their invaluable secretarial assistance.

Finally, I wish to recognize the efforts of Mr. Guzzardi and his staff of the Hotel Villa del Mare in enhancing the warm atmosphere of the Workshop.

P. Dario

Pisa, January 1988

TABLE OF CONTENTS

XI

APPENDIX

SECTION 1

THE BIOLOGICAL MODEL:
THE HUMAN BODY AND ITS SENSORY SYSTEMS

The Central Nervous System
as a Low and High Level Control System

James S. Albus
Chief, Robot Systems Division
Center for Manufacturing Engineering
National Bureau of Standards

The division of the central nervous system into high and low level control systems has long been recognized. How these two systems are interconnected and how they influence each other is one of the great mysteries of modern science. Recent attempts to produce intelligent behavior in robots and computer integrated manufacturing systems have produced insights as to how high level goals can be decomposed into low level actions, and how knowledge about the environment can be acquired, stored, and accessed by task decomposition processes to produce sensory-interactive goal directed behavior.

Research on computer brain models has also shown how networks of neuron-like elements can learn patterns and motor skills and generalize from one task to another. This paper proposes a model of how the high level understanding, evaluating, goal selection, planning and reasoning functions commonly associated with the mind are tied into the lower level sensing, filtering, recognizing, task execution, and servo control functions that are commonly associated with the mechanisms of the body.

The Brain as a Hierarchy

Specific neurological theories that assume the brain (the seat of both the mind and the motor system) to be hierarchically structured are well over a hundred years old (Jackson, 1931). In the past three decades, a large number of neurophysiological experiments have shown that the brain processes sensory information through a number of distinct hierarchical levels. It is now well established that specific functions are performed by specific neuronal structures at a variety of hierarchical levels in a number of neuronal pathways. For example, the retina of the eye is known to perform edge enhancement and spatial filtering operations on visual images produced by the optics of the eye. The lateral geniculate performs stereo matching and gating operations. The visual cortex (area 17) detects edges and corners, and measures the position and orientation of such features in the visual field. Cell clusters in the visual association areas recognize three dimensional objects, and areas in the frontal cortex perform spatial reasoning operations.

Hierarchically structured computing modules have been observed and studied not only in the ascending sensory systems but in the descending motor control pathways as well. It has been demonstrated (Evarts and Tanji 1974) that neurons in the cerebellum, thalamus, and motor cortex alter their firing rates at various intervals prior to either movement or feedback, as motor commands propagate down the task decomposition hierarchy.

NATO ASI Series, Vol. F43
Sensors and Sensory Systems
for Advanced Robots
Edited by P. Dario
Springer-Verlag Berlin Heidelberg 1988

There is much evidence that the descending pathway influences, and is influenced by, the ascending pathways. Downward flowing motor commands are known to be capable of significantly modifying the interpretation of sensory data. At many different levels, internally generated expectations are matched against experientially observed objects, relationships, and temporal events. Both correlations and differences between expectations and observations are computed.

Similarly, ascending sensory input significantly modifies downward flowing motor control signals by providing feedback at a variety of hierarchical levels. Thus, the brain does not have simply a vision system, but a vision-attention system. There is not simply a speech generating system, but a hearing-speech system. The brain is, at least at the lower levels, a strongly cross-coupled hierarchical sensory-motor system.

The Brain as a Control System

It is important to realize that the principal function of the brain is not to reason and plan, or even to sense and recognize, but to select goals and control behavior. Reasoning and planning are evolutionarily recent properties of only a tiny fraction of the brains that have ever existed. But all brains control behavior. Effective control of behavior is crucial to survival and reproduction. Behavior either succeeds or fails in achieving the ultimate goal of all living creatures: <Propagate-Genes>.

This suggests that the best approach to the study, and hopefully the artificial production, of intelligent behavior would be to build machines that must produce behavior in the real world, and then to equip these machines with sensors and control systems that enable them to carry out sensory-interactive goal-directed behavior.

Goal Selection

Any autonomous creature, be it a robot, a human, a bird, or an insect, must have some internal mechanism for selecting between alternative possible behaviors. In animals, the highest level, longest term, goals are selected by hard-wired (or PROM memory) routines, called instinct. Instinctual goals are known to be triggered into play by hormonal action, which itself is synchronized with long term external events, such as changes in the seasons, by means of sensory input.

It is also known that there are areas in the human and animal brain, primarily in the limbic regions, that evaluate the goodness or badness of situations and events. These are the neural areas that human subjects report produce feelings of pleasure-pain, joy-horror, hope-despair, love-hate, curiosity-fear, craving-revulsion, etc. The values of these "feeling variables" provide the basis for choosing between approach behavior and flight, between acceptance of other creatures as friends or rejection of them as enemies; between fighting, feeding, resting, sleeping, and caring for others; between caring for off-spring, family, peer group, or those outside the peer group. The emotional evaluation functions provided by the limbic system affect both the planning and execution of behavior at many different levels.

Goal selecting and situation evaluating functions can be provided to robots through decision-making methodologies taken from the fields of operations research, game theory, cybernetics, and artificial intelligence. Goal selection typically requires:
1. A search through the space of possible future actions
2. An evaluation of those futures considered
3. A selection of the best course of action
Clearly this process requires both a means for modeling and predicting the effect of future actions, as well as a set of evaluation functions with which to compute the cost, benefit, risk, and payoff of various alternative courses of action. Both natural and artificial autonomous creatures must be able to evaluate the costs and benefits of actions, both while they are being planned and while they are being executed.

Natural creatures that possess superior high level decision making processes and low level skills tend to be more successful than their competitors in selecting and executing behavior which propagates genes. This presumably is the mechanism of natural selection which produced natural intelligence in the first place.

Workshop Goals

The goals of this workshop are to:

1. Explore what is known in a wide diversity of fields about sensors and sensory processing systems, and
2. Understand how sensory information can be applied to advanced robotics.

The objective of this session on the "biological model" is to correlate what is know from biology with what is known from robotics and artificial intelligence.

A Unifying Hypothesis

This paper attempts to formulate a unifying hypothesis of how the high level understanding, evaluating, goal selection, planning and reasoning functions commonly associated with the mind are tied into the lower level sensing, filtering, recognizing, task execution, and servo control functions that are commonly associated with the mechanisms of the body. Hopefully, this model will focus discussion on unifying principals that form a common thread through the wide diversity of experimental data and theoretical methodologies that apply to this most challenging and interesting problem; the study of the nature of intelligence, both natural and artificial.

The model which I am proposing is derived from neurophysiological theory and experiments as well as results from CIM (Computer

Integrated Manufacturing) experiments (Albus 1982). CIM is the attempt to integrate robots, machine tools, inspection machines, and material inventory and distribution systems into totally automated factories.

The proposed model has the general structure shown in Figure 1. It consists of a hierarchy of levels, each of which is composed of task decomposition, world modeling, and sensory processing modules. Commands flow vertically downward in a task decomposition tree; sensory information flows vertically upward in a sensory processing tree; and world model information flows horizontally between task decomposition, world modeling, and sensory processing modules at each level of the hierarchy. Computing modules at all levels of the hierarchy are "data-flow" machines, each of which itteratively executes the following control cycle:

1. Read a set of input variables,
2. Compute a mathematical function on these inputs, and
      (after a computational delay)
3. Produce a set of output variables.

The functions computed depend both on the input variables and on state variables internal to the modules. If such a system is modeled as a discrete-time sampled-data system, the computational modules can be represented as finite-state automata.

There are three ways that such dataflow computational modules can be interpreted:

1. If the input to any computational module is treated as a vector of variables S, the output is a vector of variables P computed by the mathematical function H, i.e.,

$$P = H(S) \qquad (1)$$

2. If the input S is treated as an "IF premise", the output P is a "THEN consequent" such as is computed by an expert system rule, i.e.,

$$IF(S) \; THEN \; (P) \qquad (2)$$

(Here the function H is defined by the entire set of rules in the expert system.)

3. If the input S is treated as an address of a location in memory, then the output P is the contents of the address S, i.e.,

$$P = contents(S) \qquad (3)$$

(Here the function H is defined by a look-up table where for each address S there is stored a value P.)

In all three cases, the input defines an input vector S on an input space which is mapped by the function H into an output vector P on an output space.

The Functional Levels

Figure 2 shows the type of functions performed, and the type of output produced, at each level in the proposed hierarchy . The lowest level (level 0) is the servo level. At this level position, velocity, and force information is sensed, scaled, filtered, and compared with commanded values of position, velocity, and force. In a robot, the level 0 task decomposition modules compute the correct drive to the joint actuators to null the difference between commanded and observed values.

In animals, level 0 consists of the circuitry shown in Figure 3. Level 0 contains alpha and gamma motor neurons, muscles, stretch receptors, tendon tension sensors, and sensory ganglia neurons that make up the stretch reflex.

At level 0 all computations are performed in joint or muscle coordinates. Output values are typically updated every few milliseconds.

Level 1 of the proposed hierarchy computes the kinematic transformations necessary to translate from a convenient coordinate system (world, tool, or part coordinates) into joint coordinates. Commands at level 1 are defined in a coordinate system in which it is convenient to express problems of manipulation or locomotion. Sensory information is transformed into the same coordinate system so that observed positions, velocities, accelerations, and forces can be easily compared with commanded values. Level 1 commands typically are executed in a few hundreths of a second.

Level 2 in the proposed hierarchy accepts input commands defined in terms of "keyframe poses", or "key knot points", on a manipulation or locomotion elemental-move trajectory. It computes a dynamically efficient pathway, or smooth trajectory, through the keyframe poses in space/time. Level 2 also coordinates motions between closely related body parts so as to accomplish dynamically efficient movements of arms, legs, hands, and fingers. Level 2 commands typically are executed in a few tenths of a second.

In animals, level 1 and 2 functions have been hypothesized to be computed in the cerebellar cortex and in the cerebral motor cortex.

Level 3 in the proposed hierarchy transforms input commands expressed in terms of symbolic names of elemental-movements into key frame poses along trajectories in the chosen coordinate system of levels 1 and 2. Elemental-move commands are typically of the form <REACH>, <APPROACH>, <GRASP>, <LIFT>, <MOVE-TO>, etc. Level 3 modules control individual subsystems such as a single manipulator, a mobility unit, or a single machine tool. Obstacle avoidance, kinematic singularities, and limits on joint motion

are handled at this level. Level 3 commands are typically
accomplished in periods of a few seconds.

Level 4 of the proposed hierarchy accepts input commands
expressed in terms of tasks, or operations, to be performed by an
individual animal or machine on specific objects, or with respect
to other individuals. Level 4 decomposes these tasks into
sequences of elemental-moves that can be executed by manipulation
or locomotion subsystems. Coordination between various
subsystems is performed at this level: for example, manipulation-
locomotion coordination, eye-hand coordination, or coordination
between multiple arms takes place at this level. Precedence
constraints on elemental-move sequences, as well as cost/benefit
trade-offs between alternative elemental move action sequences
may be computed at this level. In the manufacturing environment,
this is the equipment level, i.e. the level of individual
machines or stand-alone equipment on the factory floor. Level 4
commands are typically accomplished in periods of a few seconds.

Level 5 takes input commands expressed in terms of tasks to be
performed on groups of objects by small tightly coupled groups,
or families, of machines. Input commands often specify only the
set of tasks and priorities without specific instructions as to
the sequence in which the tasks must be done. Coordination
between actions of various individuals, such as coordination
between robots, machine tools, automatic clamping and material
handling systems takes place at this level. In the
manufacturing environment, this is the work station level. Level
5 commands are typically accomplished in periods of minutes to
hours.

Level 6 input commands are expressed in terms of goals to be
accomplished by cells, or groups of workstations. Commands at
this level typically consists of production requirements and due
dates for the next day to a week. Level 6 schedules production
and performs batching and routing of parts and tools so that the
workstations can efficiently accomplished their assigned tasks
with a minimum of waiting. In the manufacturing environment,
this is the cell control level. Level 6 commands are typically
accomplished in a day to a week.

Level 7 takes input commands expressed in terms of goals to be
accomplished by entire manufacturing shops consisting of groups
of cells. Level 7 commands typically contain the entire backlog
of production orders, together with priorities and due dates.
Level 7 functions maintain inventory at levels necessary to meet
production schedules, and assign machine tools and other
production resources so as to meet promised deliveries in an
efficient manner. Level 7 deals with control problems that have
lead times of weeks to months.

Level 8 is where decisions are made about the product line and
the engineering design of parts to be manufactured. Decisions
made at this level must be based on predicted needs and
requirements months to years into the future.

At each level in the model described above, there exists a task decomposition, a world modeling, and a sensory processing module. At each level there is a feedback loop whereby the world model is updated and the task decomposition process can be modified to take into account previously unknown information about the external world. The information contained in the world model is thus a synthesis of apriori knowledge and sensed information.

Time

In the generation and control of behavior, time is a crucial variable. Behavior is produced in the instant of the present, at t = 0. Feedback is derived from sensory data collected during the past. Plans are made for action in the future. How far into the past is sensory data relevant to current behavior? How far into the future is it reasonable to plan? The hierarchical control model in Figure 2 suggests answers to these questions. Each level plans only how to decompose its current (or next) command into a sequence of subcommands to the next lower level. Sensory data at each level is integrated over a historical time period that extends only as far into the past as the planning algorithms at that level look into the future. Hence, modules at low levels in the hierarchy plan only for the immediate future and need sensory data integrated only over the immediate past. At successively higher levels, the planning horizon extends exponentially further into the future, and the relevant historical trace extends exponentially further into the past.

The task decomposition and sensory processing systems are duals of each other. Each level in the task decomposition hierarchy produces both a spatial and temporal decomposition of tasks and goals into subtasks and subgoals. Spatial decomposition involves assigning task subelements to subordinate computing modules. Temporal decomposition requires that each subordinate computing module decompose its task subelement into a temporal sequence of activities.

Similarly, each level in the sensory processing hierarchy produces both a spatial and temporal integration of patterns and sequences into objects and events. Spatial integration involves the collection of pixels into features, features into regions, regions into objects, etc. Temporal integration involves the collection of motion trajectories (of points, features, regions, and objects) into events, sequences, and trends.

Neuronal Computational Modules

How does the brain organize nerve cells to perform the functions of the H, M, and G modules? No one really knows. Nevertheless, a number of attempts have been made to demonstrate how a neuronal net can be organized so as to filter and process sensory data, recognize patterns, remember events, make predictions, learn, generalize, decompose tasks, and control actions (Albus 1981).

Among the earliest works in brain modeling were the simple neuron analogues of McCulloch and Pitts (1943). The concept of cell assemblies, which learn to compute behavioral functions, was introduced by D. O. Hebb (1949) as part of a neurophysiological theory of the organization of behavior. In (1961) Rosenblatt proposed a mathematical construct called the "perceptron" to model the neurodynamic functions of learning, generalization, and pattern recognition. Minskey (1967) demonstrated the ability of neural nets to compute any computable function. Later (1969) Minsky and Papert published a treatise which so effectively demonstrated the limitations of perceptrons that it had the effect of halting almost all work on brain modeling for a decade. Recently, interest in associative memory (Kohonen 1978) and brain modeling (Palm 1982) has revived.

The 1960's brought advances in the technology of single cell recording and electron microscopy. These made possible a set of experiments by Eccles and others (1967) that identified the functional interconnections between the principal components in the cerebellar cortex. This experimental data provided the basis for a theoretical model developed by Marr (1969) and Albus (1971) which suggested how a layer of cerebellar cortical tissue could learn, generalize, and compute arithmetic functions on input variables carried by a set of mossy fiber inputs. As shown in Figure 4, this model also suggested how a single precise value, such as the angle of an elbow joint, could be encoded by a multiplicity of imprecise neuronal channels.

Figures 5 and 6 show how the Marr-Albus model decodes input variables so as to select a set of synaptic weights. The model then computes by table look-up the value of functions such as described in formula (3) above. Figures 7 and 8 illustrate how the cerebellar model can produce the properties of learning and generalizing. Figure 9 illustrates how such modules can be arranged in a hierarchical architecture so as to produce sensory-interactive goal directed behavior.

Implications for Advanced Robotics

It is clear from the papers presented at this workshop that the sensors for advanced robots are being built. Sensors are being developed for proprioception (joint position, velocity, force), touch, temperature, haptic perception, vision, hearing (vibration, speech, sonor), taste, smell, and a number of senses not available to biological creatures (capacitive, inductive, and ultrasonic).

Work is proceeding rapidly in computer graphics and image understanding on methods for modeling the geometric, topological, and structural properties of objects. Much is known about how to use this information to plan, schedule, and control machines and processes to manipulate and manufacture such objects.

It seems clear that we soon will see the integration of this knowledge and these capabilities into advanced robots that are capable of successful goal selection and task decomposition in complex and unpredictable environments. Such advanced robots will provide the tools for quantitative experiments into the fundamental mystery of intelligent behavior. Only through such experiments will we ever really understand the relationship between the high and low levels of control within the central nervous system.

References

Albus, J.S., (1982) "An Architecture for Real-Time Sensory-Interative Control of Robots in a Manufacturing Facility", 4th International Federation of Automatic Control Symposium, Gaithersburg, MD.

Albus, J.S., (1971) "A New Approach to Manipulator Control: The Cerebellar Model Articulation Controller (CMAC)", Journal of Dynamics Systems, Measurement and Control

Evarts, E., and Tanji, J. (1974) "Gating of Motor Cortex Reflexes by Prior Insturction", Brain Research 71: 479-494

Eccles, J.C. (1967) "The Cerebellum as Neuronal Machine", New York: Springer-Verlag

Marr, D. (1969) "A Theory of Cerebellar Cortex", Journal of Physiology 202: 437-470

Jackson, J. (1931) "Selected Writings of John Hughlings Jackson", Edited by J. Taylor, London

Minksy, M. (1967) "Computation: Finite and Infinite Machines", Prentice-Hall, Engelwood Cliffs, NJ

Hebb, D.O. (1949) "The Organization of Behavior", Wiley, NY

Kohonen T. (1978) "Associative Memory", Springer, Berlin, Heidelberg, New York

McCulloch W.S. and Pitts W.H. (1943) "A logical calculus of ideas immanent in nervous activity",

Minsky M. and Papert S. (1969) "Perceptrons MIT Press", Cambridge, Massachusetts and London

Palm G. (1982) "Neural Assemblies. An Alternative Approach to Artificial Intelligence.", Springer, Berlin, Heidelberg, New York

Rosenblatt F. (1961) "Principles of neurodynamics: Perceptrons and the theory of brain mechanisms.", Spartan, Washington D.C.

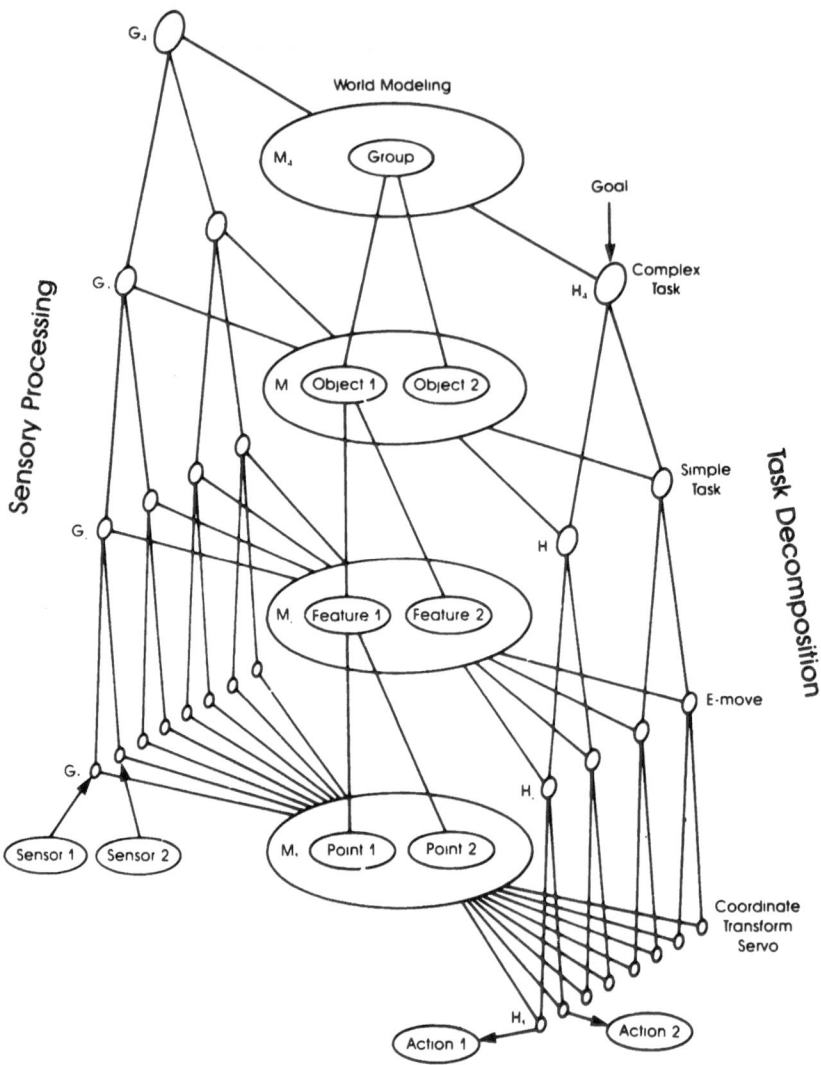

FIGURE 1

13

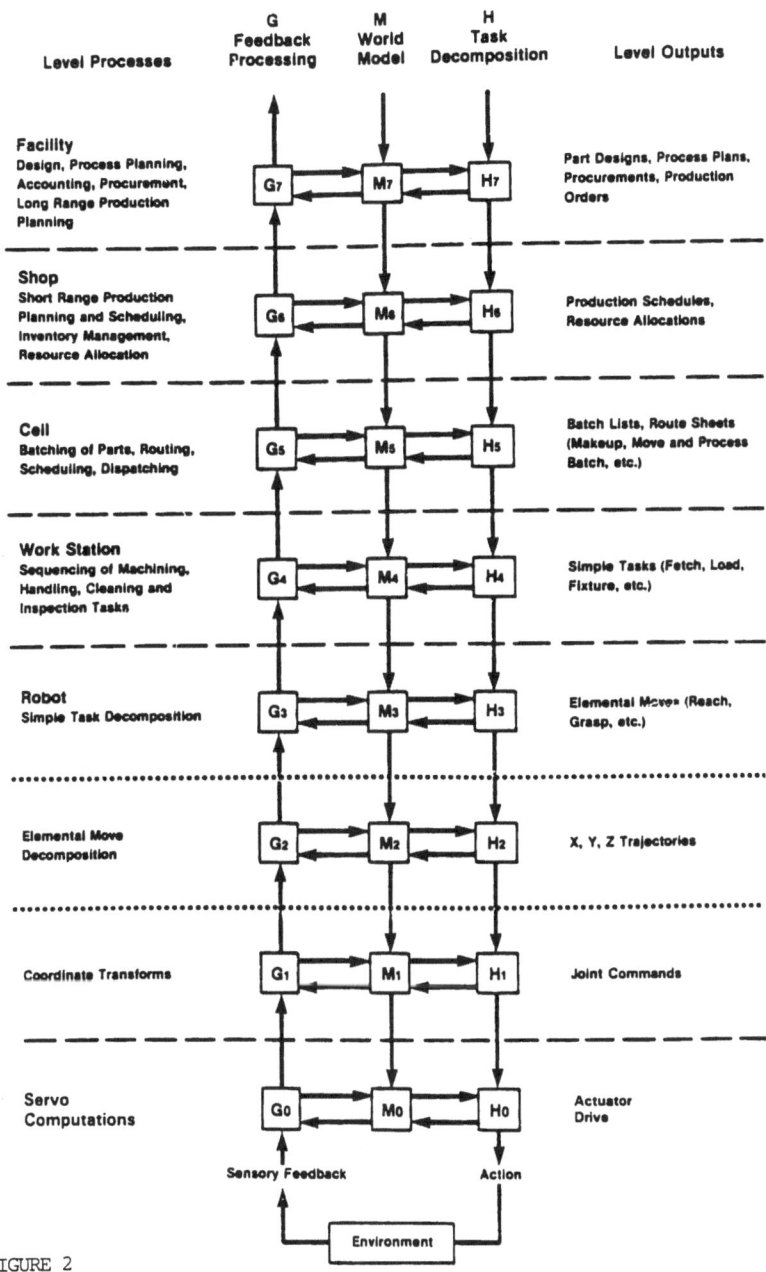

FIGURE 2

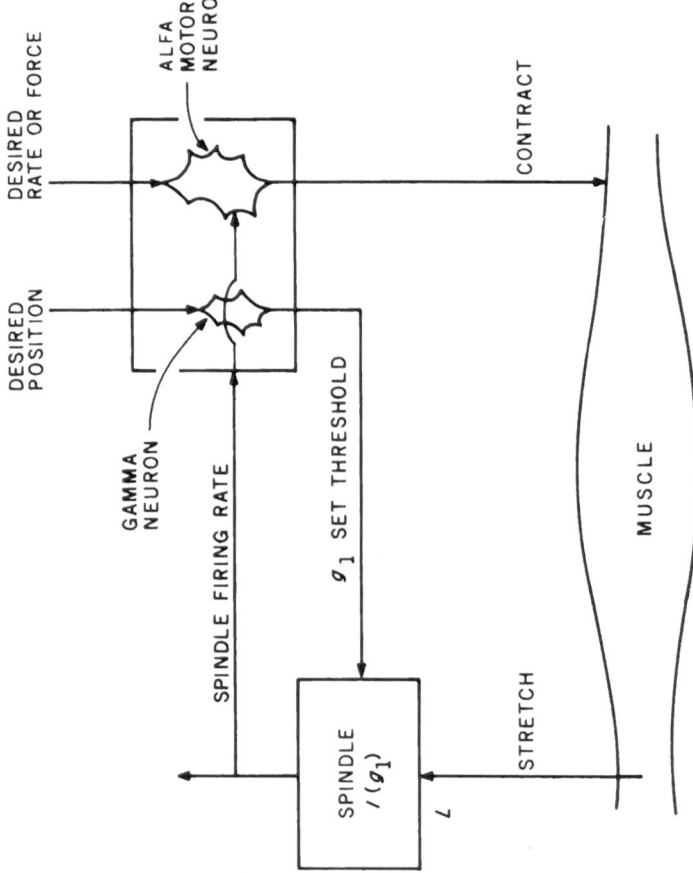

Figure 3: *A schematic diagram of the relationship between the motor neurons, muscles, stretch sensors, and input commands from higher motor centers.*

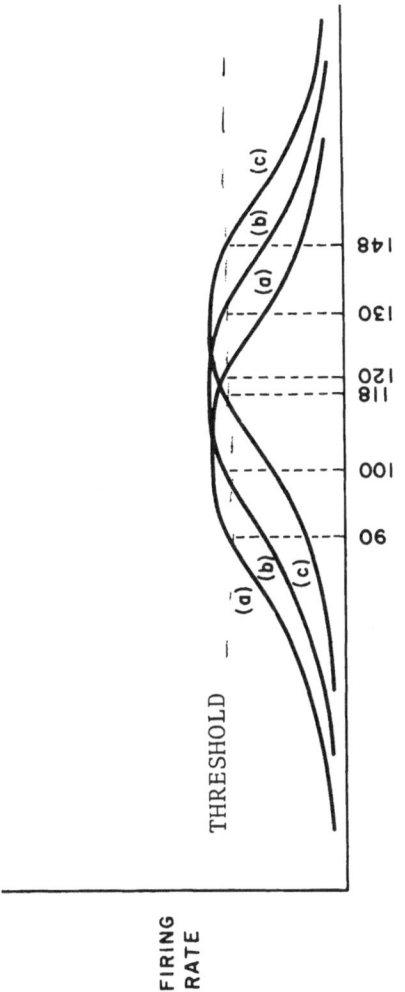

Figure 4: *Three different mossy fibers encoding a single sensory variable (elbow position). All three fibers maximally active simultaneously indicate that the elbow lies between 118° and 120°.*

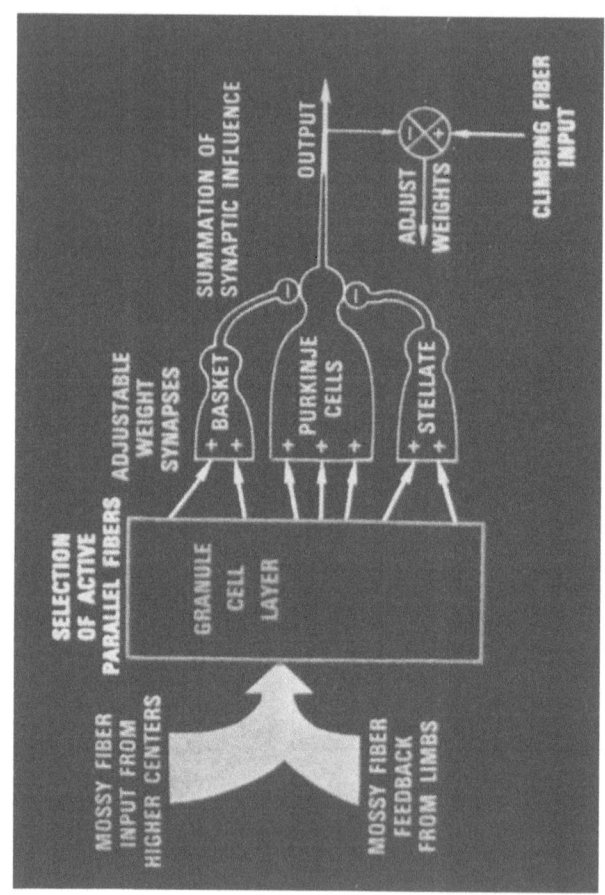

Figure 5: *A theoretical model of the cerebellum.*

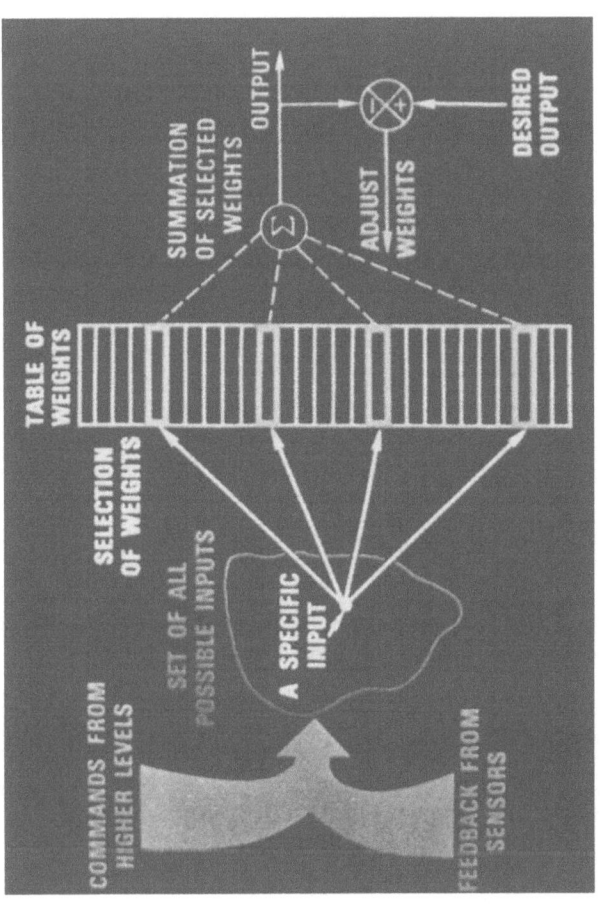

Figure 6: *A schematic representation of CMAC (Cerebellar Model Arithmetic Computer).*

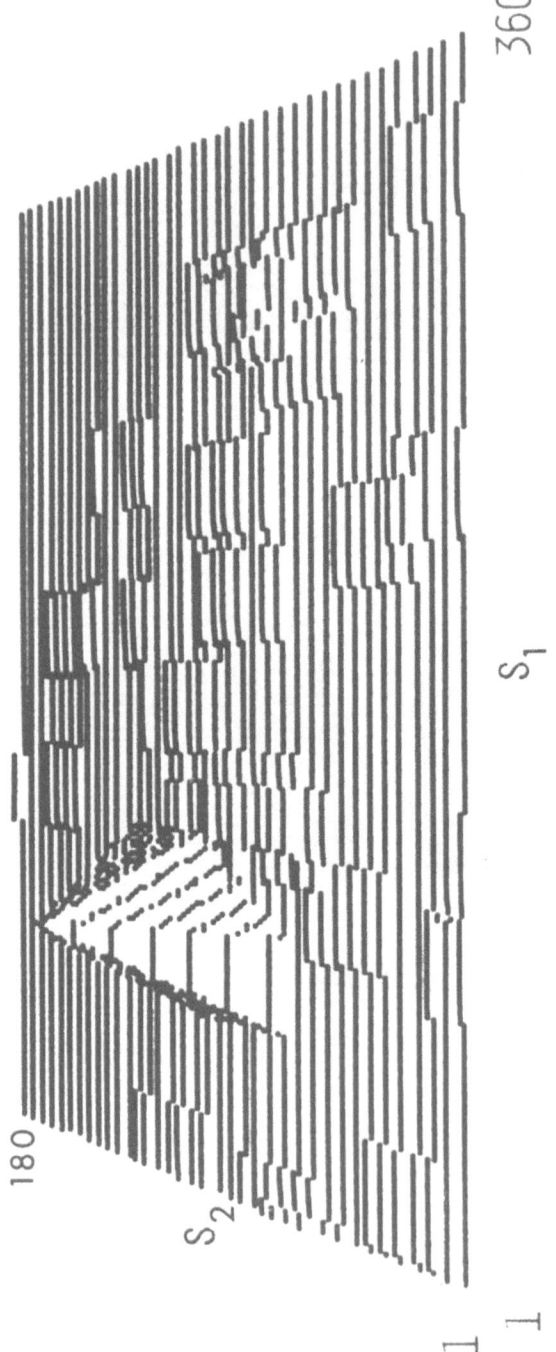

Figure 7 : *The effect of training CMAC on the function* $\hat{p} = \sin(2\pi s_1/360) \sin(2\pi s_2/360)$.
a: *One training at* $(s_1, s_2) = (90, 90)$.

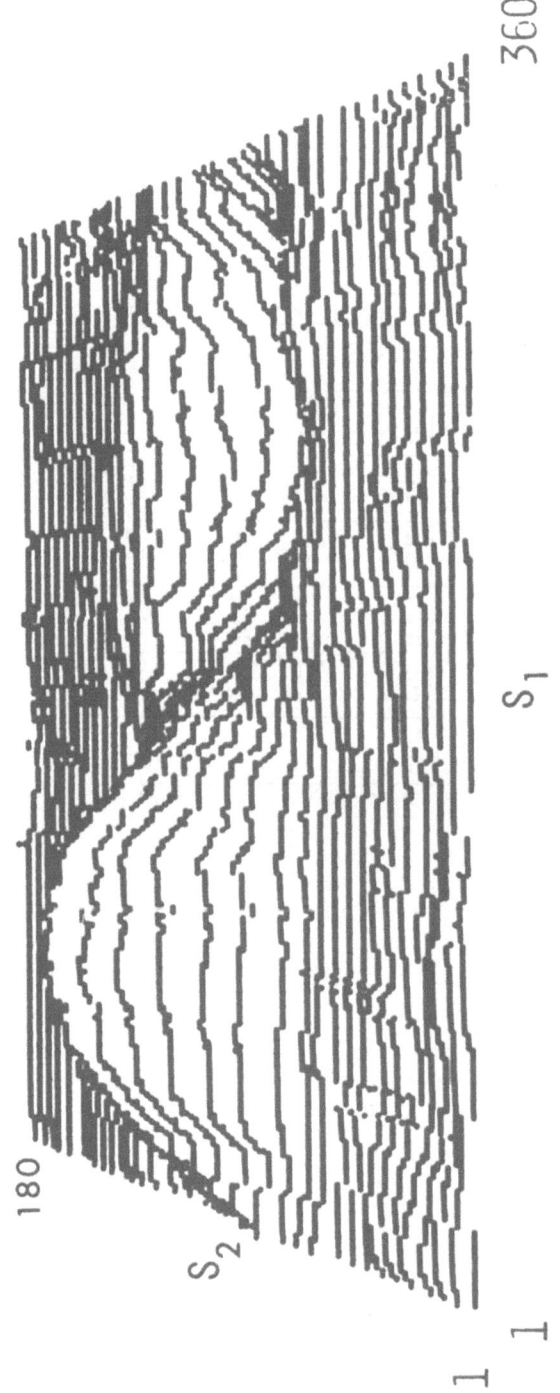

Figure 8 : *Training at 16 points along a trajectory defined by* $s_2 = 90$.

20

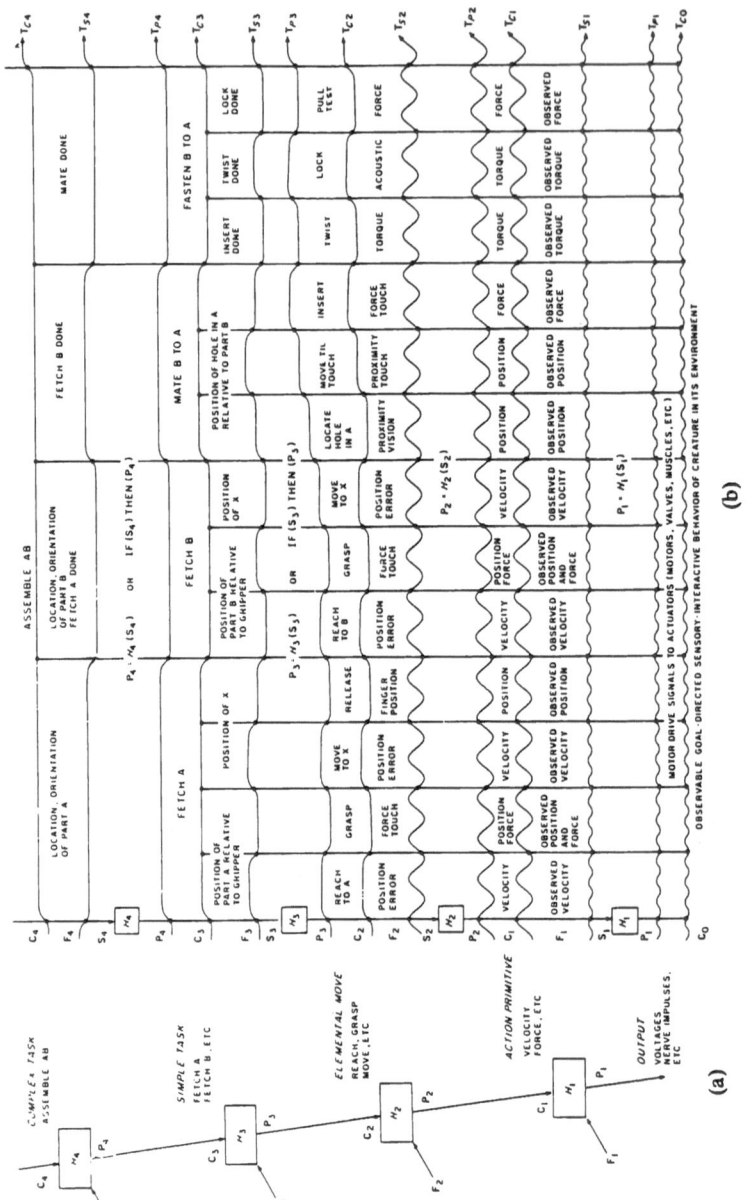

Figure 9: *A hierarchy of* H *operators produces sensory interactive goal-directed behavior. The highest level input command* C_4 *defines a goal, which in this example is* <ASSEMBLE AB>. *The feedback* F_4 *carries highly processed sensory data describing the state of environment in which the assemble command must operate, including the state of the lower level* **P** *vectors. The* H_4 *operator maps each input* S_4 *into an output* P_4. *As* F_4, *changes the goal* <ASSEMBLE AB> *is decomposed into a sequence of subgoals* <FETCH A>, <FETCH B>, <MATE B TO A>, <FASTEN B TO A>. *At each level in the hierarchy a different type of feedback data with a different rate-of-change drives the decomposition of a higher level command into a sequence of lower level subcommands. Finally, at the lowest level the* P_0 *vector consists of motor drive signals which actuate observable behavior* C_0.

(b)

(a)

PROPRIOCEPTIVE FEEDBACK FOR SENSORY-MOTOR CONTROL

Manuel Hulliger

Brain Research Institute
University of Zürich
August-Forel-Strasse 1,
CH-8029 Zürich
Switzerland

Abstract

Among the natural sensors, which provide proprioceptive feedback during movement, the muscle spindle has attracted most
interest since the sensitivity of spindle afferents to length
variations of the host muscle is subjected to efferent control
by fusimotor neurones (mostly γ-motoneurones). These are themselves controlled from numerous centres of the nervous system.
Functionally, the perhaps most important feature of γ-action
is that the fusimotor system, with its static (γ_S, sensitivity
reducing) and dynamic (γ_D, sensitivity enhancing) components,
possesses the potential of providing flexible gain (or sensitivity) control of spindle feedback. One of the most relevant
pending issues in the field of peripheral control of movement
is whether, and under which circumstances, this potential is
used during physiological motor performance.

This paper gives a brief overview of important functional properties of the sensory spindle afferents and of efferent fusimotor action and it addresses the question of fusimotor function during natural movement. It is pointed out, that strategies of fusimotor action, which have been encountered in reduced laboratory preparations, appear not be be adhered to during
normal motor performance. An alternative concept of fusimotor
function, the notion of <u>fusimotor set</u>, is presented. According
to this, key features of fusmotor action are

- largely <u>tonic</u> γ_S and γ_D firing patterns (even during rhythmic movements, featuring rhythmic skeletomotor α-activity),
  suggesting independent central control of the skeletomotor
  and fusimotor systems,
- a dramatic <u>switch</u> from predominant and low-key static action (during routine motor performance) to predominant and
  powerful dynamic action (during unfamiliar motor tasks),
- gradual <u>resetting</u> of fusimotor drive of a given type during
  periods of motor adjustment.

Thus, proprioceptive feedback from spindle afferents is not
invariant. It appears to be adapted to specific requirements
of motor tasks, conceivably to optimize motor control.

NATO ASI Series, Vol. F43
Sensors and Sensory Systems
for Advanced Robots
Edited by P. Dario
© Springer-Verlag Berlin Heidelberg 1988

Introduction

Physiological sensory systems differ from technical sensors in more than one respect. For instance, instead of providing easy to read analogue signals they transmit sensory information in the form of impulse trains, using pulse frequency modulation to encode their messages. Further, physiological sensory signals often are noisy, and it is obvious that a high - largely random - variability of the afferent firing is bound to limit the information capacity of individual neurones (Matthews & Stein, 1969b). Examples of noisy discharge patterns are illustrated in Fig. 1 (A-E). Under certain experimental conditions (see legend to Fig. 1) the responses of muscle spindle primary afferents (see below) to a physiological mechanical stimulus (stretch of the parent muscle) can exhibit so much variability of firing rate that details of the stimulus are easily obscured. In Fig. 1 the stimulus was a triangular length change of large amplitude, with additional small sinusoidal movements superimposed (H). However the responses of two of the afferents (B, D) fail to reveal any sinusoidal modulation of firing rate, the responses of other afferents (A, C) merely give a faint hint of the presence of a sinusoidal component of stretch, and only the afferent in E seems capable of - crudely - monitoring the time course of the small sinusoidal movement. The responses of Fig. 1 also illustrate a second source of variability. It is evident that even the component of response to the large triangular stimulus varies substantially between afferents, not only with respect to its size (depth of modulation; cf B with D) but also as regards the relative weight given to position (triangular) and velocity (rectangular) components, the latter being largest in B and C and smallest in E. Finally, such biological sensory responses can show distortions which reflect some form or other of non-linear behaviour. In particular in Fig. 1.B, but also in A and D the afferent firing rate tended to be held at fixed levels during muscle shortening (plotted downwards in H). In the present example this was due to the concomitant excitation of the afferent neurones by static fusimotor efferent fibres (see below, muscle spindle), which were tonically activated at a fixed rate and which, within limits, were capable of driving the afferent discharge at their own rate of firing (or at a submultiple).

In mammalian nervous systems such limitations can to some extent be compensated for by the large numbers of afferents (for instance from joints and muscles) which converge onto spinal or ascending tract neurones. Owing to such convergence at synaptic relay stations information may be recovered by spatial averaging of parallel afferent input (for a more detailed discussion of these matters see e.g. Jansen & Walloe, 1970). In order to illustrate the potential power of such spatial averaging the responses of 16 primary afferents, including those of Fig. 1 (A-E), to the same stimulus (shown in H) were summed as if they had occurred simultaneously. The resulting averaged response is illustrated in F. It can be seen that the degree

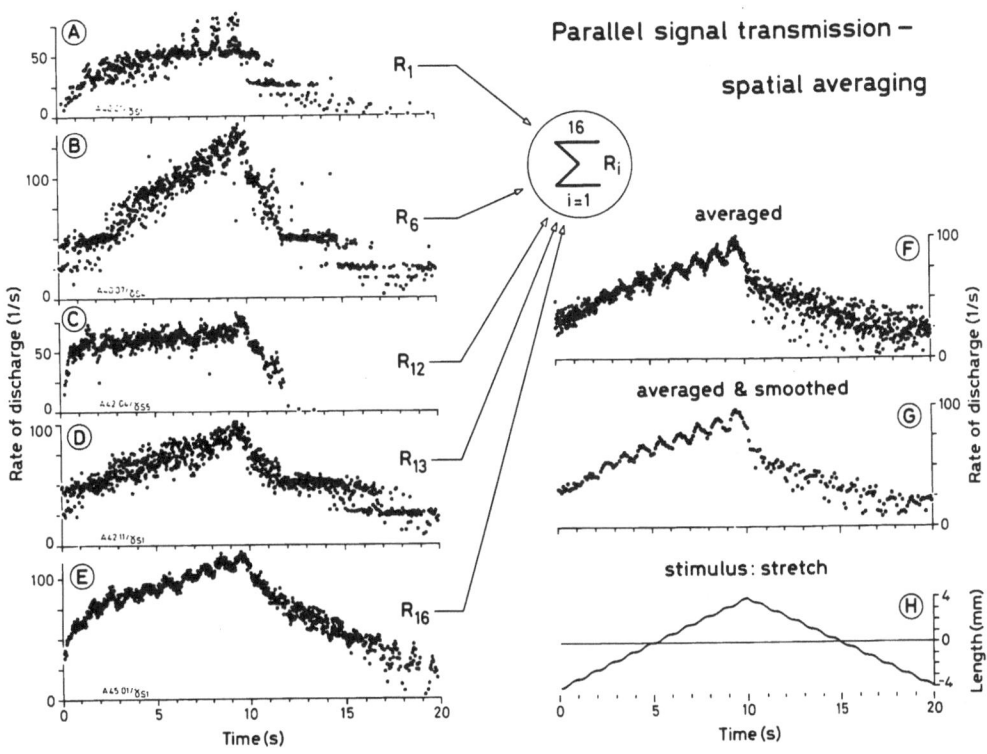

Figure 1. Parallel feedback from proprioceptors and effects of spatial averaging. In A to E, single cycle average frequency display (see Matthews & Stein, 1969a) of spindle primary afferent responses to combined sinusoidal (1 Hz, 100 µm) and triangular (0.05 Hz, 4 mm) stretch (H) of the host muscle. In all cases mechanical stimulation was accompanied by tonic electrical stimulation of static γ-fibres (all at 50/s). In F, effect of averaging across 16 parallel sensory channels. The responses from 16 combinations of a Ia afferent and a γ_S efferent (partly from separate experiments, but all subjected to identical mechanical and fusimotor stimulation) were pooled to give an averaged ensemble response. In G, the effects of possible low-pass filtering upon synaptic transmission were simulated with a moving window average algorithm (window width 80 ms). The high variability of discharge (A-E) was largely due to the presence of static γ-action, since in passive control responses (no γ-action) variability was five to ten times smaller. For further detail, see text. (From Emonet-Dénand, Hulliger & Sonnenberg, previously unpublished)

of variability is strongly reduced during muscle lengthening, that the distortions arising from efferent fusimotor driving during muscle shortening have disappeared, and that a sinusoidal modulation of the discharge rate is now clearly recognizable during the rising phase of triangular stretch. Upon

synaptic transmission afferent input signals may be subjected
to further smoothing, provided the post-synaptic membrane time
constant is long enough to confer effective low-pass filter
properties upon the summing neurone. This was simulated by
filtering the summed response (using a moving window average).
The result is shown in Fig. 1.G and reveals additional reduction of variability and enhancement of resolution.

Also, in contrast to simple technical devices, biological sensors frequently exhibit extremely <u>non-linear</u> properties, whose
functional significance at best is subject of speculation (see
below, Theories of Fusimotor Function). For instance, in the
case of the mammalian muscle spindle these non-linearities are
such that so far they have defied all attempts of comprehensive mathematical description (for review of these matters see
Hulliger, 1984). The best which so far has been achieved for
this particular system still fails to adequately account for
the complexity of muscle spindle properties (Hasan, 1983).

Proprioceptive Feedback during Movement

In addition to feedback from spindle afferents, proprioceptive
information is also provided by joint, Golgi tendon organ, and
fine (slowly conducting group III, and unmyelinated) afferents. Compared with spindle afferents their sensory messages
are of lesser complexity. <u>Tendon organs</u> are particularly sensitive to force generated actively by contraction even of individual motor units (for review see Proske, 1981). Therefore
their likely physiological role is to provide force feedback
no matter whether this affects (via central pathways) only the
relatively small muscle compartment, from which the feedback
arose, or whether such compartmentalized feedback is lost in
the course of central averaging. Be this as it may, recordings
from freely moving animals strongly support the view that during natural movements feedback from Golgi tendon organs indeed
closely monitors active contraction force (Appenteng & Prochazka, 1984). The role of <u>joint afferents</u> is more controvers-
ial, since there is the possibility that in some joints they
might act as extreme-range detectors (conceivably activating
safety stops), whereas in other joints they could operate as
continuous position sensors (cf. Burgess & Clark, 1969, Clark
& Burgess, 1975, with Carli et al., 1981; but see also Rossi &
Grigg, 1982). Moreover, 'joint afferents' do not form a homog-
enous group, much as 'muscle afferents' are a non-homogenous
group. In either case these populations encompass a substantial fraction of fine (including unmyelinated) afferents whose
main function appears to be their contribution to nociception
(see for instance Schaible & Schmidt, 1983, 1985).

Muscle Spindle

Among muscle receptors the muscle spindle has attracted most interest, since the sensitivity of its sensory afferents to length changes is subjected to <u>central control</u>, which is mediated by <u>fusimotor neurones</u>. The fusimotor supply of the spindle is mostly provided by the exclusively fusimotor γ-motoneurones, but to some extent also by the mixed skeleto- and fusimotor β-motoneurones (see Matthews, 1972, 1981a, 1981b; Laporte et al., 1981; Hulliger, 1984; general reviews). Fusimotor efferents (both γ and β) are commonly further subdivided into two categories, static (γ_S, β_S) and dynamic (γ_D, β_D), according to whether their activation decreases (S) or increases (D) the sensitivity of muscle spindle primary (Ia) afferents to dynamic changes of host muscle length (Matthews, 1962).

Examples of typical <u>fusimotor effects</u> are illustrated, in Fig. 2, in the responses of Ia afferents to ramp and hold stretches (bottom traces) applied to the soleus muscle of the cat. Compared with the control responses during fusimotor silence (so called 'passive' spindles, upper traces), electrical stimulation of a dynamic γ-axon at a fixed rate (70/s) caused a pronounced increase in mean and peak discharge rate (centre, left) during the dynamic phase of the stretch, whereas the activation of a static γ-axon nearly completely abolished the dynamic response component (centre, right). It may also be noted that static, but not dynamic, action was capable of maintaining appreciable Ia firing during muscle shortening (Fig. 2, see also Fig. 1). This observation has been repeatedly confirmed and it has been very influential for the formulation of various concepts of fusimotor function.

The size of the fusimotor effects in Fig. 2 gives a qualitative indication of the range over which the dynamic sensitivity of spindle primary afferents can be varied by fusimotor action. Systematic quantitative studies with controlled electrical stimulation of γ-fibres (as in Fig. 2) have clearly indicated that the fusimotor system with its two contrasting components (γ_S, γ_D) was equipped with the potential of providing <u>flexible</u> sensitivity or <u>gain control</u> of spindle feedback. This conclusion was based on the finding that Ia sensitivity could be continuously graded between an upper and lower limit, which were set by pure dynamic and by pure static action, respectively, intermediate values of sensitivity being determined by the relative weight of activation of the two efferent components (Hulliger et al., 1977a,b).

Since γ-motoneurones in turn are subjected to <u>central control</u> from numerous higher motor centres (see Hulliger, 1984; review), and since some of the descending pathways involved act quite selectively on either dynamic (Appelberg, 1981) or static (Schwarz et al., 1984) fusimotor neurones, the possibility could be considerd that the central nervous system was capable not only of varying the sensitivity of spindle feedback over a wide range but also of tuning it in fine gradations.

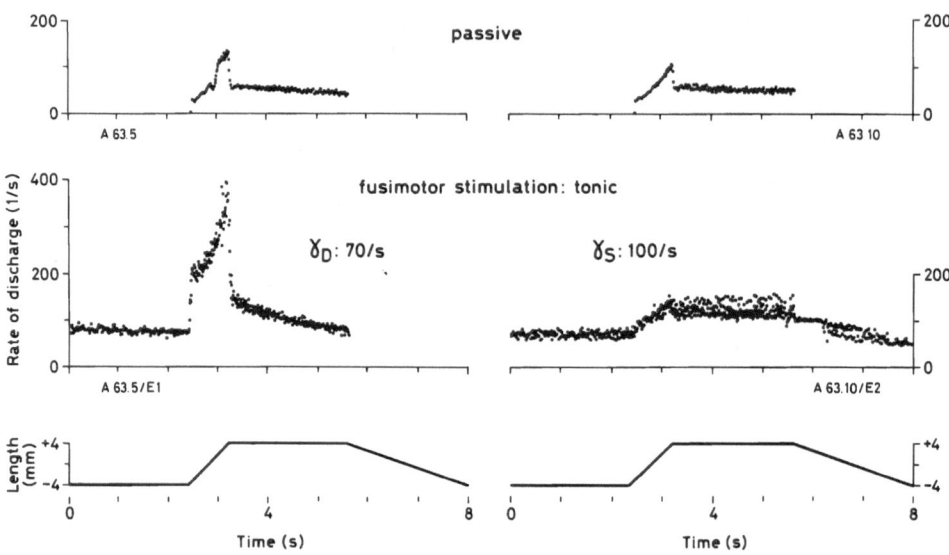

PRIMARY SPINDLE AFFERENTS

(single cycle responses, average frequency display)

Figure 2. Dynamic (γ_D) and static (γ_S) fusimotor action on Ia response to trapezoidal length changes applied to the parent muscle. From top to bottom, passive Ia control responses, responses of the same Ia afferents (as for the passive controls, above) during dynamic (left) and static (right) γ-stimulation, and time course of length changes. Illustration of two separate Ia afferents (left vs right). For further detail see text. (From Hulliger, 1984, with permission)

The knowledge about central control and action properties of fusimotor neurones was exclusively derived from investigations on severely reduced (i.e. anaesthetized, spinal, or decerebrate) preparations, the use of such reduced nervous systems being dictated by technical and ethical constraints. However, it is commonplace that in such reduced states the function of structurally preserved subunits of the nervous system may be severely altered, compared with their physiological behaviour, when they are embedded in the intact nervous system and subjected to higher order control. Thus, as regards the fusimotor system the knowledge gained from reduced preparations revealed a great deal about its functional potential, but it left open the question as to what use might be made of this potential under physiological conditions, e.g. during normal motor performance. This issue will be further pursued below.

Among the sensory neurones of the muscle spindle the large and rapidly conducting primary (Ia) afferents can be distinguished from the smaller and more slowly conducting secondary (group II) afferents. Whilst group II afferents only receive static fusimotor input, Ia afferents are subjected to both static and

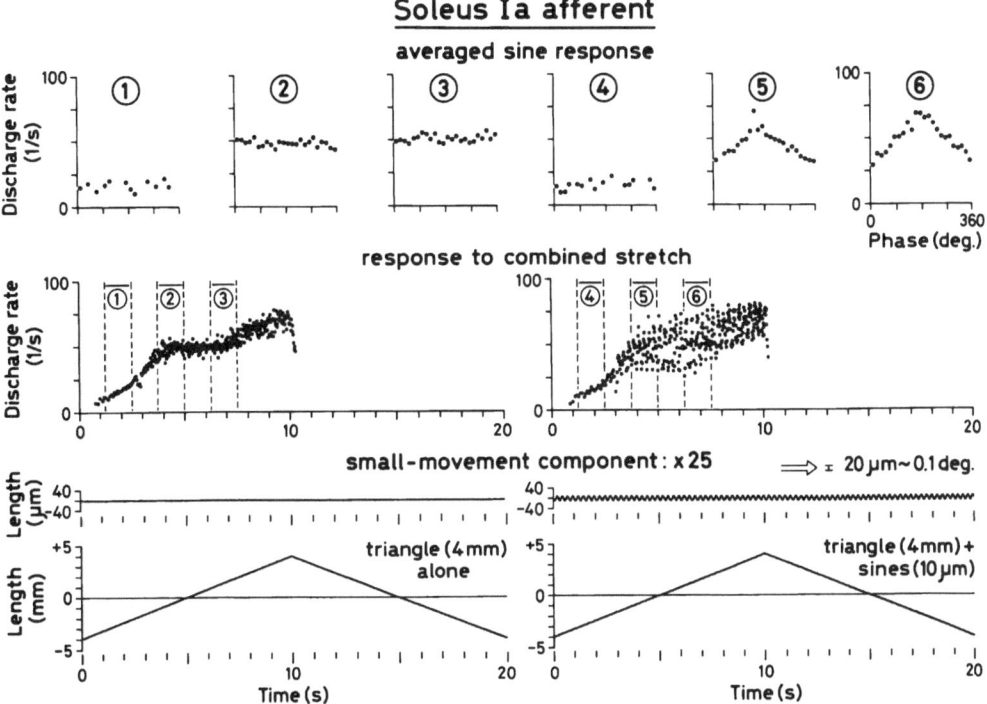

Figure 3. Small-movement responsiveness of primary spindle afferents. Centre panel left, response of a Ia afferent to triangular stretch (bottom panel, left). Centre panel right, response of the same afferent to combined sinusoidal (4 Hz, 10 μm half peak to peak) and triangular stretch (bottom right), in either case single cycle (triangle) average frequency display (see legend to Fig. 1). In the upper row sine cycle histograms are plotted, to demonstrate that the variability of firing on the right was mainly due to a sinusoidal modulation of the afferent's discharge rate, and to show the absence of comparable sinusoidal components of response on the left. The sine cycle histograms were constructed by averaging the afferent response over five successive sine cycles within the windows indicated and numbered in the centre panels. Note that the small sinusoidal oscillation is not recognizable in the plot of combined sinusoidal and triangular movement (bottom right). Therefore, the small-movement component was also plotted separately on twenty-fivefold magnification (next to bottom panels). In order to give an indication of the scale of Ia sensitivity, it may be noted that the size of the small disturbance would correspond to a peak to peak excursion of about 0.1 degree at the metacarpo-phalangeal joint of a human finger. (From Hulliger & Sonnenberg, 1985, previously unpublished)

dynamic control (for further detail see general reviews, above). Therefore the firing patterns of primary afferents can provide indirect information on the central control of these two groups of motoneurones, especially under conditions where direct recordings are feasible, technically, only from afferents but not from fusimotor efferents (see below). Yet given the dual (mechanical and neural) input to the spindle it is not surprising that the information carried by its sensory afferents often is far from straightforward.

One of the - non-linear - properties of primary spindle afferents, which has attracted particular interest, is that they are much more sensitive (by a factor of up to 100) to small than to large movements (Matthews & Stein, 1969a; Goodwin et al., 1975; Hulliger, 1984, review). A striking example of this pronounced __small-movement sensitivity__ is illustrated in Fig. 3. The middle traces show two responses of one and the same Ia afferent to large (4 mm) triangular stretches (bottom panels) which, by the look of it, are indistinguishable. Yet the afferent's responses differ appreciably: whilst the profile on the left only reveals the normal noisiness of Ia discharge (cf. Fig. 1), the response on the right contains an additional rhythmic component. This was due to a minute (10 μm) sinusoidal oscillation which was superimposed on the large triangular movement (bottom right; note separate, blown-up, display of small-movement component). These added oscillatory movements were so small as to be invisible, not only in the bottom display of Fig. 1 but also in reality, during the experiment, when they could not be detected by eye. Yet the receptor responded quite vigorously to the small stretch components. The size of the oscillatory response was roughly one forth of the triangle response, whilst the components of movement differed by a factor of 400.

Theories of Fusimotor Function

Especially primary spindle afferents are sensitive not only to the position, but also to the velocity and acceleration components of movement. Thus passive Ia afferents tend to be silenced as soon as their host muscle shortens, and so they cease to provide useful feedback on the time course of the movement at issue (see e.g. Fig. 3). Therefore, since the excitatory action of single γ-motoneurones was first described, __maintenance of__ some __spindle firing__ during muscle shortening was seen as a central feature of fusimotor function.

Initially, when the spindle was looked at as a straightforward stretch receptor, the benefit ascribed to this was to simply keep the afferents in a suitable __working range__. According to this the purpose of fusimotor excitation was to provide some __bias__ to enable the afferents to faithfully monitor the full range and time course of movements, including segments of muscle shortening. Given the ability of static, but not of dynamic, fusimotor efferents to strongly excite Ia afferents during muscle shortening (above, and Fig. 2) a participation of the static system seemed indispensable for such biasing action.

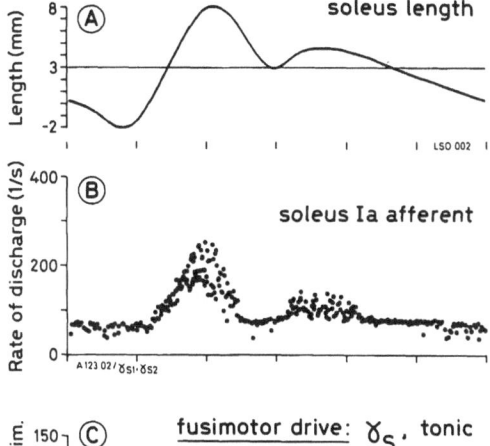

Figure 4. Maintained Ia firing during muscle shortening owing to tonic static fusimotor action. Two functionally single γ_S axons were stimulated electrically at 75/s (C). The movement imposed on the soleus muscle of an anaesthetized cat (A) was an idealized step which was derived from measurements of soleus length changes during normal locomotion (walking; Goslow et al., 1973; Fig 11). The responses of a Ia afferent (B) to three successive movement cycles were pooled and are displayed as average-frequency cycle histogram (see Fig. 1, legend).

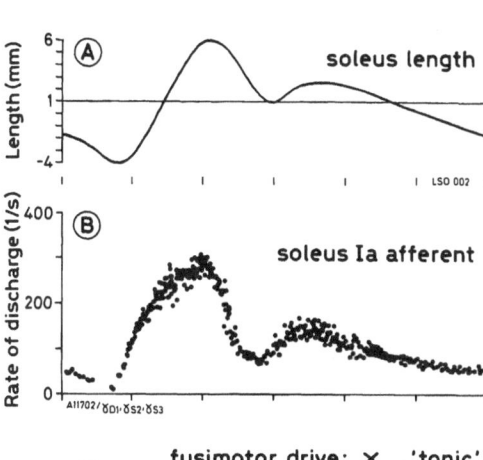

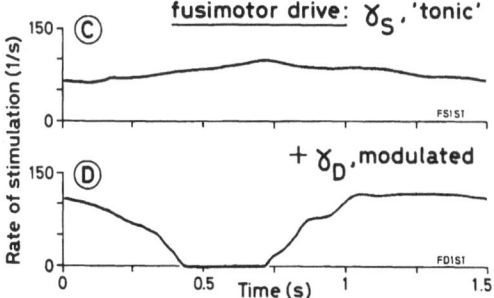

Figure 5. Ia firing maintained throughout a step cycle during concomitant weakly modulated γ_S and deeply modulated γ_D action. Same movement as in Fig. 4. Two γ_S axons acting on the afferent under study were activated in parallel, using rate modulated stimulation driven by the profile in C. In addition, a γ_D axon (operating the same spindle) was activated simultaneously, the rate of stimulation being modulated by the profile in D. The Ia response shown in B was averaged over three successive stimulation cycles (as in Fig. 4). Stimulation profiles from published records of γ-firing during locomotion in decerebrate cats (see Figs. 9.C and 11.B).

Yet apart from that, this notion was not discriminative, as a number of different patterns of γ_S and/or γ_D action, achieving this goal, can easily be conceived. This is illustrated in Figs. 4 and 5, in responses of soleus Ia afferents to an experimentally simulated idealized step. In Fig. 4 maintained, if modulated, spindle firing was achieved by simple tonic γ_S action accompanying the movement. In Fig. 5 a qualitatively similar effect (maintained modulated discharge) was provoked by a more complex pattern of fusimotor drive, when essentially tonic γ_S stimulation was paired with deeply modulated γ_D stimulation. The particular profiles of γ-action simulated in Fig. 5 were taken from the literature and are dealt with in more detail below (cf. Figs. 9.C and 11.B). What matters in the present context is, first, that in either case some form of static fusimotor drive was required in order to maintain spindle discharge throughout an entire step cycle and, second, that even strongly modulated dynamic action on its own was not normally capable of achieving this (not illustrated). However, the presence (in Fig. 5) of additional dynamic drive was not without effect, since it led to much more faithful Ia monitoring of the length changes imposed on the parent muscle (cf. Fig. 5 with 4).

Later, when theories of servo control began to make their impact felt, the advantage of maintained spindle firing was seen in the provision of an adequate and approximately steady carrier frequency to enable the afferents to monitor small disturbances (to which they are particularly sensitive, see above, 'small-movement sensitivity') rather than the overall course of a movement. This notion, which was first formulated as the concept of servo assistance of movement (Matthews, 1972; also Stein, 1974), is illustrated schematically in Fig. 6. Fusimotor drive (A.3) was thought to be modulatd and adjusted so as to give approximately steady Ia discharge (A.2) in spite of the concomitant movement, but only as long as movements proceeded as intended by the central nervous system (Fig. 6.A). If, however, movements were perturbed (Fig. 6.B.1), e.g. by unpredictable variations of external load, the deviations from the intended movement would be monitored as fluctuations of the spindle discharge (centered around the fixed carrier frequency; Fig. 6.B.2). Thus spindle (in particular Ia) afferents were seen as misalignment detectors. In the schematic drawing of Fig. 6.B.2 the Ia afferent is illustrated as operating at a fixed value of small-movement sensitivity. Although this was never spelled out explicitly, it is obvious that for operational simplicity such constant-sensitivity behaviour would be highly desirable. However, as is pointed out below (see also Fig. 8), muscle spindle properties appear not to be suited to meet this goal.

A variation of the theme of servo assistance was the proposal (Taylor & Appenteng, 1981; see also Appenteng et al., 1980) of modulated γ_S action forming a 'temporal template' (Fig. 6.3) of the intended movement (to provide some background Ia discharge), combined with tonic γ_D action to set incremental (i.e. small-movement) sensitivity of Ia afferents to suitable values. In Fig. 7 it is shown that patterns of modulated γ_S activation (C) can indeed be found which are capable of offsetting the effects of the length changes during movements of physiological amplitude (A), so that a Ia afferent 'responds' with an approximately constant rate of discharge (B). The par-

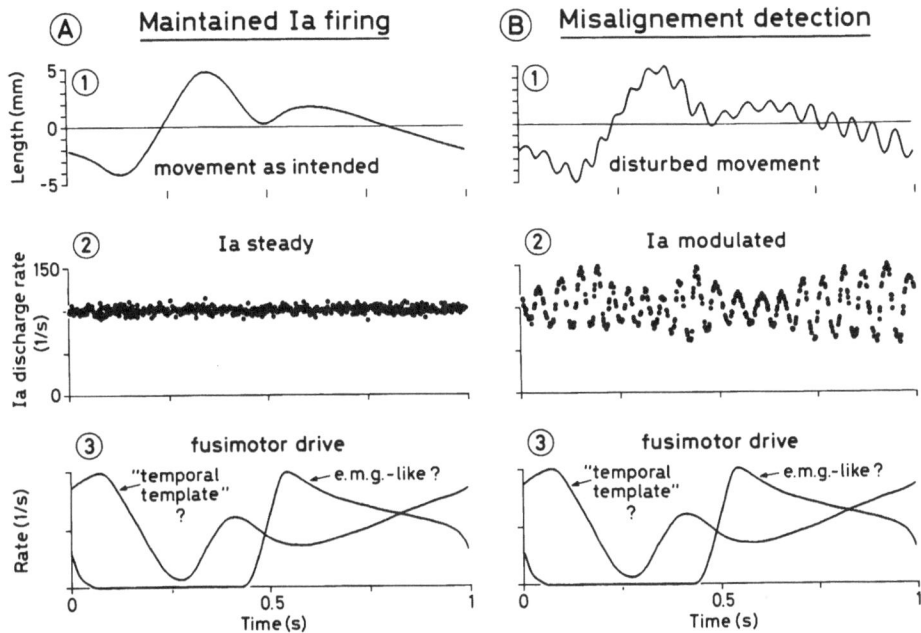

Figure 6. Schematic illustration of the concept of servo assistance. From top to bottom: movement, Ia discharge, and fusimotor activity profiles. In A, idealized length changes during unperturbed locomotion (stepping), with concomitant modulated γ-action (A.3), adjusted to offset the effects of the length variations, so as to lead to steady Ia firing (A.2). In B, the same movement with superimposed disturbances, accompanied by fusimotor action of the same time course (B.3). On this view the afferent is expected to respond to the disturbances (B.2) but not to the length variations of the underlying intended movement. Putative profiles of γ-action providing such compensatation for planned length variations are shown in the bottom panels (3). Note that the 'temporal template' (drawn as inverted and phase advanced movement) is not unlike the profile of Fig. 7.C, which indeed elicited steady Ia firing during concomitant length variations. In contrast, e.m.g.-like patterns of γ-drive have so far never been found to provoke servo-assistance like Ia discharge (unpublished observations).

ticular profile of Fig. 7.C was generated in the experiment using an iterative on-line procedure (see Hulliger, this volume) to produce pre-defined Ia target responses (here a constant rate of discharge). It bears emphasis that this modulated compensatory γs profile (C) bore no resemblance to the e.m.g. (reflecting modulated ensemble skeletomotor α-activity) as recorded during normal stepping (D). Thus, although 'temporal template' like γs action was modulated, it was by no means tightly coupled with extrafusal α-activity, suggesting that rigid linkage of α- and γ-activity would not be suited to provide servo-assistance like Ia discharge patterns.

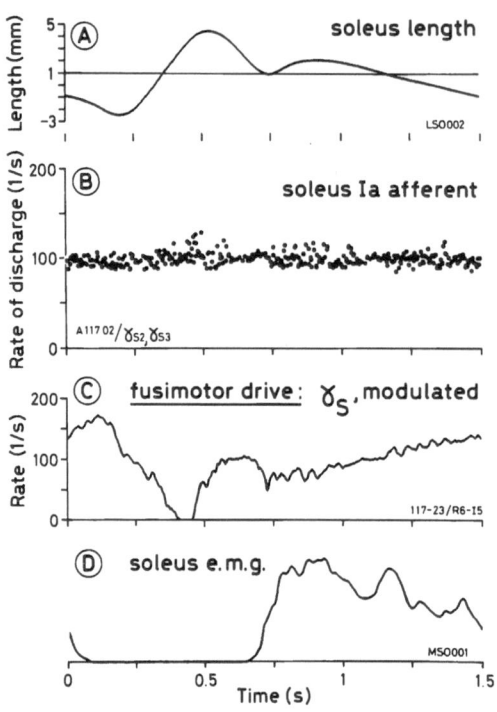

Spindle Ia response to simulated step

(A) soleus length LSO002

(B) soleus Ia afferent A11702/γ_S2, γ_S3

(C) fusimotor drive: γ_S, modulated 117-23/R6-15

(D) soleus e.m.g. MSO001

Figure 7. Steady Ia firing during large movement, elicited by compensatory modulated γ_S action. In A, length changes during idealized step (same as in Fig. 4). In B, Ia firing averaged over three successive cycles of stimulation. In C, time course of rate modulated γ_S stimulation. In D, e.m.g. during walking (data from Walmsley et al., 1978, Fig. 2, low-pass filtered and smoothed by eye). Same primary afferent and static efferents as in Fig. 5. The fusimotor driving signal in C was generated with an iterative on-line technique (see text and Hulliger, this volume).

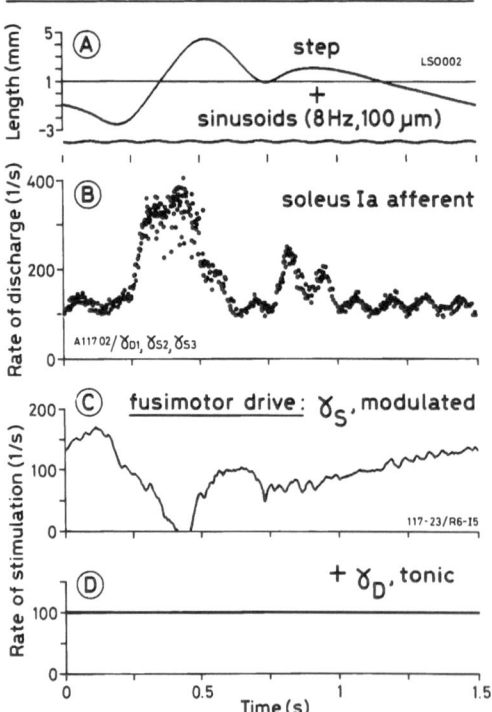

Ia response to simulated step with "noise"

(A) step LSO002 + sinusoids (8 Hz, 100 μm)

(B) soleus Ia afferent A11702/γ_D1, γ_S2, γ_S3

(C) fusimotor drive: γ_S, modulated 117-23/R6-I5

(D) + γ_D, tonic

Figure 8. Experimental assessment of the notion of servo assisting spindle feedback as illustrated in the scheme of Fig. 6.B. The experimental design was the same as in Fig. 7, apart from the small sinusoidal disturbances which were added to the large length changes of the idealized step (in A; note sines plotted separately), and apart from the additional tonic activation of a dynamic γ-axon (D). Same afferents and efferents as in Figs. 5 and 7. For further detail see text.

Whilst Fig. 7 illustrates 'temporal template' like γ_S action in isolation, Fig. 8 shows a complete test of the notion of Taylor & Appenteng (above). A response of the same Ia afferent as in Fig. 7 is displayed in B. However, the mechanical stimulus was a simulated step of large amplitude (same as in Fig. 7) combined with small superimposed sinusoidal disturbances, which, for clarity, are plotted separately in A. Moreover, the response of Fig. 8.B was obtained during modulated and 'temporal template' like γ_S action (C; same profile as in Fig. 7) which was accompanied by tonic γ_D drive (D). It can be seen that such combined γ-action disproportionally enhanced small-movement sensitivity during the lengthening phases of the step cycle (plotted upwards in Fig. 8.A). Moreover, steady background discharge was no longer present, since γ_D action also strongly and selectively enhanced the Ia response to the large components of stretch (cf. Fig. 7 with 8). As to underlying receptor mechanisms these effects are explicable in terms of strongly non-linear γ_S / γ_D interactions, e.g. by the well known occlusion of dynamic by static action during muscle shortening (Hulliger et al., 1977b; Hulliger, 1984). For functional considerations the important conclusion is that tonic γ_D action combined with modulated γ_S drive, the latter adjusted to offset the effects of large length changes, obviously failed to provide the constant small-disturbance sensitivity (throughout an entire movement cycle), which might be expected of a simple servo-like feedback signal.

Concepts of Fusimotor Function Based on Afferent Recordings

In spite of some similarity between the concepts described in the preceding section (e.g. as regards the desirability of maintained spindle firing), agreement could not be reached in detail on the patterns of static and/or dynamic fusimotor activity, which should be expected (on theoretical grounds) to take place during natural movements.

Technical obstacles have up until now - in spite of repeated attempts - ruled out direct recordings from γ-efferents during natural movements. This applies both for freely moving laboratory animals and for human subjects, the main difficulties being that proper classification of efferents (conduction velocity in γ-range) and type-identification (γ_S vs γ_D) proved beyond present techniques (for further detail see Prochazka & Hulliger, 1983). Notions of fusimotor function have therefore often been based on qualitative inferences made from recordings of spindle afferent discharge. Such concepts have been difficult to evaluate, all the more as studies on different paradigms in different preparations have led to different conclusions. From the compilation of the most important notions in Table 1 it is evident that some of these are mutually exclusive (cf. for instance A-E with L, also F with G). However such contradictions may be more apparent than real, given the indirect and qualitative nature of some of the evidence (C, D,

TABLE 1

	YS	Y	YD	Movement	Preparation	Species	Recording	Reference
A	?	α-coupled	?	(fictive) locomotion	spinal decerebr.	cat	γ,Ia	Sjöström & Zangger(1976); Severin(1970), Severin et al.(1967)
B	?	α-coupled	?	respiration	anaesth.	cat	γ,Ia,II	Critchlow & Von Euler(1963), Eklund et al.(1964), Sears(1964)
C	?	α-coupled	?	tracking isometric	alert	man	Ia,II	Vallbo et al.(1979), Burke(1981), Hagbarth(1981)
D	?	α-coativated	?	chewing	alert	monkey	Ia,II	Goodwin & Luschei(1975)
E	α-coupled flexors		α-coupled extensors	spontaneous stepping	decortic.	cat	Ia	Perret & Berthoz(1973), Cabelguen(1981)
F	α-coupled		tonic	chewing	anaesth.	cat	γ,Ia,II	Appenteng et al.(1980), Gottlieb & Taylor(1983)
G	tonic		α-coupled	locomotion treadmill	decerebr.	cat	γ,Ia,II	Murphy et al.(1984), Taylor et al.(1985)
H	———		phasic	biting	alert	monkey	Ia,II	Larson et al.(1981)
I	?	α-independent	?	tracking	alert	man Monkey	Ia	Vallbo & Hulliger(1981), Schieber & Thach(1985)
K	? swing phase biartic. m.	stance phase extensors swing phase biartic. m.	swing phase extensors stance phase biartic. m.	walking treadmill	alert	cat	Ia	Loeb & Hoffer(1985), Loeb et al.(1985)
L	tonic, weak		tonic,strong imposed mvt.	walking imposed mvt.	alert	cat	Ia	Hulliger et al.(1985), Prochazka et al.(1985)

Table 1. Overview of concepts of fusimotor function. For details see text. Abbreviations: **anaesth.**, anaesthetized; **biartic. m.**, biarticular muscles; **decerebr.**, decerebrated; **decortic.**, decorticated; **mvt.**, movement.

I, K), and considering the distinct possibility that to some extent such contradictions may be attributable to genuine differences between reduced preparations and normal alert individuals (see Table 1, Preparation). Finally, there is no compelling reason that fusimotor strategies should be identical for different muscle groups (Loeb, 1984; E and K in Table 1) or different motor tasks (L in Table 1).

Concepts of Fusimotor Function from Reduced Preparations

In contrast to unrestrained animals and human volunteers, direct recordings from γ-efferents have been achieved in reduced preparations (anaesthetized, spinalized, or decerebrated animals). These have led to three main concepts of fusimotor function which are illustrated schematically in Fig. 9 and which, for rhythmic movements, all entail some modulated or even α-coupled fusimotor outflow:

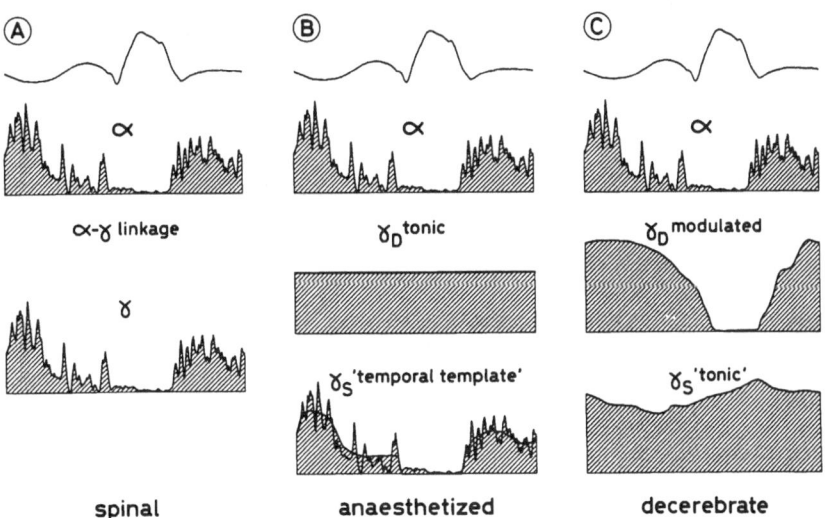

Figure 9. Concepts of fusimotor function derived from reduced preparations. For details see text. For references see, for A, Table 1 (A); for B, Table 1 (F); and for C, Table 1 (G). The length variations (top traces) and e.m.g. records (α, second row; γ in A, and $γ_S$ in B, bottom panels) were redrawn from Prochazka et al. (1985). The $γ_D$ and $γ_S$ firing profiles in C were redrawn from Murphy et al. (1984, Fig. 3). (From Hulliger, 1986, with permission)

First, the concept of **α-γ linkage** (Fig. 9.A; Table 1, A), based on early reflex studies (Granit, 1955) envisages a relatively close coupling of skeletomotor (α) and fusimotor (γ) efferents. As to locomotion the strongest evidence in its favour derives from Sjöström & Zangger's (1976) finding of tightly linked α- and γ-firing patterns during fictive locomotion in spinal cats. Note that in the studies listed in Table 1 (A) no attempt was made to differentiate between $\alpha-\gamma_S$ and $\alpha-\gamma_D$ linkage (cf. also Fig. 9.A with B and C).

Second, the notion of **modulated γ_S** combined with α-independent or even **tonic γ_D** activity (Fig. 9.B) restricts α-related behaviour to static efferents. On its own this might guarantee maintained Ia firing during muscle shortening, whilst additional γ_D drive would maintain suitable Ia sensitivity to disturbances (see servo assistance, above). For rhythmic movements this view is supported by the observation of simultaneously occurring α-coupled (γ_S) as well as tonic (γ_D) γ-firing patterns during reflexly induced jaw movements in anaesthetized cats (see Table 1, F).

Third, the concept of **modulated γ_D** combined with largely **tonic γ_S** action (Fig. 9.C; see also Fig. 5) is based on recordings of γ-discharge during treadmill locomotion in decerebrate cats (Murphy et al., 1984; Table 1, G). Although it is the reverse of the notion above, this pattern too might maintain spindle firing during shortening whilst restricting Ia sensitization to the stance phase of a step cycle.

Assessment of Theories in Simulation Experiments

Recently an experimental **simulation technique** has been introduced, which permits quantitative testing of explicit theories of fusimotor function, and which has been used to assess the applicability (to normal movements) of the concepts of fusimotor function which were derived from direct recordings from γ-efferents in reduced preparations (above, Fig. 9, and Table 1, A, F, G). For a detailed description of this approach see Hulliger & Prochazka (1983) and Hulliger (this volume). Briefly, normal movements, which were originally recorded in chronically implanted cats (Prochazka, 1984; also Fig. 10, left), are reproduced in nerve-muscle preparations of acutely anaesthetized animals (Fig. 10, right). The responses of spindle Ia afferents to these simulated movements are then recorded from dorsal root filaments whilst, at the same time, static and/or dynamic γ-efferents (isolated in ventral root filaments) are activated electrically, using rate modulated stimulation in order to test the effects of various patterns and constellations of the time course of fusimotor activation. These simulated Ia responses are then compared and assessed for goodness of match with target Ia responses, which in the chronic experiments had been recorded at the same time as the particular segment of movement under study.

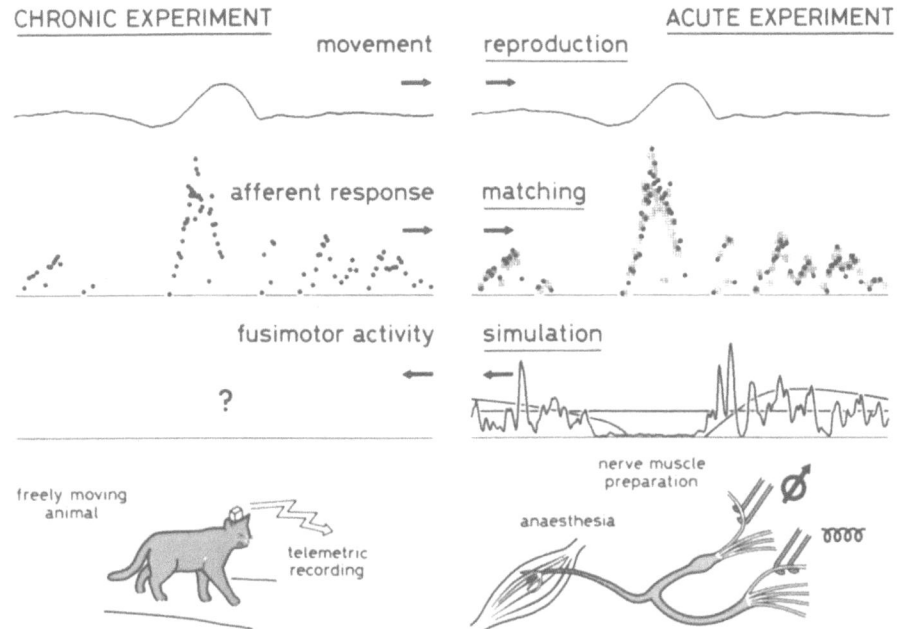

CHRONIC EXPERIMENT   ACUTE EXPERIMENT

movement   reproduction

afferent response   matching

fusimotor activity   simulation

?

freely moving
animal

telemetric
recording

nerve muscle
preparation

anaesthesia

Figure 10. Schematic illustration of the experimental simulation method which combines chronic with acute recordings (see also text). Note that the chronically recorded afferent response (left) is replotted on the right as target response (grey raster display) which is to be matched by simulated responses (right, black dots). The bottom traces on the right are examples of driving functions as used for rate modulated electrical stimulation of fusimotor efferents which were isolated in fine ventral root filaments. (From Hulliger et al., 1986, with permission)

Using this simulation technique the three concepts of fusimotion, which were derived from reduced preparations (preceding section), were experimentally evaluated by assessing their ability to account for Ia firing patterns which were chronically recorded during normal **self-paced locomotion** (gentle **walking**). It emerged that none of the three notions appeared to apply to Ia behaviour during physiological motor performance.

First, when reproduced stepping movements were accompanied by strictly **α-linked γ**-activity (stimulation rate modulated by e.m.g. envelope) simulated responses consistently failed to reproduce chronically recorded Ia discharge patterns (Hulliger et al., 1985, 1986; Prochazka et al., 1985). This was true irrespective of both type (γ_S, γ_D, $\gamma_S + \gamma_D$) and strength (modulation depth) of γ-action, and regardless of the number of efferents used for simulations.

Second, when the reproduction of the length variations of normal steps was combined with e.m.g. **modulated γ_S-** and **tonic γ_D-**

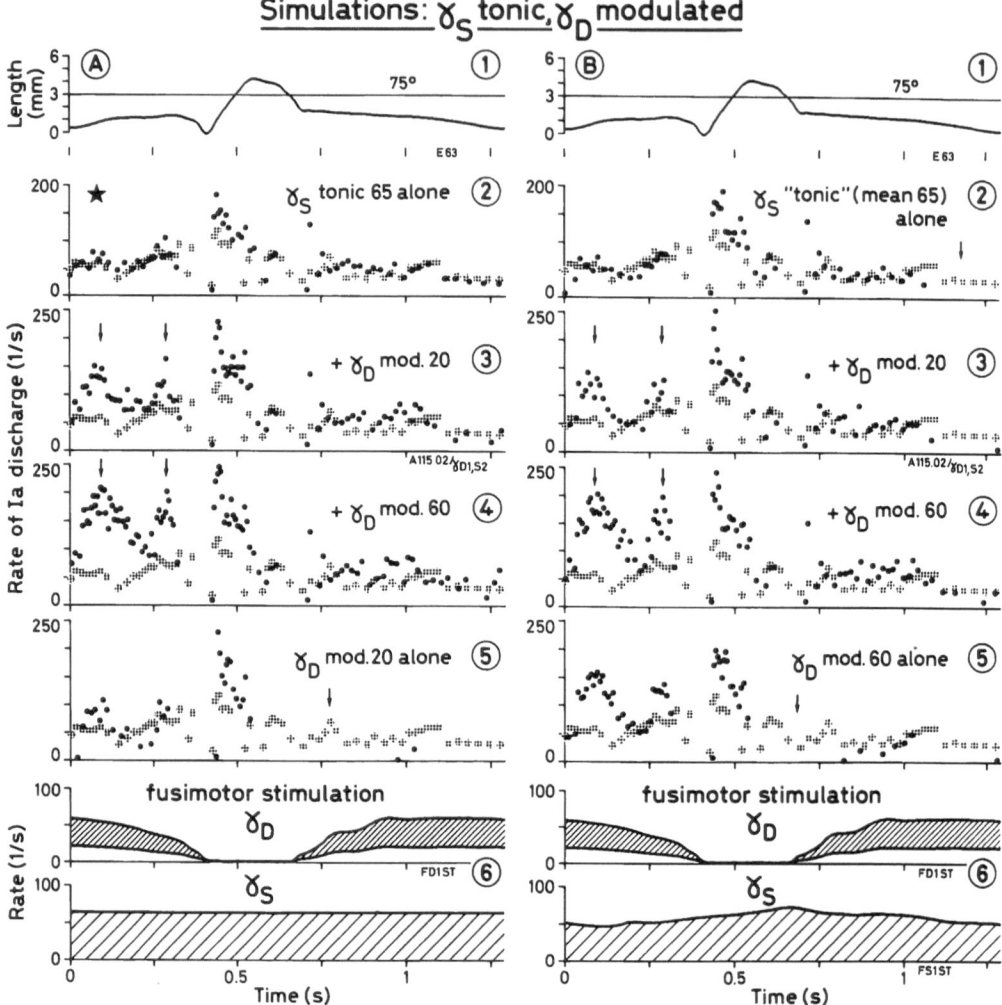

Figure 11. Tests of the notion of modulated γ_D action accompanied by largely tonic γ_S action (Murphy et al., 1984; see also Table 1, G). In A, strictly tonic γ_S drive was combined with rate modulated γ_D stimulation at two different values of modulation depth (20/s in 3, 60/s in 4, zero to peak values; see contours of densely hatched stimulation profiles in 6). In B, the same patterns of deeply modulated γ_D action were combined with weakly modulated γ_S action, the mean rate of stimulation being the same as in A. The stimulation profiles of B.6 are smoothed versions of the γ-discharge patterns recorded by Murphy et al. (1984, Fig. 3). In A.2 and B.2, effects of γ_S action alone; in A.5 and B.5, effects of modulated γ_D action alone. Note the satisfactory match of the target by the simulated response in A.2 (asterisk, tonic γ_S action alone). Arrows indicate segments of distinct deviations of the simulated responses from the target. (Data from Hulliger et al., 1986, with permission)

action, the chronic target responses were almost invariably poorly matched (Hulliger et al., 1986). However, in rare exceptional cases a high level of dynamic drive was required for successful matching of chronic recordings (see e.g. Hulliger et al., 1985). This tended to occlude concomitant modulated γ_S action, even when it was of intermediate strength, so that its presence in the original recording could not be excluded.

Third, when tonic or largely **tonic γ_S**-action was combined with deeply **modulated γ_D**-action (stimulation rates modulated by the γ_S- and γ_D-profiles of Murphy et al., 1984), simulated responses regularly failed to match chronic target responses, unless γ_D modulation was very small. Examples are illustrated in Fig. 11. Note that in this and the following illustrations chronic target responses are plotted as raster background profiles, whereas simulated responses are shown as superimposed black dots. In the clear majority of cases Ia responses during stepping were best matched by simple tonic γ_S action alone (similarity of raster and black-dot profiles in Fig. 11.A.2, asterisk; see also Fig. 12). Adding γ_D action of increasing modulation depth (profile in A.6) rapidly led to distinct mismatches between simulated and target response (arrows in A.3 and A.4). Substituting weakly modulated (B.6) for tonic (A.6) γ_S action did not make much difference. Whilst, for trials with γ_S stimulation alone, modulated action was slightly less satisfactory than tonic action (cf. B.2, arrow, with A.2), the extent of mismatch with additional modulated γ_D action was much the same in either case (cf. B.3, B.4 with A.3, A.4).

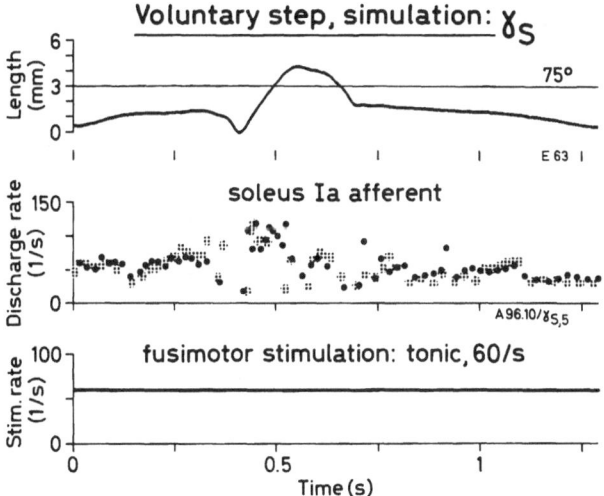

Figure 12. Simulation of chronically recorded discharge during normal walking. In the acute experiment the original movement was reproduced identically and accompanied by tonic activation of a single γ_S axon at 60/s. The chronic target is shown as raster pattern, the simulated response as a series of black dots. Same chronic data as in Fig. 11, but separate simulation experiment. (Data from Prochazka et al., 1985, redrawn)

Fusimotor set

The assessment (in simulation experiments) of chronically recorded Ia firing during <u>different types of movement</u> has led to
a new hypothesis of fusimotor function, <u>'fusimotor set'</u> (Hulliger et al., 1985; Prochazka et al., 1985), which is summarized below and schematically illustrated in Fig. 15 (note also
the list of key features in the figure legend). The main implications for general considerations of motor control are that
the balance between static and dynamic fusimotor action, and
hence the quality and information content of proprioceptive
feedback from muscle spindles, is strongly dependent on the
type of movement performed, and that this balance can be flex-
ibly adjusted to meet particular requirements of the motor
tasks which are executed.

For <u>routine movements</u>, such as normal locomotion, fusimotor
drive appears to be mainly static, tonic, and of low intensity, implying that Ia sensitivity to disturbances of planned
movements (see above, 'servo assistance') is unimpressive (unpublished observations). The evidence to support this conclusion is the finding, that Ia discharge during stepping was most
successfully reproduced when simulated movements were accompanied by tonic static γ-action. Satisfactory matches, as in
Figs. 11.A.2 and 12, were often obtained by activation of single γ_S axons at stimulation rates between 50 and 100/s. Since
muscle spindles normally receive a converging supply of between 5 and 10 static efferents (see Hulliger, 1984, review),
these optimal stimulation rates correspond, in a crude approximation, to single-fibre discharge rates of around 10/s (assuming distributed activation of static efferents). Thus, during routine movements fusimotor output appears to be of low
intensity and mainly restricted to static efferents.

In contrast, predominant activation of dynamic efferents, and
hence a striking <u>switch</u> of <u>γ-action</u> from static to dynamic,
has been deduced from recordings during separate motor paradigms, e.g. when animals resisted movements which were unexpec-
tely imposed (referred to as 'resisted stretch'). Thus there
was a dramatic change in fusimotor control of proprioceptive
feedback when <u>unfamiliar motor tasks</u> were executed. Examples
of Ia firing during resisted stretches, which were well simulated by dynamic, but not by static action (not illustrated),
are shown in Figs. 13 and 14. In Fig. 13 the two successive
peaks of the target response were best matched by only margin-
ally different, and relatively low, levels of tonic γ_D drive.
Yet in Fig. 14 two γ_D axons, both acting on the same Ia afferent, had to be activated simultaneously, with stimulation rates up two 100/s (in other cases even up to 150/s), in order to
match the very high firing rates of the chronic target. Since
the normal dynamic fusimotor supply of a spindle consists of 2
to 3 converging axons, the same estimation as above suggests
that individual dynamic efferents can be activated very power-
fully indeed (firing rates around 100/s).

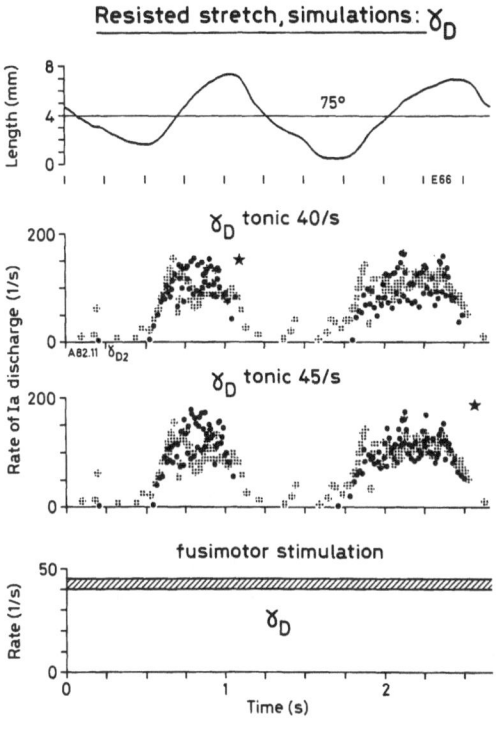

Resisted stretch, simulations: γ_D

Figure 13. Simulation of chronic primary afferent response to stretches, which were moderately resisted by the animal. The best matches were obtained with tonic dynamic action, but the optimal levels of γ_D drive were slightly different for the two peaks of the target response (asterisks). From top to bottom: reproduced movement, simulated (black dots) and chronic target (raster pattern) responses, time course of stimulation. Chronic data was from the same afferent as in Figs. 11 and 12, but simulations were performed in separate experiments. (From Prochazka et al., 1985, redrawn)

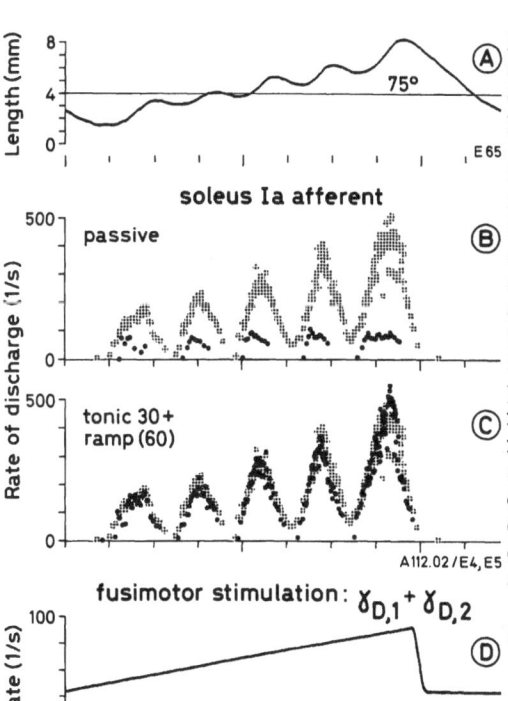

Figure 14. Gradual resetting of dynamic fusimotor drive during a sequence of unexpectedly imposed stretches (A). Best match (C) of chronic response was obtained with stimulation profile in D (slowly rising ramp superimposed on a tonic background component). Two - powerful - γ_D axons, both acting on the primary afferent of the acute simulation experiment, were stimulated simultaneously. In B, passive controls to show that mere passive reproduction of movement (absence of fusimotor action) could not possibly account for the chronically recorded firing. Same chronic afferent as in Figs. 11 to 13, separate simulation experiment. (From Hulliger et al., 1985, redrawn)

The findings of Figs. 11 to 13 put the emphasis on the occurrence of largely tonic fusimotor firing patterns during rhythmic movements (guided by rhythmic skeletomotor activity). However, this is not to imply that fusimotor activity is invariably set to fixed tonic levels. On the contrary, there are
observations which indicate that gradual changes of fusimotor
outflow may occur which, over a number of movement cycles,
lead to a _resetting_ of fusimotor activity of a given type from
one level of drive to another. For the target response of Fig.
14 levels of tonic γ_D drive could be found which gave satisfactory simulations, but only of individual peaks and not of
all peaks simultaneously in a single trial. More successful
matching was achieved (C) when a gradually increasing component was superimposed on a tonic background level of dynamic
drive (D).

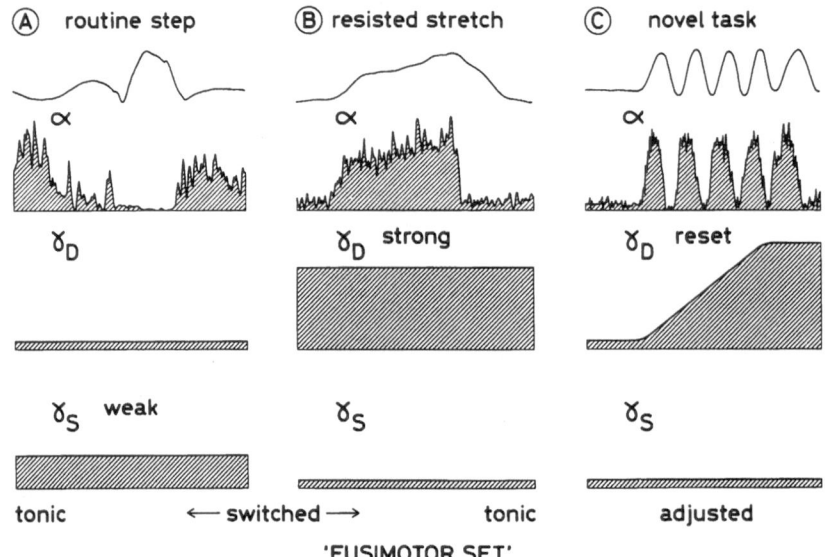

Figure 15. Schematic illustration of **'fusimotor set'**. From top
to bottom, time course of movement, skeletomotor (α), dynamic
fusimotor (γ_D), and static fusimotor (γ_S) activity. The notion
describes task-dependent fusimotor setting and resetting:
1. Under **steady state** conditions, and for movements of a given
   category, fusimotor activity is largely **tonic** (A, B), even
   when α-activity shows deep rhythmic modulation.
2. Fusimotor activity may **switch** from predominantly **static** and
   a low level of drive, during **routine tasks** (stepping, A),
   to predominantly **dynamic** and a high level of drive, during
   novel tasks (resisted stretch, B), the switch depending on
   the type of motor task performed.
3. During phases of motor adjustment fusimotor activity of a
   given type may also be subjected to gradual changes over
   several movement cycles (C). This leads to **resetting** from
   one level of drive to another, and it may entail a redistribution of the relative weight given to γ_D and γ_S action.
(From Hulliger, 1986, with permission)

These observations have been generalized to provide the concept of **'fusimotor set'** (Hulliger et al., 1985; Prochazka et al., 1985). On this view central control of fusimotor output can be described in terms of largely tonic γ_S- and/or γ_D-activity (Fig. 15 A,B), to 'set' Ia sensitivity, and of striking changes in γ_S/γ_D balance, to 'reset' Ia sensitivity according to the demands of a given motor task. Such resetting may be abrupt, between two successive cycles of movement (not illustrated), or it may take place over a number of seconds and movement cycles (as in Fig. 14; also Fig. 15.C) leading to gradual adjustments of the static/dynamic balance.

It has long been known that the central nervous system can gate or weight sensory feedback, to either ignore it or treat it with special attention. This is achieved by the action of descending pathways from higher centres of the brain, which control synaptic transmission at the sensory relay stations (such as the dorsal column nuclei or the thalamus). The notion of fusimotor set now suggests that such weighting of proprioceptive feedback can already occur in the periphery, at the level of individual sensors by flexible adjustment of their sensitivity. Thus, the concept of 'fusimotor set' may be a special case of the general control principle characterized by the notions of 'motor set' and 'sensory set' (Evarts et al., 1984), and it might well constitute a peripheral link between these two.

Acknowledgements: This work was supported by grants from the Swiss National Science Foundation (3.157.81, 3.071.84), the Jubiläumsstiftung and Forschungsstiftung of the University of Zürich, the Dr. Eric Slack-Gyr Foundation, and the European Science Foundation (ETP). Thanks are extended to Drs. F. Emonet-Dénand and R. Sonnenberg for permission to use previously unpublished material (Figs. 1 and 3) and to Dr. A. Prochazka for permission to use some of his original recordings for the illustrations redrawn for the present paper (Figs. 12 to 14).

References

Appelberg B (1981) Selective central control of dynamic gamma motoneurones utilised for the functional classification of gamma cells. In: Taylor A, Prochazka A (eds) Muscle receptors and movement. Macmillan, London, pp 97-107

Appenteng K, Prochazka A (1984) Tendon organ firing during active muscle lengthening in awake, normally behaving cats. J Physiol(Lond) 353:81-92

Appenteng K, Morimoto T, Taylor A (1980) Fusimotor activity in masseter nerve of the cat during reflex jaw movements. J Physiol(Lond) 305:415-431

Burgess PR, Clark FJ (1969) Characteristics of knee joint receptors in the cat. J Physiol(Lond) 203:317-335

Burke D (1981) The activity of human muscle spindle endings in normal motor behavior. Int Rev Physiol 20:91-136

Cabelguen JM (1981) Static and dynamic fusimotor controls in various hindlimb muscles during locomotor activity in the decorticate cat. Brain Res 213:83-97

Carli G, Fontani G, Meucci M (1981) Static characteristics of muscle afferents from gluteus medius muscle: comparison with joint afferents of hip in cats. J Neurophysiol 45: 1085-1095

Clark FJ, Burgess PR (1975) Slowly adapting receptors in cat knee joint: can they signal joint angle? J Neurophysiol 36: 1448-1463

Critchlow V, von Euler C (1963) Intercostal muscle spindle activity and its gamma motor control. J Physiol(Lond) 168: 820-847

Eklund G, von Euler C, Rutkowski S (1964) Spontaneous and reflex activity of intercostal gamma motoneurones. J Physiol (Lond) 171:139-163

Evarts EV, Shinoda Y, Wise SP (1984) Neurophysiological approaches to higher brain functions. Wiley, New York

Goodwin GM, Luschei E (1975) Discharge of spindle afferents from jaw closing muscles during chewing in alert monkeys. J Neurophysiol 38: 560-571

Goodwin GM, Hulliger M, Matthews PBC (1975) The effects of fusimotor stimulation during small amplitude stretching on the frequency response of the primary ending of the mammalian muscle spindle. J Physiol(Lond) 253:175-206

Goslow GE, Reinking RM, Stuart DG (1973) The cat step cycle: hind limb joint angles and muscle lengths during unrestrained locomotion. J Morphol 141:1-41

Gottlieb S, Taylor A (1983) Interpretation of fusimotor activity in cat masseter nerve during reflex jaw movements. J Physiol(Lond) 345:423-438

Granit R (1955) Receptors and sensory perception. Yale University Press, New Haven

Hagbarth K-E (1981) Fusimotor and stretch reflex functions studied in recordings from muscle spindle afferents in man. In: Taylor A, Prochazka A (eds) Muscle receptors and movement. Macmillan, London, pp 277-286

Hasan Z (1983) A model of spindle afferent response to muscle stretch. J Neurophysiol 49:989-1006

Hulliger M (1984) The mammalian muscle spindle and its central control. Rev Physiol Biochem Pharmacol 101:1-110

Hulliger M (1986) The role of muscle spindle receptors and fusimotor neurones in the control of movement. Electroenceph Clin Neurophysiol (in press)

Hulliger M (1986) An iterative and interactive simulation method for reconstruction of unknown inputs contributing to known outputs of neuronal systems (this volume)

Hulliger M, Prochazka A (1983) A new simulation method to deduce fusimotor activity from afferent discharge recorded in freely moving cats. J Neurosci Methods 8:197-204

Hulliger M, Sonnenberg R (1985) Does the paradoxical γ_D-mediated reduction of Ia sensitivity to small stretches persist during larger background movements? Neurosci Letters Suppl 22:S595

Hulliger M, Matthews PBC, Noth J (1977a) Static and dynamic fusimotor action on the response of Ia fibres to low frequency sinusoidal stretching of widely ranging amplitude. J Physiol(Lond) 267:811-838

Hulliger M, Matthews PBC, Noth J (1977b) Effects of combining static and dynamic fusimotor stimulation on the response of the muscle spindle primary ending to sinusoidal stretching. J Physiol(Lond) 267:839-856

Hulliger M, Zangger P, Prochazka A, Appenteng K (1985) Fusimotor "Set" vs. α-γ linkage in voluntary movement in cats. In: Struppler A, Weindl A (eds) Electromyography and evoked potentials. Adv Appl Neurol Sci 1: 56-63

Hulliger M, Prochazka A, Zangger P (1986) Fusimotor activity in freely moving cats. Tests of concepts derived from reduced preparations. In: Grillner S, Stein PSG, Forssberg H, Herman RM, Stuart DG, Wallén P (eds) Neurobiology of vertebrate locomotion. Macmillan, London (in press)

Jansen JKS, Walloe L (1970) Signal transmission between successive neurons in the dorsal spinocerebellar pathway. In: Schmitt FO (ed) The neurosciences, second study program. Rockefeller University Press, New York, pp 617-629

Laporte Y, Emonet-Dénand F, Jami L (1981) The skeletofusimotor or β-innervation of mammalian muscle spindles. TINS 4:97-99

Larson CR, Smith A, Luschei ES (1981) Discharge characteristics and stretch sensitivity of jaw muscle afferents in the monkey during controlled isometric bites. J Neurophysiol 46:130-142

Loeb GE (1984) The control and responses of mammalian muscle spindles during normally executed motor tasks. Exercise Sports Sci Rev 12:157-204

Loeb GE, Hoffer JA (1985) Activity of spindle afferents from cat anterior thigh muscles. II. Effects of fusimotor blockade. J Neurophysiol 54:565-577

Loeb GE, Hoffer JA, Pratt CA (1985) Activity of spindle afferents from cat anterior thigh muscles. I. Identification and patterns during normal locomotion. J Neurophysiol 54:549-564

Matthews PBC (1962) The differentiation of two types of fusimotor fibre by their effects on the dynamic response of muscle spindle primary endings. Q J Exp Physiol 47:324-333

Matthews PBC (1972) Mammalian muscle receptors and their central action. Arnold, London

Matthews PBC (1981a) Evolving views on the internal operation and functional role of the muscle spindle. J Physiol(Lond) 320:1-30

Matthews PBC (1981b) Muscle spindles: their messages and their fusimotor supply. In: Brookhart JM, Mountcastle VB, Brooks VB (eds) Motor control, part 1. American Physiological Society, Bethesda, pp 189-228 (Handbook of physiology, sect 1, vol 2)

Matthews PBC, Stein RB (1969a) The sensitivity of muscle spindle afferents to small sinusoidal changes of length. J Physiol(Lond) 200:723-743

Matthews PBC, Stein RB (1969b) The regularity of primary and secondary muscle spindle afferent discharges. J Physiol (Lond) 202:59-82

Murphy PR, Stein RB, Taylor J (1984) Phasic and tonic modulation of impulse rates in γ-motoneurons during locomotion in premammillary cats. J. Neurophysiol 52:228-243

Perret C, Berthoz A (1973) Evidence of static and dynamic fusimotor actions on the spindle response to sinusoidal stretch during locomotor activities in the cat. Exp Brain Res 18: 178-188

Prochazka A (1984) Chronic techniques for studying neurophysiology of movement in cats. In: Lemon R (ed) Methods of neuronal recording in conscious animals. IBRO handbook series: methods in the neurosciences 4. Wiley, Chichester, pp 113-128

Prochazka A, Hulliger M (1983) Muscle afferent function and its significance for motor control mechanisms during voluntary movements in cat, monkey, and man. In: Desmedt JE (ed) Motor control mechanisms in health and disease. Adv Neurol 39:93-132

Prochazka A, Hulliger M, Zangger P, Appenteng K (1985) "Fusimotor set": new evidence for α-independent control of γ-motoneurones during movement in the awake cat. Brain Res 339:136-140

Proske U (1981) The Golgi tendon organ. Properties of the receptor and reflex action of impulses arising from tendon organs. Int Rev Physiol 25:127-171

Rossi A, Grigg P (1982) Characteristics of hip joint mechanoreceptors in the cat. J Neurophysiol 47:1029-1042

Schaible H-G, Schmidt RF (1983) Responses of fine medial articular nerve afferents to passive movement of knee joint. J Neurophysiol 49:1118-1126

Schaible H-G, Schmidt RF (1985) Effects of an experimental arthritis on the sensory properties of fine articular afferent units. J Neurophysiol 54:1109-1122

Schieber MH, Thach WT (1985) Trained slow tracking. II. Bidirectional discharge patterns of cerebellar nuclear, motor cortex, and spindle afferent neurons. J Neurophysiol 54: 1228-1270

Schwarz M, Sontag K-H, Wand P (1984) Non-dopaminergic neurones of the reticular part of substantia nigra can gate static fusimotor action onto flexors in cat. J Physiol(Lond) 354: 333-344

Sears TA (1964) Efferent discharges in alpha and fusimotor fibres of intercostal nerves of the cat. J Physiol(Lond) 174:295-315

Severin FV (1970) The role of the gamma motor system in the activation of the extensor alpha motor neurones during controlled locomotion. Biophysics 15:1138-1145

Severin FV, Orlovskii GN, Shik ML (1967) Work of the muscle receptors during controlled locomotion. Biophysics 12:575-586

Sjöström A, Zangger P (1976) Muscle spindle control during locomotor movements generated by the deafferented spinal cord. Acta Physiol Scand 97:281-291

Stein RB (1974) Peripheral control of movement. Physiol Rev 54:215-243

Taylor A, Appenteng K (1981) Distinctive modes of static and dynamic fusimotor drive in jaw muscles. In: Taylor A, Prochazka A (eds) Muscle receptors and movement. Macmillan, London, pp 179-192

Taylor J, Stein RB, Murphy PR (1985) Impulse rates and sensitivity to stretch of soleus muscle spindle afferent fibers during locomotion in premammillary cats. J Neurophysiol 53:341-360

Vallbo ÅB, Hulliger M (1981) Independence of skeletomotor and fusimotor activity in man? Brain Res 223:176-180

Vallbo ÅB, Hagbarth K-E, Torebjörk HE, Wallin G (1979) Somatosensory, proprioceptive and sympathetic activity in human peripheral nerves. Physiol Rev 59:919-957

Walmsley B, Hodgson JA, Burke RE (1978) Forces produced by medial gastrocnemius and soleus muscles during locomotion in freely moving cats. J Neurophysiol 41:1203-1216

PHYSIOLOGY AND PSYCHOPHYSICS IN TASTE AND SMELL

K.C. Persaud, J.A. DeSimone and G.L. Heck

Dept. of Physiology and Biophysics, Medical College of Virginia, Virginia Commonwealth University, Box 551, MCV Station, Richmond, Va 23298, U.S.A.

1. INTRODUCTION

The ability to recognize chemical stimuli in the external environment is a feature of living organisms ranging from bacteria to man. It has played a crucial role in the search for food, detection of adverse environments, and sexual behaviour. The transduction process of chemoreception in both lower organisms and higher vertebrates consists of binding of chemical stimuli to a receptor membrane, giving rise to a receptor generator potential. This potential either triggers a chemotactic response in lower organisms or is transmitted in the form of neural impulses to signal processing centres in higher animals. In vertebrates, chemical stimuli in external environments are received at gustatory and olfactory receptor cells. These cells detect chemicals, encode information about the intensity, duration and quality and transmit it to higher processing centres. This chapter discusses some aspects of current research in the physiology and psychophysics of taste and smell in vertebrates.

Both olfactory and gustatory stimuli are presented to the sensory membranes as aqueous solutions. Airborne chemicals are partitioned from a gas to a liquid phase in the case of olfaction. Prior to detection and transduction of a chemical stimulus, the chemical must be convected to the vicinity of the receptors and diffuse through barriers of mucus in the case of olfaction or through taste pores in the case of gustation.

Chemoreceptory quality is multidimensional - there are correlations between size, shape, charge and hydrophobicity of stimulatory molecules to the sensation perceived (Boelens, 1974). The sensitivity of the olfactory system is of the order of 10^{-9} M for many odorants and the recognition capabilities are immense. There are likely to be many different types of odorant receptor sites on the olfactory mucosa, each with rather broad specificity, or alternatively, each sensory cell could contain a mixture of several receptor types of higher specificity (Beets, 1978). This gives rise to many different qualities of perception. Where the sense of taste is concerned, little is also known about the nature of the gustatory receptors. The sensitivity of the gustatory system is lower than the olfactory system being of the order $10^{-4} - 10^{-5}$ M for most compounds. The

NATO ASI Series. Vol. F43
Sensors and Sensory Systems
for Advanced Robots
Edited by P. Dario
© Springer-Verlag Berlin Heidelberg 1988

classical view has been that there are four taste qualities
- sweet, sour, salty and bitter, the Japanese adding a fifth
one called "umami" describing the taste perceived with
monosodium glutamate (Arai *et al*, 1973). This implies that
there are only four or five taste "receptors". This view is
likely to be simplistic since many of these stimuli interact
with each other producing different taste perceptions. To
this effect, Fring (1948), Erickson (1963) and Schiffman &
Erickson (1971) have suggested that the common four tastes
are points of familiarity in a continuous taste spectrum.

The fundamental problems facing any sensory system are :-

(a) Is anything there ? - the detection problem;

(b) What is it ? - the problem of recognition;

(c) How much of it is there ? - the problem of scaling;

(d) Is one stimulus different from another ? - the problem
of discrimination.

In order to determine how a living organism solves these
problems the researcher has to understand :-

(a) The nature of the chemoreceptors;

(b) The mechanism of transduction;

(c) The neural coding mechanisms for intensity and quality;

(d) The nature of the higher information processing
pathways.

Although the chemical senses have been socially and
economically important to man throughout the centuries, it
has only been in the last two decades that multidisciplinary
research into the nature of these senses has given any clues
as to the nature of chemoreception.

2. GUSTATION

2.1 The Chemo-sensors

The sensors of the gustatory system are the taste bud
cells which are concentrated over the upper surface of the
tongue, though they are also found on the palate and
pharynx. The taste buds are located in groups within raised
specialised papillae, each bud consisting of 40 to 50
transducer cells grouped together in a flask-shaped
structure (Fig. 1). There is an opening at the top called
the taste pore, through which the chemical stimulus
diffuses. Microvilli extend from the cells into the taste
pore and are in contact with saliva. These are thought to be
chemosensitive. Each cell is innervated by fine nerve fibres
conducting taste information to the brain (Beckstead &

Norgren, 1979). Although the individual cells within the
taste bud are continually replaced, the integrity of the
taste bud is constantly maintained – the taste nerve endings
continuously move and innervate new taste cells (Guth,
1971). Morphologically different papillae are observed. The
fungiform papillae in the anterior two thirds of the dorsal
tongue surface contain 1-5 taste buds and are innervated by
the chorda tympani nerve. Several larger circumvallate
papillae in the posterior part of the tongue contain
hundreds of taste buds which are innervated by the
glossopharyngeal nerves. Foliate papillae on the sides of
the tongue also contain many taste buds and these may be
innervated by both the glossopharyngeal and chorda tympani
nerves. Little is known about the central taste pathways.
From the tongue, the nerves project to the thalamus, and at
least two cortical taste areas, but gustatory information
also reaches many other brain areas such as the hypothalamus
and amygdala (Beidler, 1980).

From the above description, the gustatory system can be
regarded as a convergent neural network, with a large number
of chemo-sensors in a pseudo-parallel configuration sending
information to a multilevel processing system, with outputs
at different levels of signal processing.

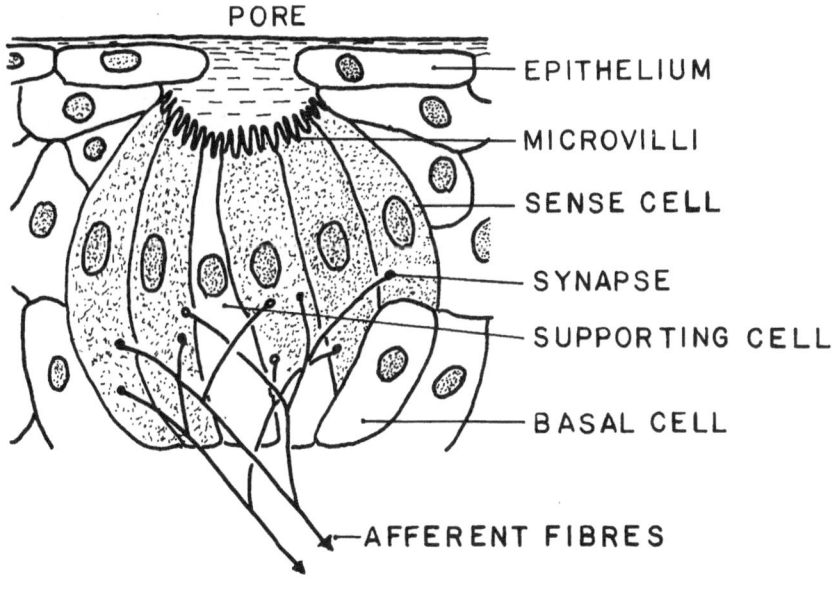

Fig. 1 Diagram of a taste bud.

2.2 Methodology in Gustation

Until comparatively recently, there were very few probes available for understanding how the chemical senses work. The classical approach has been to ask questions of human subjects such as

(a) What does this taste of ?

(b) How much is there ?

(c) How different is it from something else ?

This enters the field of psychophysics where an observer is required to

(a) partition a given segment of a sensation continuum into a predetermined number of subjectively equal intervals;

(b) make a direct judgement of apparent sensation magnitudes.

By suitable scaling it is possible to create a multi-dimensional space where groups of compounds clump together depending on how closely they are identified to each other (Schiffman, 1974).

One can ask similar questions of animals using the technique of aversive conditioning. If an unpleasant tasting compound is given to an animal, it will avoid it as well as mixtures of that compound with others. One can then ask the animal how close is another compound in taste to one of the compounds the animal is aversive to (LeMagnen, 1971).

This methodology helps to delineate the type of information that is produced by a taste stimulus. It has been demonstrated that precisely designed molecules in defined classes such as sugars or dihydrochalcones will give defined taste quality e.g. sweet. By chemically modifying these molecules stepwise, the taste perceived can also be changed stepwise to bitter (Birch, 1981). This structure-activity relationship may give some information about the nature and number of different types of the taste receptor binding sites. It is now possible to categorize certain molecular parameters such as size, shape, hydrophilicity, lipophilicity, chirality or chemical substitution which define a particular taste effect. In spite of this there is still very little understanding of vertebrate receptor structure at the molecular level.

Some progress has been made using electrophysiological techniques to measure either the integrated neural responses from the chorda tympani and glossopharyngeal nerves or the responses from individual nerve fibres to taste stimuli placed on the tongue. Both nerves respond to all stimuli, but there are differences in sensitivity between the two towards different compounds. Thus the glossopharyngeal nerve is exquisitely sensitive to the bitter taste of quinine, while the chorda tympani exhibits high sensitivity to sodium chloride (Frank & Pfaffmann, 1969). Single fibers respond to more than one taste stimulus but in varying degrees, so that the chemical specificity of single fibres is relative. However there is a quantitative difference in the electrical activity produced by gustatory stimuli among fibres and several categories of fibres may be identified (Sato, 1971). The frequency of impulse discharge immediately after gustatory stimulation is expressed by a power function of stimulus strength. Many nerve fibres also give responses to thermal stimuli placed on the tongue, and it is probable that this may also be encoded for taste perception.

The electrophysiological results suggest that sensory quality does not depend on the activation or deactivation of a particular nerve fibre alone, but on the pattern of other active fibres. Thus the message describing the gustatory stimulus is expressed in terms of the relative amounts of neural activity across many neurons. Some of these across-fibre patterns have been investigated by Erickson (Erickson, 1963), and correlation matrices between pairs of stimuli have been demonstrated to give good indication of similarities and differences between compounds, in their perceived taste. These correlations have been expanded by Schiffman *et al.* (1978) using multidimensional scaling techniques. Intensity coding, on the other hand, appears to be a function of the frequency of firing rate for nerve fibres. This may also be a very complex phenomenon, since Bartoshuk (1977) showed that perceived intensity of taste when plotted against concentration can show either synergism or suppression when the taste stimulus is presented as a mixture of different compounds. Thus a number of compounds are known to enhance sweet taste, such as some surfactants, or depress sweet taste, such as gymnemic acid.

Deutsch (1967) has set out a number of models for the instinctive recognition of chemical patterns. Assuming four primary modalities of sensation, ABCD, which are nonoverlapping in specificity, then the recognition of a specific compound X can be set out in terms of

$$X_{10} = A_{84}B_0C_0D_{286}$$

where compound X has a stimulus amplitude of 10 units and contains for example 84 stimulus units of A and 286 stimulus units of D but does not contain any B or C component. Since in practice, the specific responses of four different taste buds to four primary modalities overlap each other, then a transformation matrix may be constructed using experimental data which looks like:

Effective Stimuli in Receptors

		A	B	C	D
Input Elements	A1 (salt)	0.29	0.29	0.08	0.34
	B1 (sour)	0.04	0.46	0.04	0.46
	C1 (sweet)	0.08	0.15	0.62	0.15
	D1 (bitter)	0.09	0.17	0.14	0.60

then the effective stimulus matrix for compound X_{10} is given by a matrix multiplication

[Input X] . [Transformation Matrix] = [Effective X]

Effective Stimuli in Receptors

		A	B	C	D
Input Elements	A_{84}	24.4	24.4	6.7	28.5
	B_0	0	0	0	0
	C_0	0	0	0	0
	D_{286}	25.8	48.6	40.0	171.6
	Sums	50	73	47	200

so that the effective stimuli perceived at the neural level are given by

$$X_{10} = A_{50} B_{73} C_{47} D_{200}$$

which is a unique descriptor of compound X. Such a model can be developed further in terms of neural processing networks in a biological system, and as more raw experimental data becomes available it may be possible to construct more realistic models of the pattern recognition systems in the chemical senses.

2.3 Salt Taste Transduction

This section discusses a specific example of current research into the transduction mechanism responsible for salt taste. Here, the input from both electrophysiological techniques as well as psychophysical techniques have complemented each other in the elucidation of the events surrounding the excitation of mammalian taste-bud cells by salt.

It was formerly thought that the dorsal lingual

epithelium acts as an impermeable barrier against ions and other taste stimuli and that the interaction between stimulus ions and taste buds was adsorption to a receptor (Beidler, 1971). DeSimone *et al.*, 1981 were the first to utilize an *in vitro* Ussing preparation (Ussing and Zerahn, 1951) of the canine lingual epithelium, and subsequently the rat lingual epithelium, and were able to show that the dorsal and ventral surfaces of the tongue are actively engaged in ion transport. When Krebs-Henseleit buffer was placed symmetrically across the tissue, ouabain, a potent blocker of sodium / potassium ion cotransport, caused the short circuit current to decay to zero. When hyperosmotic sodium chloride was placed on the dorsal surface of the tongue an electrical response was recorded. Both the short circuit current as well as the open circuit potential of this response increase nonlinearly, and a sigmoid saturating curve results with increasing sodium chloride concentration. The integrated chorda tympani neural response in the rat is also sigmoidal in shape. The diuretic amiloride, which is a potent inhibitor of sodium ion transport in a variety of epithelial tissues, was applied to the dorsal lingual surface. Fig. 2(a) shows the open circuit potential resulting from 0.5 M NaCl solutions placed on the mucosal surface of the *in vitro* canine tongue preparation. The response was substantially attenuated when 10^{-4} M amiloride was added to the mucosal adapting solution prior to stimulation with sodium chloride. Fig. 3 shows the open-circuit response to 0.5 M NaCl before and after application of 10^{-4} M amiloride in 10^{-3} M NaCl for 10 minutes. Both the rapid rise and the steady state voltage were reduced, and this effect was selective for NaCl over KCl. It has been shown that the amiloride sensitive currents are coupled to early events in gustatory transduction and subsequent neural events since it is possible to reversibly and specifically block the neural response to NaCl by rinsing the rat's tongue in amiloride. The neural response to KCl is unaffected.

In a recent and elegant improvement in technology, Heck *et al*, (1986) have been able to record simultaneously both the short circuit current response from the lingual epithelium as well as the integrated chorda tympani response from an *in vivo* rat preparation in response to a gustatory stimulus. The simultaneous recordings to a 1M NaCl stimulus are shown in Fig. 3. When the tongue is treated with amiloride, both the short circuit current as well as the integrated chorda tympani response are substantially reduced (Mierson, *et al.*, 1986).

The validity of these findings have been confirmed by human taste psychophysical techniques (Schiffman, 1983). In human subjects, 500 μM amiloride was delivered to one half of the tongue and a filter paper soaked in deionised water was applied to the other side to serve as a control. A standard concentration of a salt solution

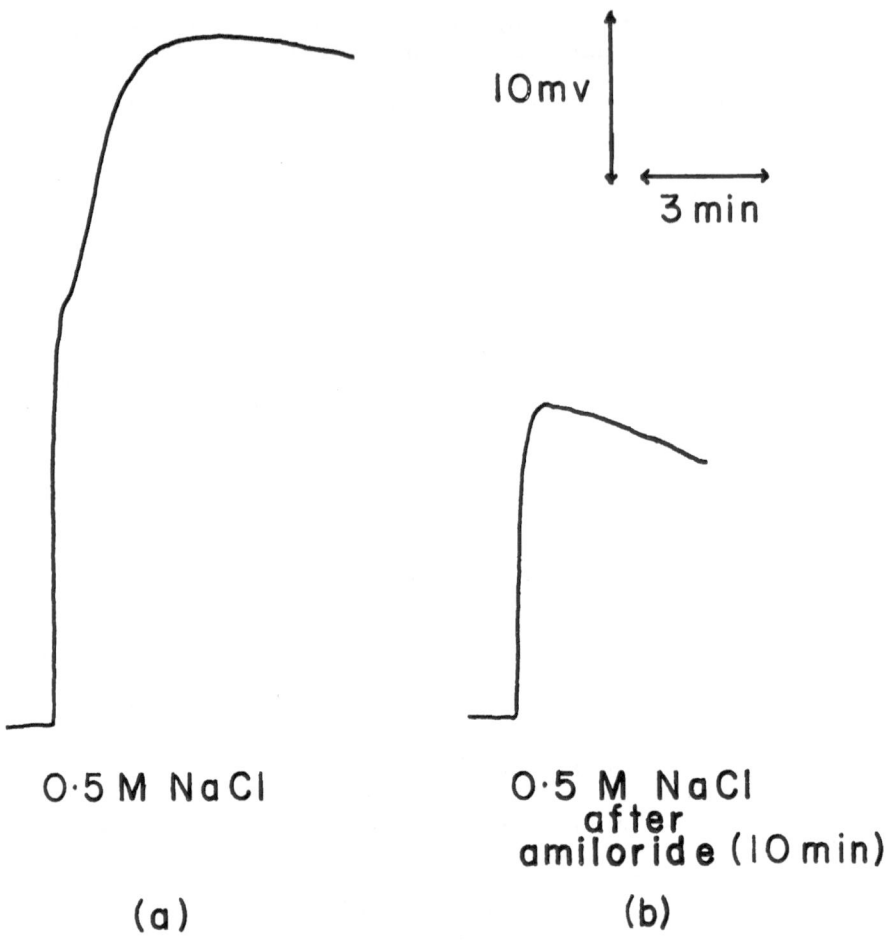

0·5 M NaCl

0·5 M NaCl
after
amiloride (10 min)

(a)

(b)

Fig. 2. Open circuit response to NaCl using the *in vitro* canine lingual preparation.

(a) The response to 0.5 M NaCl showing a characteristic two component response – an initial fast component followed by a slower second component which eventually reaches a steady state.

(b) The response to 0.5 M NaCl after the tissue had been treated with 10^{-4} M amiloride for 10 min. The baseline current represents 1 mM NaCl.

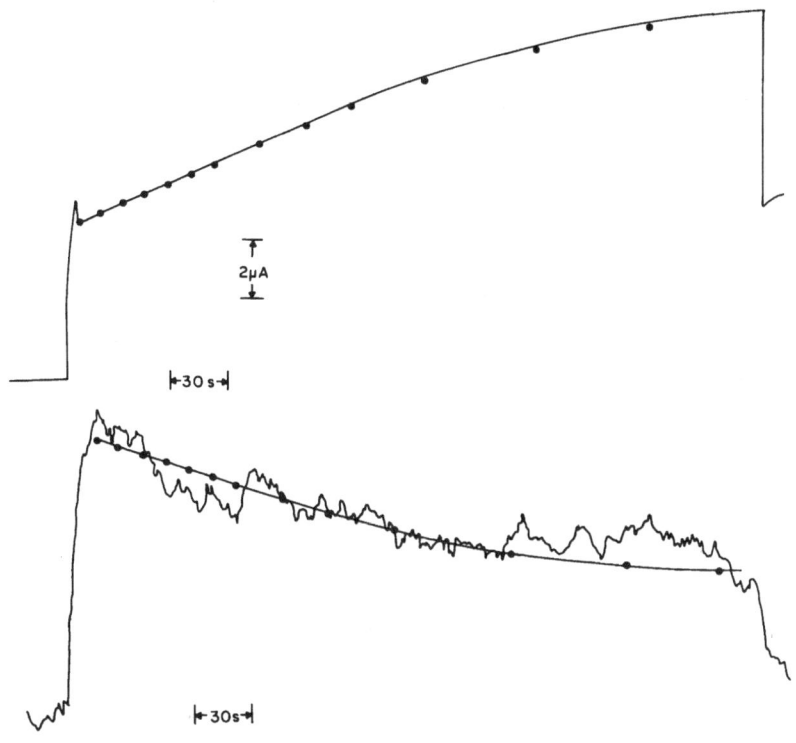

2µA

|←30 s→|

|←30s→|

<u>**Fig. 3**</u> Simultaneous recording of short circuit current response in the lingual epithelium and the chorda tympani neural response.

The upper trace shows the short circuit response from an *in vivo* rat lingual preparation in response to 1M NaCl. The lower trace shows the simultaneous integrated recording of the chorda tympani response to the stimulus. The exponential time course of the slower secondary component of the lingual response has been scaled and superimposed on the secondary component of the chorda tympani response showing that both follow the same exponential time course.

impregnated in circles of chromatography paper was placed on the side of the tongue to which amiloride had been applied. Test stimuli were then applied to the other side of the tongue which had not been treated with amiloride. The concentrations of the test stimuli were then adjusted to match the perceived intensity of the standard. It was found that amiloride reduced the taste intensity of both sodium and lithium salts. The perceived intensity of 0.20 M, 0.40 M and 0.60 M NaCl decreased by 50.0%, 57.5% and 56.7% respectively. The amiloride had no effect on the taste perception of potassium or calcium salts but had a moderate effect on the perception of choline chloride.

Schiffman *et al.* also undertook neurophysiological recordings of single neurons from the nucleus tractus solitarius of the medulla where afferent nerve fibres from the tongue are terminated. The results indicated that amiloride placed on the tongue suppresses the the responses to NaCl but leaves the response to KCl relatively unaffected.

Thus it has been shown that the taste bud cells contain an amiloride sensitive sodium ion transport pathway which plays an important role in NaCl taste transduction. This pathway serves simultaneously as a specific Na recognition site and the means by which the current that causes depolarization is transmitted. The failure of the response to potassium ions to be affected indicates that there are different pathways involved in the transduction of other salt stimuli. The salt taste response to sodium chloride is due to an ion transport system resembling other epithelial transport systems, such as those found in the gut and kidneys.

The response to sweet taste is also affected by amiloride. The transduction system for sweet taste may involve a receptor site which modulates an amiloride sensitive ion channel. Effects on the bitter or sour modalities are still to be investigated.

3. OLFACTION

The fundamental problems outlined in Section 1 are very applicable to the olfactory sense. Very little is known about the nature of the olfactory receptors, the mechanism of transduction, or the neural coding mechanisms for intensity and quality.

3.1 The odour-sensor structures.

The sensory mucosa in vertebrates consists of three major cell types, sensory, supporting and basal (Fig. 4). The sensory neurons are bipolar nerve cells with slender rod-like processes or dendrites extending to the surface of the epithelium. These dendrites form knob-like endings of

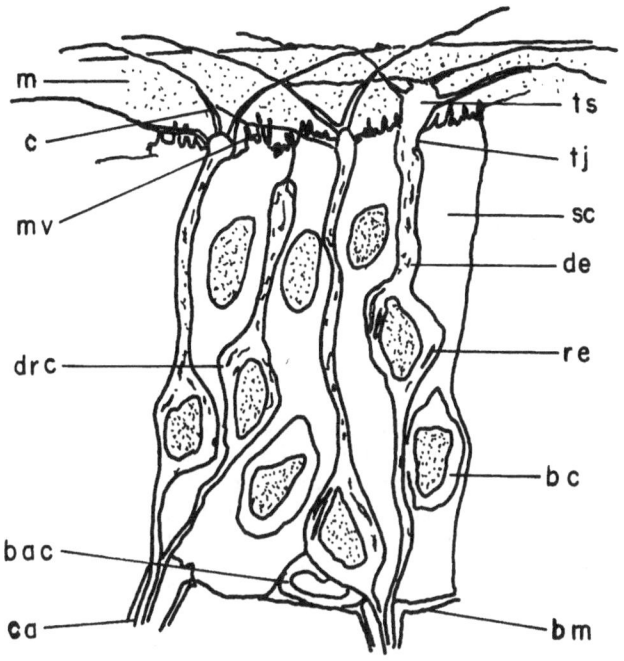

Fig. 4. Transverse Section of a generalised vertebrate olfactory epithelium (adapted from Andres, 1966).

m; mucus layer, c; cilia, ts; terminal swelling, mv; microvilli, tj; tight junction, sc; supporting cell, de; dendrite, re; receptor cell, drc; developing receptor cell, bc; blastema cell, bac; basal cell, bm; basement membrane, oa; olfactory axons.

diameter 1-2 μm at the surface of the epithelium. Long olfactory cilia arise from these structures. The axons of the olfactory neurons extend to the brain where they synapse with the dendrites of the mitral cells in the glomeruli of the olfactory bulb. The axons are nonmyelinated and are among the thinnest fibres of the body. As a result their rate of electrical conduction is slow (0.2 m/sec) (Moulton & Beidler, 1967). In vertebrates, olfactory receptor neurons are arranged between a superficial layer of supporting cells above a layer of basal cells lying on the basal lamina.

Much ultrastructural research has concentrated on the characterization of the olfactory cilia since, by the nature of their location, these are likely to contain the odorant receptor sites of the olfactory system. Menco *et al* (1976) using electron micrograph freeze-fracture techniques have shown that large numbers of particles occur in the membranes of the olfactory cilia of the cow. These have been subsequently shown to occur in several animals (Menco, 1978, Usukura & Yamada, 1978). These particles may have a function in the sensory process and they may be the odorant receptor sites of the olfactory receptor cell (Menco, 1977).

The supporting cells are approximately cylindrical in shape with a diameter of 4-5 μm. These cells possess large numbers of microvilli which project into the mucus coating which covers the olfactory epithelium. The cells have a folded appearance, and the folds apparently surround the receptor cells and are important in the electrical isolation of adjacent dendrites from each other (Shepherd and Ottoson, 1978). The cells may be essential accessories to the receptor cells and may have secretory and nutritional functions in addition to mechanical support functions.

The basal cells are arranged in an irregular manner on a thin basement membrane. These cells form a protective sheath around the axons of the olfactory neurones as they extend into the basement membrane. Among these cells are blastema cells which may differentiate into new olfactory neurons or supporting cells.

Below the epithelium, two distinctive features are bundles of olfactory nerve fibres and Bowman's glands. The latter are tubular glands with a secretory function. Their secretions are discharged on the surface of the epithelium.

The many axons of the olfactory receptor cells enter the olfactory bulb, where they converge to form glomeruli. Each glomerulus contains the endings of about 25,000 axons and makes contact with about 25 second-order neurons called mitral cells. The axons of the mitral cells enter the olfactory tubercle, the prepiriform cortex and periamygdaloid area of the brain. From these areas, olfactory messages may be relayed to the thalamus and hypothalamus to influence feeding and reproductive behaviour.

Thus, resembling in some ways the gustatory system, the olfactory system is an organised, convergent structure. It consists of layered sheets of interacting, interconnecting neurones with common afferent and efferent projections. A large number of odour sensors in a pseudo-parallel configuration send information to a multilevel processing system.

3.2 Methodology in Olfaction

This section discusses some psychophysical and physiological experiments which have contributed to our present understanding of the olfactory sense.

The size and shape of an odorant molecule, together with the distribution of polar groups determine the odour description (Amoore, 1970). The psychophysical techniques outlined in Section 2 have been widely applied to the olfactory sense. However, the exact structural requirements specifying a particular odour type are still ill-defined. It seems likely that the most important molecular parameters of an odorant which determine the olfactory response are; (a) adsorption and desorption energies of the molecule from an air / water -lipoprotein interface; (b) partition coefficients between water and lipoprotein; (c) electron donor / acceptor interactions which depend on the polarisability of the molecule; (d) molecular size and shape (Boelens, 1974, Getchell *et al*, 1984).

Electrophysiology has given much basic information. When an electrode is placed on the surface of the olfactory epithelium and an odourous stimulus is applied, a slow potential change of up to several millivolts is observed (Fig. 5). This phenomenon, first observed by Hosoya and Yoshida (1937) in canine olfactory epithelium has been observed in many other animals, ranging from fishes to man. This potential change is now called the electro-olfactogram (EOG) (Ottoson, 1956), and is generally accepted as representing the summated generator potentials of the olfactory neurons. The EOG precedes single cell spike activity and olfactory nerve discharge indicating that it is the primary event for a chain of activity at higher levels (Doving, 1966). The amplitude of the EOG is proportional to the logarithm of the stimulus concentration as in gustation. Electrical stimulation of the olfactory nerve does not result in potential changes at the level of the olfactory epithelium (Ottoson, 1959). This observation indicates that the EOG is not due to the activity of the nerve fibres. The psychophysical response to odorants has been correlated with the EOG response in humans. The EOG response is now thought to consist of three components; (a) a negative generator potential, due to the influx of Na^+ with a corresponding increase in membrane permeability to K^+ ions; (b) a positive hyperpolarising positive component mainly due to the movement of Cl^- ions, with a contribution from K^+ ions; (c) a slow positive potential, accompanied by vigourous secretory activity in the supporting cells (Takagi *et al.*,1969, Okano & Takagi, 1974).

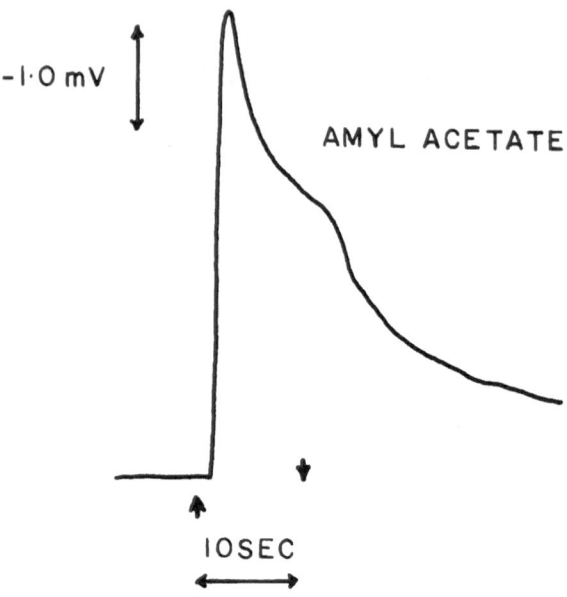

-1·0 mV

AMYL ACETATE

10SEC

Fig. 5. Electro-olfactogram elicited to a pulse of amyl acetate vapour on the olfactory mucosa of the frog.

The relationship between olfactory stimulus magnitude and peripheral nerve spike activity has been investigated by O'Connell and Mozell (1969). The frequency of neural spike activity also obeyed a power law relationship to odorant concentration. This dynamic range compression is similar to that found in gustation and other sensory modalities.

Information generated at the peripheral level of the olfactory system is processed by the higher centres leading to discrimination between odours. A pattern or code is generated at the level of the olfactory epithelium when an odorant interacts with the receptor sites. Adrian (1954) proposed that receptors which are highly selective towards particular odorants would signal how close a fit there is between odorant and receptor and that a unique pattern for a particular odorant would result as different receptors respond at various levels to each other. Another mechanism proposed is that there is spatiotemporal patterning whereby the physicochemical characteristics of the odorant determine the distribution and rate of migration across the epithelium. A space-time activity pattern would be generated across the olfactory epithelium which may be unique to a particular odorant.

Amoore (1962) attempted to identify the fundamental odour receptor sites and the *primary* odours which interact with these sites. Information was collected about the odours of different chemicals and these were grouped into common general classes. The shapes of these molecules were correlated with the odour type. Seven *primary* odours were identified. These were classified as ethereal, camphoraceous, musky, floral, pepperminty, pungent and putrid. In a modelling approach it was illustrated how a complex odour like *almond* may interact with the proposed camphoraceous, floral and pepperminty receptor sites.

The phenomenon of smell blindness or specific anosmia to the odour of verbena flowers was reported by Blakeslee (1918). This was subsequently investigated by Guillot (1959) who found that small portions of the human population could not smell the odours of violets, benzyl salicylate, ambergris, thymol, sclareol, farnesol, lewisite, scatol, musks and steroids. Amoore recognised that this phenomenon may be due to genetic defects at the level of the receptor sites and he undertook a systematic study of the sweaty fatty acid anosmia in order to determine the structural limits of *primary* odours. The olfactory thresholds of a homologous series of fatty acids were measured and the level of anosmic defects found were correlated with the molecular size of the fatty acid chains. The study revealed that the level of anosmic defects was greatest for fatty acids with 5 carbon chains. This observation was taken to reflect the degree of specificity exhibited by the receptor site. Similar studies have also been undertaken with the spermous, fishy, malty, minty, musky and urinous odours (Beets, 1978). These results provide evidence that certain primary odour receptor sites exist in the human olfactory epithelium and that odorant interactions between these sites may provide a peripheral coding mechanism. However it is not known whether the molecular receptor types correspond directly to psychophysically defined odours.

There is also electrophysiological evidence that quality coding in the olfactory system may occur through odorant interaction with selective receptor sites. Gesteland *et al.* (1963), recording the electrical activity of single units from the olfactory bulb of the frog, found that a differential response occurred to various odorants. It was demonstrated that single olfactory receptors in the frog have a highly individual, but wide selectivity. Eight overlapping groups of receptors were categorised.

There is a topographic organisation of the olfactory nerve projections from the olfactory epithelium to the olfactory bulb (Clark, 1951, Costanzo & O'Connell, 1978). The projection from the olfactory epithelium to the dorsal part of the olfactory bulb is very sharp, but becomes more diffuse in the anterioposterior region. It appears that any regional organisation of receptors on the olfactory epithelium would be reflected at the bulbar level. Boulet *et al.* (1978) have made a detailed study of qualitative and quantitative odour discrimination by mitral cells as compared to the anterior olfactory nucleus cells of the olfactory bulb. The responses to odour stimulation of 61 mitral cells and of 64 single anterior olfactory nucleus units were simultaneously recorded from rabbit. Six odours were delivered at two intensity levels. It was found that certain stimuli were discriminated at the level of the olfactory epithelium and these remained so at the level of the mitral cells. Other stimuli were poorly differentiated at the level of the olfactory epithelium but became separated at the mitral cell level of the olfactory bulb. Processing of information from the olfactory epithelium appeared to involve elimination of redundant information and retention of differences. It was concluded that much olfactory discrimination occurred in the mitral cells while the anterior olfactory nucleus units were more sensitive to intensity changes. It is likely that the higher centres of the brain receive information in the form of a multidimensional continuum in which each chemical substance occupies a specific position (Daval *et al.*, 1974).

The sorptive properties of the olfactory epithelium may contribute to the unique pattern of interaction generated by a particular odorant. The horizontal movement of odorants across the olfactory epithelium has been analysed in detail. When the column of a gas chromatograph was replaced by the olfactory epithelium of an intact frog, there were large differences in the rates at which different odours moved across the epithelium (Mozell & Jagodowicz, 1973). Similarly the distribution of tritiated odorants drawn across the epithelium forms a steep gradient between the anterior and posterior of the epithelium (Hornung & Mozell, 1980).

There is also an inherent spatial patterning which may be generated by the receptors themselves. These patterns may be generated not only by the odour specificities of the receptors, but also by variations in the densities of the receptors. Detailed maps of the spatial response patterns of the salamander olfactory epithelium are now available (Kubie *et al.*, 1980).

It is now clear that the unique pattern generated when an odorant interacts with the olfactory epithelium may be due to (a) interaction with receptor sites of varying specificities; (b) a spatial distribution of these sites across the olfactory epithelium; and (c) the physicochemical characteristics of the odorant molecule contributing to defined distribution patterns across the epithelium. In addition, the relative importance of each of these features will vary according to molecular structure (either flexible or rigid), hydrophobic, or hydrophilic character of an odorant molecule.

3.3 The voltage clamped olfactory epithelium.

This section describes current research in our laboratory

into the odorant induced short circuit current transients across the olfactory mucosa of the frog.

The most thoroughly characterized electrophysiological property of the vertebrate olfactory mucosa is the EOG. The rate and rise of amplitude of the EOG are related to odorant presentation, so that the EOG probably reflects certain aspects of ion translocation during the early stages of transduction. The potential measured is associated with a transcellular current and an equal and opposite return current. The return current pathways include the extracellular fluid spaces, and sustentacular cells. We have used the Ussing method (previously discussed in Section 2.3) to study the transepithelial currents in frog olfactory mucosa (Heck *et al.*, 1984). When bathed symmetrically in amphibian Ringer's the mean value of the transepithelial potential, short circuit current and resistance were -3.8 ± 0.6 mV, 56.0 ± 6.3 $\mu A/cm^2$, 73 ± 15 ohm $/ cm^2$, respectively, indicating that the olfactory mucosa supports active ion transport. This method has enabled us to record the first measurements of odorant induced short circuit currents.

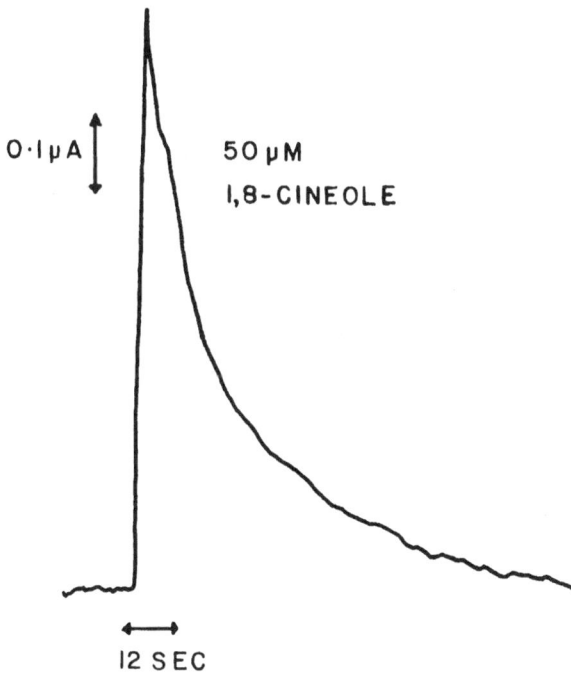

0·1 μA

50 μM
1,8-CINEOLE

12 SEC

Fig. 6. Odorant induced current response in the voltage clamped frog olfactory mucosa.

50 μM 1,8-cineole was introduced into the mucosal compartment of the Ussing chamber as a bolus of about 15 sec. The baseline current was recorded with amphibian Ringer's solution in both mucosal and serosal compartments.

Fig. 6 shows the current transient induced when a bolus of the odorant 1,8-cineole at 50 μM was presented to the short circuited olfactory mucosa. This was usually bimodal, the initial phase characaterized by a rapid onset, rise and decline of current, and a slower second phase which was either a monotonic decrease in current, a plateau followed by a decrease in current or a second maximum followed by a decrease in current.

A major difference between this measurement and the EOG measurement is that odourous stimuli are presented to the mucosa in aqueous phase in known concentrations. For EOG measurements, the stimuli are presented in the vapour phase and then have to be partitioned between the vapour phase and an aqueous phase at the level of the mucosa. Thus the exact concentrations at the receptor sites are never known. The Ussing type preparation allows different types of experiments to be designed for probing the olfactory receptors and associated ion channels. Thus it is possible to probe events in the process of olfactory transduction using pharmacological agents. We are now in the process of investigating the characteristics of the odorant gated ion channels of the olfactory mucosa using these techniques.

4.0 SUMMARY

It is clear that research into the chemical senses of gustation and olfaction is a multidisciplinary activity. Psychophysical techniques have been particularly powerful in giving data on stimuli which occupy particular groupings in gustatory or olfactory multidimensional space (Schiffman, Amoore). They indicate how some of the coding mechanisms in the chemical sensory systems may operate, as well as how many basic modalities of taste or smell exist. Physiologists have been able to take some of the psychophysical measurements and quantititate them in terms of electrical measurements at the primary levels of transduction, or at the secondary levels of signal processing (Heck *et al.*, Boulet *et al.*). This chapter has not covered biochemical investigations of the chemical senses. Progress is being made in characterization of the proteins which may function as odour receptors (Lancet, 1986, Getchell, 1986), or as taste receptors (Cagan, 1981). We can look forward to a better understanding of the molecular masis of chemical sensory transduction.

BIBLIOGRAPHY

Adrian, E.D. (1954) The basis of sensation, some recent studies of olfaction. Br. Med. J. 1, 287-290.

Amoore, J.E. (1962) The stereochemical theory of olfaction 1. Identification of the seven primary odours. 2. Elucidation of the stereochemical properties of the olfactory receptor sites. Proc. Sci. Sec. Toilet Goods Assoc. 37 (Suppl.), 1-23.

Amoore, J.E. (1970) Olfactory genetics and anosmia. In: Handbook of Sensory Physiology 1. Olfaction, Ed. Beidler, L.M. Springer-Verlag, New York, pp 245-256.

Andres, K.H. (1966) Der feinbau der regio olfactoria von macrososmatikern. Z. Zellforsch 69, 140-154.

Arai, S., Yamashita, M., Noguchi, M., Fujimaki, M. (1973) Tastes of L-glutamyl oligopeptides in relation to their chromatographic properties. Agric. Biol. Chem. 37,151-156.

Bartoshuk, L.M. (1977) Modification of taste quality. In: Sensory properties of Foods, Eds. Birch, G.G., Brennan, J.G. & Parker, K.J. Appl. Sci. Publ. London, pp 5-26.

Beckstead, R.M. & Norgren, R. (1979) An autoradiographic examination of the central distribution of the trigeminal, facial, glossopharyngeal, and vagal nerves in the monkey. J. Comp. Neurol. 184, 455-472.

Beets, M.G.J. (1978) Structure-activity Relationships in Human Chemoreception. Appl. Sci. Publ., London.

Beidler, L.M. (1980) The chemical senses: gustation and olfaction. In: Medical Physiology, Ed. Mountcastle, V.B., Mosby, St. Louis.

Birch, G.G. (1981) A model for gustatory transduction. In: Perception of behavioural chemicals Ed. Norris, D.M. Elsevier/ North-Holland Biomedical Press, Amsterdam, New York. pp 187-204.

Blakeslee (1918) Unlike reaction to different individuals to fragrance of verbena flowers. Science, 48, 298-299.

Boelens, H. (1974) Relationship between the chemical structure of compounds and their olfactive properties. Cosmetics and Perfumery 89 1-7.

Boulet, M., Daval, G. & Leveteau, J. (1978) Qualitative and quantitative odour discrimination by mitral cells as compared to anterior olfactory nucleus cells. Brain. Res. 142, 123-134.

Cagan, R.H. (1981) Recognition of taste stimuli at the initial binding interaction. In: Biochemistry of Taste and Olfaction Ed. Cagan, R.H. & Kare, M.R. Academic Press, New York, London. pp. 175-205.

Clark, W.E. Le Gross (1951) The projection of the olfactory epithelium on the olfactory bulb of the rabbit. J. Neurol. Neurosurg. Psychiat. 14, 1-10.

Costanzo, R.M. & O'Connell, R.J. (1978) Spatially organised projections of hamster olfactory nerves. Brain Res. 139, 327-332.

Daval, G., Leveteau, J., Macleod, P., Holley, A., Duchamp, A. & Revial, M.F. (1972) Studies on local EOG and single receptors response to multiple odour stimulation in the frog. In: Olfaction and Taste Vol. IV, Seewissen, 1972, Ed. Schneider, D. pp. 109-121.

DeSimone, J.A., Heck, G.L., DeSimone, S.K. (1981) Active ion transport in dog tongue: A possible role in taste. Science 214, 1039-1041.

Doving, K. (1966) Problems in the physiology of olfaction. Chemistry and physiology of flavours. Ed. Schultz et al. Avi. Publ. Comp. Conn. pp 52-94.

Erickson, R.P. (1963) Sensory neural patterns and gustation. In: Olfaction and taste, Ed. Zottereman, Y. Pergamon press, New York. pp 205-213.

Frank, M. & Pfaffmann, C. (1969) Taste nerve fibers: A random distribution of sensitivities to four tastes. Science 164, 1183-1185.

Getchell, T.V., Margolis, F.L. & Getchell, M.L. (1984) Periceptor and receptor events in vertebrate olfaction. Prog. Neurobiol. 23, 317-345.

Getchell, T.V. (1986) Functional properties of vertebrate olfactory receptor neurons. Physiol. Rev. in the press.

Gesteland, R.C., Lettvin, J.Y., Pitts, W.H. & Rojas, A. (1963) Odor specificities of the frog's olfactory receptors. In: Olfaction and Taste Ed. Zotterman, Y. Pergamon Press, Ltd. Oxford. pp. 19-34.

Guillot, M. (1953) Anosmics et paranosmics individuelles specifiques. Recherches (1953) Octobre, 26-31.

Guth, L. (1971) Degeneration and regeneration of taste buds. In: Handbook of Sensory Physiology. Chemical Senses 2. Taste. Ed. Beidler, L. Springer-Verlag, Berlin-Heidelberg, New York (1971) pp. 63-74.

Heck, G.L., Mierson, S. & DeSimone, J.A. (1981) Salt taste transduction occurs through an amiloride-sensitive sodium transport pathway. Science 223, 403-405.

Heck, G.L. DeSimone, J.A. & Getchell, T.V. (1984) Evidence for electrogenic active ion transport across the frog olfactory mucosa in vitro. Chemical Senses 9, 273-283.

Heck, G., DeSimone, J.A. & Persaud, K. (1986) Electrical properties of the frog olfactory epithelium in vitro Biophys. J. 49, 182a.

Hornung, D.E. & Mozell, M.M. (1980) Odorant molecule accessibility to olfactory receptors. Abstr. 7th Int. Symp. Olf. Taste. IRL London.

Hosoya, Y. & Yoshida, H. (1937) Uber die bioelectrischen erscheinungen an der reichschleimhaut. Jap. J. Med. Sci. III Biophys. 5 22-23.

Kubie, J.L., Mackay-Sim, A. & Moulton, D.A. (1980) Inherent spatial patterning of response to odorants in the salamander olfactory epithelium. Abstr. 7th Int. Symp. Olf. Taste. IRL London.

Lancet, D. (1986) Vertebrate olfactory reception. Ann. Rev. Neurosci. in the press.

LeMagnen, J. (1971) Olfaction and nutrition. In: Handbook of Sensory Physiology IV Ed. Beidler, L.M. Springer-Verlag, New York. pp. 465-482.

Menco, B.P.M., Dodd, G.H., Davey, M., Bannister, L.H. (1976) Presence of membrane particles in freeze etched bovine olfactory cilia. Nature, 263, 597-599.

Menco, B.P.M., Leuwissen, J.L.M., Bannister, L.H. & Dodd, G.H. (1978) Bovine olfactory and nasal respiratory epithelium surfaces. High voltage and scanning electron microscopy and cryo-ultramicrotomy. Cell. Tiss. Res. 193, 503-524.

Mierson, S. DeSimone, J.A., Heck, G.L. & DeSimone, S.K. (1986) Ion transport mechanisms in salt taste transduction. Biophys. J., 49, 182a.

Moulton, D.G. & Beidler, L.M. (1967) Structure and function in the peripheral olfactory system. Physiol. Rev. 47, 1-52.

Mozell, M.M. & Jagodowicz, M. (1973) Chromatographic separation of odorants by the nose: retention times measured across the in vivo olfactory mucosa. Science 191, 1247-1249.

O'Connell, R.J. & Mozell, M.M. (1969) Quantitative stimulation of frog olfactory receptors. J. Neurophysiol. 32, 51-63.

Okano, M. & Takagi, S.F. (1974) Secretion and electrogenesis of the supporting cell in the olfactory epithelium. J. Physiol. 242, 353-370.

Ottoson, D. (1956) Analysis of the electrical activity of the olfactory epithelium. Acta. Physiol. Scand. 35 Suppl. 122, 1-83.

Ottoson, D. (1959) Comparison of slow potentials evoked in the frog's nasal mucosa and olfactory bulb by natural stimulation. Acta. Physiol. Scand. 47, 149-159.

Sato, M. (1971) Neural coding in taste as seen from recordings from peripheral receptors and nerves. In: Handbook of Sensory Physiology. Chemical Senses 2. Taste. Ed. Beidler, L.M. Springer-Verlag, New York. pp. 116-147.

Schiffman S. & Erickson, R.P. (1971) A psychophysical model for gustatory quality. Physiol. Rev. 7, 617-633.

Schiffman, S.S. (1974) Contributions to the physicochemical dimensions of odors: A psychophysical approach. Ann. N.Y. Acad. Sci. 237, 164-183.

Schiffman, S.S., Lockhead, E. & Maes, F.W. (1983) Amiloride reduces the taste intensity of Na^+ and Li^+ salts and sweetners. Proc. Natl. Acad. Sci. USA. 80, 6136-6140.

Shepherd, G. & Ottoson, D. (1978) Olfactory Physiology. Physiological Society Monograph Series.

Takagi, S.F., Kitamura, H., Imai, K. & Takeuchi, H. (1969) Further studies on the roles of sodium and potassium in the generation of the olfactogram: effects of mono-, di-, and trivalent cations. J. Gen. Physiol. 53, 115-130.

Ussing, H.H. & Zerahn, K. (1951) Active transport of sodium as the source of the electric current in the short-circuited isolated frog skin. Acta Physiol. Scand. 23, 110-127.

Usukura, J. & Yamada, E. (1978) Observations on the cytolemma of the olfactory receptor cell in the newt. 1. Freeze replica analysis. Cell Tiss. Res. 188, 83-98.

THE PHYSIOLOGY AND PSYCHOPHYSICS OF TOUCH

S.J. Lederman† and R.A. Browse ‡ †
†Department of Psychology
‡Department of Computing and Information Science
Queen's University, Kingston, Ontario
Canada K7L 3N6

We have selected material for this presentation by structuring the discussion around a series of critical questions that might be asked with a view toward modelling a biological tactile sensing system.

1. What is touch used for?

In our everyday lives, there are a number of functions which we regularly perform with touch. We use cutaneous information to grasp and manipulate tools and other objects with considerable dexterity. Without this sensory input, we drop things and frequently hurt ourselves. In the absence of vision, we use touch for purposes of object identification, and for learning about the spatial layout of objects and surfaces near the body. In this way, for example, we may reach for a glass of juice by our bed in the dark without accident. We also use touch to assess various perceptual attributes of objects, particularly surface texture, hardness, thermal qualities, weight, and movement of an object's parts. Of course, it is also possible to determine shape, size, and function of objects.

Touch is particularly useful in sensing vibration, for the most part operating within the space delimited by our outstretched arms. Vibrations arising from events that occur at greater distances may be sensed too, for example a passing truck, but this is done at a very coarse level. Finally, touch is invaluable in locating and exploring holes .

Normally vision, touch, and proprioception operate together to explore the environment, with smooth transitions among the perceptual systems occurring frequently, though often unconsciously. A comprehensive understanding of tactual perception clearly must include the study of intersensory processing.

NATO ASI Series, Vol. F43
Sensors and Sensory Systems
for Advanced Robots
Edited by P. Dario
© Springer-Verlag Berlin Heidelberg 1988

2. What are the primary units that are sensitive to mechanical stimulation?

As the majority of tactual perception studies focus on the hairless (glabrous) skin of the hand, we will similarly limit our discussion. Also because of time and space constraints, we will only discuss mechanoreceptors, although there are also separate units that selectively respond to small increases or decreases in skin temperature. The interested reader is referred to papers by Johnson, Darian-Smith, & LaMotte (1973) and Johnson, Darian-Smith, LaMotte, Johnson, & Oldfield (1979).

Beneath the surface of the skin can be found the sensory neurons whose responses constitute the peripheral, first stage in the neural processing of cutaneous stimulation. These neurons terminate in special endings that render them especially sensitive to light mechanical deformation. It is likely that the particular properties of the specialized ending of an individual neuron determine its response. In the discussion that follows, we will use the term "tactile unit" to refer to the peripheral sensory neurons with their specialized endings. Note that the receptor and neuron functions are performed within the same tactile unit; however, the details of the transformation process are not clearly understood.

It has been estimated that there are about 17,000 tactile units innervating the hairless skin of the human hand (Johansson & Vallbo, 1979). Although the cell bodies of all tactile units lie outside the spinal cord, the first synapse occurs in the dorsal column nucleus of the spinal cord.

There are many different levels at which the performance of the tactile system may be examined. In Sections 2.0 - 4.0, we consider important anatomical and neurophysiological aspects of touch; in Sections 5.0 - 7.0, we study the system at the behavioural level.

2.1. What are the response characteristics of these tactile units?

First, it is possible to obtain a two-dimensional sensitivity profile for an individual tactile unit by moving a point probe over the skin surface and recording the unit's response. The two-dimensional sensitivity profile is represented as the unit's "receptive field". Not surprisingly, the sensitivity is highest directly above the specialized ending(s) for the tactile unit; it tails off as the distance from the centre of the receptive field increases. The receptive field is usually circular or slightly oval in shape (Johansson & Vallbo, 1983). Its size is influenced by several structural factors that are discussed next; it also varies dynamically with stimulus intensity.

Second, it is possible to measure a tactile unit's temporal response by examining its adaptation to both sustained and dynamic stimulation.

2.2. What are the factors that determine a unit's response characteristics?

The response characteristics of the primary tactile units are believed to be affected by at least the following factors, whose relative importance and interactions are not well understood.

These factors are:

The structure of fingertip skin

The number, distribution, and placement of the specialized ending(s) for a single tactile unit

The morphology of the specialized endings

A tactile unit may be classified into one of four groups on the basis of its response characteristics and the above factors. In what follows, we consider each of these factors and how it influences the unit's response. In the process, therefore, we will elaborate the accepted taxonomy of tactile units (Vallbo & Johansson, 1983).

2.2.1. The structure of fingertip skin: All human skin is composed of two major layers, each with its own sub-layers: the outer layer is called "epidermis" and the layer beneath, with which it is firmly interlocked, is known as the "dermis". A layer of subcutaneous tissue which lies beneath the skin, coextensive with it and loosely attached to both the dermis above and to muscles and bone below, should also be considered in any discussion of skin function.

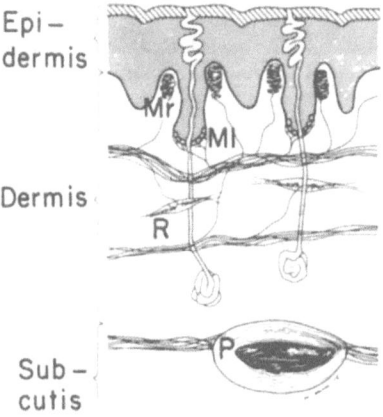

Figure 1. The structure of fingertip skin. Ml: Merkel's neurite complex; Mr: Meissner corpuscle; R: Ruffini cylinder; P: Pacinian corpuscle.

The characteristic patterns of ridges (fingerprints) on the hairless skin of the fingertips are derived from the underlying structure of the interface between epidermis and dermis. The epidermis projects into the dermis in a regular series of pegs, deeply-projecting pegs ("intermediate ridges") alternating with shallowly-projecting pegs ("limiting ridges"). Each fingertip ridge seen with the naked eye overlies one of the deep pegs; each groove on the fingertip lies above a shallow peg. While the shallow pegs are firmly attached to the dermis, the deeper ones are essentially free-floating structures.

For purposes of modelling, Johnson & Phillips (1981) have assumed that for small surface deflections, the mechanical properties of the skin may be approximated as a homogeneous, elastic, isotropic medium. This property alone predicts that the receptive field of an individual tactile unit will be circular, with decreasing sensitivity for distant stimuli. As the depth of the specialized endings from the skin surface increases, presumably the size of the receptive field must also increase: when a point on the skin is deformed, units whose end-organs are adjacent to the line of force should be more intensely stimulated the deeper they lie. Therefore deep end-organs which lie laterally further from the line of force than shallower ones may also be activated.

2.2.2. Number, distribution, and depth from the skin surface of the specialized ending(s) for a tactile unit:

Number: Tactile units appear to vary in the number of specialized end-organs in which they terminate, i.e. 1, 4 to 7, or 12 to 17 (Johansson, 1978). Johansson found that when a unit ends in more than one specialized ending, the sensitivity of the receptive field remains relatively uniform across the corresponding area above the endings, compared to when there is only one ending per unit.

Distribution of multiple specialized endings: Tactile units vary in terms of whether the multiple end-organs for a single unit are densely packed or more broadly distributed (Johansson, 1978). The more closely together the endings of a unit lie, the smaller is the receptive field.

Depth from the skin surface of specialized endings: It is evident in Figure 1 that the presumed endings of the tactile units lie at different depths from the skin surface. The specific details will be discussed in Section 2.2.3. For the moment, we simply wish to emphasize that the closer the end-organ is to the skin surface, the steeper is the decline in sensitivity toward the periphery of the receptive field.

The effect of indenting the skin is to distribute a force across many specialized endings embedded at different distances from the surface. Can the sensitivity profiles obtained by Johansson (1976) be mathematically modelled? This seems a little difficult to do in the cases where the profiles are flattened, i.e. when a unit terminates in a number of sensory end-organs. Another possibility is to model the response of each of the specialized endings, such as with a gaussian distribution. The sensitivity profile for a unit may be better described by something like the sum of the individual responses for the specialized end-organs. The space constants of the individual gaussians would be determined by the depth of the end-organ from the surface. In this way, we suggest it might be possible to model the unit's sensitivity profile in three, not just two dimensions, as is the case when using the receptive field as the unit's response property.

2.2.3. The morphology of the specialized endings: Each tactile unit is presumed to end in one of four types of specialized endings that may be morphologically distinguished from each other (Figure 1). In many cases, it would seem that the structure and/or location of an ending determines in part the unit's specific signal transmission characteristics.

Meissner's corpuscles are embedded close to the surface of the skin, in the dermal cavities between the limiting and intermediate ridges. They are ovoid in shape, and are almost always aligned with their principal axis at right angles to the skin surface. This structure probably determines their sensitivity to forces acting normal to the skin surface. There is usually only one Meissner corpuscle associated with each of the dermal papillae.

Merkel's cell complexes are situated a little further from the surface, being associated with the tip of the intermediate ridges. A single tactile unit will terminate in a tightly clustered group of these end-organs. These specialized endings also seem to be well positioned for responding to forces acting normal to the surface, since the intermediate ridge with which they are associated, is free to move when stimulated from above.

The spindle-shaped Ruffini endings are located more deeply in the dermis. Each ending is hooked up to dermal structures with collagen fibres extending from its poles (Chambers, Andres, During, & Iggo, 1972), a morphological feature that seems to render it particularly sensitive to lateral skin stretch.

Lastly, the Pacinian corpuscles lie furthest from the skin surface, usually buried within the sub-cutaneous tissue. They consist of a thin capsule surrounded by a lamellated onion-shaped structure of non-neural tissue. It has been demonstrated by Loewenstein & Skalak (1966) that the response characteristics of the outer structure may be approximated by a linear filter characterized by high-pass filtering properties.

It is not known whether the morphological properties affect other aspects of the tactile units' adaptation responses to sustained and transient stimulation.

2.3. Classification of tactile units

Table 1 summarizes the properties of the four different groups of tactile units which are commonly referred to in the neurophysiological literature (Johansson & Vallbo, 1983): FAI (RA or QA, as they were known previously), FAII (or PC), SAI and SAII. These units are typically distinguished by the size of their receptive fields, and by their adaptation to sustained stimulation. The table also includes a summary of the values for the additional factors discussed in this section. Note that the specialized endings for the four different groups of tactile units have not been confirmed in human glabrous skin. The proposed relations shown in Table 1 are based on indirect evidence from separate morphological and physiological studies in human, and from experiments that have combined the study of both structure and function in non-human species.

Summary of Fingertip Tactile Units and Their Properties				
property	FAI	FAII	SAI	SAII
specialized endings	Meissner	Pacinian	Merkel	Ruffini
location of endings	shallow	deep	shallow	deep
number of endings per unit	12-17	1	4-7	1
unit's endings clustered?	no	-	yes	-
adaptation	fast	fast	slow	slow
mean receptive field (mm^2)	12.6	101.0	11.0	59.0

3. What are the specific features of the mechanical stimulus to which individual tactile units selectively respond?

3.1. Sustained deformation normal to skin surface

Only SA units respond to sustained stimulation of the skin, normal to its surface (Johnson & Lamb, 1981.

3.2. Spatially patterned deformation (normal to or scanned across fingertip skin)

Recent neurophysiological work with humans and monkeys implicates the SA units, particularly SAI, as well as FAIs in coding the spatial aspects of stimulation (e.g., Johnson & Lamb, 1981). At the single unit level, we are probably talking about edges.

3.2.1. Edge enhancement: When a stationary edge is applied to the skin, following the initial transients, there is no temporal variation in the input, and only the SA units respond; the FA units are silent. When the stimulus is moved across the skin, however, all units may potentially fire. Johannson, Landstrom, & Lundstrom (1982) observed that the majority of FAI and SAI units innervating the glabrous skin of the human hand which they explored were most responsive to edges placed over the receptive field. This was particularly true of the SAI units. Edge detection may be particularly important for spatial analysis during object manipulation, exploration, and identification.

As all of the mechanoreceptor units synapse in the spinal cord, there are no lateral interconnections among single units in the periphery. Hence, there is no lateral inhibition at this level. Edge enhancement may be due to mechanical rather than neural factors affecting the response of first-order SAI units, as elegantly demonstrated in the continuum mechanics model of the skin proposed by Johnson & Phillips (1980).

3.3. Transient deformation, i.e. vibration (normal to skin surface)

The FA units (I and II) are particularly sensitive to mechanical vibration applied normal to the skin surface. Low frequency stimulation is mainly coded by the FAI units, which are maximally tuned to frequencies ranging from 8 to 64 Hz. High frequencies, those above 64 Hz are mainly coded by the FAII units, which are maximally tuned to about 250 Hz. There is some suggestion that the SA units may respond to frequencies below about 5 Hz; however, the specific role of SA units in coding vibration is still unclear (for human data, see Johansson, Landstrom, & Lumdstrom, 1982; the monkey data obtained much earlier by Talbot, Darian-Smith, Kornhuber, & Mountcastle, 1968 are comparable in many respects, although there appears to be a difference of interpretation concerning the contribution of the SA fibres to vibration).

3.4. Temporal vs spatial tradeoffs

One of the interesting things to note about the response properties of the various mechanoreceptor units is the apparent trade-off between spatial and temporal resolving power in at least three of the four types. The SAI units are capable of resolving the finest spatial details, and to the extent that they encode vibratory frequency, are tuned to the very bottom end of the vibrotactile frequency spectrum of the skin. The FAI units are somewhat poorer than SAI units at resolving spatial details, but also somewhat better at resolving vibrotactile patterns. The FAII units cannot code spatial details at all, but are sensitive to the highest portion of the frequency spectrum. This is somewhat reminiscent of the sustained vs transient distinction (X vs Y cells, respectively) made in vision (see chapter by Maffei).

The status of the SAIIs in this comparison is somewhat less clear. Note that it is not always possible to differentiate between SAI and SAII units in subhuman species. In humans, they appear to be limited spatially as well as temporally. Their particular sensitivity to lateral skin stretch (see Section 3.5), however, suggests that their contribution may not be particularly important in coding spatial or temporal changes in deformation normal to the skin surface.

3.5. Sustained deformation lateral to the skin surface, i.e. skin stretch

In addition to responding to sustained deformation normal to the skin surface, the SAII units are particularly sensitive to lateral stretching of the skin, and often show selectivity to the direction of stretch.

4. Other issues/parameters that affect the response of populations of tactile units

Given that mechanical stimulation most often extends beyond the receptive field of an individual tactile unit, it is clear that we must further consider the response of populations of tactile units with the same individual response characterstics.

4.1. What is known about the population distribution of the four types of tactile units in the fingertip?

Johansson & Vallbo (1979) have analyzed the absolute densities of the four tactile unit types on the palms, and on the proximal and distal phalanges of the fingers. The absolute densities were calculated on the basis of an estimate of the total number of myelinated nerve fibres in the median nerve at the wrist. FAI (RA) and SAI units are particularly numerous in the distal portion of the fingertips (140 and 70 units per cm squared, respectively); their densities decrease in the more proximal phalanges, and still more in the palm area. The FAI and SAI unit gradients are similar, although the absolute density of the SAI units is somewhat lower. The SAII and FAII (PC) units are more evenly distributed across the entire glabrous skin area. For the fingertip, the densities are 50 and 20 units per cm squared, respectively.

4.2. What is known about convergence in the cutaneous system beyond the first synapse?

Beyond the first synaptic level, convergence definitely occurs; however, it is apparently restricted to units originating in the same mechanoreceptor populations, i.e. FAI, FAII, SAI and SAII.

4.3. A neural alternative to mechanical edge enhancement

In this section, we speculate on an alternative to the mechanical enhancement of edges, proposed by Phillips & Johnson (1981). We propose that as in vision, edge detection is performed beyond the first synapse by neural centre-surround mechanisms, which may be approximated mathematically by difference of gaussian operators (DOGs). Such operators applied somatotopically, point by point, provide second derivative information from which the extraction of zero-crossings will yield edges. Our logic is as follows.

One way to model the response profile of a tactile unit is by using a gaussian function, with the space-constant providing an estimate of the radius of the receptive field. This model seems reasonable because most receptive fields are roughly circular, and response decreases as the distance to the stimulus increases (Johansson, 1976). In order to estimate the space-constants for the response distributions of the four tactile units, we begin with the mean receptive field areas reported by Vallbo & Johansson (1984). The areas are converted into radii of receptive fields, which are taken as the space-constant approximation as follows: FAI=2.00 mm, FAII=5.67 mm, SAI=1.87 mm, SAII=4.33 mm. For both the fast and slow-adapting populations, there are two gaussians, with one small and one large space constant. To obtain the second derivative (i.e., edge information) of a spatial pattern using a DOG, it is necessary to have one small and one large gaussian, the difference of which represents the actions of a centre-surround mechanism. In this way, it is possible to mathematically model the response of both the fast- and slowly-adapting systems. For the fast-adapting system, the large FAII gaussian is subtracted from the small FAI gaussian. For the slowly-adapting system, the large SAII gaussian is subtracted from the small SAI gaussian. In both cases, the subtraction process is carried out somatotopically, point by point. This edge detection process must be carried out beyond the first synapse, because convergence is required. In support of this edge detection process is the fact that centre-surround receptive fields have actually been found in the somatosensory cortex (e.g., Laskin & Spencer, 1979).

There are certain interesting parallels to be noted between the above analysis and the mathematical modelling of visual edge detection by Wilson & Bergen (1979). In vision, Wilson & Bergen have also worked with gaussians with varying space constants. They have modelled the responses of two "sustained" and two "transient" channels as differences of gaussians. It is interesting to note that the ratios of gaussain space-constants which we derive for the tactile slowly-adapting and fast-adapting space constants all lie within the range of values for vision. The ratio of small to large space constants for the slowly-adapting channel was 1:2.3 (cf 1:1.75 for the sustained channels in vision) and for the fast-adapting channel was 1:2.8 (compared 1:3.0 for the transient channels in vision). Though far from perfect agreement, it is worth pointing out these similarities in the two modalities.

4.4. Which populations of tactile units are involved in coding complex spatial patterns applied to the fingertip?

Research by Johnson & Lamb (1981) indicates that only the SA fibres, and to a lesser extent the FAI fibres, resolve complex spatial patterns such as braille-like dot patterns. However, there is a limit to the spatial acuity of these units; as low as about 2 mm (centre-to-centre) inter-dot spacing, the responses of SA and possibly FAI units preserve the spatial information in the stimulus pattern. Below that value the acuity decreases, until at about 1 mm separation, the spatial representation is considerably reduced in all units. The critical value at which spatial representation of surface details is lost is determined by the spatial patterning in the response of the SA units. Moving the stimulus across the skin clarifies these results, but does not alter the conclusions in any other fundamental way. The corresponding behavioural data are considered in Section 5.4.1.

4.5. Which mechanoreceptor populations are involved in coding textures moved across the fingertip?

The information above clearly indicates that very fine spatial details are not registered in the response profiles of any single population of afferent unit. Phillips and Johnson (1985) propose that information about textured surfaces with spatial detail finer than the acuity limits imposed by the SA units can be coded by the relative activity in the SA, FAI, and FAII mechanoreceptor populations (cf colour vision). The corresponding behavioural data are considered in Sections 5.4.2 and 5.4.3.

5. What is known about the human response to stimulus properties at the behavioural level?

In this section, we consider the stimulus properties that humans attend to when detecting, discriminating, and recognizing very small stimuli (i.e., no more than fingertip size) that contact the fingertip. Unfortunately, the field has been severely hampered to date by the lack of an effective technology with which to produce the stimuli required for careful psychophysical study. Accordingly, we consider the parameters whose effects we would like to know about, and simply indicate when certain information is not yet available.

We divide the material first in terms of the physical parameters of stimulation; within each section, the material is further organized by the perceptual task performed with these stimulus parameters. For additional information, the reader should consult new review chapters by Sherrick & Cholewiak (1986) and Loomis & Lederman (1986).

5.1. Point or unpatterned probes

5.1.1. Pressure thresholds: Studies of the pressure thesholds for detecting a step function input (nylon monofilaments) to the skin have been performed most recently by Weinstein (1968). Thresholds tend to vary widely with body locus: Face and torso are most sensitive to pressure, followed by the fingers and lower extremities. The threshold at the fingertips is 1-2 mg.

5.1.2. Perceived intensity of static stimuli: Studies have determined that the perceived intensity of stimulation is affected by both depth of penetration (e.g., Harrington & Merzenich, 1970; Knibestol & Vallbo, 1980) and by rate of skin indentation (Poulos, Mei, Horch, Tuckett, Wei, Cornwall, & Burgess, 1984). Actually intensity judgments are more closely correlated with stimulus force than with indentation. The work of Greenspan and his colleagues (Greenspan, 1984; Greenspan, Kenshalo, & Henderson (1984) may also be consulted.

5.1.3. Point localization: A common test of spatial ability is the accuracy with which a point applied to the skin may be localized, i.e. point localization. The most recent values on the fingertip have been supplied by Weinstein (1968) and by Loomis (Loomis & Collins, 1978; Loomis, 1979). Weinstein found that observers could localize the position of a fine nylon monofilament to within about 1 mm. However, using more refined methodologies, Loomis has demonstrated a form of tactile hyperacuity for absolute localization: a shift of .1 mm can be detected. This form of spatial acuity is probably related to the considerable overlap of receptive fields on the fingertip.

5.1.4. Vibrotactile thresholds: Many studies have systematically evaluated human response to temporal sinusoidal stimulation of the skin, using small probes with smooth surfaces. The skin is sensitive to a range of sinusoidal frequencies from about 2 Hz up to well over 500 Hz, although the top end has not been precisely determined. Verrillo (1971) demonstrated that vibrotactile thresholds on the fingertip are relatively high and uniform up to about 40 Hz. The threshold function then becomes U-shaped, decreasing with a slope of 12 dB per doubling of frequency to about 250Hz, and increasing in threshold again up to 700Hz, the highest frequency tested. There is considerable psychophysical evidence (e.g. Verrillo & Gescheider, 1979; Gescheider & Verrillo, 1979; Lederman, Loomis, & Williams, 1982) to suggest that there are at least two separate systems that code vibration, one tuned to low-frequencies (up to about 40 Hz), and one tuned to frequencies above that (maximally tuned to around 200-300 Hz). Recalling the earlier discussion of tactile unit specifications, it appears that the psychophysical and neurophysiological data tend to compliment each other reasonably well.

Only the high frequency system appears to summate over space (Verrillo, 1963) and time (Gescheider, 1976). These temporal and spatial factors clearly set limits on tactile pattern recognition.

5.1.5. Detecting single irregularities on a surface: With active exploration by the observer, Johansson & LaMotte (1983) found that as the pulse width of a single protruberance is increased, a lower height is required for detection. We are remarkably good at this task, e.g. for a pulse edge with infinite diameter, observers detected an irregularity of .85 micron! For diameters of 602, 231, and 40 microns, the threshold for detecting a single asperity increased to 1.09, 2.94, and 5.97 microns, respectively. However, it is important to note that such fine detection only occurs with lateral motion between skin and surface. This is probably because the papillary ridges on the fingertip are deformed as they catch on the edges of the protruberances during lateral motion.

5.2. Edge properties

Although the effects of edges have been systematically studied and modelled in human vision, technological difficulties in producing the fine spatial variations required have hampered progress in the field of tactile psychophysics. What is needed is careful evaluation of the effects of edge length and the slope of the edge that penetrates the skin. There has also been very little assessment of the effect of fine changes in edge orientation on perceived orientation (Loomis, 1979; Johnson & Phillips, 1981).

5.3. Stimulus Width

5.3.1. Measurement of the spatial discriminative capacities of the skin: Spatial discriminative capacities of the skin have been evaluated in a number of different ways. It is important to emphasize that the estimates of spatial acuity vary considerably, both with body locus (Weinstein, 1968) and with task.

The classic two-point touch threshold (the minimum separation between two dots which results in the sensation of two, not one, points on the skin) is generally between 2 and 3 mm on the fingertip (Weinstein, 1968; Loomis, 1979). This task assesses spatial resolution acuity, and defines the upper limit of spatial frequencies that the skin can resolve. It is probably determined by the use of lower spatial frequency information that varies when one vs two points on the skin are touched; Loomis (1979) demonstrated that when this low spatial frequency information was equated, the threshold increased to about 3.4 mm.

Note, however, that other tasks have resulted in much finer estimates of the skin's spatial capacities. Consider the thresholds for detecting misalignment of dots in a multi-dot configuration and for spatial interval discrimination. In the latter task, the observer must say whether a fixed pair of lines, with spacing of 5 mm (bar centre to bar centre), was bigger or smaller than a pair of lines with variable separation. Thresholds in both tasks were less than 20% of the two-point touch threshold. (Recall also that the point localization task above yielded very low thresholds. Moreover, a similar finding was obtained earlier by Vierck & Jones (1969) for size discrimination tasks performed on the forearm.) Loomis has suggested that such tasks (with the exception of two-point touch) warrant the label "hyperacuity". Presumably, the discriminations may be made on the basis of lower spatial frequency information, although it is not claimed that the skin must perform a spatial frequency analysis. Presumably, it is not just the registration of the local elements which permits such good performance, since the spatial separation of the elements affects performance so strongly.

More recently still, Johnson & Phillips (1981) obtained similar estimates of spatial acuity by having subjects discriminate between bars with and without central gaps of varying width, and between two orthogonal, but otherwise identical gratings of varying spatial periods applied to the skin. In both cases, the gap thresholds (half the spatial period of the gratings) were about .9 mm.

5.4. Patterns

5.4.1. Pattern recognition: Johnson & Phillips (1981) have examined the skin's capacity to recognize arabic letter patterns applied to the fingertip. They found that letter recognition varied with character height, recognition threshold corresponding to a letter height of 4.5 mm. Note that at this height, the structural elements of the letters are separated by about .9 mm. (The resolution limit was only slightly improved to .7 mm when observers were allowed to scan the letters freely.) As it is estimated that the spacing between receptors (both SA and FAI units in the fingertip) is about 1 mm, the investigators conclude that the limits of spatial resolution and of pattern recognition seem to be determined by the spacing of the specialized endings in the fingertip, the area of maximum spatial acuity.

Both Johnson & Phillips (1981) and Loomis (1981) have demonstrated a strong correlation between tactual and visual letter recognition. They note the fact that vision too appears to be limited by the spacing of the cones, rather than by central factors. Loomis (1981) has suggested that touch works much like blurred vision, and has shown that both systems when equated for spatial acuity, show superior performance for recognizing braille characters over arabic letters. He suggested that this superiority may be explained by the greater dissimilarity among the braille characters in their low spatial frequency information. He further demonstrated similar effects of character height on recognition by the two systems.

There has also been work done in the area of vibrotactile pattern recognition, in which patterns are created by vibrating a spatially distributed array of pins. For a review and analysis of the major findings of this area, the review chapter by Loomis & Lederman (1986) is recommended. Not surprisingly, size again affects recognition accuracy, presumably due to the limits imposed by the spatial bandwidth of cutaneous processing at the fingertip.

It is important to recognize that the detectability or identifiability of a vibrotactile pattern may be hampered by the variation of other stimulus events that are contiguous in space ("lateral masking"). When two patterns occupy the same location at different times, patterns (e.g., letters) presented very close together in time can interfere with the recognition of one another. There is more backward masking (the mask follows the target pattern) than forward masking (the mask precedes the target pattern). Finally, there is a type of masking known as metacontrast, in which two patterns overlap neither in space or time. Research by Weisenberger & Craig (1982) indicates that maximum interference occurs, not when there is zero asynchrony between mask and target, but when the mask onset followed the target onset by about 50 ms. Visual and tactile masking results share a number of similarities. For further details and discussion, the reader is referred to Loomis & Lederman (1986).

There have also been a number of studies that examine the effect of presenting two-dimensional vibrotactile patterns to the fingertip in different modes, e.g. static mode (where the entire stimulus is turned on and off without lateral motion), scan mode (where the pattern is moved across the tactile display), and a variety of sequential modes (e.g., Times Square). The recognition accuracy obtained with different display modes varies with the size of the patterns presented to the fingertip (for details, see discussion in Loomis & Lederman chapter).

5.4.2. Roughness perception: Lederman (e.g., Lederman, 1983; Taylor & Lederman, 1975; Lederman, Loomis, & Williams, 1982) has systematically assessed the parameters affecting the perceived magnitude of the roughness of rectangular metal gratings. The research clearly indicates that perceived roughness grows with increases in groove width (inner edge to inner edge); it is relatively unaffected or if anything, decreases with increases in ridge width; groove width-to-ridge width ratio and spatial period do not influence perceived roughness, nor does the coefficient of friction between skin and surface. The experiments clearly demonstrate that the temporal frequency of vibrations set up in the skin by the relative motion between skin and surface was not used to perceive roughness. Taylor & Lederman (1975) have developed a model of human roughness perception based on skin mechanics using the data base from Lederman (1974).

The perception of the roughness of two-dimensionally varying surfaces appears to be influenced by many of the same factors, whether the raised elements are organized in a grid configuration, or randomly jittered about these locations (Stevens & Harris, 1962; Lederman, Thorne & Jones, 1986).

It is important to note that without lateral motion between skin and surface, it is impossible to perform the fine texture discriminations of which we are clearly capable (Katz, 1925; Lederman, 1975); however, it makes little difference whether the movement is effected by the observer or by an external agent moving the patterns across the observer's skin (Lederman, 1981); movement velocity is also relatively unimportant (Lederman, 1974; 1983).

5.4.3. Texture discrimination: Morley, Goodwin, & Darian-Smith (1983) have assessed observers' ability to discriminate among gratings varying in spatial period (which covaried with groove width). In this task, observers examined the surfaces freely, but were not directed to attend to a particular dimension of texture, e.g. roughness. The results indicated a difference of 5% in spatial period could be differentiated. Lamb (1983) found 2% variation in spatial period of two-dimensional dot patterns could be discriminated.

It would be interesting to compare texture discrimination under conditions involving lateral motion vs motion normal to the fingertip. It is possible that lateral motion will prove superior because of the additional stimulation caused by the skin catching on irregularities of the surface.

5.5. Surface topology

We are not aware of any studies that have assessed human ability to determine the topology, e.g. planarity, radius of curvature, etc. of surfaces applied to the skin. However, Gordon & Morison (1982) recently measured absolute and difference thresholds for detecting curvature. Their observers were capable of detecting a curve with a base-to-peak height of only .09 mm in a 20 mm strip by freely moving their hands back and forth across the surface. Furthermore, observers were able to discriminate between two curved (20 mm long) surfaces with a base-to-peak height difference of .11 mm. Their research indicated that the gradient of curvature (i.e. base-to-peak height/half the surface length), rather than radius of curvature, was the parameter observers used in judging curvature (see also Davidson, 1972).

We do not know if local curvature information can be detected without lateral motion.

5.6. Detecting Slip for Controlling Precision Grip

Slip produced by lateral motion of objects across the skin probably contributes important cutaneous information for precision gripping (e.g. Westling & Johansson, 1984). Lateral forces must be known to assess the coefficient of friction, which humans use so successfully to produce and maintain a stable precision grip. It appears to us therefore, that the design of tactile sensors should include measurement of lateral forces (e.g. the Lord sensor). A second argument in favour of this recommendation is that our exquisite sensitivity to microscopic irregularities on a surface appears to involve or be mediated by shear forces resulting from lateral motion between skin and surface.

6. How is cutaneous information used to understand objects?

The previous section dealt with the perception of very small patterns. However, most things we come into contact with are considerably larger than a single fingertip. It is therefore necessary to integrate the cutaneous information from multiple finger contacts to form a coherent perception. This integration, in turn, requires the localization of all cutaneous inputs relative to some world coordinate system. Presumably, the latter is accomplished by means of additional inputs from proprioception. How the integration is performed is still unclear.

Under such circumstances, we are using the haptic system , a perceptual system which uses cutaneous and kinesthetic input to derive information about objects or patterns, and about their properties (Loomis & Lederman, 1986).

6.1. Limitations on cutaneous integration

It should not be assumed that the cutaneous integration process is without difficulties. The following limitations must be recognized initially.

6.1.1. Integrating across multiple finger inputs:
Several studies have indicated a serious limitation on tactual processing that is not shared by vision; the skin is apparently unable to read more than one pattern (e.g., a letter, or braille character) moved across the finger at a time. Studies which have attempted to increase the tactile "field of view" across fingers have generally failed (e.g., Craig, 1986; Hill, 1974; Lappin & Foulke, 1973). There is improvement, although only very

slight, when the patterns are spread over homologous fingers of the two hands.

Lederman & Browse (submitted) have further demonstrated that the cutaneous system is limited in its ability to process complex inputs to separate fingers in parallel. Observers were required to actively search among items that lay under the fingers of both hands for the presence of a conjunction of simple features, i.e. a bar which was rough AND oriented vertically. They chose to perform this task using a serial processing strategy, by examining the items one finger at a time until they located the target. Only when observers were required to detect a simple target, e.g., a bar that was either rough OR vertical, did they seem to integrate the information across the fingers in parallel. Lederman & Browse therefore proposed that texture and oriented bars may be reasonable dimensions to use as "primitives" for computational models of machine touch.

6.1.2. Temporal integration More often than not it is impossible to obtain a complete view of the object or pattern in a single "glance". It is therefore necessary to explore the object sequentially with one or more fingers, and to construct a representation over time. The exploration of structure is particularly constrained by the considerable memory demands which are placed on the haptic system. This is probably not the case when extracting information concerning surface texture, possibly because the observer assumes, at least initially, that the local input concerning texture is representative of the entire surface. (This is also likely true when perceiving the thermal qualities of objects.)

6.2. The role of hand/arm movements

As Gibson has observed (1962), purposive exploration, i.e. touching, is required for spontaneous perception of the external world of objects and their properties, and of surface layout. When the observer is passively stimulated, i.e. touched, it is more common to experience the events as separate local sensations. For example, when an object is impressed against our fingertips, we tend to experience five local pressure sensations rather than a single object "out there". With active touch, we experience object constancy. Although it is possible to gain some information about the external world by passively stimulating the skin, it is rarely as natural a process as with active exploration.

Purposive exploration involves the use of both cutaneous and kinesthetic inputs to derive information about objects and their properties. The perceptual system that uses such information is referred to as the haptic system (Gibson, 1962; Loomis & Lederman, 1986).

Does purposive exploration guarantee good perception and recognition by the haptic system? The answer is complex.

6.2.1. Raised 2-dimensional displays Purposive haptic exploration does not ensure veridical experience. For example, current displays which use raised symbols to produce graphics for the blind, e.g. maps, pictures, etc. are very poorly apprehended by the haptic system. Blindfolded and blind observers are usually prohibitively slow and highly inaccurate at perceiving and identifying such raised patterns. We note that these displays typically present very limited cutaneous information (e.g. little if any variation in the third dimension) and no thermal variation. The kinesthetic input from hand movements is limited to variation in the plane of the display; in fact, such displays offer little information that can be apprehended effectively by any kind of hand movement (Lederman & Klatzky, submitted).

6.2.2. Apprehension and recognition of common 3-dimensional objects In contrast, we are remarkably skilled at identifying common three-dimensional objects haptically. For example, Klatzky, Lederman, & Metzger (1985) demonstrated that blindfolded observers identified 100 common objects with over 96% accuracy, in only 1 to 2 seconds for most objects.

Lederman et al. consider this finding to be an existence proof that haptics can perform at a very high level. Our (Lederman & Klatzky) long-term goal is a process model of human haptic object apprehension (i.e., deriving information about object properties and how they combine to form the whole) and recognition (i.e., assigning a label to an object). We have begun by focussing on the hand movements people choose to execute during object exploration. We argue (Lederman & Klatzky, submitted) that hand movements are guided by the type of object knowledge that is desired. We have demonstrated a number of stereotypical patterns of hand movement, which at a certain level (regardless of the specific hand configuration or end effectors) are strongly linked to obtaining a specific type of object knowledge, e.g. texture, hardness, mass, etc.

The research indicates that these links are probably forged because the particular hand movements provide the best information sought. The movement patterns are often necessary, or else the optimal (i.e. most accurate and fast) of several sufficient ways for obtaining a particular type of object knowledge. They vary, however, in terms of the degree of specialization with which they provide knowledge about objects. Specifically, a two-handed "enclosure", in which the fingers mould to the object's contours, is the least specialized of movement patterns; with a single quick grasp, it is possible to obtain sufficient information about many different object dimensions simultaneously, for purposes of matching objects along some defined dimension. In contrast, exerting pressure normal to the object surface (by poking, pressing, tapping, squeezing) is the most highly specialized of movements, providing information primarily about the perceived hardness of the object. The particular type of task will clearly affect the appropriate sequence of hand movements. Such information may prove valuable in programming a robot hand to extract information about an object for various tasks.

7. Other issues to consider

Although time (and space) restrictions prevent us from going into detail, there are several additional questions that are important to raise.

7.1. What coordinate system(s) do we use to code the cutaneous/kinesthetic input?

Do we begin by perceiving the cutaneous input to each finger within its finger-centred coordinate system? Do we next switch to a body-centred coordinate system, and finally to an object- centred coordinate system, as Marr (1982) has done when modelling human vision? If this is the case, what is the nature of each coordinate system, e.g. polar? At which point do the shifts in coordinate system occur? Is there a single cutaneous egocentre, or are there multiple egocentres, defined by the part of the body used, e.g., individual fingers, hand, arm. Corcoran (1977) showed that perception of patterns drawn on the hand is influenced by the perceived static posture of the body surface upon which the drawing occurs; pattern orientation appears to be represented in terms of distal space rather than in terms of a body-centred system. Such orientation phenomena also occur when the observer actively examines raised letters (Oldfield & Phillips, 1983).

7.2. How does the human system integrate perceptual and manipulative demands during haptic exploration?

The sensory and motor systems are clearly both integrally involved in the apprehension/recognition and manipulation of objects. It is critical that we understand the control processes involved in guiding the interplay between the two systems for these tasks. There has been little systematic investigation of this issue as yet in psychology; the motor control and hand rehabilitation fields should also be consulted.

7.3. How do haptic and visual inputs become integrated?

As mentioned initially, touch and vision usually work together in a highly cooperative and effective manner during the manipulation, apprehension and identification of objects. There has been a wealth of research on human perception and animal physiology that has begun to articulate important issues in this field and to tackle questions concerning the nature of multimodal perception and intersensory integration. The interested reader should consult Freides (1974), Walk & Pick (1981), and Welch & Warren (1980).

8. References

Chambers, M.R., Andres, K.H., During, M. von, & Iggo, A. (1972). The structure and function of the slowly adapting type II mechanoreceptor in hairy skin. *Quarterly Journal of Experimental Physiology, 57*, 417-445.

Corcoran, D.W. (1977). The phenomena of the disembodied eye or is it a matter of personal geography? *Perception, 6*, 247-253.

Craig, J.C. (1986). Attending to two fingers: Two hands are better than one. *Perception & Psychophysics, 38,* 496-511.

Davidson, P.W. (1972). Haptic judgments by blind and sighted humans. *Journal of Experimental Psychology, 93,* 43-55.

Freides, D. (1974). Human information processing and sensory modality: Cross-modal functions, information complexity, memory and deficit. *Psychological Bulletin, 81,* 284-310.

Gescheider, G.A. (1976). Evidence in support of the duplex theory of mechanoreception. *Sensory Processes, 1,* 68-76.

Gescheider, G.A., & Verrillo, R.T. (1979). Vibrotactile frequency characteristics as determined by adaptation and masking procedures. In D.R. Kenshalo (Ed.), *Sensory Functions of the Skin of Humans.* New York: Plenum Press.

Gibson, J.J. (1962). Observations on active touch. *Psychological Review, 69,* 477-490.

Gordon, I.E., & Morison, V. (1982). The haptic perception of curvature. *Perception & Psychophysics, 31,* 446-450.

Greenspan, J.D. (1984). A comparison of force and depth of skin indentation upon psychophysical functions of tactile intensity. *Somatosensory Research, 2,* 33-48.

Greenspan, J.D., Kenshalo, D.R., & Henderson, R. (1984). The influence of rate of skin indentation on threshold and suprathreshold tactile sensations. *Somatosensory Research, 1,* 379-393.

Harrington, T., & Merzenich, M.M. (1970). Neural coding in the sense of touch: Human sensations of skin indentation compared with the responses of slowly adapting mechanoreceptive afferents innervating the hairy skin of monkeys. *Experimental Brain Research, 10,* 251-264.

Hill J.W. (1974). Limited field of view in reading lettershapes with the fingers. In F.A. Geldard (Ed.), *Cutaneous Communication Systems and Devices.* Austin, Tex.: The Psychonomic Society.

Johansson, R.S. (1976). Receptive field sensitivity profile of mechanosensitive units innervating the glabrous skin of the human hand. *Brain Research, 104,* 330-334.

Johansson, R.S. (1978). Tactile sensibility in the human hand: receptive field characteristics of mechanoreceptive units in the glabrous skin area. *Journal of Physiology, London, 281,* 101-123.

Johansson, R.S., & LaMotte, R.H. (1983). Tactile detection thresholds for a single asperity on an otherwise smooth surface. *Somatosensory Research, 1,* 21-31.

Johansson, R.S., Landstrom, U., & Lundstrom, R. (1982). Responses of mechanoreceptive afferent units in the glabrous skin of the human hand to sinusoidal skin displacements. *Brain Research, 244,* 17-25.

Johansson, R.S., & Valbo, A.B. (1979). Tactile sensibility in the human hand: Relative and absolute densities of four types of mechanoreceptive units in glabrous skin. *Journal of Physiology, 286,* 283-300.

Johansson, R.S., & Valbo, A.B. (1983). Tactile sensory coding in the glabrous skin of the human hand. *Trends in Neuroscience, 6,* 27-32.

Johnson, K.O., Darian-Smith, I., & LaMotte, C. (1973). Peripheral neural determinants of temperature discrimination in man: A correlative study of responses to cooling skin. *Journal of Neurophysiology, 36,* 347-370.

Johnson, K.O., Darian-Smith, I., LaMotte, C., Johnson, B., & Oldfield, S. (1979). Coding of incremental changes in skin temperature by a population of warm fibres in the monkey. *Journal of Neurophysiology, 42,* 1332-1353.

Johnson, K.O., & Lamb, G.D. (1981). Neural mechanisms of spatial tactile discrimination: Neural patterns evoked by braille-like dot patterns in the monkey. *Journal of Physiology, 310,* 117-144.

Johnson, K.O., & Phillips, J.R. (1981). Tactile spatial resolution. I. Two-point discrimination, gap detection, grating resolution, and letter recognition. *Journal of Neurophysiology, 46,* 1177-1191.

Katz, D. (1925). Der aufbau der tastwelt. *Zeitschrift fur Psychologie,* Leipzig: Barth.

Klatzky, R.L., Lederman, S.J., & Metzger, V.A. (1985). Identifying objects by touch: An "expert system". *Perception & Psychophysics, 37,* 299-302.

Knibestol, M., & Valbo, A.B. (1980). Intensity of sensation related to activity of slowly adapting mechanoreceptive units in the human hand. *Journal of Physiology, 300,* 251-267.

Lamb, G.D. (1983). Tactile discrimination of textured surfaces: Psychophysical performance measurements in humans. *Journal of Physiology, 338,* 551-565.

Lappin, J.S., & Foulke, E. (1973). Expanding the tactual field of view. *Perception & Psychophysics, 14,* 237-241.

Laskin, S.E., & Spencer, A. (1979). Cutaneous masking. II. Geometry of excitatory and inhibitory receptive fields of single units in somatosensory cortex of the cat. *Journal of Neurophysiology, 42,* 1061-1082.

Lederman, S.J. (1974). Tactile roughness of grooved surfaces: The touching process and effects of macro- and micro surface structure. *Perception & Psychophysics, 16,* 385-395.

Lederman, S.J. (1981). The perception of surface roughness by active and passive touch. *Bulletin of the Psychonomic Society, 18,* 253-255.

Lederman, S.J. (1983). Tactual roughness perception: Spatial and temporal determinants. *Canadian Journal of Psychology, 37,* 498-511.

Lederman, S.J., & Browse, R. (submitted). Haptic feature integration.

Lederman, S.J., & Klatzky, R.L. (submitted). Hand movements: A window into haptic object recognition.

Lederman, S.J., Loomis, J.M., & Williams, D.A. (1982). The role of vibration in the tactual perception of roughness. *Perception & Psychophysics, 32,* 109-116.

Lederman, S.J., Thorne, G., & Jones, B. (1986). The perception of texture by vision and touch: Multidimensionality and intersensory integration. *Journal of Experimental Psychology: Human Perception and Performance, 12.*

Loewenstein, W.R., & Skalak, R. (1966). Mechanical transmission in a pacinian corpuscle. *Journal of Physiology, 182,* 346-378.

Loomis, J.M. (1979). An investigation of tactile hyperacuity. *Sensory Processes, 3,* 289-302.

Loomis, J.M. (1981). Tactile pattern perception. *Perception, 10,* 5-27.

Loomis, J.M., & Collins, C.C. (1978). Sensitivity to shifts of a point stimulus: An instance of tactile hyperacuity. *Perception & Psychophysics, 24,* 487-492.

Loomis, J.M., & Lederman, S.J. (1986). Tactual perception. In K.Boff, L. Kaufman, & J. Thomas (Eds.), *Handbook of Perception and Human Performance.* New York: John Wiley and Sons.

Marr, D. (1982). *Vision.* San Francisco: Freeman.

Morley, J.W., Goodwin, A.W., & Darian-Smith, I. (1983). Tactile discrimination of gratings. *Experimental Brain Research, 49,* 291-299.

Oldfield, S.R., & Phillips, J.R. (1983). The spatial characteristics of tactile form perception. *Perception, 12,* 615-626.

Phillips, J.R., & Johnson, K.O. (1981). Tactile spatial resolution. II. Neural representation of bars, edges, and gratings in monkey afferents. *Journal of Neurophysiology, 46,* 1192-1203.

Phillips, J.R., & Johnson, K.O. (1985). Neural mechanisms of scanned and stationary touch. *Journal of the Acoustical Society of America, 77,* 220-224.

Poulos, D.A. Mei, J., Horch, K.W., Tucket, R.P., Wei, J.Y, Cornwall, M.C., & Burgess, P.R. (1984). The neural signal for the intensity of a tactile stimulus. *The Journal of Neuroscience, 4,* 2016-2024.

Sherrick, C.E., & Cholewiak, R.W. (1986). Cutaneous sensitivity. In K. Boff, L Kaufman, & J. Thomas (Eds.), *Handbook of Perception and Human Performance.* New York: John Wiley and Sons.

Stevens, S.S., & Harris, J.R. (1962). The scaling of subjective roughness and smoothness. *Journal of Experimental Psychology, 64,* 489-494.

Talbot, W.H., Darian-Smith, I., Kornhuber, H.H., & Mountcastle, V.B. (1968). The sense of flutter-vibration: Comparison of the human capacity with response patterns of mechanoreceptive afferents from the monkey hand. *Journal of Neurophysiolgy, 31,* 3301-3344.

Taylor, M.M., & Lederman, S.J. (1975). Tactile roughness of grooved surfaces: A model and the effect of friction. *Perception & Psychophysics, 17,* 23-36.

Valbo, A.B., & Johansson, R.S. (1984). Properties of cutaneous mechanoreceptors in the human hand related to touch sensation. *Human Neurobiology, 3*, 3-14.

Verrillo, R.T. (1963). The effect of contactor area on the vibrotactile threshold. *Journal of the Acoustical Society of America, 35*, 1962-1966.

Verrillo, R.T. (1971). Vibrotactile thresholds measured at the finger. *Perception & Psychophysics, 9*, 329-330.

Verrillo, R.T., & Gescheider, G.A. (1979). Psychophysical measurements of enhancement, suppression, and surface gradient effects in vibrotaction. In D.R. Kenshalo (Ed.), *Sensory Functions of the Skin of Humans*. New York: Plenum Press.

Vierck, C.J., & Jones, M.B. (1969). Size discrimination on the skin. *Science, 158*, 488-489.

Walk, R.D., & Pick, H.L. (1981). *Intersensory Perception and Sensory Integration*. New York: Plenum.

Weinstein, S. (1968). Intensive and extensive aspects of tactile sensitivity as a function of body part, sex, and laterality. In D.R. Kenshalo (Ed.), *The Skin Senses*. Springfield, Ill.: Charles C. Thomas.

Weisenberger, J.M., & Craig, J.C. (1982). A tactile metacontrast effect. *Perception & Psychophysics, 31*, 530-536.

Welch, R.B., & Warren, D.H. (1980). Immediate perceptual response to intersensory discordance. *Psychological Bulletin, 88*, 638-667.

Westling, G., & Johansson, R.S. (1984). Factors influencing the force control during precision grip. *Experimental Brain Research, 53*, 277-284.

Wilson, H.R., & Bergen, J.R. (1979). A four mechanism model for threshold spatial vision. *Vision Research, 19*, 19-32.

SECTION 2

THE ARTIFICIAL COUNTERPART:
ROBOTS AND THEIR SENSORY FUNCTIONS.
GENERAL REMARKS

Steps toward making robots see

Michael Brady
Robotics Research Group,
Department of Engineering Science,
University of Oxford,
Oxford OX1 3PJ,
U.K.

1 Abstract

This paper reports on recent progress in Computer Vision by the Oxford Robotics Research Group. We discuss in particular: edge and corner finding; shape from contour; parallel algorithms for computing shape representations; parallel architectures for computer vision; and the application of truth maintenance systems to recognise variable geometry objects in cluttered images. Model-based vision and data-directed vision are discussed as extreme cases of architectures for vision systems.

2 Introduction

Computer vision is our most powerful sense. Using vision, we are able to determine the positions of three-dimensional objects as well as their orientations; we can identify objects, inspect them for flaws, determine how to grasp them, or figure out how to navigate around them. Vision is our most stable sense insofar as it is least sensitive to slight shifts in viewpoint.

Vision is also our most complex sense, so that severe limitations need to be imposed on vision systems if they are to work *quickly* and *reliably*. These are the two principal reasons why vision has had limited application in industry. First, vision inherently demands staggering amounts of memory and processing. A single brightness image, with eight bits-per-pixel (since currently-available cameras can deliver eight bits, a unit of storage that exactly fits a byte of computer memory), and with a spatial resolution of 512 pixels square, occupies a quarter of a megabyte of memory. Adding colour to a monochrome image at least triples the storage required; stereo further doubles it; and motion multiplies it by the temporal sampling frequency (say 10 Hz). Advanced algorithms can require tens of thousands of operations to be performed at each pixel, implying billions per image. Evidently, "real time" vision can only be made possible using the parallel architectures that have become available because of the development of VLSI. Parallel architectures specially designed for vision are discussed in Section

NATO ASI Series, Vol. F43
Sensors and Sensory Systems
for Advanced Robots
Edited by P. Dario
© Springer-Verlag Berlin Heidelberg 1988

8. Vision machines are still relatively expensive and this is one factor retarding their deployment in industry.

However, a second, and by far the more important, reason why advanced vision algorithms have had limited application in industry to date is that they do not, in general, perform reliably on real data. For example, few, if any, of the hundreds of published motion algorithms have been adequately tested on real image sequences. The situation is somewhat better with respect to stereo (Grimson 81) and shape (Grimson and Lozano-Pérez 86), (Bolles and Cain 82), (Brady and Asada 84), (Connell and Brady 85) but algorithm testing falls a long way short of meeting industrial testing requirements. Reliability is much more important than efficiency: garbage that is computed in one picosecond is still garbage.

For this reason, the bulk of research in computer vision has concentrated on developing its theoretical and practical base in: geometry and representations of geometric structures; signal processing; and algorithms, particularly those that support local parallel computations. As we discuss further in Section 7, there is a fledgling theory of parallel algorithms based on the propagation of local constraints. Other approaches to computer vision are sometimes mistakenly called "Artificial Intelligence" approaches because of their emphasis on the active role to be played by domain specific knowledge, a topic to which we return in the next Section.

In this paper, we review recent work in our Laboratory in the following areas: image surface descriptions; (ii) shape from contour; (iii) geometric reasoning: the interpretation of mechanical drawings and the interpretation of cluttered images involving variable geometry shapes using a truth maintenance system; (iv) parallel architectures for early vision; and (v) parallel algorithms for computing shape descriptions. The overall framework for our research is described in Section 2.

3 Prior knowledge, models, and constraint propagation

For the past few years, research in computer vision has emphasised signal processing, geometry and representations of shape, as well as the development of specific algorithms for individual modules such as stereo or motion. Most implemented programs operate in a strictly data-driven manner: in a typical setting, an image is processed to yield edges, the edges are then grouped to form continuous (often closed) boundaries, from which shape or texture descriptions are computed, so that finally an object is recognised by an appropriate matching operation (perhaps in a "feature" space). Marr (Marr 82) proposes the model of human vision shown in Figure 1. The image is unintelligently processed, that is to say without deploying information specific to a particular domain, to yield an information-preserving representation of the "significant" intensity changes in the image(s). The intensity change representation(s) is then operated upon by "shape from" processes such as stereo, motion, contour interpretation, texture gradients, to yield a sparse map of depths or surface normals. The map is sparse because the intensity changes "significantly", and generally these do not form a dense set. Following a process

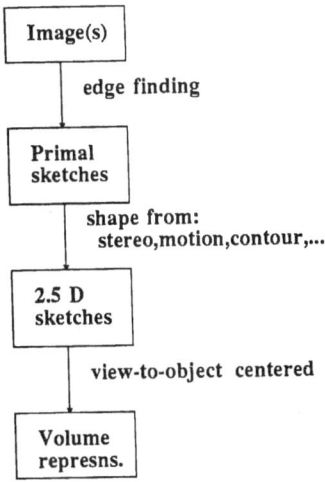

Figure 1: Marr's model of human vision, a sequence of elaborated representations.

of surface reconstruction that interpolates a dense surface using the sparse set of given depth values as "weak" constraints (Grimson 81; Terzopoulos 83; Blake and Zisserman 87), shape representations are computed (Ponce and Brady 87; Brady, Ponce, Yuille, and Asada 85) and objects are recognised. The theoretical framework of Marr has supported many substantial advances in our understanting of computer vision.

It is frequently objected, however, that human perception cannot be organised as Marr suggests, and that machine perception shouldn't be either. Some of the reasons given for objecting to the strictly data-driven organisation of the vast majority of extant programs are:

1. images are noisy, so dumb routines that know nothing of the context in which they are applied must inevitably make mistakes. Edge detection and edge grouping are particularly unintelligent processes that regularly make mistakes on real images. Errors computed early in the visual pathway make errors at subsequent (sequential) processing stages more likely;

2. the real world is too complex to process exhaustively. Surely it is more efficient to process blindly as little data as possible in order to invoke some stored model, whose structure can be used to selectively guide subsequent image processing;

3. real objects such as trees, people, and houses are infinitely varied in their geometry. Surely it is unrealistic to suppose that unintelligent processes can accommodate such variations;

4. Rubin's vase (Figure 2) reminds us that ambiguity is rife in perception. A single

Figure 2: Rubin's vase reminds us that ambiguity is rife in perception.

signal is *interpreted* in two different ways. Surely, perception is primarily a process of seeking after meaning in which signal processing and geometry are only concerned with the "relatively uninteresting" initial processing;

5. the man in the moon, figures in the fire, and the illusory figures studied by Kanisza are taken as further evidence for the constructive nature of visual perception in which descriptions, culled from stored models, are imposed upon a scene.

To varying extents, these are cogent arguments against the strictly data-driven approach to computer vision with which Marr is associated. (Actually, Marr was often at pains to point out that his principal concern was to identify a number of distinct *representational* stages in visual processing, with no commitment to a particular *processing architecture*: if the algorithms he and his followers constructed as computational experiments were data-driven that was primarily for reasons of parsimony.) This in turn raises the issue of what an alternative might be.

In contrast to data-driven processing, *model-directed* or *top-down vision* is characterised by the idea that models be "imposed" upon images. The idea is that a stored model should be identified early in the perceptual process and used to guide subsequent processing. An early proponent of this view was Guzman (Guzman 71), who coined the slogan "if you have found a leg, you know where to look for a foot". It is important to realise that an inherently sequential account of perception is implicit in this view: why not search simultaneously for *both* the foot and the leg, since the interpretation of the leg scene structure as a leg is as much constrained by the presence of the foot event and its interpretation as a foot as vice versa.

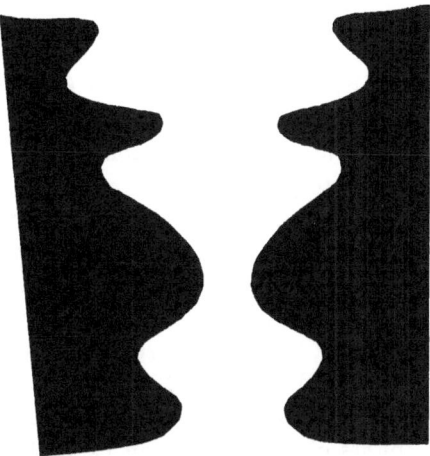

Figure 3: A Rubin-like figure lacking obvious semantic interpretation.

Not surprisingly, top-down vision has been much concerned with esoteric software process architectures such as blackboards. In an early system, Freuder (Freuder 75) developed a program to locate hammers in images. The performance of his program should be contrasted (unfavourably) with that of (Grimson and Lozano-Pérez 86). Brady and Wielinga (Brady and Wielinga 78) developed a top-down (called "heterarchical" to distinguish it from the more conventional "hierarchical") program that read Fortran coding sheets; but they considered the heterarchical programming approach to studying vision too limiting (Brady 81; Brady 82). In particular, they drew attention to the unwarranted, though rigid, distinction made in top-down programs between what does and what does not constitute a model.

To this end, consider that Rubin's vase is particularly compelling because both interpretations have *clear semantic interpretations*: as a vase or as two faces. Figure 3 shows that the same reversal of figure and ground occurs despite the absence of obvious semantic interpretations. There is no *single* representational level in the visual system that constitutes "model". An edge finder implicitly models the world as a patchwork quilt of uniformly grey patches (Binford 81), or as low-curvature loci of rapidly changing intensities. Stereo matching can not handle images of snowstorms; instead, it models the world as composed of piecewise smooth surfaces (the smoothness requirement may be quite weak, as in the Lipschitz condition imposed by PMF (Pollard, Porrill, Mayhew, and Frisby 85)). The further one progresses from generic routines such as these, the more specialised and more consciously-available become the models. The job of vision is to invoke models *at the appropriately highest level as quickly as possible, as reliably as possible, and so as to impose maximal constraint.*

This suggests a refined version of model-based vision, in which relationships within a model are used to constrain matching between scene fragments $\alpha, \beta, \ldots$ and corresponding model fragments $A, B, \ldots$ based upon relationships computed between the scene fragments. This approach was explored initially by Bolles and Cain (Bolles and Cain 82; Bolles and Horaud 86) and by Faugeras and Hebert (Faugeras and Hebert 86). Lozano-Pérez and Grimson (Grimson and Lozano-Pérez 86) have explored constraint-propagation as constrained search. The model-based constraints provably cut-off the expected combinatorial explosion of interpretations. Recently, (Murray, Castelow, and Buxton 87) has extended Lozano-Pérez and Grimson's approach to handle parametric constraints typified by the depth-scale ambiguity in computing visual motion.

Other work has explored the application of truth-maintenance systems to the problem of visual interpretation (Herman and Kanade 87; Bowen and Mayhew 86; Provan 87). Provan, working in our Laboratory, has applied truth-maintenance techniques to a problem originally studied by (Hinton 76) in his PhD thesis. Hinton's aim was to study the role that relaxation processes might play in constructing structured representations corresponding to perceptual accounts of a scene. As an example, he developed a program to find one or more "optimal" instances of a "puppet" figure in a sea of rectangles. The notion was to deliberately ignore the early processing and concentrate on interpretation. The key issue is that the puppet model is only holonomically constrained: legs are allowed to swing in restricted ways about knees and thighs, arms swing about elbows and shoulders, and so forth. In Hinton's approach, interpretation is cast as an integer programming problem. The description of how a particular scene can be viewed as containing puppets occurs as a second non-trivial pass.

Provan's program, on the other hand, *explicitly computes the structural description of the puppet at the time that it computes the interpretation*. Figure 4 shows an example interpretation computed by Provan's program. The program initially interprets a small number of rectangles as possible heads or trunks because they overlap an appropriate number of rectangles that have the right relative aspect ratios to serve as a neck or as upper arms or thighs. These initial interpretations are called "seeds". The holonomic puppet model then constrains the interpretation of further, intrinsically ambiguous, rectangles as the remaining parts of the puppet. The program can recognise puppets in a variety of postures in highly cluttered scenes. If parts of the puppet are missing, for example an upper leg, the program will supply a "default" of the right size and in the right position to join to a putative lower leg. If a more satisfactory upper leg is later added to the image, the interpretation of the lower leg can change automatically. Though the imagery considered by Provan may be considered somewhat artificial, it is important to realise that the *interpretation process* utilised in Herman and Kanade's (Herman and Kanade 87) seemingly more realistic imagery is considerably weaker than Provan's. Coincidentally, Provan's original motivation for constructing his program was not to study vision but to investigate the large scale behaviour and asymptotic complexity of TMS algorithms. Vision is a particularly challenging application for TMS algorithms because of the large number of "contexts" or partial interpretations that need to be generated.

A more technical presentation of the ideas in this section can be found in (Brady 87).

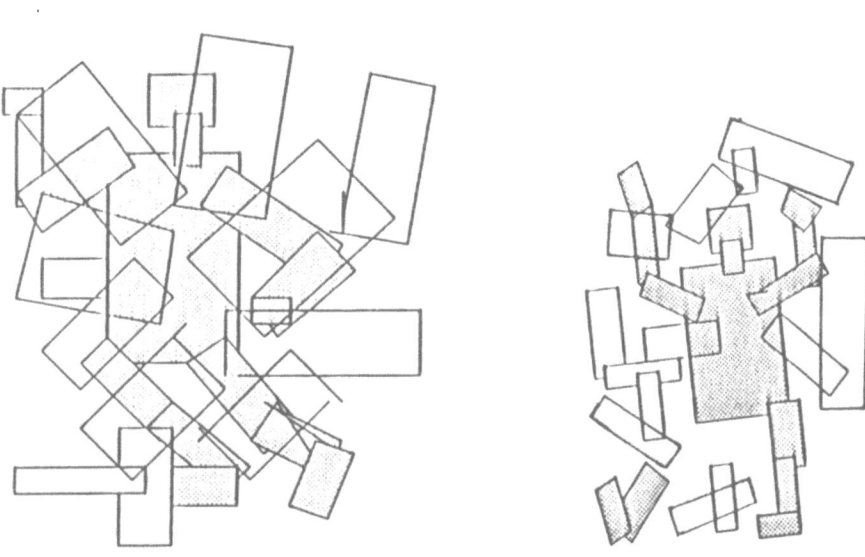

Figure 4: Examples of the interpretations of puppets in cluttered scenes found by Provan's TMS-based vision program. See text for further details.

4   Describing intensity changes

The human brain is primarily a detector of change. It responds slowly or not at all to slow changes in irradiance across the visual field. There is a good reason why it should be like this: absolute intensity values vary considerably with illumination and viewpoint from image to image (Horn 86) while irradiance changes usually correspond to "geometric and reflectance events" that persist from image to image. Of course, this observation need not apply to industrial vision systems where the illumination can be carefully controlled; but it has often proved difficult in practice to control lighting carefully enough for image-based operations such as thresholding to be robust and effective. The intensity level stored in a pixel is a function of: the illumination, and the reflectance and local surface normal of the corresponding scene point (Horn 86). Rapid changes in illumination correspond to shadows or highlights (Healey and Binford 87). Rapid changes in reflectance typically correspond to changes in surface material. Rapid changes in local surface normal correspond to "edges", which may in turn be occluding boundaries or creases (Noble 87a; Noble 87b; Koenderinck and van Doorn 87). These observations suggest edge finding as a key, largely autonomous, process of early vision whose goal is

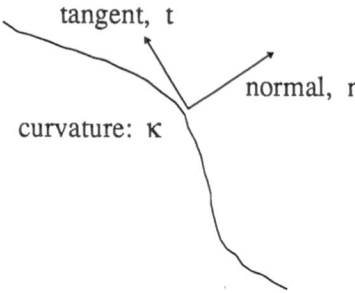

Figure 5: The (two-dimensional) curvature of the edge as projected into the image plane.

to elicit a description of all the "significant" changes in the intensity surface (Marr 76; Marr 82).

Edge finding is the most thoroughly studied process in vision. Even so, we still know remarkably little about it. Attention has almost entirely concentrated on detecting and describing *monochrome, step edges of low curvature that are spatially separated sufficiently* (the two-dimensional curvature of the edge as projected into the image plane) (Figure 5). This is an important though restricted class of edges which are closely related to step changes in one-dimensional signals. We return to this point below. A step change in intensity can be detected either as a peak in the intensity derivative, or as a zero-crossing of the second derivative. The human visual system appears to use zero-crossings of second derivatives insofar as we find it relatively difficult to detect a low gradient ramp superimposed on a step edge (so-called "lateral inhibition"). On the other hand, simple numerical considerations would suggest using first derivatives on the grounds that finite differences of slowly changing signals amplify noise. For this reason, *ad hoc* finite difference operators that approximate first and second intensity derivatives work poorly and unpredictably. (We shall see below how a difference of two appropriately smoothed versions of an image can approximate a second derivative.)

There have been three major approaches to developing reliable edge detection. The first, pioneered by Prewitt (Prewitt 70), is based on approximating each local window of the image by an analytic surface. In her technique, there are three choices to be made: the size of the approximating window, the parametric form of the approximating analytic surface, and the precise form of the goodness of fit of the surface to the image data. For example, choosing the window size to be three pixels square, the approximating parametric form to be the planar surface $ax + by + c$, and the goodness of fit process to be least squares error, leads to the familiar finite difference approximation for first derivatives of the image. This approach has been developed by Haralick (Haralick 80; Haralick 84; Haralick 85) in his "facet model", and later in the "topographic primal sketch" (Haralick 85). An important innovation in Haralick's work on the planar facet is his treatment of noise. Assuming Gaussian noise, Haralick (Haralick 80) showed that the gradient estimator has an associated significance level determined by a F^2 distribution. Extending the treatment of noise to more complex surface geometries is an unsolved problem currently receiving attention in our laboratory (Noble 87a; Noble 87b). The second approach to a theory of edge detection is based on the design of *optimal filters*. Canny's (Canny 83; Canny 86) work is a notable example of this approach and surveys previous work. Boie, Cox, and Rehak (Boie, Cox, and Rehak 86), Shen and Castan (Shen and Castan 86), and Medioni (Heurtas and Medioni 86) have developed the approach and developed efficient implementations using recursive filter designs. One of the author's graduate students, Spacek (Spacek 87), suggested adding a further noise reducing term to the optimisation of the detection and localisation originally suggested by Canny. He found that a cubic is then the best "simple" approximation to his optimal filter. Spacek's work is more notable for its analysis of edge curvature, however, as described below.

The final approach to a theory of edge detection attempts to model computationally various aspects of the human visual system. Low contrast spatial frequency studies have shown that the human visual system has a set of approximately bandpass filters at each retinal location. Roughly, there are four families of filters: two have sustained responses and give accurate localisation of the positions of edges; two are transient and give more accurate information regarding motion than position. Wilson and Bergen (Wilson and Bergen 79) suggest that each family of retinal cells compute a convolution of the image with a difference-of-Gaussians (DOG) $G_{\sigma_1} - G_{\sigma_2}$ where the ratio of the space constant σ_2 of the negative surround to that of the positive center $(\sigma_2/\sigma_1) \approx 1.6$. These findings are in remarkable agreement with those reported for tactile sensing by Ledermann in these proceedings. Several authors (Hawken and Parker 86; Young 86) have suggested that the DOG model do not fit closely empirical data on simple cells, and have identified the large negative surround G_{σ_2} as the culprit. Instead, they suggest difference-of-offset-Gaussian (DOOG) models, in which the large negative surround is replaced by a set of smaller negative Gaussians symmetrically placed about the mode of the positive Gaussian.

In any case, it has frequently been observed that DOG and DOOG filters approximate the second derivative of a Gaussian. Indeed, Marr and Hildreth (Marr and Hildreth 80) suggested that an image should first be smoothed with a Gaussian of some appropriate size (scale, in current jargon), after which zero-crossings of the Laplacian are found. This "zero-crossings of $\nabla^2 G_\sigma$" theory of edge detection has supported some impressive work in computer vision, including the Marr-Poggio-Grimson theory of stereo vision and Hildreth's work on visual motion. With few exceptions, however, such work is not critically reliant upon the $\nabla^2 G_\sigma$ theory. Note that the fact that some simple cells carry DOGs, DOOGs, or $\nabla^2 G_\sigma$ *do not imply* that edges are found by some simple operation applied to their values, such as by marking zero-crossings. Several researchers have suggested that information other than zero-crossings are important (Mayhew and Frisby 81; Watt and Morgan 83). (Koenderinck and van Doorn 87) and others (Young 86) have noted that second, third, fourth, and possibly as high as seventh derivatives may be computed by the human visual system.

A major difference between Marr and Hildreth's Laplacian of a Gaussian operator and that proposed by Canny is that the former is non-directional while the latter is directional. Refer to Figure 6. If the unit normal to the edge (more generally, intensity level crossing) is given by $\mathbf{n}$, the directional derivative is $d\|\nabla I\|/d\mathbf{n}$. Spacek (Spacek 87) has noted that a version of the Navier-Stokes equation can be used to relate the directional derivative and the (non-directional) Laplacian of a Gaussian:

$$\frac{d\|\nabla I\|}{d\mathbf{n}} = \nabla^2 I + \kappa\|\nabla I\|$$

Torre and Poggio (Torre and Poggio 86) have derived a related result. It follows that for straight edges in ideal images (for which the two-dimensional curvature κ is zero), the zero-crossings of the directional and non-directional operators are the same. However, Canny (Canny 86) has pointed out that the Laplacian sums contributions from both the $\mathbf{n}$ and the $\mathbf{n}^\perp$ directions; the latter contributes only to noise and not to the edge signal. For this reason, directional edge finders are to be preferred in practice for low curvature edges. Noble (Noble 87a; Noble 87b) reviews recent work on directional and non-directional edge operators by Berzins, Clark, and by Haralick.

Earlier, we noted that edge detection has concentrated almost entirely upon a limited class of intensity changes, namely *monochrome, step edges of low curvature that are sufficiently spatially separated* (Noble 87b). Figure 7 shows the edges computed by the implementation of the Canny edge finder used in the implementation of the PMF stereo algorithm (Pollard 85; Pollard, Porrill, Mayhew, and Frisby 85). Observe that it regularly misses corners and junctions (often T-junctions) at which several surfaces meet. Similar examples can be found, for example, in Gennert (Gennert 86). One might object that corners are inherently noisy and difficult to detect. Indeed they are, but they are also the tightest source of constraint in images (see (Brady 87) for a discussion of corners as "seeds of perception").

Alison Noble, of our Laboratory, has recently (Noble 87b) reviewed work aimed at corner detection. Early approaches were based on computing the (projected) two-dimensional curvature of an edge using one of the analytic forms familiar from the calculus. As we noted above, Spacek (Spacek 87) has shown how to relate directional edge operators to Marr and Hildreth's Laplacian of a Gaussian. The curvature of an edge can be estimated from the responses to *both* operators. Spacek discusses a number of techniques for estimating the curvature in this way and has implemented a technique that seems to work well (Figure 8). The corners are used as maximally constraining features in an implementation of Ullman's minimal mapping theory (Ullman 79). Figure 8 shows stereo or motion information that is computed by Spacek's program.

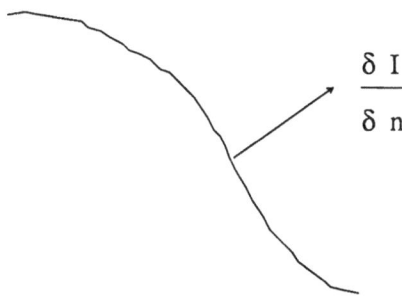

$$\frac{\delta\ I}{\delta\ n}$$

Figure 6: Directional and non-directional edge operators. The unit normal to the edge (more generally, intensity level crossing) is given by **n**, so that the directional derivative is $d\|\nabla I\|/d\mathbf{n}$.

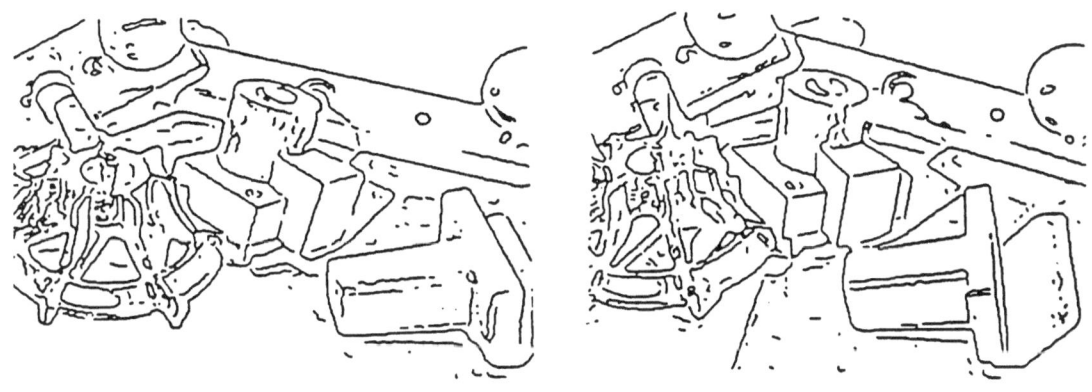

Figure 7: The edges computed by the implementation of the Canny edge finder used in the implementation of the PMF stereo algorithm.

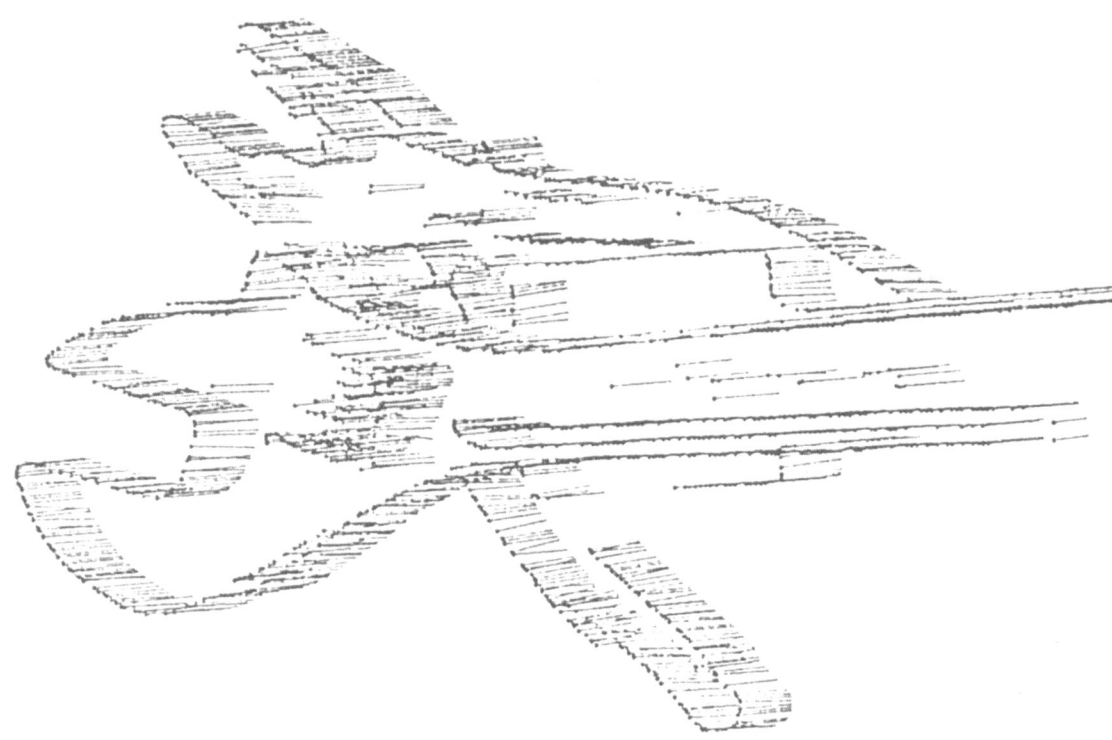

Figure 8: Stereo range computed by Spacek's program that extracts corners.

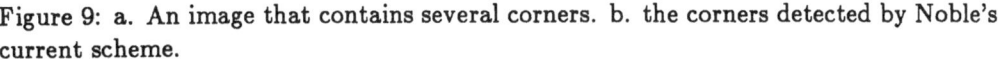

Figure 9: a. An image that contains several corners. b. the corners detected by Noble's current scheme.

Image corners, T-junctions, X-junctions, and the like correspond to complex geometric events in the image considered as a surface. It is natural, therefore, to base a corner detection scheme upon an analysis of the image surface. Haralick's (Haralick 85) "topographic primal sketch" was an early attempt at such an analysis. More recently, in our Laboratory, Noble has developed the differential geometrical analysis of images as well as techniques for extracting corners from images. Figure 9a shows an image that has several corners and Figure 9b shows the corners detected by Noble's current scheme.

Margaret Fleck, working in our Laboratory, has noted (Fleck 87) that, formally, differential geometry imposes a number of conditions to be met in order that the first and second fundamental forms exist. The conditions are *not* met, for example, at the X-junctions on a chess board. This implies that one has to be extremely careful in applying differential geometric analysis to images. Strictly, the considerations do not apply to the work of Ponce and Brady (Ponce and Brady 87) on representing surface changes because they first convolve the surface function with a Gaussian, and this makes the surface infinitely differentiable. Gaussian convolution has the nice property that it may smooth an image sufficiently to support local differential geometric analysis. However, it has the disadvantage that it *changes the local geometry and topology of the image.*

This is particularly marked for densely textured images. Instead, Fleck has proposed a hierarchy (akin to scale space) of finite resolution representations called an *adjacency structure*. An early application of Fleck's theory has been to edge detection. Figure 10 shows an example. The reliability with which corners and T-junctions can be computed for real images by Fleck's edge finder is in marked contrast to Figure 7. Further details of Fleck's edge finder can be found in Fleck (Fleck 87) and in a recent paper that describes progress on the development of a mobile robot system at Oxford (Brady, Cameron, and et. al. 87).

5   Shape from contour

In conventional computer vision, edge finding is followed by a number of independent "shape from" processes, including stereo, motion, shape from contour, shape from texture, and shape from focus. In this section, we report briefly on progress we have made recently on the determination of shape from contour. Recent work on motion and stereo is reported in Brady (Brady 87).

Shape from contour is one of the most powerful cues available to the human visual system for determining the orientation of surfaces, especially those beyond the effective range of stereo vision. Brady and Yuille (Brady and Yuille 84) suggested a variational principle for computing the perceived orientation of planar surfaces indicated by their bounding contours. According to Brady and Yuille's variational principle, the surface orientation is chosen as that which maximises the area divided by the square of the perimeter of the rotated contour. Recently, Horaud and Brady (Horaud and Brady 87) have combined the work of Brady and with (i) Malik's (Malik 87) approach to labelling image contours as either bounding or surface creases; and (ii) an assumption that the object viewed is a generalised cone whose axis is a straight line. For a given planar contour, such as that shown in Figure 11, Horaud and Brady plot the area/perimeter-squared variational measure proposed by Brady and Yuille's as a surface indexed by the possible orientations of the planar contour. Superimposed upon this variational surface is a constraint on the orientation of the generalised cone determined from bounding contours. In general, these two constraints cannot both be met exactly. Horaud and Brady suggest choosing the surface orientation that maximally satisfies both constraints.

6   Geometric reasoning

Figure 12a shows a geometric drawing of an object that is displayed isometrically in Figure 12b. Figure 12b is also the interpretation of the three drawings shown in Figure 12a after they have been input to a program written by Meng Er (Er 86), one of the author's graduate students. The key contribution of Er's analysis is a reformulation of *dual space*, in which planes are transformed to points (two components of the unit surface normal and the offset from the origin); points are transformed to planes (the duals of all the planes that contain the point); and lines are transformed to lines.

Figure 10: An example of the performance of Fleck's phantom edge finder.

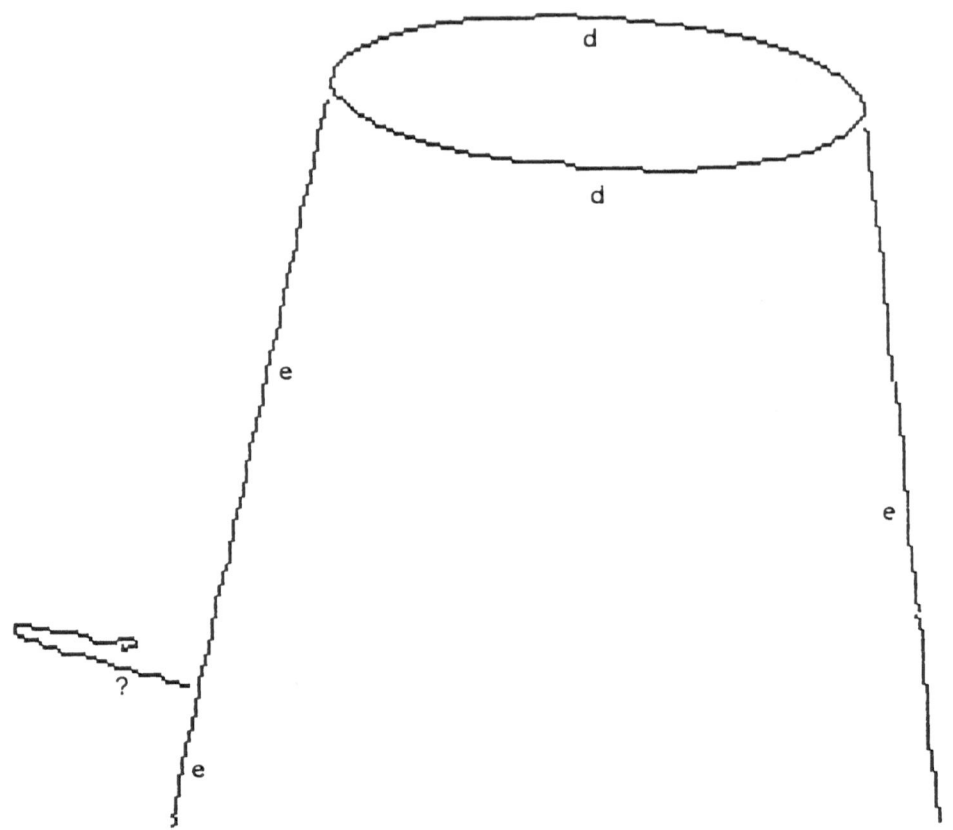

Figure 11: Constraint integration in Horaud and Brady's work on shape from contour.

The notion of dual space was introduced into computer vision by (Huffman 71) and used by Draper (Draper 81) in his analysis of the visual competence of a number of computer programs. Draper concluded from his analysis that dual space was inherently limited in what kinds of visual reasoning it could support. Er's reformulation of dual space avoids the problems discussed by Draper. He has shown how dual space can support reasoning about "impossible figures" such as that shown in Figure 13 . In a second application of his formulation of dual space, Er has developed a fast hidden-line algorithm for computer graphics.

Another graduate student, Greg Provan has investigated Truth Maintenance Systems (TMS) and their applicability to high level vision. Most of Provan's thesis is concerned with a complexity analysis that provides a precise foundation for the continuing debate in Artificial Intelligence regarding the competences of justification-based vs assumption-based TMS. Provan shows that published claims regarding the efficiency of ATMS algorithms need to be treated with care, especially in applications in which large numbers of "contexts" need to be generated. To explore these issues, Provan has applied TMS algorithms to high level visual interpretation. The notion is that partial perceptual interpretations correspond to contexts: the expectation is that complex images generate many contexts because ambiguity, especially local ambiguity, is rife in high level vision. This was pointed out earlier in the discussion of Rubin's vase.

Consider the puppet figure shown in figure 14a. The task is to find the puppet in cluttered environments such as Figure 14b. The geometry of the puppet is deliberately

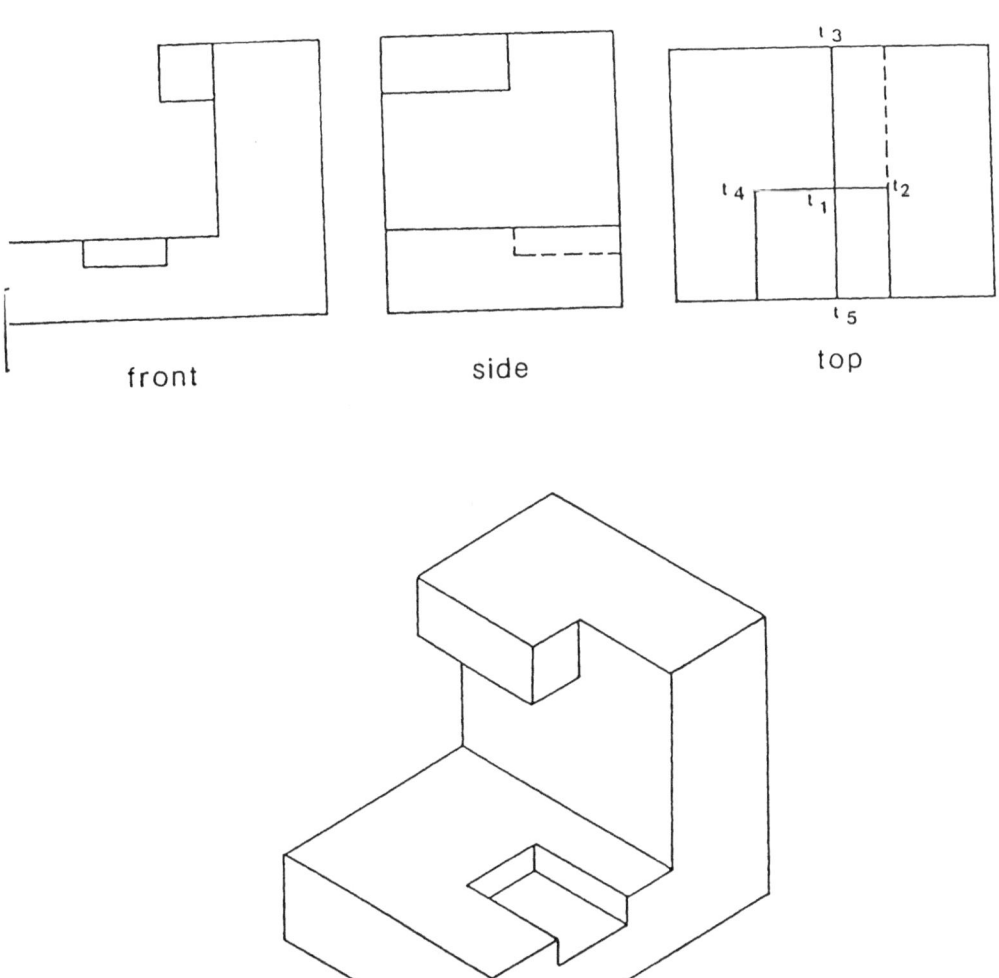

front   side   top

Figure 12: a. A geometric drawing of an object, corresponding to front, side, and plan elevations. b. An isometric projection of the object interpreted by Er's program given the drawings in a.

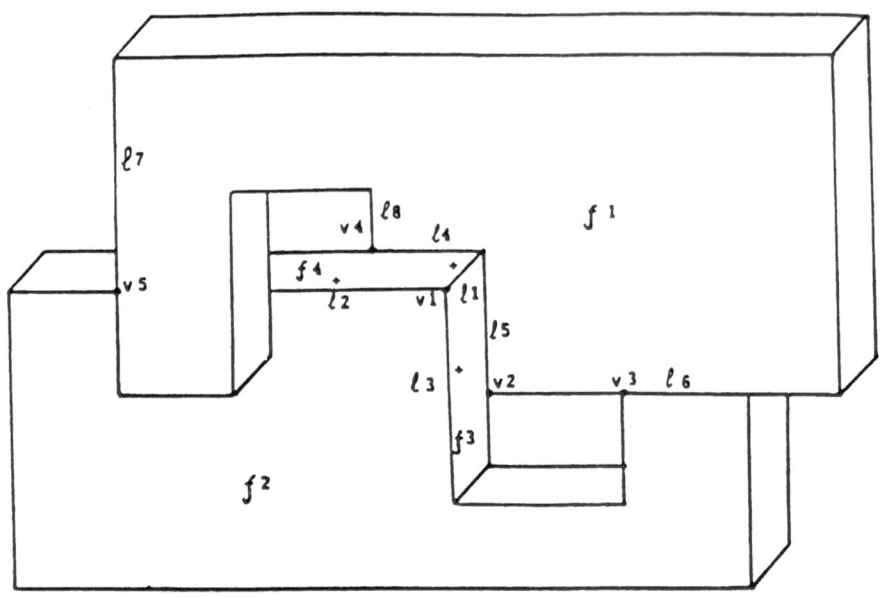

Figure 13: An "impossible" figure that is recognised as such by Er's program. The figure illustrates the visual competence of Er's formulation of dual space.

trivialised to focus attention upon the interpretive higher levels of visual analysis. Hinton (Hinton 76) originally introduced the puppet task, and it was to compare and contrast the performance of Provan's TMS based program with that of Hinton that the task was chosen. Hinton's program was based on relaxation using integer programming to search for an optimal puppet. The perceptual structure of the puppet fragment "found" by Hinton's program was implicit in the coding of the vertices of the integer programming space; a separate post-processor had to convert the vertex coding into a puppet labelling. Provan's program reasons *explicitly* about the puppet structure, thus: I believe this rectangle could be a trunk since rectangle A could be a head, and rectangle B a neck and A is attached to B and B is attached to the putative trunk, and the aspect ratios and position constraints of heads, necks, and trunks are satisfied.

Provan (Provan 87) discusses how his program is capable of recognising puppets that have parts missing. For example, in Figure 15a, the program proposes the isolated rectangle as a lower leg and it supplies a default as the (invisible) upper leg. If, later, two additional rectangles are supplied (Figure 15b) that correspond to a more reasonable interpretation as a leg, then: (i) all the perceptual structure of the puppet aside from that leg is preserved; and (ii) the proposed lower leg and the default upper leg are no longer considered part of the optimal interpretation.

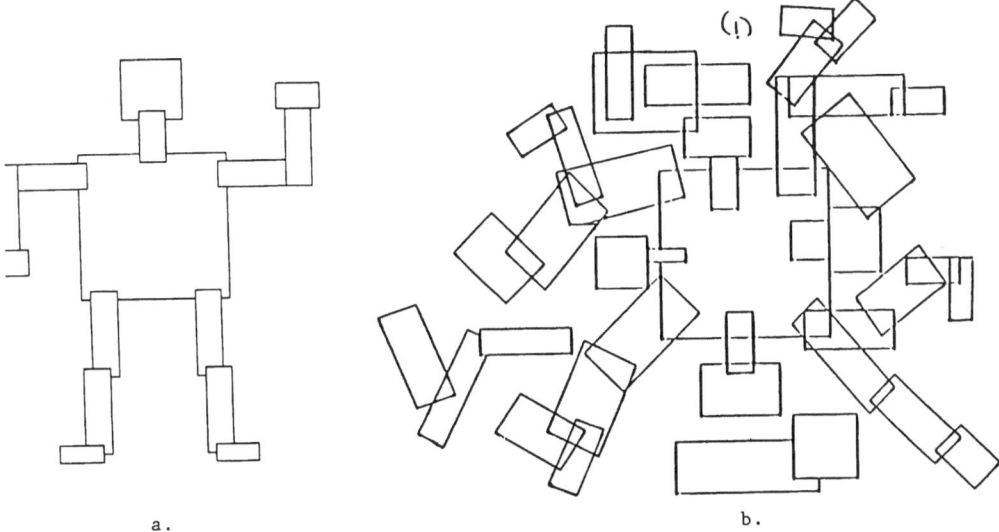

a.   b.

Figure 14: a. A puppet figure. Note that the puppet has variable geometry as the legs and arms bend, and variable postures. This complicates the interpretation process as the constraints are parametrically described rather than fixed. b. The puppet task is to find the best puppet(s) in a cluttered figure such as that shown.

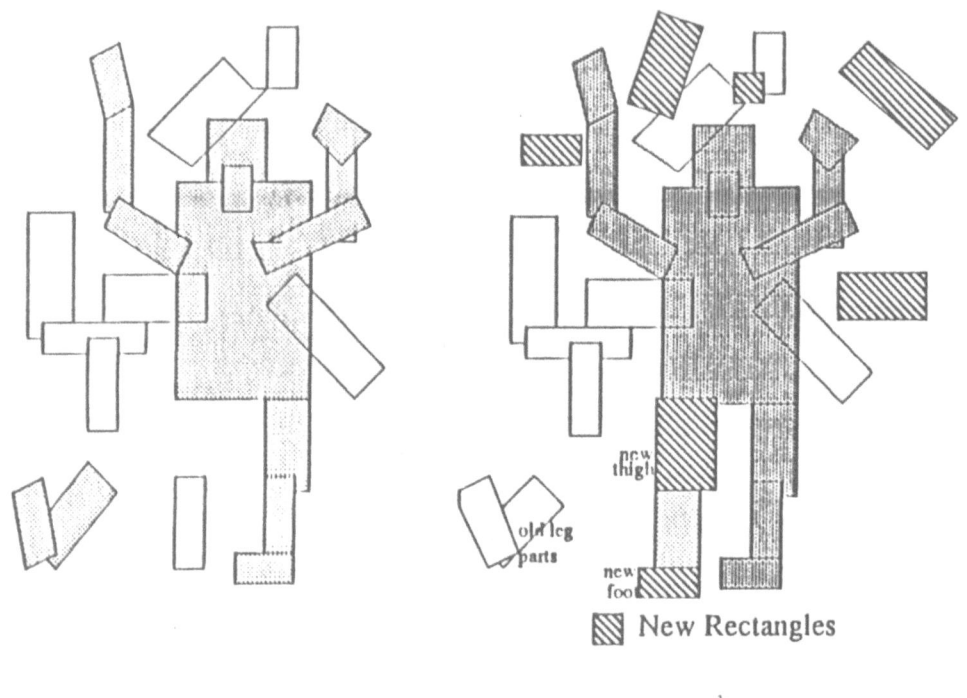

a.   b.

Figure 15: a. Failing to find a right lower leg to complete the interpretation of the puppet, the program proposes the isolated lower leg and then imagines a default upper leg to complete the figure. b. Later, the proposed right leg is replaced by a new pair of rectangles that are evidently better.

7   Vision machines

We noted in the Introduction that advanced vision algorithms demand billions of cycles corresponding to tens of thousands at each pixel location. Serial machines can't keep pace. Vector machines such as the Cray family of computers and a variety of vector processing peripherals optimistically cut processing time by $O(n)$, where the image has $n * n$ pixels. Array processors potentially offer $O(n * n)$ speed up and with it the prospect for real time vision including algorithms such as stereo, motion,and reliable edge detection. There are major practical difficulties involved in building vision computers. First, to date, it is still quite expensive to build machines with one processor per pixel for images of size $n = 256$ or $n = 512$. Second, the relatively small size of individual chips (as opposed to wafer scale integrated systems) means that the pixel processors need to be distributed over many chips, so that inter-chip communication limits the speed of the overall machine.

Third, current processor architectures lie at one or other extreme with respect to synchronisation. Most available vision machines are exclusively SIMD in that the behaviours of the individual pixel processors are strictly synchronised to the ticks of a central clock. This may be too restrictive for higher level processes. However, the main alternative is asynchronous communication between cooperating sequential processes (Hoare 79; Jones 87). Primitives to support such communication are provided by the languages Ada or Occam (Jones 87) and its associated hardware realisation the Transputer. Page (Page 87a) presents an excellent overview of parallel architectures for computer vision, as well as a representative sample of parallel vision algorithms.

Fourth, in our Laboratory, Foster (Foster 87) has recently analysed the time required to convolve an image, an operation which, though linear, is a key component of many non-linear intermediate vision computations. He has shown that the time to communicate an image to and from a serial device such as a TV camera or a TV monitor substantially exceeds the processing time for carrying out the convolution. In serial computers, pipelining is the standard technique for increasing information throughput. Foster has designed and implemented a convolver that incorporates a variety of pipelining techniques to increase the overall speed by a factor of six.

A colleague, Ian Page (Page 87b) has implemented a mixed SIMD/MIMD system. The SIMD part of the processor is the ARROW processor (Page 87b) which was originally designed as a computer graphics engine. ARROW directly implements RasterOp, otherwise known as Bitblt, which was identified as a generic operation for computer graphics by Newman and Sproull (Newman and Sproull 73). RasterOp takes two bit maps SRC and DEST as arguments and computes

$$DEST := SRC \; logop \; DEST$$

in one cycle, where *logop*, one of sixteen Boolean operations on two arguments, is applied to each pair of corresponding bits of the bit maps. Figure 16 shows a portion of

Canny's edge finding algorithm (Canny 83; Canny 86). The routine shown in the figure is the hysteresis thresholding algorithm described by Canny. We have implemented a portion of Canny's algorithm on ARROW, the speed up being about 256, ARROW's array of processors being 16*16. Recently, an array of 42 transputers has been added to the system to provide a MIMD capability. To date, Page, and his graduate student Phil Winder, have implemented a variety of graphics algorithms on the mixed SIMD/MIMD processor, nicknamed the DisPuter, including the Mandelbrot set, an Eulerian fluid flow, and a display of overlapped two-dimensional shapes. Currently, we are implementing a version of Fleck's algorithm on the DisPuter.

```
(defun EXTEND_CONTOURS(x_slope y_slope contours_1 contours_2 count
@optional (x0 0) (y0 0) width height)
(or width  (setq width  (- (array-dimension-n 1 contours_1) x0)))
(or height (setq height (- (array-dimension-n 2 contours_1) y0)))
(let* ((dims (list (- (+ 31 width) (\ (1- width) 32)) height))
       (coords (list x0 y0))
       (x>y (make-array dims ':type art-1b))
       (x<y (make-array dims ':type art-1b))
       (temp_1 (make-array dims ':type art-1b))
       (temp_2 (make-array dims ':type art-1b)))
  (map_2d #'m_abs (list x-slope coords y-slope coords) (list x>y))
  (map_2d #'m-plusp (list x-slope coords y-slope coords) (list temp_1))
  (map_2d #'m-plusp (list x-slope coords) (list x<y))
  (bitblt xor width height temp_1 0 0 x y 0 0)
  (do ((i 0 (1+ i)))
      ((= i count))
    (bitblt seta width height contours_2 x0 y0 temp_1 0 0)
    (bitblt and  width height x>y 0 0 temp_1 0 0)
    (bitblt ior  width (1- height)temp_1 0 1 temp_1 0 0)
    (bitblt ior  width (- height 1)temp_1 0 1 temp_1 0 1)

    (bitblt seta width height contours_2 x0 y0 temp_2 0 0)
    (bitblt andca width height x>y 0 0 temp_2 0 0)
    (bitblt ior  (1- width) height temp_2 1 0 temp_2 0 0)
    (bitblt ior  (- width 1) height temp_2 0 0 temp_1 1 0)
    (bitblt ior  width height temp_2 0 0 temp_1 0 0)

    (bitblt seta width height contours_2 x0 y0 temp_2 0 0)
    (bitblt and width height x>y 0 0 temp_2 0 0)
    (bitblt ior  (1- width) (1- height) temp_2 1 1 temp_2 0 0)
    (bitblt ior  (- width 1) (- height 1) temp_2 0 0 temp_2 1 1)
    (bitblt ior  width height temp_2 0 0 temp_1 0 0)
```

```
(bitblt seta width height contours_2 x0 y0 temp_2 0 0)
(bitblt andca width height x>y 0 0 temp_2 0 0)
(bitblt ior  (- width 1) (1- height) temp_2 0 1 temp_2 1 0)
(bitblt ior  (1- width) (- height 1) temp_2 1 0 temp_2 0 1)
(bitblt ior  width height temp_2 0 0 temp_1 0 0)

(bitblt and width height contours_1 x0 y0 temp_1 0 0)
(bitblt ior  width height temp_1 0 0 contours_2 0 0))))
```

Figure 16. The contour extension algorithm in the original LISP machine implementation of Canny's algorithm. Note the preponderance of calls to the function bitblt (alias rasterop).

8 Parallel algorithms for shape representation

This final section reports work aimed at the development of parallel algorithms for computing representations of two-dimensional shape. It is described more fully in (Brady and Scott 87). Some time ago, we introduced a representation of two-dimensional shape that we called *smoothed local symmetries* (SLS). It was based upon an analysis of a number of problems with the influential Symmetric Axis Transform (SAT) representation (Brady 85; Brady and Asada 84). As noted in (Brady 85), the reflectional symmetry axes and rotational symmetry centres uncovered by the SLS constitute *intrinsic coordinate frames* for describing the shape. This is a powerful idea, which is the basis for our work on *intrinsic surface patches* for representing visible surfaces (Brady, Ponce, Yuille, and Asada 85). Scott (Scott 87) has explored a similar notion in proposing "snake-like" representations for curves.

Vaious parallel algorithms have been developed for computing shape representations such as the SAT and the SLS. The first such was the "grassfire" algorithm (Blum 73). More recently, Crowley (Crowley and Parker 84) has published a similar algorithm which seems to be distinguished from the grassfire algorithm only insofar as the shape is blurred by a series of Gaussian (more generally low pass) filters. His algorithm marks peaks and ridges in the blurred shapes. Brady and Scott (Brady and Scott 87) recount their experience with a prototype implementation of the SLS on the Connection Machine (Hillis 85). They point out that the major problem stems from the inadequate mathematical basis for the shape computation, which means that it is not possible to guarantee the correctness of an implementation nor is it possible to make precise statements for any given shape, say a rectangle containing an elliptic hole, what will in fact be computed as the representation.

Brady and Scott identify and unify three ideas that are at the heart of the SLS and parallel algorithms for computing shape representation. First, as noted above, reflectional and rotational symmetries in the SLS or SAT act as local coordinate frames that "organise" the description of the shape. This is reminiscent of the familiar mathematical

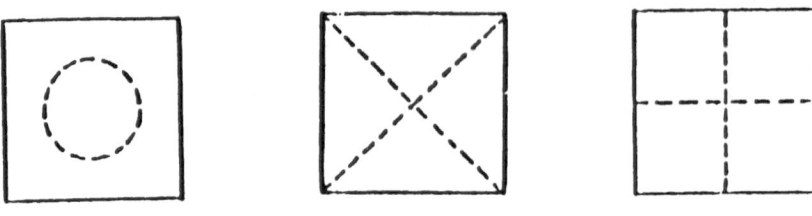

Figure 17 : Modes for a number of shapes considered as thin plates.

notion of orthogonal basis functions for describing a complicated function. Second, the idea of computing shape descriptions at a set of spatial scales is further reminiscent of an orthogonal function decomposition using familiar mathematical bases such as Fourier, Hermite, or Walsh functions. Finally, algorithms such as Crowley's (and Pizer's recent work in a similar vein) that blur the shape and extract structure from it, are reminiscent of the computation of modes, or eigenfunctions, of an appropriate differential equation. For example, Figure 17 shows the modes for a number of shapes considered as thin metal plates. Note that the reflectional and rotational symmetry descriptions are simply computed as different modes. Note also for the hole in the rectangle that the presence of the hole is ignored in certain low modes. For Gaussian blurring, for example, the appropriate differential equation is the diffusion equation. The decomposition of a shape according to a orthogonal function basis is related to the modal or eigenfunction solutions to a differential equation via Sturm-Liouville theory. Brady and Scott discuss a number of possible differential equations and show the results of initial computational experiments with the ideas.

9 Acknowledgements

We thank our several colleagues, particularly those whose work has been highlighted in this paper: Meng Er, Margaret Fleck, David Forsyth, David Foster, Radu Horaud, Alison Noble, Ian Page, and Guy Scott. Conversations with Hugh Durrant-Whyte and Ron Daniel were very helpful.

References

T. O. Binford, 1981. Inferring surfaces from images. *Artificial Intelligence*, 17:205–245.

A. Blake and A. Zisserman, 1987. *Visual Reconstruction*. MIT Press, Cambridge, Mass.

H. Blum, 1973. Biological shape and visual science (part 1). *Journal of Theoretical Biology*, 38:205–287.

R. A. Boie, I. J. Cox, and P. Rehak, 1986. On optimum edge recognition using matched filters. In *Proc. Conf. Pattern Recognition Image Processing*, pages 100–108, Miami Beach, FL.

R. C. Bolles and R. A. Cain, 1982. Recognising and locating partially visible objects. *International Journal of Robotics Research*, 1(3):57–82.

R. C. Bolles and P. Horaud, 1986. 3DPO: A three-dimensional part orientation system. *International Journal of Robotics Research*, 5(3):3–26.

J. B. Bowen and J. E. W. Mayhew, 1986. Consistency maintenance in the REV graph. Sheffield University AIVRU 020.

J. M. Brady, 1981 (). Dangerous behaviour. In *Proc. AISB Conf.*, Amsterdam, Netherlands.

Michael Brady, 1982. Computational approaches to image understanding. *Computing Surveys*, 14(1):3–63.

Michael Brady, 1985 (). Criteria for representations of shape. In *Human and Machine Vision (Beck, Hope, and Rosenfeld eds.)*, page , Academic Press, NY.

J.M. Brady, 1987. Seeds of Perception. In *Proc. Alvey Vis. Conf.*, page , Cambridge, UK.

M. Brady and H. Asada, 1984. Smooth local symmetries and their implementation. *International Journal of Robotics Research*, 3(3).

Michael Brady and G. L. Scott, 1987 (). Parallel algorithms for shape representation. In *Parallel Architectures and Computer Vision (I. Page editor)*, page , Oxford University.

J. M. Brady and B. J. Wielinga, 1978. Reading the writing on the wall. In *Computer Vision Systems (Hanson and Riseman eds.)*, pages 283–302, Academic Press, New York.

Michael Brady and A. Yuille, 1984. An extremum principle for shape from contour. *IEEE Trans. Pattern Anal. Machine Intell.*, PAMI-6:288–310.

Michael Brady, S. A. Cameron, and et. al., 1987. Progress towards a system that can acquire pallets and clean warehouses. In *Proc. Fourth Int. Symp. Rob. Research (Bolles, R. and Roth, B. eds.)*, page , MIT Press.

M. Brady, J. Ponce, A. Yuille, and H. Asada, 1985. Describing surfaces. *Comput. Graphics Image Processing*, 32:1–28.

J.F. Canny, 1983. *Finding Edges and Lines*. Technical Report Tech.Rep. 720, Massachusetts Inst. Technol.

J. Canny, 1986. A computational approach to edge detection. *IEEE Trans. Pattern Anal. Machine Intell.*, PAMI-8:679–698.

J.H. Connell and J.M. Brady, 1985 (). Learning shape descriptions. In *Int. J. Conf. Artif. Intell.*, Los Angeles, Ca.

J. L. Crowley and A. C. Parker, 1984. A representation for shape based on peaks and ridges in the difference of low-pass transform. *IEEE Trans. Pattern Anal. Machine Intell.*, PAMI-6:156–169.

S. W. Draper, 1981. The use of gradient and dual space in line-drawing interpretation. *Artificial Intelligence*, 17:461–508.

Meng Chiau Er, 1986. *Computer interpretation of engineering drawings*. PhD thesis, Essex University.

O. D. Faugeras and M. Hebert, 1986. The representation, recognition, and locating of 3-d objects. *International Journal of Robotics Research*, 5:27–52.

Margaret M. Fleck, 1987 (). The phantom edge finder. In *Proceedings of the Alvey Vision Conference*, Cambridge, England.

David J. Foster, 1987. *The role of pipelining in machine vision*. PhD thesis, Oxford University.

E. C. Freuder, 1975. *A computer vision system for visual recognition using active knowledge*. PhD thesis, MIT Artificial Intelligence Laboratory.

A. Gennert, Michael, 1986 (). Detecting half-edges and vertices in images. In *Proc. CVPR*, pages 552–557, Miami, Fla.

W. E. L. Grimson, 1981. *From Images to Surfaces*. MIT Press, Cambridge, Mass.

W. Eric L. Grimson and Tomás Lozano-Pérez, 1986 (). Search and sensing strategies for recognition and localization of two-and three- dimensional objects. In *Third Symp. Robotics Research*, pages 73–82, Gouvieux, France.

A. Guzman, 1971 (). Analysis of curved line drawings using context and global information. In *Machine Intelligence 6 (Meltzer and Michie eds.)*, pages 325–375, Ellis Horwood.

R.M. Haralick, 1980. Edge and Region Analysis for Digital Image Data. *Comput. Graphics Image Processing*, 12:60–73.

R.M. Haralick, 1984. Digital Step Edges from Zero-crossings of Second Directional Derivatives. *IEEE Trans. Pattern Anal. Machine Intell.*, PAMI-6(1):58–68.

R.M. Haralick, 1985. Second Directional Derivative Zero-crossing Detector using the Cubic Facet Model. In *Proc. Conf. Pattern Recognition Image Processing*, pages 672–677.

M. J. Hawken and A. J. Parker, 1986. Spatial properties of neurons in the monkey striate cortex. *Proc. Roy. Soc. London*, B 231:(in press).

G. Healey and T. O. Binford, 1987 (). Local shape from specularity. In *Proceedings of the First International Conf. Comp. Vision.*, pages 151–160, London, England.

Martin Herman and Takeo Kanade, 1987. Incremental reconstruction of 3-d scenes from multiple complex images. *Artificial Intelligence*, 31:.

A. Heurtas and G. Medioni, 1986. Detection of Intensity Changes with subpixel accuracy using Laplacian-Gaussian Masks. *IEEE Trans. Pattern Anal. Machine Intell.*, PAMI-8(4):651–664.

W. Daniel Hillis, 1985. *The Connection Machine*. MIT Press, Cambridge, Ma.

G. H. Hinton, 1976. *Relaxation and its role in computer vision*. PhD thesis, Edinburgh University.

C. A. R. Hoare, 1979. Communicating sequential processes. *Comm. A. C. M.*

Radu Horaud and Michael Brady, 1987 (June). On the geometric interpretation of image contours. In *Proceedings of the First International Conf. Comp. Vision.*, pages 374–382, London, England.

B. K. P. Horn, 1986. *Robot Vision*. MIT Press, Cambridge, Ma.

D. A. Huffman, 1971. Impossible objects as nonsense sentences. *Machine Intelligence 6*, 295–323.

Geraint Jones, 1987. *Programming in Occam*. Prentice Hall, Englewood Cliffs, NJ.

J. J. Koenderinck and A. J. van Doorn, 1987. Dynamic shape. *Biol. Cyb.*

J. Malik, 1987. Interpreting line drawings of curved objects. *International Journal of Computer Vision*, 1:73–103.

D.C. Marr, 1976. Early Processing of Visual Information. *Phil. Trans. Roy. Soc. London B*, 275:483–524.

D. Marr, 1982. *Vision*. Freeman, San Francisco.

D. Marr and E. C. Hildreth, 1980. Theory of edge detection. *Proc. Roy. Soc. London*, B207:187–217.

J. W. Mayhew and J. P. Frisby, 1981. Psychophysical and computational studies toward a theory of human stereopsis. *Artificial Intelligence*, 17:349–387.

D. W. Murray, D. A. Castelow, and B. F. Buxton, 1987. From an image sequence to a recognised polyhedral object. In *Proc. Alvey Vis. Conf.*, Cambridge, UK.

W. M. Newman and R. F. Sproull, 1973. *Principles of Interactive Computer Graphics*. McGraw-Hill, New York.

J.A. Noble, 1987a. Finding corners. In *Proc. Alvey Vis. Conf.*, Cambridge, UK.

J.A. Noble, 1987b. The Geometric Structure of Images. M.Sc. report.

Ian Page, 1987a. *Parallel Algorithms and Computer Vision*. Oxford University Press, Oxford.

Ian Page, 1987b. The DisPuter. In *Parallel Architectures for Computer Vision (Page, Ian ed.)*, page , Oxford University Press.

S. B. Pollard, 1985. *Identifying correspondences in binocular stereo*. PhD thesis, Sheffield University.

Stephen B. Pollard, John Porrill, John E. W. Mayhew, and John P. Frisby, 1985 (). Disparity gradient, lipschitz continuity, and computing binocular correspondences. In *The Third International Symposium on Robotics Research*, pages 19–26, Gouvieux, France.

J. Ponce and M. Brady, 1987. Towards a surface primal sketch. In *Three-dimensional vision (Kanade ed.)*.

J. M. S. Prewitt, 1970 (). Object enhancement and extraction. In *Picture Processing and Psychopictorics*, pages 75–149, .

G. M. Provan, 1987. *The complexity of Truth Maintenance Sytems and their application to vision*. PhD thesis, Oxford University.

G. L. Scott, 1987. The alternative snake – and other animals. In *Proc. Alvey Vis. Conf.*, Cambridge, UK.

Jun Shen and Serge Castan, 1986. An optimal linear operator for edge detection. In *Proc. CVPR*, pages 109–114, Miami, Fla.

Libor A. Spacek, 1987. Edge detection and motion detection. *Image and Vision Computing*, 4:43–56.

D. Terzopoulos, 1983. The role of constraints and discontinuities in visible-surface reconstruction. In *Proc. 7th Int. J. Conf. Artif. Intell.*, pages 1073–1077, Karlsrühe.

V. Torre and T. Poggio, 1986. On edge detection. *IEEE Trans. Pattern Anal. Machine Intell.*, PAMI-8:147–163.

Shimon Ullman, 1979. *The Interpretation of Visual Motion*. MIT Press, Cambridge, Ma.

R.J. Watt and M.J. Morgan, 1983. The Recognition and Representation of Edge Blur: Evidence for Spatial Primitives in Human Vision. *Vision Res.*, 23(12):1465–1477.

H. R. Wilson and J. R. Bergen, 1979. A four mechanism model for spatial vision. *Vision Research*, 19:19–32.

Richard A. Young, 1986 (). Simulation of human retinal function with the gaussian derivative model. In *Proc. CVPR*, pages 564–569, Miami, Fla.

An Overview of Local Environment Sensing in Robotics Applications

B. ESPIAU

I.R.I.S.A.

Campus de Beaulieu

35042 RENNES Cedex, France

Abstract

This paper presents an overview of the basic technical solutions for local environment sensing, and gives some examples of realizations. A first section sets the problem of local sensing through three concepts : relativity of the informations, necessity of compactness, fastness of the related control loops. Section 2 gives a comparative presentation of the possible local sensors and describes contact, magnetic and ultrasonic sensors. Section 3 is devoted to optical sensors, including local vision and optoelectronic systems : triangulation, phase shift and reflectance sensors. Section 4 deals with multisensory systems and local control. The paper ends with the presentation of some realizations and gives some remarks about the future of local environment sensing.

1 Introduction

Most classical industrial robots have no real – time closed – loop control with regard to the environment : the only existing feedback is provided by joint sensors, and the use of vision systems is limited to some high – level functions (sorting, recognition, inspection) which are activated from time to time. As mentionned in [GOOD], this leads to inability to detect and correct errors, thus to a lack of adaptivity, which may be easily overcome by using end – effector sensing. For example, in space applications ([BEJCZY 80]), local sensing may compensate for terminal errors due to the flexibilities of a large light structure. In an other way, it is often easier, in many industrial applications, to compensate for relative positionning errors by moving the robot than the workpiece ; further, the precise clamping of a part at a reference position is sometimes difficult, and, finally, the deformations or uncertainties on the workpiece itself may lead to the need of a final relative adjustment : among the most known industrial examples, let us recall the mounting of a windshield in car industry or the tracking of a seam in arc welding.

An other important problem which requires local sensing mainly in non manufacturing applications, is the one of on – line collision detection and obstacle avoidance. That problem is encountered when the environment is not well – known, unperfectly modelled, or liable to unexpected modifications : most of mobile robots are concerned, especially outdoor systems, or rescue robots.

NATO ASI Series, Vol. F43
Sensors and Sensory Systems
for Advanced Robots
Edited by P. Dario
© Springer-Verlag Berlin Heidelberg 1988

This problem is also important in the applications of teleoperation, mainly in hostile worlds : undersea or nuclear environments. In all cases, fast detection and control/decision algorithms are required, which proves again the need of efficient local sensing.

From the analysis of the main kinds of problems which requires local environment sensing, some of them were described above, we may select the essential features of local exteroceptive sensing :

1. The aim is to provide a set of *relative* informations with regard to the assigned task. These informations will generally be of geometric kind, and thus will depend in a way on the relative position/orientation of a local sensor and of the related target. This is why local sensing is more concerned with low – level measurements, like target's distance and orientation, than with high level informations (recognition).

2. Due to the requirement of relativity, it is obvious that local sensors are « *on – board* » sensors : they have to be located either on the robot, or... on the workpiece itself, or its carrier. In the first case (the most frequently encountered), all the elements of the system, i.e. all the links for a manipulator, may be concerned, and a classical sensing distribution has to allow to prevent the links from collision, while aiding the end effector in its task achievement. In industrial applications or in teleoperation, a more general point of view is to consider that an and effector is a *tool* rather than a gripper, and that each removable tool has to be equipped with the adequate local sensing capabilities.
An obvious consequence is that the *size* of the sensors is a basic parameter for choice and evaluation. This point is of course more crucial for manipulators than for mobile robots or legged vehicles.

3. As the informations are low – level, the related actions or motions are generally of « *reflex* » kind rather than « reflective ». If we compare with the human behaviour, a simple interpretation is that local sensory – based motions are close to virtual force – controlled (or tactile) actions, even providing a « remote – touch » feeling in some applications ([ESPIAU 84a], [ANDRE 85a]). The immediate consequence is the introduction of constraints for the response – time of local loops. These real – time constraints induce three main requirements : the local sensors themselves must have a low response time ; the related communication channels and protocoles have to be fast enough ; the programming and software tools have to allow to close the local loops with the required bandwith.

These three parameters : size, nature of the provided output, and response time mainly characterize the possible technical solutions for local sensing which will be presented in the following.

Local sensing is thus an essential complement to off – line robot programming and high level sensing and reasoning. The aim of this paper is to give an overview of some existing local sensors, through experiments reported in the robotics literature, with an emphasis on optical techniques. The paper is organized as follows : we begin Section 2 with some general informations about the various physical principles which are used. Then, we present a survey of non – optical techniques. Section 3 is devoted to optical sensors, including reflectance, triangulation or phase measurements. The paper ends in Section 4 with some remarks on the need of multisensory systems and on the control methodology related to local sensing. Supplementary drawings and figures related to this paper may be found in [ESPIAU 86].

2 Physical Principles of local sensors

2.1 Generalities on the available technologies

Several criteria may be used in order to classify local sensors :

Passive/active :   a sensor is « active » if it emits an information which has to interact with the environment before being analyzed.

binary/non – binary :   many industrial sensors (presence sensors for example) directly provide a logical output, through an adjustable threshold. Analog sensors have a continuous output, and discrete sensors provide a digital signal with a given quantization.

nature of the information :
Some sensors directly provide an information of distance. We call them « range » sensors. Some others are able to give further geometric parameters, like the orientation. The sensors the output of which do not *only* depend on geometric relations are usually called « proximity » sensors.

The Table of figure 1 presents the characteristics of the main existing local sensors through the above classification.
Figure 2 gives a comparizon of the performances and of the conditions of use for the various local sensing principles including criterias given in the introduction.

2.2 Some examples of the use of non – optical sensors in robotics applications

2.2.1 Contact sensors

We do not here consider tactile or force sensing, but only mechanical contact sensors. The most known electromechanical distance sensor is the linear variable differential transformer, which consists of magnetically coupled coils through which a permeable rod segment may move according to the contact distance. An other kind of electromechanical distance sensor is the magnetoacoustic sensor, the principle of which may be found in [KOENIGSBERG 83].

However, robotics applications of local sensing do not generally use contact *distance* sensors, but rather *Proximity* mechanical sensors. The principle is to combine a simple industrial microswitch with a mechanical device (probe, needle, plate, ring,...) adapted to the application. An interesting example of mechanical local sensing is provided in [HIROSE 84], where « whisker » sensors, made of shape memory alloy, are used for local sensing at the feet of a walking machine.

Close to the principle of contact sensor is the pneumatic sensor : air pressurized escapes through a small hole, and when an object lies into the airstream, close to the orifice, the inside pressure increases. This system is used in a multidigital gripper, in [BERTIN 82].

2.2.2 Capacitive and magnetic sensors

The principle of capacitive sensors is to measure the change of a capacitance induced by the presence of an object into the sensitive field of the sensor. In some cases, one of the electrodes is the object itself, and the sensor output depends on its position. These sensors are not widely used in robotics applications.

Much more frequently encountered are the magnetic sensors. The more simple systems are reed microswitches or Hall effect switches, but the most used systems in robotics are based on electromagnetic inductive principles, mainly eddy currents generation. The basic principle consists in creating a magnetic field using appropriate coils, a core with high permeability, and an oscillator with a frequency excitation high enough to minimize the penetration of the field inside a conductive material. Such sensors are presented in [CLERGEOT 84], from which are taken figures 3 and 4, and are mainly used in seam – tracking applications ([DETRICHE 81], [NICOLO 80]). An other example of magnetic sensor may be found in [HORNAK 83]. The main problems related to magnetic sensors are their high size/range ratio, the sensitivity to temperature, the difficulty of reliable distance measurement in general cases.

2.2.3 Ultrasonic sensors

The basic principle of these sensors is to measure the time taken by a sound wave from its emission to its reception, through a conductive medium, and after having been reflected by a target. Otherwise, using Doppler effect allows to measure the velocity of a target. A well – known example of acoustic sensor is the Polaroid range finder, used in automatic focus cameras. A well – suited acoustic sensor for robotics is presented in [CANALI 81].

In robotics, ultrasonic sensors are widely used for mobile robots (example in [KANAYAMA 84], [TACHI 84], [BAUZIL 82]), in local navigation tasks. Some other applications are seam tracking ([GUNNARSSON]), object recognition [SASAKI 84] or real – time trajectory correction ([ANDRE 85a]). Although they give a distance measurement, ultrasonic sensors present some drawbacks : first, at the used wavelengths, the targets are mirror – like objects, and there is no significant reflected signal when the angle between the normal to the object surface and the sensor direction is larger than about ten degrees. An other problem is the *minimal* detectable range, which is usually greater than 10 cm, due to the measurement principle. Significant improvment is got when using separate emitter and receiver ; [MILLER 84] presents a method of active damping which greatly increases the performances of acoustic sensors.

A last drawback of ultrasonic sensors is their overall size, which reduces the possibility of integration within an end effector.

2.2.4 Microwaves and Radiowaves sensors

The basic principle is to transmit an electromagnetic signal through an antenna, and to measure the round trip time after reflection on a target. Usually, the practical range of such systems is very large, and they mainly concern outdoor autonomous robots or space systems. However, some short – range radars have been realized by RCA Company, and a system measuring the back reflected energy instead of the time of flight is presented in [ANDRE 83].

2.2.5 Miscellaneous sensors

Some industrial sensors measure distance or flow by sensing scattered or absorbed radiations using gamma generators and cameras. We do not know present applications in robotics, but the emission of nuclear rays by a target might be used in some cases of intervention, for example for homing a gripper to find and catch a gamma source. The more general underlying idea is to use the specific features of an application to design an « ad – hoc » local sensor. For example, thermal infrared measurement has been used in seam tracking applications.

3 Optical sensors

We will distinguish two main classes of optical sensors :

- The sensors which use a classical imaging system (camera), in a « local vision » approach.
- The optoelectronic sensors.

3.1 Local vision

The basic idea here is : « Eye in the hand ». The availability of geometrically fiber optic bundles has improved the possibilities of reducing the size of a sensitive head, which classically includes a lens and a solid – state imaging array. Two examples of fiber – optic image sensors may be found in [AGRAWAL 83] and [TAYLOR 83]. From our point of view, the vision sensors become local sensors if they may be used in a closed loop way, as in reference [SANDERSON 83], where an on – board camera is used to correct positions in real – time. Another example of closed – loop control under passive vision may be found in [SHOHAM], where a simple four – quadrants optical detector is used to correct robot position and orientation in a seam tracking experiment.

An other possibility is to combine local vision with a device allowing range measurement. In reference [ORROCK], a CID camera is coupled with two range sensors. Close to the principle of optical triangulation (see section 3.2), is the idea of adding a laser spot emitter to the camera, and by measuring its XY coordinates in the image plane, to get a relative position measurement (an example is the I – SIGHT 32 system, from Electronic Automation Ltd, which provides both a local vision sensor and a laser device).

Moreover, this principle of local active vision may be improved by adding a sweeping mechanism, to extend the number of measured 3D points. This is for example used in some seam tracking applications (ex. in [HUBER 84] and [BAMBA 81]. Of course, an other natural way of realizing local active vision is to project given patterns onto the objects, and to process the image in order to recover 3D relative informations. In [VANDERBRUG 79], a light plane is used and the processed informations provide data about range, orientation and edges of the object. In [AGIN 85], a light – stripe projector with a cylindrical lens is used. More complex patterns may be projected, as in [KINOSHITA 85], where a ring is used for range finding and surface tracing.

However, all these sensors may be considered as local 2D or 3D visual sensors, but their use as local environment sensors satisfying the requirements given in Section 1 is yet doubtful.

3.2 Optoelectronic sensors

This class of sensors includes all the system which associate emitting and receiving optoelectronic components, like LEDs, Laser diods, PIN diods, Phototransistors, etc. Several principles of measurement may be used :

- the triangulation sensors use the geometric distribution of projected spots to compute distance and orientation

- the phase sensors indirectly measure the time of flight of the emitted light after reflection onto the target, or use Lambert's law to compute geometric phase shift.

- the reflectance sensors measure the back – diffused energy from a target.

3.2.1 Triangulation sensors

The basic principle of these sensors is quite simple : a photoreceiver (point, line or plane) senses the image of the intersection of an emitted light beam and a target. By knowing the position of the spot on the receiver, and the geometry of the emitter/receiver device, the distance may be computed. By using a mobile emitter, or several emitters in conjunction with a 2D receiver, supplementary informations are provided : local shape of an object, orientation of a target. However, the trend is presently to avoid mechanical scanning devices, in order to increase the reliability of the device, and to reduce its overall size.

Let us give some example of triangulation sensors :

Systems with mobile sensors

A first idea is to search for the maximal intensity received by scanning emitter and receivers, and to compute the distance and the orientation of a plane target using a photometric model (ex. in figure 5). A more efficient approach is to use a set of reflectance sensors (as the ones described in section 3.2.3), and to search for the maximum reflected intensity in each sensor. In [YAMADA 83], two sensors with infrared emitters allow to locate a cylindrical target. A more compact realization is mentionned in [ANDRE 83].

A second class of scanning triangulation device is the well – known mobile spot, which is mainly used in 3D active imaging and in seam tracking applications. The most classical system uses a scanning mirror, with one or two rotations. For example, in [BAMBA 81], the sensor includes a LED, a mobile mirror, and a 2D photodiod receiver. The same author, in [BAMBA 84] uses a Laser Diod, with a linear PIN and the scanning is performed by rotating the whole head around the torch. Many other references on that class of sensors are available through the literature. However, all these sensors may be considered as local sensors in our sense only if they satisfy the constraints given in section 1. This generally requires a large reduction of their overall size, which is not presently easily feasible.

Systems without mobile mechanical parts

The principle here is to replace mechanical scanning by « electrical » scanning. In [OKADA 82], a unique laser light source is used with a linear array of 512 photosensors through pinholes. According to the principle given figure 6, if the maximal intensity value is located at the position X_i, the range of the target is obtained from :

$$D = \frac{HX_O}{X_i - X_O} \tag{1}$$

A photometric study is provided in the mentionned paper together with experimental results.

Some other interesting realizations are described in [FUHRMANN 84] and [KANADA 83] : a first sensor uses 6 LED and a planar PIN photodiod Hamamatsu ; the 6 LEDs are aimed toward the same point along the optical axis. The obtained precision is 0.07mm and 1.5° over the range of 40mm to 50MM and −30 ° to +30 °. The orientation is estimated using a least – squares fitting of a plane to the 6 measured points. A second version improves the performances by focusing three sources toward one point and three other sources toward a second point on the optical axis. A last proposed sensor uses multiple cones of 18 light sources with expected accuracy of 0.05mm and 1 °, and sensor diameter of 11cm.

The same principle of measurement is found in [NAKAMURA 83] : 30 infrared LEDs are arranged circularly around the axis of a position sensitive device Hamamatsu. All the beams converge conically at the same point on the optical axis. The depth range is 30mm to 50mm. Let us notice that a microcomputer system is integrated inside the sensory device, without too much increasing the overall size.

Based upon the same triangulation principle, other realizations may be found in the literature, some of them using fiber optics instead of on – board LEDs.

The main problems in triangulation methods lie in :

- the errors due to the distorsion or the separation of the spot when the beam is not an ideal line.

- the blind areas which may occur in abrupt discontinuities.

3.2.2 Optical Phase Sensors

We may distinguish two kinds of sensors according to the used measurement principle :

Phase shift from photometry and geometry

This kind of sensor is described in [MASUDA 81] and [MASUDA 85] : the principle of distance measurement is the following (fig. 7): the two light sources are modulated such as the luminous intensities are :

$$LED_1 : G_1 = A \sin \omega t \qquad (2)$$

$$LED_2 : G_2 = B \cos \omega t \qquad (3)$$

The brightness of Point P is :

$$L_P = C \left[\frac{G_1}{h_1^2} \cos \theta_1 + \frac{G_2}{h_2^2} \cos \theta_2 \right] \qquad (4)$$

with the following hypothesis :

- the object surface is Lambertian with parameter C (perfect diffusivity)
- the intensity of light sources is not affected by the angles of emission.
- the receiver is narrow – beam.

Then, (4) gives :

$$L_P = Cz \left[\frac{A}{(a^2 + z^2)^{3/2}} \sin \omega t + \frac{B}{(b^2 + z^2)^{3/2}} \cos \omega t \right] = D \sin (\omega t + \phi) \qquad (5)$$

with

$$\tan \phi = \frac{B}{A} \left[\frac{a^2 + z^2}{b^2 + z^2} \right]^{3/2} \qquad (6)$$

Knowing A, B, b, and measuring ϕ then gives z via (6).

More complex combinations of emitters allow to measure target inclination and position. The practical range is 10 to 60 mm with accuracy ± 1 mm and ± 30° with accuracy ±2°.
A very close system with similar performances is proposed in [TROUNOV 84].

Phase shift from time – of – flight ([CHAPPELL])

The principle here is to build an optical radar system. However, as the practical desired working distance is small (less than 1500 mm) and as we need a rather good accuracy, it is not possible to directly measure the time – of – flight of a light pulse (1 nanosecond for 300 mm). A more convenient way is to modulate the flight beam and to measure the phase shift due to the distance ([CHAPPELL]). The phase shift of a signal with frequence f and velocity c on a distance $d = ct$ is :

$$\phi = 2\pi ft = \frac{\omega d}{c} \qquad (7)$$

thus :

$$\frac{\Delta d}{d_{max}} = \frac{\Delta \phi}{\phi_{max}}$$
(8)

for example with a desired accuracy in distance of 1/1800 the required accuracy in the measurement of ϕ is 0.1° on a 180° range.

From (7), we obtain :

$$f = \frac{C}{2\pi} \frac{\Delta \phi}{\Delta d}$$
(9)

Thus

$$f = 10^8 \frac{1}{1,2\Delta t(mm)} Hz$$
(10)

The order of magnitude of f thus is from 50 to 200 Mhz according to the desired range and accuracy.

The principle of a phase sensor is given figure 8. The light source is generally an infrared laser diod, with a rather low output power (some mW or tenth of mw). A fraction of the emitted beam constitutes the reference signal. A transposition signal ω_t allows to work at the frequency $\omega_m - \omega_t$, much lower than ω_m. The measurement of ϕ may be performed from $s(t)$ either by lowpass filtering or counting.

Two examples of such systems may be found in [PAGE] and [SLAVIK].

3.3 Reflectance sensors

Among the optoelectronic sensors, this kind of sensor is the simplest one : as shown in figure 9, it includes an emitter and a receiver with appropriate focusing devices. The output signal represents the diffused light intensity from the target to the receiver, which depends not only on the distance but also on the photometric properties and the local orientation of the object. The optical parts are generally lenses or optical fibers. To obtain a good signal to noise ratio, two measurement methods are used :

- synchronous modulation

- pulse intensity measurement

The last solution allows to emit short pulses of large intensity, and, by measuring the difference between the received level and the steady state value to increase the SNR. However, a resting time between successive pulses is needed.

Typical outputs of a reflectance sensor facing a plane target with albedo λ are presented figure 10. If the geometric and optical parameters of the sensor are well – tuned, the range section from the origin to the maximum presents a linear area with high sensitivity, which is sometimes used for accurate positionning. A complete photometric model of a reflectance sensor may be found in [ESPIAU 80]. However, by making the emitter and the receiver as close as possible with quasi – parallel axis, it may be shown that the output y tends to :

$$\tilde{y} = \frac{K}{d^2} \qquad (11)$$

where d is the distance to the plane target, and K a global parameter depending on the target's characteristics. This may be verified on the experimental results presented figure 11, approximatively when the distance is greater than the one corresponding to the maximal intensity value.

Some realizations of reflectance sensors are reported in [JOHNSTON 77], [WAMPLER 84], [BALEK 85]. Sensors with optical fibers made by SAGEM are described in [ESPIAU 80]. More recent sensors are presented in [ESPIAU 85a] : they simply consist in a set of selected couples of LEDs and photoreceivers with the appropriate geometric arrangements (fig. 12). The emitters are driven by pulse generators, with a typical pulse width of 20 µs and a resting time of 200 µs.

Before ending this section, let us recall the main problems related to optical sensors :

- Sensitivity to the nature of the target :
  Much sensors use the property of diffusivity. Of course, some objects (metallic for example) look like mirrors, and there exists a risk of light saturation in specular reflection, while no significant signal occurs for any other relative orientation. Further, highly absorbing objects (« black bodies ») may lead to a bad signal – to – noise ratio.

- Sensitivity to the environment :
  Several elements may disturb an optical sensor : the medium may be polluted by dust of smoke ; the ambient light may modify the received signal ... In this last case, modulation or pulse driving are two ways of avoiding the influence of external light sources. A supplementary precaution is to use narrow band emitters with the associated filters on the receivers. This method is more efficient with laser diodes which have a very thin light band – width.

An other problem is specific of reflectance sensors : the direct measurement of range and/or orientation is not possible when some characteristics of the target are unknown. However, it is possible to overcome this difficulty if the relative **motion** of a sensor and its target is known. Recursive algorithms which simultaneously estimate K and d in equation (11) are given in [ESPIAU 85a].

4 Multisensory systems ans local control

4.1 Introduction : Multisensory systems are needed

The basic techniques for local sensing have been presented above ; some of them provide rather fine informations about the local geometry of an object : distance, orientation, 3D relative position of some points, etc... However, in practice, the choice of a local sensory system does not depend only on the performance of the candidate sensors but also on the particular constraints assigned by the task. For example, the size requirements for the integration of sensors within a gripper may lead to prefer small refectance sensors to more accurate triangulation sensors. This has two main consequences :

a)   Achievement of non trivial tasks with simple local sensors will require the use of **several** sensors to extract the needed informations.

b)   As various technologies may be used to perform a same kind of task depending on environmental conditions, the constraints of **flexibility**, and **transparency** from the control point of view become of prime importance.

For these reasons, and recalling that **cooperation** between sensors is also an important trend in robotics, the need of systems able to manage and process several local sensors clearly appears. An interesting example of such a system is given in [ANDRE 85b] : the multiproximity sensor system (MSP) presented fig. 13 allows to combine various sensor modules, with high hardware and software modularity. The main components of the MSP are presently :

a)   A set of small sensory heads which can be arranged on a robot arm or mounted on an end effector : Infrared reflectance sensors (fig. 12), ultrasonic transducers (POLAROID or MASSA E 188), and magnetic sensors.

b)   The associated remote fast electronic modules, generally located near the sensory heads.

c)   An acquisition and processing system, including the dedicated interfaces and a microcomputer for local real – time processing and external communications. Several channels are available for each kind of sensor (up to 16 optical sensors).

An other multiproximity system is presented in [CROSNIER] : the MULTICOFO system is able to process up to 56 fiber optics detectors, and an example of gripper control is described.

4.2 Control and applications :

The third feature of local exteroceptive sensing proposed in section 1 is the need of real time sensory feedback. This problem has been underlined by several authors, as in [ALBUS 80], [HIRZINGER 85b], [NITZAN 83], [NITZAN 81], [WAMPLER 84]. Briefly, this problem has to main aspects : the algorithmic point of view and the implementation/programming aspect. Concerning the first one, a general possible approach is to use so – called «Hybrid Control» [CRAIG 82], where the force – controlled part may be driven by local sensors producing either virtual or actual forces. As mentionned in [LEBORGNE 84], [ESPIAU 84b] and [ANDRE 85a], this idea may be applied with proximity sensors by associating elementary actions to elementary sensors, building action primitives looding like virtual contacts between solids ([ESPIAU 84c] and computing the associated screw to be introduced in the control scheme (fig 14) :

$$f = \sum_i \vec{V_i} \tag{12}$$

$$\tau = \sum_i \vec{Ox_i} \wedge \vec{V_i} \tag{13}$$

More difficult is the practical problem of introducing these control variables into the programming system, with the guarantee that the real – time constraints will be satisfied by the sensory feedback loop. A first idea is to compute (12) (13) in parallel with the main controller, as it is done in the MSP [ANDRE 85b], in order to simplify and to normalize the transmitted data between the sensory system and the control device. However, this is not sufficient for making easy the use of sensory – based loops, and much work remains to be done at the level of programming tools and languages.

In spite of these limitations, some applications of local sensing in control have been reported in the literature : the main work in space applications has been done at JPL (see for example [BEJCZY 80]), with optical sensors. In [WAMPLER 84], a gripper with 10 infrared reflectance sensors is described. In Figure 15, a bidigital gripper including 3 reflectance sensors using optical fibers, designed by SAGEM is presented. In [BALEK 85], a parallel – jaw gripper is equipped with 7 infrared reflectance sensors, for bin – picking applications. Some other realizations are presented in [LUO 84], [DILLMANN 82], [BENI 83], and [ESPIAU 86]. Other examples of simple sensitive grippers may be found in [BARDIAUX 85], [BARDIAUX 84], [WITWICKI 79]. Let us end this survey of applications with the infrared devices presented in [ESPIAU 85a], which have been used in several applications of teleoperation ([VERTUT 84], [ESPIAU 84a]): a bidigital gripper, with up to 12 reflectance sensors allows automatic grasping and obstacle avoidance, and a planar 4 – sensors system is dedicated to surface – following tasks. An other version, with ultrasonic sensors, has been used for the compensation of the oscillations of a teleoperator on a moving carrier (figure 16, from [ANDRE 85a]).

4.3 Concluding remarks

This paper has presented an overview of local sensing techniques, with an emphasis on optical systems, and has given some examples of applications. A question now is : what is the future of local sensing in robotics? To that question are associated some problems which remain to be solved, in relation with the three main characteristics of local sensing given in section 1 :

a)   The first problem is the integration of sensors into robotics devices : the most part of the existing sensors has to be reduced in size if we want to use them in a wide class of applications. A related problem is the one of data and energy transmission : decreasing the number of wires is a necessary challenge, which may be partially won by realizing local processing and « smart sensors ».

b)   The second problem is the reliability of the provided informations with regard to the task. This problem is not only a technical one, but also an algorithmic one : for example, we may overcome the sensitivity of reflectance sensors to the nature of objects by using normalized error signals, or recursive estimation schemes. Some studies remain to be done in the field of exploiting the time – redundancy of the informations during a closed – loop motion.

c)   Finally, the industrial future of local sensing techniques is linked both to the availability of robust, reliable, and accurate sensors and to the easiness of using them in a robotics system. This last point demands to reconsider the existing hardware and software tools (interfaces, languages), in order to authorize the design and the implementation of efficient local sensory – based loops.

References

[AGIN 85] **AGIN G.J.** Calibration and use of a light stripe range sensor Mounted on the hand of a robot
IEEE Conference on Robotics and Automation, St Louis, March 1985.

[AGRAWAL 83] **AGRAWAL A.** Robot Eye – in – hand using fiber optics
ROVISEC 3, Cambridge Ma, Nov. 1983.

[ALBUS 80] **ALBUS J.S., BARBERA A.J., FITZGERALD M.L., NAGEL R.N., VANDERBRUG G.J., WHEATLEY T.E.** A measurement and control model for adaptive robots
10th ISIR, 1980.

[ANDRE 83] **ANDRE G.** Conception et Modélisation de systèmes de perception proximétrique. Application à la Commande en Téléopération
Thesis, University of Rennes, Oct. 1983.

[ANDRE 85a] **ANDRE G., FOURNIER R.** Generalized End effector Control in a computer aided teleoperation system with application to motion coordination of a manipulator arm on an oscillating carrier
1985 International Conference on Advanced Robotics, Tokyo, Sept. 1985.

[ANDRE 85b] **ANDRE G.** A Multiproximity sensor system for the guidance of robot end effectors
5th ROVISEC, Amsterdam, Oct. 1985.

[BALEK 85] **BALEK D.J., KELLEY R.B.** Using gripper Mounted Infrared Proximity Sensors for robot feedback control
IEEE Conference on Robotics and Automation, St Louis, March 1985.

[BAMBA 81] **BAMBA T., MARUYAMA H., OHNO E., SHIGA Y.** A visual Sensor for arc welding robots
11th ISIR, Tokyo 1981.

[BAMBA 84] **BAMBA T., MARUYAMA H., KODAIRA N., TSUDA E.** A visual seam tracking system for arc – welding robots
14th ISIR, Gothenburg, Sweden, Oct. 1984.

[BARDIAUX 84] **BARDIAUX J.C.** Implantation de Capteurs extéroceptifs simples sur un mini – robot
Conférence CIAME « Capteurs 84 », Paris, June 1984.

[BARDIAUX 85] **BARDIAUX J.C.** A simple low – cost intelligent sensor for recognition and localization of moving workpieces
15th ISIR, Tokyo, Sept. 1985.

[BAUZIL 82] **BAUZIL G., BRIOT M., RIBES P.** A Navigation Sybsystem using ultrasonic sensors for the mobile robot HILARE
2nd ROVISEC, 1982.

[BEJCZY 80]   BEJCZY A.K. "Smart sensors for smart hands", from « Remote Sensing of Earth from Space : Role of Smart Sensors »
R.A. Brecken ridge Ed., Vol. 67 of Progress in Astronautics and Aeronautics, 1980.

[BENI 83]   BENI G., HACKWOOD S., HORNACK C.A., JACKEL J.L. Dynamic sensing for robots : an analysis and implementation
International Journal of Robotics Research, Vol. 2, no 2, Summer 1983.

[BERTIN 82]   BERTIN Cy. Pince équipée de capteurs de proximité pneumatiques pour la saisie en vrac
Journal « Le Nouvel Automatisme », France, March 1982.

[CANALI 81]   CANALI C., de CICCO G., MORTEN B., PRUDENZIATI M., TARONI A. An Ultrasonic sensor operating in air
Sensors and Actuators, 2(1981/82) 97 – 103.

[CHAPPELL]   CHAPPELL A. ed. Optoelectronic range finder, in « Optoelectronics, Theory and Practice »
pp 389 – 397, Mc Graw – Hill Book Company.

[CLERGEOT 84]   CLERGEOT H., PLACKO D., MONTEIL F. Flexible Eddy Current Sensors for Industrial Applications
4th ROVISEC, London, Oct. 1984.

[CRAIG 82]   CRAIG J.J., RAIBERT M.H. Hybrid Position/Force control of manipulators.
Journal of Dynamic Systems, Measurement and Control, Vol. 102, June 1982.

[CROSNIER]   CROSNIER J.J. Sens du toucher aritficiel par fibres optiques pour organes préhenseurs de robots industriels in « Etat de la Robotique en France »
Company Souriau, 3 av. du Mal Devaux, Paray – Vieille – Poste, France.

[DETRICHE 81]   DETRICHE J.M., MARECHAL P., CORNU J. Self – adaptive Arc Welding by mean of an automatic Joint following system
4th Symposium ROMANSY, Warsaw, Poland, Sept. 1981.

[DILLMAN 82]   DILLMAN R. A sensor – Controlled gripper with tactile and non tactile sensor environment.
2nd ROVISEC, Stuttgart, 1982.

[ESPIAU 80]   ESPIAU B., CATROS J.Y. Use of Optical Reflectance Sensors, in Robotics Applications
IEEE Transactions on SMC, Vol. SMC – 10, no 12, Dec. 1980.

[ESPIAU 84a]   ESPIAU B., ANDRE G. Sensory – Based Control for Robots and Teleoperators
5th ROMANSY, Udine, Italy, June 1984.

[ESPIAU 84b]   ESPIAU B. Closed Loop Control of robots with local environment sensing : principles and applications
2nd International Symposium on Robotics Research, Kyoto, Aug. 1984.

[ESPIAU 84c]   ESPIAU B. Note sur l'interprétation de primitives d'action proximétriques en termes de liaisons cinématiques fictives
IRISA Technical Report no 242, Rennes, Nov. 1984.

[ESPIAU 85a]   **ESPIAU B.** Use of Optical Reflectance Sensors in :
Recent Advances in Robotics, John Wiley Publ., New – York, 1985.

[ESPIAU 85b]   **ESPIAU B., BOULIC R.** Collision avoidance for redundant robots with proximity sensors
3d International Symposium on Robotics Research, Gouvieux, France, Oct. 1985.

[ESPIAU 86]   **ESPIAU B.** Etat de l'art en capteurs d'environnement local pour la robotique
Internal Report, IRISA, Rennes, France, June 1986

[FUHRMANN 84]   **FUHRMANN M., KANADE T.** Optical Proximity Sensor using multiple cones of light for measuring surface shape
Optical Engineering, Vol. 23, no 5, Sept/Oct. 1984.

[GOOD]   **GOOD M.C., SWEET L.M.** Structures for sensor – based robot motion control
Corporative Research and Development, General Electric Company, Shenectady, USA.

[GUNNARSSON]   **GUNNARSSON K.T., PRINZ F.B.** Ultrasonic sensors in robotic seam tracking
Dept of Mechanical Eng., Carnegie Mellon Univ., Pittsburgh, USA.

[HIROSE 84]   **HIROSE S., MASUI T., KIKUCHI H., FUKUDA Y., UMETANI Y.** TITAN III : A Quadruped Walking Vehicle
2nd International Symposium on Robotics Research, Kyoto, Aug. 1984.

[HIRZINGER 85a]   **HIRZINGER G.** Sensory Feedback in the external loop
IFAC Symposium on Robot Control, Barcelona, Nov. 1985.

[HIRZINGER 85b]   **HIRZINGER G.** Robot Learning and Teach – in based on sensory feedback
3d International Symposium on Robotics Research, Gouvieux, France, Oct. 1985.

[HORNACK 83]   **HORNACK L.A., HACKWOOD S., BENI G.** Reentrant Loop Magnetic Effect Proximity Sensor for robotics.
3rd ROVISEC, Cambridge, Nov. 1983.

[HUBER 84]   **HUBER C.** Sensor – based tracking of large quadrangular weld seam paths
14th ISIR, Gothenburg, Sweden, Oct. 1984.

[JOHNSTON 77]   **JOHNSTON A.R.** Proximity Sensor Technology for manipulator and effectors.
Mech. Machine Theory, 12, 1977.

[KANADE 83]   **KANADE T., SOMMER T.M.** An Optical Proximity Sensor for Measuring Surface Position and Orientation for Robot Manipulation
Technical Report TR – 83 – 15, Carnegie Mellon Univ., Pittsburgh, 1983.

[KANAYAMA 84]   **KANAYAMA Y., YUTA S., KUBOTERA Y.** A sonic range finding module for mobile robots
14th ISIR, Gothenburg, Sweden, Oct. 1984.

[KINOSHITA 85]   KINOSHITA G., IDESAWA M. Optical Range finding system by projecting
                     ring beam patterns
                     1985 International Conference on Advanced Robotics, Tokyo, Sept. 1985.

[KOENIGSBERG 83]   KOENIGSBERG W.D. Non contact Distance Sensor Technology
                     3d ROBISEC, Cambridge MA, Nov. 1983.

[LEBORGNE 84]   LEBORGNE M., ESPIAU B. Modelling and Closed – loop Control of Robots
                     in local operating space
                     IEEE Conference on Decision and Control, Las Vegas, Dec. 1984.

[LUO 84]   LUO Ren Chyuan, GRANDE D. Servo – Controlled Gripper with Sensors for
                     flexible assembly
                     IEEE Conference on Robotics, Atlanta, March 1984.

[MASUDA 82]   MASUDA R., HASEGAWA K. WEITING GONG Total Sensory System for
                     Robot control and its design approach
                     11th ISIR, Tokyo, Oct. 1981.

[MASUDA 85]   MASUDA R. Multifonctionnal Optical Proximity Sensor by using phase
                     information
                     1985 International Conference on Advanced Robotics, Tokyo, Dec. 1985.

[MILLER 84]   MILLER G.L., BOIE R.A., SIBILIA M.J. Active Damping of Ultrasonic
                     Transducers for Robotic Applications
                     IEEE Conference on Robotics, Atlanta, March 1984.

[NAKAMURA 83]   NAKAMURA Y., HANAFUSA H. A new optical proximity sensor for three
                     dimensionned autonomous trajectory control of robot manipulators
                     1983 International Conference on Advanced Robotics, Tokyo, 1983.

[NICOLO 80]   NICOLO V., MOREGGIA V., VARRONE P.G., VENTURELLO G. Industrial
                     Robots with Sensory feedback ; Application to continuous arc welding
                     10th ISIR, Milano, 1980.

[NITZAN 81]   NITZAN R. Assessment of robot sensors
                     1st ROVISEC, Stratford on Avon, 1981.

[NITZAN 83]   NITZAN D., BARROUIL C., CHEESEMAN P., SMITH R. Use of sensors in
                     robot systems
                     ICAR 83 Conference, Tokyo, Dec. 1983.

[OKADA 82]   OKADA T. Development of an Optical Distance sensor for Robots
                     International Journal of Robotics Research, Vol. 1, no 4, Winter 1982.

[ORROCK]   ORROCK J.E., GARFUNKEL J.H., OWEN B.A. An Integrated Vision/Range
                     sensor
                     pp 263 – 269, Honeywell Inc., Technology Strategy Center, Roseville,
                     Minnesota, U.S.A.

[PAGE]   PAGE C.J., HASSAN M. Non contact Inspection of complex components
                     using a rangefinder vision system
                     ROVISEC, pp 245 – 254.

[SANDERSON 83]   SANDERSON A.C., WEISS L.E. Adaptative Visual Servo – Control of
                     Robots, in « Robot Vision »
                     Ed. Alan Pugh, 1983, IFS Publ. Ltd, U.K. and Springer – Verlag.

[SASAKI 84]   SASAKI K., TAKANO M., ONO K. Development of Ultrasonic Robot
                       Sensor.
                       No 3 – B 14 International Symposium on Design and Synthesis, Tokyo,
                       1984.

[SHOHAM]   SHOHAM M., FAINMAN Y., LENZ E. An Optical Sensor for Real – time
                       Positionning, Tracking and Teaching of Industrial Robots
                       1983. Faculty of Mechanical Engineering, Technion, Israël Institute of
                       Technology Haifa, Israël.

[SHOLL 84]   SHOLL P., LOUGHLIN C. A practical solution to real – time path control of
                       a robot
                       ROVISEC 4, London, Oct. 1984.

[SLAVIK]   SLAVIK J. Three – Dimensional Optical Sensor design for industrial robots
                       ROVISEC, pp 255 – 262.

[TACHI 84]   TACHI S., KOMORIYA K. Guide Dog Robot
                       2nd International Symposium on Robotics Research, Tokyo, Aug. 1984.

[TAYLOR 82]   TAYLOR P.M., SELKE K.K.W., TAYLOR G.E. Closed – loop Control of an
                       industrial robot using visual feedback from a sensory gripper
                       12th ISIR, Paris, June 1982.

[TAYLOR 83]   TAYLOR W.K., LAVIE O., ESAI I.I. A Curvilinear snake arm robot with
                       gripper axis fiber optic image processor
                       Robotica 83, Vol. 1, 1983.

[TROUNOV 84]   TROUNOV A.N. Application of Sensory Modules for Adaptive Robots
                       4th ROVISEC, London, Oct. 1984.

[VANDERBRUG 79]   VANDERBRUG G.J., ALBUS J.S., BARKMEYER E. A vision System for
                       real – time control of robots
                       9th ISIR, Washington, 1979.

[VERTUT 84]   VERTUT J., FOURNIER R., ESPIAU B., ANDRE G. Sensor – aided and/or
                       Computer aided Bilateral Teleoperator System
                       5th ROMANSY, Udine, Italy, June 1984.

[WAMPLER 84]   WAMPLER C. Multiprocessor Control of a Telemanipulator with Optical
                       Proximity Sensors
                       International Journal of Robotics Research, Vol. 3, no 1, Spring 1984.

[YAMADA 83]   YAMADA Y., MATSUDA F., TSUCHIDA N., UEDA M. Control of an
                       Industrial Arm to grasp moving objects with the aid of a two – dimensional
                       range sensor.
                       1983 International Conference on Advanced Robotics, Tokyo, 1983.

[WITWICKI 79]   WITWICKI A.T. A method of non – positioned workpieces taking
                       9th ISIR, Washington, 1979.

F I G U R E S

Sensor Technology \ Criterias	active/passive	binary/ non binary	Nature of the provided output
Mechanical sensor (microswitch)	passive	binary	presence (contact)
Pneumatic (air system)	active	binary	presence (non contact)
Acoustic (ultrasonic)	active	non binary	distance (time of flight)
Optical (reflectance)	active	analog	f(distance, albedo, orientation)
Optical (triangulation)	active	analog	distance, sometimes orientation
Optical (Phase/Time of flight)	active	analog	distance
Capacitive	active	often binary	presence
Inductive/ Eddy Currents	active	analog	f(range, magnetic properties)
Microwaves/Radiowaves	active	analog	distance (time of flight)
2D local vision	passive	2D digital image	2D reflectance/luminance
Local stereovision	passive	2x2D digital image	2x2D reflectance/luminance to be processed
Active local vision with Laser spot	active	digital	distance
Active local vision with projection	active	digital	distance and local geometric features

Figure 1 : Classification of local sensors

Sensor Technology / Characteristics	Practical range	Size	Distance measurement	Accuracy	Response Time	Sensitivity to the nature of the target	sensitivity to the geometry and orientation to the target	Mechanical Robustness	sensitivity to environmental noises and disturbances	Easiness of Implementation	Cost of data Processing	Physical Principle usable in: Undersea	Nuclear	Space
Mechanical sensor (micro switch)	some centimeters	small	no	/	low	no	no	fair	no	variable	low	yes	yes	yes
Pneumatic (air system)	some mms	small	no	/	medium	no	yes	fair	no	easy	low	no	yes	no
Acoustic (ultrasonic)	from some cms to several meters	medium	yes	fair	high	no	yes	fair	wind	medium	low	yes	yes	no
Optical (reflectance)	from some mm to 20 cm	very small	not easy	good	very low	yes	yes	medium	low	easy	low to medium	difficult	yes	yes
Optical (triangulation)	from some mm to 20 cm	medium	yes	medium	very low	low	low	medium	low	medium	rather low	difficult	yes	yes
Optical (Phase of Time flight)	from some mm to 1 meter	very small	yes	fair	very low	low	low	medium	low	complex	rather low	difficult	yes	yes
Capacitive	some mm	medium	no	poor	very low	yes	yes	good	no	variable	low	yes	yes	no
Inductive/Eddy currents	some cm	medium	difficult	good	very low	yes	yes	very good	no	very easy	low	yes	yes	yes
Microwaves/Radiowaves	from several cms to kms	large	yes for large distances	variable	variable	yes	yes	medium	medium	not easy	rather low	no	perhaps	yes
2D local vision	some meters	small to large	no	/	medium to high	yes	yes	medium	medium	rather easy	high	(yes)	yes	yes
Local stereovision	up to 50 cms	medium to large	not directly	may be poor	medium to high	yes	yes	poor	medium	not so easy	very high	(yes)	yes	yes
Active local vision with Laser spot	up to 50 cms	small to large	yes	variable	low to medium	low	low	medium	low	rather easy	rather low	(yes)	yes	yes
Active local vision with projection	up to 1 m.	medium to large	yes but not directly	variable	low to medium	very low	provides geometric information	medium	low	not easy	high	(yes)	yes	yes

Figure 2   Comparizon of local sensors.

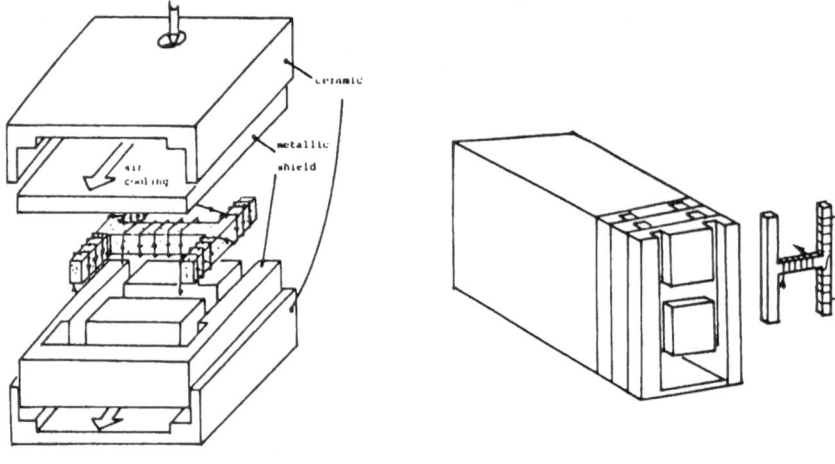

figure 3   figure 4

Array of eddy currents sensors (CLERGEOT 84)

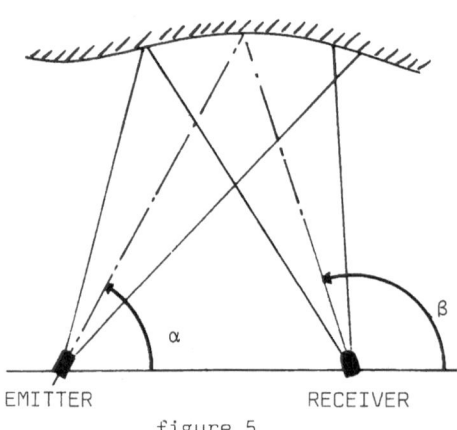

figure 5

A simple scanning system

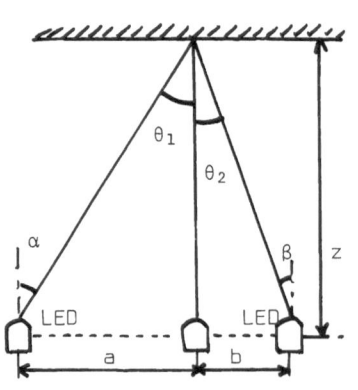

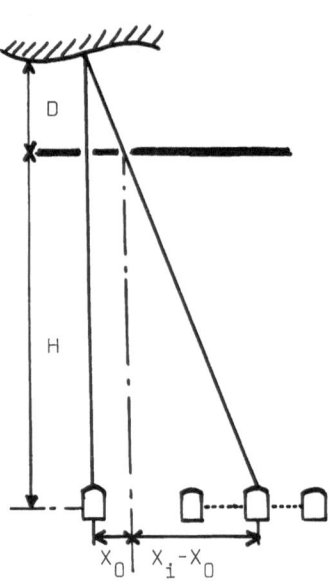

figure 6

A principle of optical
triangulation

figure 7

Principle of a phase shift
sensor

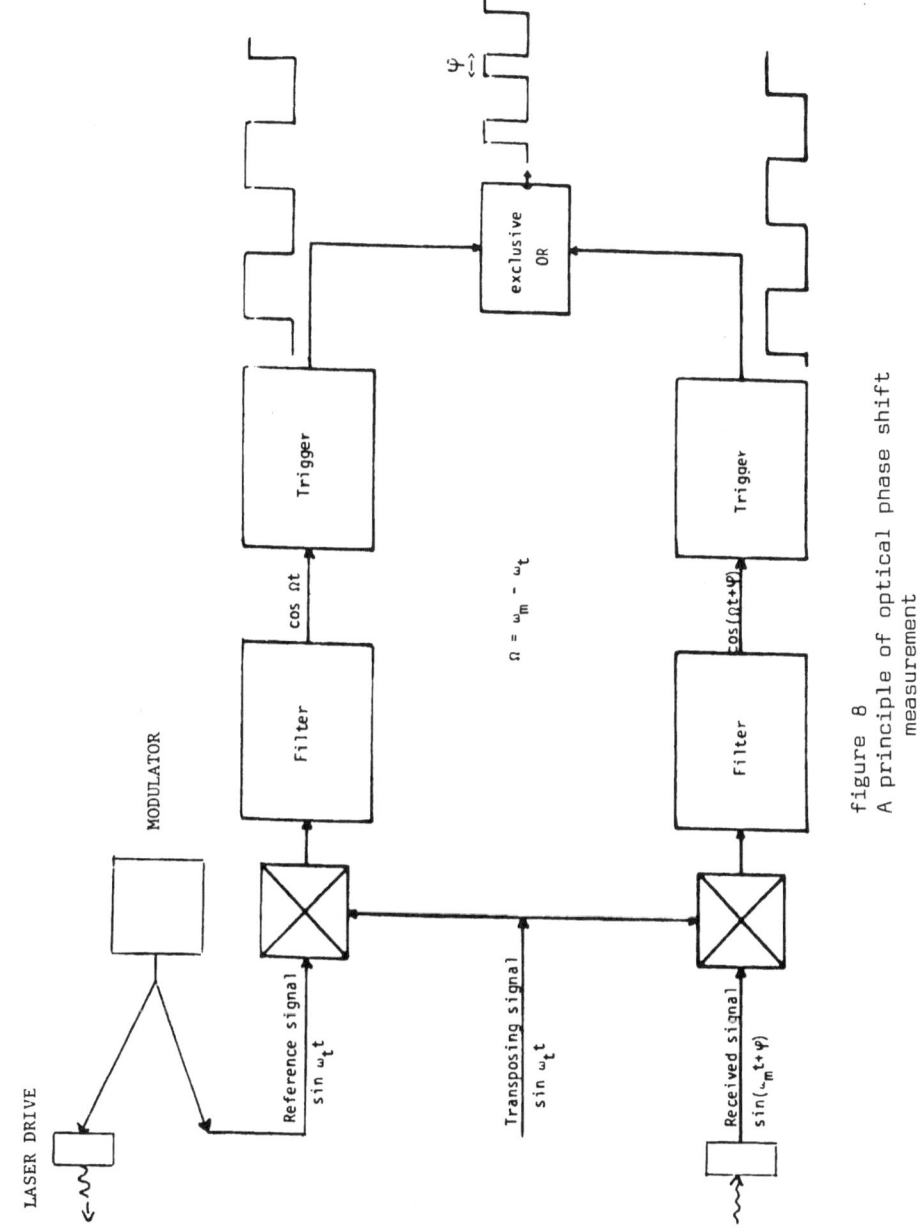

figure 8
A principle of optical phase shift
measurement

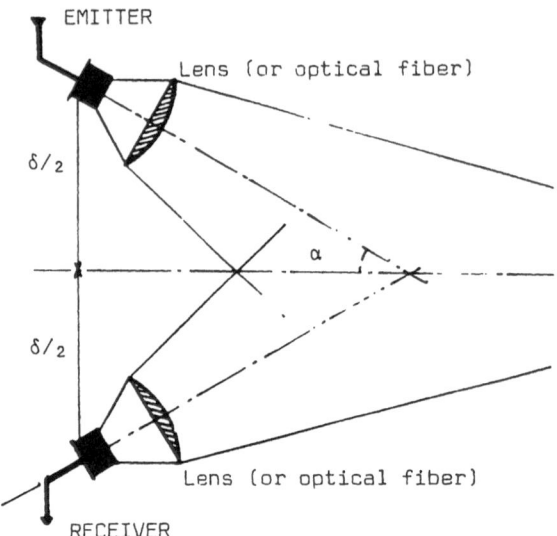

figure 9
Principle of optical reflectance sensors

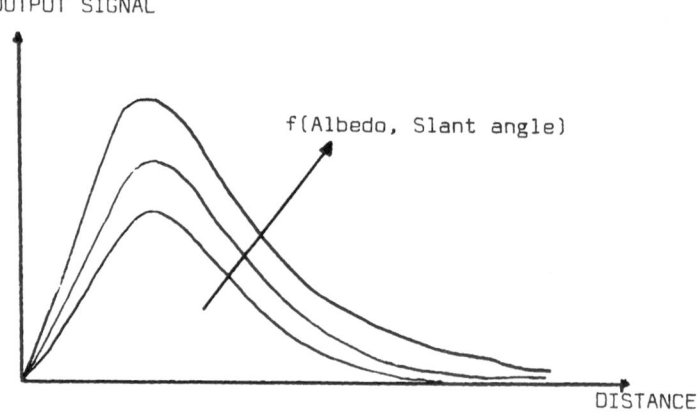

figure 10
Typical outputs of optical reflectance sensors

149

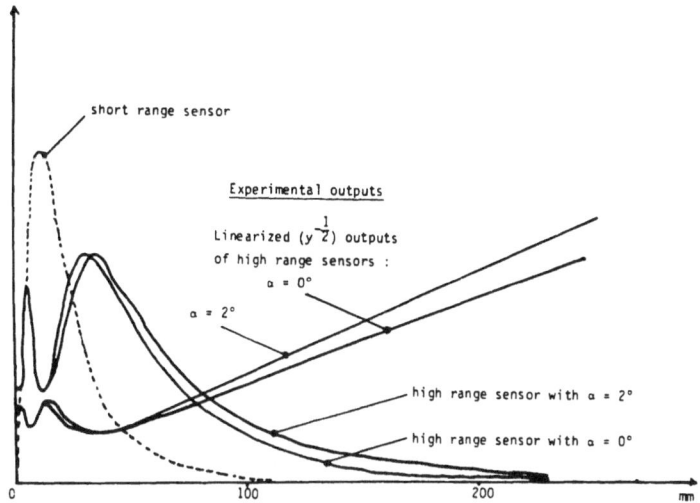

figure 11
Experimental outputs of infrared
reflectance sensors

	Emitter	Receiver	α(deg)	δ(mm)	Practical maximal range (mm) with a white plane target orthogonal to sensor axis	field: ϕ(deg)
1	Siemens LD 242	Siemens BP 104 or BP 103	0	8	120	± 60
2	Siemens LD 271 or TiL 33	Texas TiL 99	4	7	200	$\phi_E = \pm 25$ $\phi_R = \pm 10$
3	Texas TiL 31	Texas TiL 81	6	9	250	$\phi_E = \pm 5$ $\phi_R = \pm 10$
4	Siemens LD 261 (Small size)	Siemens BPX 81	0	3	80	$\pm 10^{\cdot}$

figure 12 [ESPIAU 85a]
Some arrangements of optoelectronic sensors

150

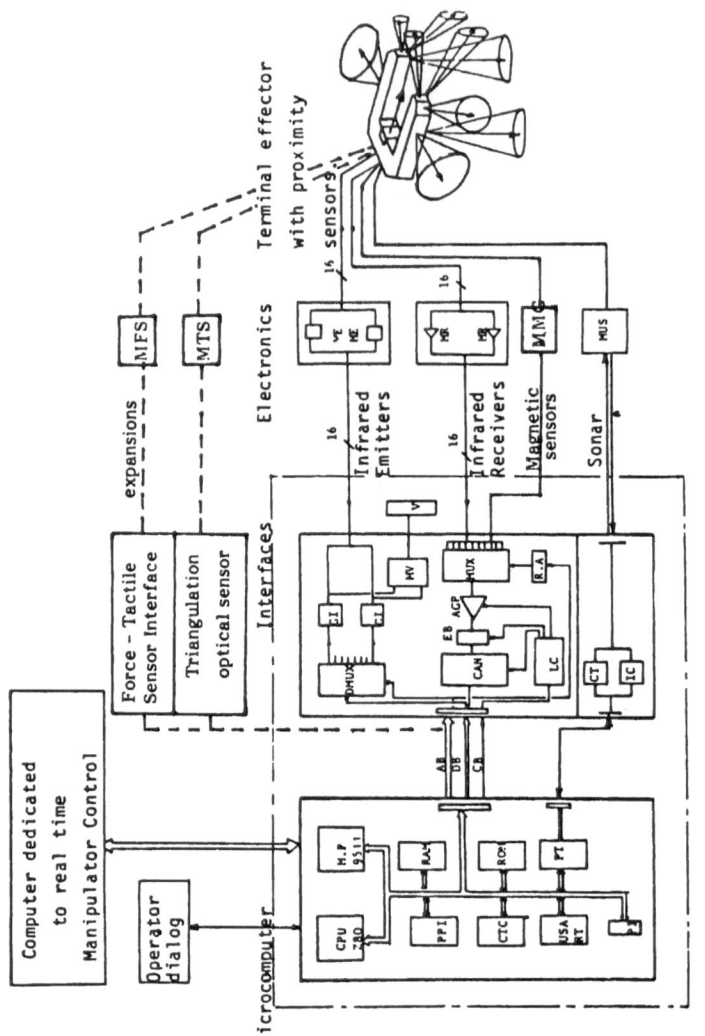

figure .13 ([ANDRE 85b])
A multisensor processing system

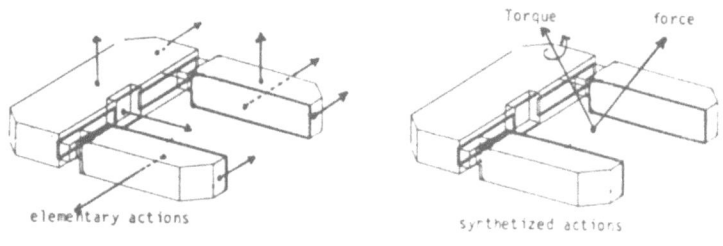

figure 14
Synthesis of control

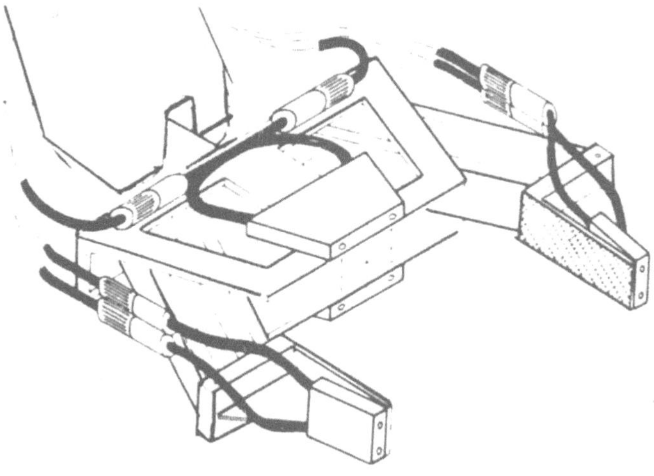

figure 15
A bidigital gripper with 3 optical-
fiber sensors, by SAGEM

figure 16 (|ANDRE 85a|)
Compensation for the oscillations of a
telemanipulator with proximity sensors

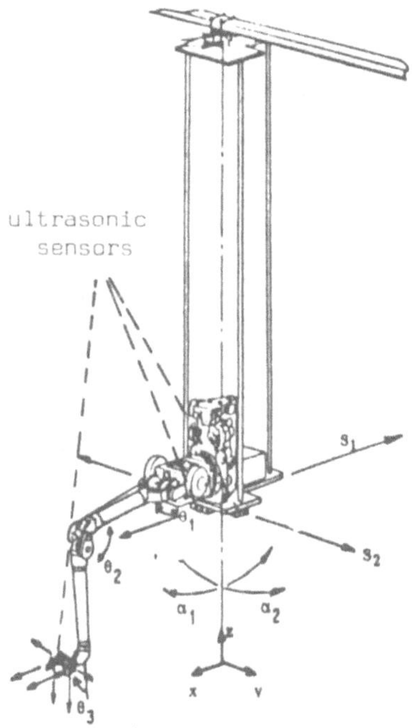

FORCE AND TACTILE SENSING FOR ROBOTS

P. Dario, M. Bergamasco, A. Fiorillo
Centro "E. Piaggio"
University of Pisa
Via Diotisalvi, 2
56100 Pisa
Italy

ABSTRACT

This paper deals with methods and techniques for sensing and controlling contact forces originated by the physical interaction of a robot with the environment.

Contact sensing includes force and tactile sensing. A distinction between these two sensing modalities is that force sensing refers primarily to the measurement of the resultant mechanical effects of contact, while tactile sensing involves the detection of a wide range of local parameters (physical and chemical) affected by contact.

This paper is comprised of a section in which force sensing techniques and devices usable in different classes of robots are discussed, and a second section dealing with true tactile sensing. In each section the motivations and the general problems encountered in each sensing modality are analyzed first, then the state of the art of related sensing technologies is surveyed.

In a third section, a coherent approach is presented to the problem of replicating human tactile sensing capabilities in an artificial robotic system. The advantage of this approach is to allow a comprehensive analysis of the sensory, motor, and control components of an advanced robot system, as well as to illustrate the roles of force and tactile sensing, and the importance of their coordination, in artificial tactile perception.

1. INTRODUCTION

The ability to sense and control parameters related to physical contact with the external environment will be a fundamental part of the sophisticated sensory capabilities required of future robots in order to adaptively interact with their surroundings. Peculiar to contact sensing, compared to other robotic sensory modalities, are both the variety of possible data extraction procedures and the richness of information obtainable by the tactile interaction of the robot with the environment.

In this paper the term "contact sensing" implies the generality of effects which take place when the robot end effector is brought into contact with an object (although tactile sensing could even be viewed as

NATO ASI Series, Vol. F43
Sensors and Sensory Systems
for Advanced Robots
Edited by P. Dario
© Springer-Verlag Berlin Heidelberg 1988

degenerate range sensing at or near zero range (Brown 1986), in the following we shall not consider this possible conceptual association of contact and non-contact sensing).

Contact sensing includes force and tactile sensing. The difference, rather subtle in some cases, between those two sensing modes can be better understood by analyzing the effects originated by the contact between the robot end effector and an object. At macroscopic level, a range of forces and moments is transmitted through the contact. At microscopic level, those forces and moments result from a distribution of stresses inside the covering of the end effector, which originate from the intimate interpenetration of the two contacting surfaces. However, this intimate contact may also produce other physical (e.g. temperature variation) and chemical (e.g. chemical reactions) modifications in the contacting bodies.

We define "force sensing" as the measurement of the global mechanical effects of contact, and "tactile sensing" as the detection of the wide range of local parameters affected by contact.

Contact forces and moments, or collectively, forces, are transmitted to the chain of serially connected links forming the mechanical structure of the robot manipulator where they can be measured by means of appropriate force sensing techniques (Salisbury 1985). The analysis of such techniques and related devices forms the core of the next section of this paper. The second section deals with tactile sensing and discusses various approaches for sensing parameters relative to intimate, local contact. The introduction to each section provides further justification to the distinction we have adopted and points out the areas of application in which the two concepts overlap. As an example of integration of force and tactile sensing, the third section of the paper describes the system design approach and the implementation of a laboratory robot system dedicated to the investigation of artificial tactile perception.

In the field of robot tactile sensory systems, research on contact sensing is inextricably related to research on other topic areas such as end effector and manipulator design, hierarchical control, data handling and pattern recognition (Harmon 1984). The emphasis of this paper is concerned with the analysis of the force and tactile transducer components of robot systems.

2. FORCE SENSING

A robot manipulator should be able to sense and control the resultant forces and moments generated at the end effector in order to perform compliant motion tasks. "Compliance", defined as the ability of a manipulator to react to contact forces or tactile stimuli as the motion proceeds (Mason 1982), is a key requirement both for a robot system designed for assembly operations in industry, and for an advanced robot intended to operate in unstructured environments. Compliance, in fact, is fundamental in many high level robot tasks. Tactile exploration is an example of a task in which the position of the end effector is constrained by task geometry; accomplishing tactile exploratory tasks requires the control of both position and force.

In general, controlling forces is much more reasonable than controlling position in all those operations in which there is uncertainty in the position of the parts the robot has to manipulate, or errors in the position servo of the manipulator (Craig 1986). Unfortunately, true force control does not exist today in industrial robots: this deficiency is one of the factors currently limiting the range of applications of robot systems to spot welding, spray painting and pick and place operations.

Hybrid position/force controllers able to control position/orientation along specified degrees of freedom and independently force/torque along the remaining degrees of freedom, are a solution to the problem of active robot compliance. However, their practical implementation is hampered by such problems as the amount of computation required, the difficulty of speci- fying appropriate position/force strategies and the lack of rugged force transducers. Despite all of these problems, general methods and techniques for robot hybrid control have been proposed by a number of investigators (Mason 1981, Zhang and Paul 1984, Whitney 1985); a few force sensor-based strategies have also been described for industrial applications (Van Brussel and Simons 1978, Cutkosky and Wright 1984, Cutkosky 1985).

As anticipated in the introduction, in this section we devote our attention to the transducer aspect of the force control of robot systems. Extending the classification proposed by Shimano (1978), that refers to the placement of the force transducer relative to the manipulator with which it is used, we shall consider the following categories of force sensing devices: force pedestal, joint torque sensors, wrist sensors, gripper sensors and fingertip sensors.

2.1. Force pedestal sensing

The force exerted between a robot and the environment can be measured by an instrumental platform equipped with sensors. This approach is equivalent to sensorizing, to some extent, the environment, instead of sensorizing the robot. The platform can be static, or it can incorporate actuators in order to cooperate with the robot in performing insertion operations.

A static platform of the first kind has been proposed by Watson and Drake (1975) for assembly operations. The platform incorporates sensors which detect the horizontal and vertical components of the resultant force acting on the platform, from which the point of application of the resultant can be calculated. In a typical peg-in-hole assembly operation, schematically depicted in Fig. 1, the platform carries the recipient part while the robot gripper holds the part to be inserted. In this, as in many other similar situations, possible insertion strategies involve force feedback accommodation procedures, consisting of commanding motions of the manipulator in certain directions, with the constraint that given force threshold values are not exceeded (Coiffet 1983).

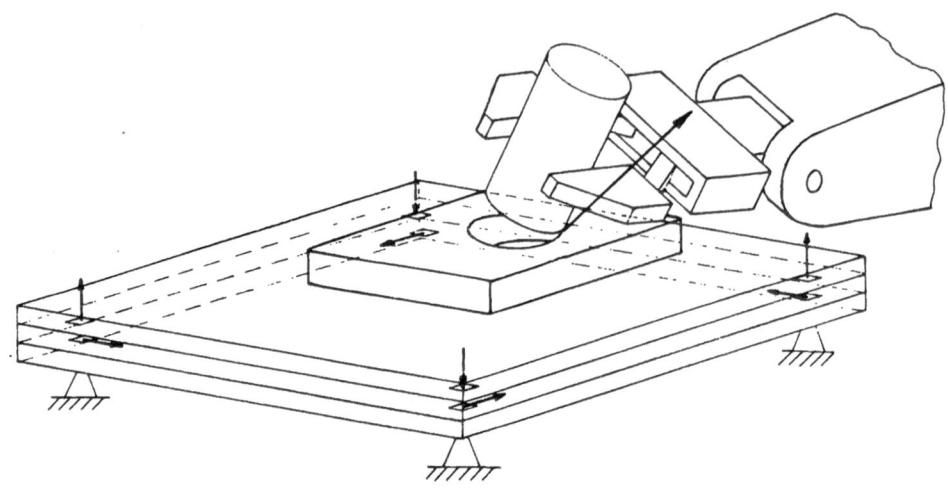

Fig. 1. A sensorized platform for the detection of contact force.

In a second interesting approach described by a group of Japanese investigators (Kasai et al. 1981), the platform has the twofold function of sensing forces and commanding accommodation movements. The proposed assembly system comprises the active sensory table, a positioning robot without sensors and a microprocessor-based control unit.

The platform incorporates both a semiconductor strain gauge-based transducer capable of resolving the magnitude of forces in the direction of the x, y and z axes and the moments around these axes, and a six-degree-of-freedom fine motion driver. The authors demonstrated that the system can be "trained", i.e. the time required for insertion operations can be gradually reduced, by preadjusting insertion parameters using sensor information provided by preceding trials.

It is interesting to observe that the proposed assembly system is functionally very similar to the now increasingly investigated case of two cooperating robot manipulators. In this context, one could observe that it might be possible and convenient to concentrate all the sensory and sensor-based control capabilities in one "intelligent" robot, and to leave an essentially passive, coarse positioning function to the second robot.

The main advantages of force pedestal devices are high structural stiffness and good sensitivity; a major disadvantage is the need for some object positioning and holding system, which renders the platform strongly object-dependent. An additional disadvantage is the time required and the complexity of converting the force signal detected by the platform into information useful for servoing the robot joint motors.

2.2. Joint torque sensing

Sensing the torques produced at each of the manipulator's joints is an old sensing technique: in fact it has been used successfully for many years in master-slave manipulators. Joint torque sensing has the advantage of not only detecting force and moments applied at the hand, but also those applied at other points of the manipulator. Furthermore, a joint sensor is protected by the inertia of the robot links and by the stiffness of the arm. The main disadvantage of joint torque sensing, particularly serious in manipulators with large weight and inertia, is that forces and moments at the robot hand are not measured directly but have to be inferred. In particular, the effects of gravity (and possibly dynamic forces) should be modelled in order to extract force information at the hand from joint torque measurements.

Joint torque can be sensed either directly, by monitoring the motor currents, or by dedicated joint torque sensors. An example of strain gauge-based joint torque sensor has been recently reported by Wu (1985), who also proposed a compliance control method exploiting the feedback provided by such a sensor. The joint torque servo adopted by Wu has a number of potential advantages. The first advantage is that joint torque sensing needs only simple, linear equations to resolve the torque for each servo joint of a robot, instead of the more complicated methods of resolving the sensing force signals from wrist or finger sensors. Thus, joint sensing has a much faster response. A second advantage is that the joint torque servo system is closed around the joint motor and gear reductions, and hence it is independent of manipulator configuration. Based on the free-joint method (Shimano 1978, Paul 1981), in his paper Wu also presents a new joint selection method for force sensed joints, according to which the joints with most freedom and less frictional effect are selected to achieve the required compliance.

A special case in which joint torques are also sensed (although indirectly) for force control is in some recently developed articulated end effectors (hands), whose fingers are usually driven through cables (Salisbury and Craig 1982, Jacobsen et al. 1984). For example, both in the Stanford/JPL hand and in the UTAH/MIT hand, semiconductor strain gauge-based tendon tension sensors are located on each cable, where it enters one of the fingers. Also the exploratory finger we have developed in our laboratory (described in the last section of this paper) incorporates tendon tension sensors, but they are located in each joint (Dario et al. 1985). The primary role of these tendon tension sensors is to allow applying appropriate combinations of cable tensions in order to exert arbitrary torques about joint axes, and ultimately to apply forces accurately with the fingertip (Salisbury 1985). Being placed after the major sources of friction in the actuation and force transmission system, tendon tension sensors can be used in a servo loop which encloses these non-linearities and minimizes their effect on system performance; moreover, forces applied to the fingertip can be sensed rather accurately by monitoring the cable tensions. These force data, along with those relative to joint position and velocity, measured by other sensors, are fundamental for the implementation of control architectures for active position and force control of grasped objects.

2.3. Wrist sensing

In order to reduce the uncertainty in measuring and controlling forces generated at the end effector (gripper, hand), these forces should be isolated from those produced by the links and joints. A method for accomplishing this is to mount a force sensing device between the last powered joint and the end effector, i.e. at the manipulator wrist. Typically, a wrist sensor consists of a mechanical structure deforming under the applied force. If appropriate sensing means are provided to detect the deflection of the elastic elements, the transducer can measure from three to six components of the force/torque vector acting on the end effector.

A number of force sensing wrists have been developed, based on different sensing techniques (strain gauge, potentiometric, electro-optical). Surveys of instrumented wrists have been presented by Bejczy (1977) and Pugh (1986), while Shimano (1978) has discussed general criteria for the design of optimal force sensing wrists for which high stiffness, compact design, good linearity, and low hysteresis and internal friction are required.

Strain gauges have proven to be the smallest, easiest to use, cheapest and most reliable sensing elements for measuring the deflection of the elastic elements of the force wrist. Mainly for these reasons, all the existing commercial force sensing wrists incorporate strain gauges.

A problem in strain gauge-based wrist (but also gripper and fingertip) sensing is how to convert into useful force vector information the readings of the strain gauge bridges. A particular solution to this problem (valid when the strain gauge signal vector contains six components) involves the use of a decoupling matrix, whose form reflects the quality of the design of the force sensor (Van Brussel et al. 1985). For instance, a force sensing wrist developed at SRI mechanically decouples the effects of the various force components, so that forces and moments can be detected separately. In this case, the simple form of the decoupling matrix allows extremely simple, and fast, conversion routines. Otherwise, experimental transformation/calibration matrices must be obtained.

Two interesting evolutions of the concept of force sensing wrist are worthy of mention. The first is a 6-axis force sensor with onboard microprocessor (Barry Wright Co. 1984a). In this configuration the sensing wrist becomes an integrated, "smart" measuring instrument which includes the transducer structure, strain gauges, and a built-in microprocessor which can resolve in real time the six force components about any coordinate system specified by the operator.

The second device is the Instrumented Remote Center Compliance (IRCC), that is basically the well known Remote Center Compliance (RCC), in which some or all the internal deflections have been instrumented by means of optical sensors (De Fazio et al. 1984). The RCC part of the IRCC is a multi-axis compliant interface between the robot and its gripper or tool. If the IRCC's complete stiffness matrix is known, applied forces and torques can be determined from the measured lateral and angular displacements of the deflecting parts of the IRCC.

An increasing number of force sensing wrist-based, active control strategies are being proposed, mostly for the execution of deburring, 3-D contour tracking and assembly operations. One could conclude that, with the exception of some problems still encountered in withstanding harsh industrial environments, wrist force sensors per se are not limiting progress in force control-based robot applications, and that, rather, more effort is still needed to identify and solve basic theoretical problems (Whitney 1985).

2.4 Gripper finger sensing

A force sensing wrist has some limitations. For instance, while a force sensing wrist tends to be more accurate than a joint torque sensor in detecting forces at the end effector, a wrist sensor is less sensitive than a finger sensor. In fact, the sensitivity of a wrist sensor may not be sufficient for some very delicate manipulation tasks. Moreover the weight of the end effector should always be considered when processing the readings of a wrist sensor. Finally, a wrist sensor is not capable of measuring "internal" forces, i.e. the force that each jaw, or finger, of the end effector exerts on the grasped object. A possible solution to these problems is to incorporate a force transducer in each jaw of a robot gripper or in the tip of each finger of an articulated hand. In this paragraph the case of a force sensing finger for a robot gripper is considered.

Like the force pedestal illustrated at paragraph 2.1, a force sensing gripper finger associating a 6-axis force transducer with a planar active contact surface on which distributed forces are exerted, can be capable of accurately resolving the six components of the force-torque vector acting on it, and the position of the point of application of the resultant contact force.

Although this information is, in principle, powerful enough to be exploited in many different force feedback-based robot control strategies, there is not much literature reporting on applications of this type of sensorized gripper. A reason for this is the limited number of practical implementations of this class of devices (Wang and Will 1978, Purbrick 1981, Lestelle 1985, Morris 1985), which has rendered it difficult to many interested investigators to experiment new control schemes. A second, more fundamental reason is that those applications in which the high sensitivity of a gripper mounted sensor would be really necessary (for instance, the manipulation of small, delicate parts) require a degree of dexterity difficult to obtain with a one-degree-of-freedom end effector, although sensorized. In fact, the difficulty of using efficiently the force feedback information provided by local sensitive force sensors for accurately controlling large robots with massive links, has been indicated as one of the reasons that justify the development of dexterous, articulated hands (Salisbury 1984).

Despite all of these facts, a sensing gripper-based force feedback has been actually implemented for the control of a robotic system by Stokic et al. (1986), who developed an experimental gripper with simple strain gauge force transducers that measure three components of the forces in contact points between the payload and gripper. Furthermore, the recent commercial availability of an instrumented gripper (Lord Corporation 1985a) that incorporates, among other sensing features, force sensing fingers, indicates a potential interest of this approach for some industrial applications (e.g. for electronics assembly).

Since the contact between each gripper finger and the grasped object will occur, in general, over a surface, a question could be raised on the opportunity of measuring contact forces locally by means of distributed sensors (and of eventually calculating the resultant force from the readings of those individual sensors) versus measuring only their resultant by a force vector transducer. A pragmatic answer to this question is that, since all currently available force sensing arrays are affected, in practice, by serious limitations in terms of sensitivity, spatial resolution, dynamic response and hystheresis, it seems preferable to make use of a fast and reliable resultant force vector transducer for the basic function of controlling robot compliant motion. On the other hand, the information provided by a distributed sensor (that basically consists of a measure of the local normal components of the contact forces between the sensor and the grasped object) could be used more appropriately for extracting global or local tactile features, usable at high control level for the overall management of the manipulation task.

Finally, in the next paragraph another possible force sensor configuration is illustrated, which conceptually represents the "transition" between the purely force sensing devices discussed so far, and the tactile sensors, intended for high level manipulation control strategies, discussed in the next section.

2.5 Fingertip sensing

Being at the interface between force and tactile sensing, fingertip sensing offers many hints for speculating on the different functions of robotic contact sensing. For this reason, this type of sensing modality and the context in which it is being investigated most extensively (i.e. dexterous manipulation with articulated hands) are discussed here with particular emphasis.

As Salisbury (1984) pointed out, a multi-degree-of-freedom end effector could dramatically enhance the limited choice of grasping configurations allowed to current single-degree-of-freedom grippers. This would make it possible to augment the resolution, sensitivity and bandwidth of the robot arm's joints and would provide for unique sensing modalities. As true dexterity results from a combination of sophisticated robot mechanisms and high level, sensor-based control strategies, different types of sensors have been, or are expected to be, incorporated in many recently developed robot hands (Salisbury and Craig 1982, Jacobsen et al. 1984). In this paragraph the features of the contact resolving sensor developed as a fingertip for the Stanford/JPL hand are described.

A robot hand designed for dexterous manipulation should be capable of controlling and sensing the forces exerted on the grasped object, in order to mantain a stable grasp as well as to perform delicate motions. Not only should one control the external forces upon the grasped object (as in an assembly task), but also the internal forces. Controlling internal forces contributes both to grasp stability, by ensuring that the normal forces at the points of contact are large enough to support tangential forces due to friction, and to the manipulation of fragile objects. In order to obtain all these capabilities, an accurate model of the grasp kinematics is mandatory. The model used by Salisbury and Craig (1982) assumes that an object is held in the fingertips of three, three-degree-of-freedom fingers and that the contact with the object is only at the fingertips, through point contacts with friction. This type of contact permits only forces to be transmitted at each fingertip. To exert (or sense) a force vector at a fingertip the Jacobian transpose matrix (or its inverse) of that finger must be known to relate finger force to joint torques (and vice versa)(Shimano 1978). If the location of the point of contact on the last finger link is known, then the Jacobian transform is a function of the joint positions only and may be computed easily (Salisbury 1984). As manipulation is mostly dynamic, the new locations of the contact point must be deduced during relative fingertip-object motions to update properly the Jacobian matrix.

Although it might be possible in some circumstances to calculate the location of the contact points from the joint positions, a fingertip sensor like the one proposed by Salisbury and developed by Brock and Chiu (1985) provides a more direct and reliable solution to the problem of determining the location and the orientation of a contact. A diagram of the fingertip sensor, which basically consists of a hemispherical cover supported by a structure that permits sensing of all three components of the applied force and all three components of the applied moment resolved at the origin of the sensing system, is depicted in Fig. 2.

With the already mentioned assumption of contact occurring at a single point with friction, and with the additional assumptions that the shape of the fingertip contact surface is known and is convex, and that the contact exerts a force directed into the surface, the fingertip sensor (which incorporates small semiconductor strain gauges on metal flexures to measure the forces and the moments about the origin of the fingertip) is capable of reading the magnitude and direction of the resultant contact force as well as the location of the contact. These performances are obtained with few transducer elements (and wires); furthermore, virtually no nonlinearity or hystheresis exist, and negligible errors are introduced by soft fingertips exerting moments at the contact. These attractive features are only slightly moderated by some fragility of the sensor. Hence this type of fingertip sensor can be very effective for force control in dexterous manipulation.

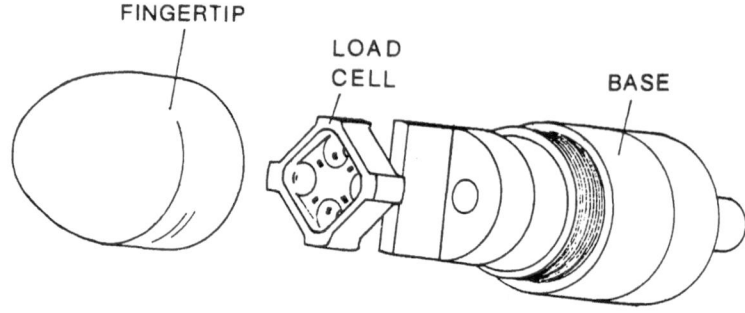

Fig. 2. Contact resolving fingertip sensor.

However, it is most interesting to observe that the force sensing ability of a contact resolving fingertip can be exploited not only at the (relatively) low control level required for manipulation, but also at the higher control level, as a tool for "perceiving" some features of the environment. In fact, some preliminary experiments reported by Brock and Chiu (1985) showed the possibility of using the fingertip sensor with an integrated sensor/hand programming system for the mapping of surface contours. These contours were reconstructed by recording the sequence of contact locations and orientations detected during the motion of the fingertip along a given object surface. This type of information can be used as raw data for an object recognition system, as proposed by Grimson and Lozano-Pérez (1984). Furthermore, the fingertip sensors can be used to determine the mass and the center of mass of a grasped object and, ultimately, to infer the stability of a grasp in different hand orientations (Brock and Chiu 1985).

A final set of observations relate to some further implications of sensing contact locations and orientations, suggested by Salisbury (1984). He considers three types of idealized contacts between rigid bodies (point contact, line contact and planar contact) with and without friction. Then for each type of contact between the fingertip sensor and the grasped object, Salisbury calculates how many components of the wrench system (Salisbury 1985) describing the collection of forces and moments that can be exerted by one body upon the other through the particular contact, must be measured in order to uniquely determine contact features. A conclusion of this analysis is that, in the case of line contact with friction, "active sensing" is required to locate the line of contact uniquely. Active sensing can be obtained by perturbing contact conditions, for instance by rolling the finger slightly or by slightly changing the direction the finger is pushing. Similarly, a friction point contact can be sensed with only the moment sensors (three instead of the six required in the general case), if the point contact is allowed to exert forces in two different directions on the fingertip, through small exploratory finger motions. In both cases, therefore, the sensing surface must cause the contact to reveal its location by causing it to exert a range of forces within its wrench system. Another interesting observation regarding the value of active force sensing relates to the possibility of using not only moment measurements, but also their derivatives to calculate the location of a point contacting the hemispherical fingertip sensor (Salisbury 1984). There are obvious analogies between these possible artificial sensing modalities and those usually observed in humans during haptic exploration. For instance, we use dynamic exploration for sensing geometrical object features, when static grasping does not provide sufficient information. Also, some receptors in the human tactile sensing system (the rapidly adapting sensory units, RA) are selectively sensitive to dynamic stimuli (Johansson 1979).

In conclusion, fingertip force sensing should be considered not only a fundamental tool for accomplishing dexterous manipulation in articulated end effectors, but also an important component of a system for active tactile exploration.

3. TACTILE SENSING

According to the definition of force and tactile sensing we have proposed in the introduction, tactile sensing is aimed at detecting all the various effects originated by the intimate contact between the robot end effector (in which, for most robot applications, tactile receptors are located) and an object. Most significant among those contact-based effects are contact stresses in the two contacting surfaces. Also, heat exchanges, field interactions and chemical reactions can be important. Hence one could immediately recognize that the amount of information on the touched object that could be extracted by an ideal sensor capable of detecting variables related to all the above effects is much larger than in the case of simple force sensing: tactile sensing can be used not only for the mere mechanistic control of robot movements, as conventional force sensing, but rather, through the artificial equivalent of perceptual processes, for creating models of the outer environment and recognizing objects.

Tactile sensing obviously requires sophisticated sensors. However, it should be pointed out immediately that the availability of these sensors would not be sufficient per se for tactile sensing. In fact, it would be also necessary to accurately control the modalities with which the tactile sensor contacts the explored object (e.g. contact force, finger position and orientation, relative velocity; in other words, the dexterity of the end effector). Thus, it is important to emphasize that force sensing and control capabilities are a prerequisite for tactile sensing.

As in the case of human touch (the only model that seems adequate for investigating high level robot touch), artificial tactile sensing would result from the operation of a system comprising kinesthetic and cutaneous sensors and a dexterous multi-degree-of-freedom end effector, coordinated by a complex control hierarchy. Hence, although a sophisticated sensor will always be the core of a robotic tactile sensing system, one should not expect that tactile sensing will result automatically from the availability of such sensor. Also, when designing the tactile sensor, the importance of the type of motor acts which will be necessary to elicit specific tactile sensations should not be underestimated.

Despite the large number of tactile sensors which have been proposed recently by many investigators, it does not seem that the above observations and their implications have been adequately considered (with a few significant exceptions) in the design of most of those tactile sensors. Although some of the concepts proposed in this section were already contained in the last published work of the late Leon D. Harmon (1984), the only criteria followed in the design of most of the existing tactile sensors derive either from the indications of a questionnaire survey (Harmon 1982) or from the confidence in the validity of the biological model (represented by the human skin).

Only recently an approach based on continuum mechanics analysis which provides quantitative insight on some phenomena occurring during contact, and derives some indications on the desirable measuring features of a tactile sensor, has been proposed (Fearing and Hollerbach 1985).

In this section, some representative tactile sensors developed so far are presented. In the next section, we shall discuss the design specifications for an ideal tactile sensor, as derived from a system approach. An example of implementation of some of the proposed concepts in a laboratory robot system will be also described.

Some of the characteristics which have usually been assumed to be desirable for a tactile sensor array and taken as a model in the design of the tactile sensors described in this section are summarized as follows (Harmon 1982) : 10 x 10 sensing elements on a 25 mm x 25 mm flexible surface; response time of each element: 1-10 ms; force measuring range (if not pointed out differently, only the normal component of the contact forces is measured): .01-10 N; the sensing elements need not be linear, but they must have low hysteresis; skin-like sensing material has to be compliant and durable. As already mentioned, however, none of the sensors which will be described fulfills completely the above specifications.

Extensive surveys on the state of the art of robotic tactile sensor technology, providing details on the performance of each type of tactile sensor, have already been presented (Harmon 1982, Ogorek 1985, Allan 1985, Dario and De Rossi 1985, Pennywitt 1986). In the following paragraphs the features of some different tactile sensors (including also a few recently developed devices), classified on the basis of the transducer technology they use, are summarized.

3.1. Piezoresistive sensors

This class of sensors is quite broad and includes a multitude of different devices and approaches, all based on materials or elements whose electrical conductivity varies as pressure is applied.

Conductive elastomers have been investigated by a number of researchers as a means to fabricate tactile sensing arrays because of their potential advantages, which can be summarized as follows: conductive elastomers are cheap and can be obtained in thin, flexible and compliant sheets; they can measure static and dynamic forces; matrix arrays can be easily fabricated and conveniently scanned with row-by-column addressing methods.

Different materials (isotropically conductive elastomers, conductive foam, carbon fibers) have been used, and a number of electrode arrangements proposed to read resistance variations at selected points of the sensor array. Among the most significant implementations of tactile sensors making use of conductive materials, one could mention the following: a 4 x 4 array in which conductive elastomer is laid over a printed circuit board etched with sixteen pairs of concentric rings: each pair of rings forms a sensing element. An appropriate circuitry is used to read the variation of resistance between the rings of each element caused by pressure on the elastomer (Snyder and St. Clair 1978); a 4-by-8-element transducer developed at the Jet Propulsion Laboratory in Pasadena (Bejczy 1983); a 16 x 16 array obtained over a 1-square-centimeter area of a flexible printed circuit board at the MIT Artificial Intelligence Laboratory in Cambridge, and capable of measuring pressures ranging from .05 to .5 N per square millimeter (Hillis 1981); a 133-element transducer developed by the Laboratoire d'Automatique et d'Analyse des Systemes in Toulouse and tested at the University of Pennsylvania in Philadelphia, whose main feature is to be shaped much like a finger (Bajcsy 1984); an original transducer made of a layer of pressure-sensitive rubber sandwiched between two layers of conductive coating plastic on which four peculiarly positioned electrodes are deposited. The sensor provides both the position of the center of mass of the distributed pressure on the sensor, and the value of the total pressure, without giving, however, details on the actual pressure distribution (Ishikawa and Shimojo 1982).

The limitations of conductive elastomers are widely recognized. In fact, conductive rubber tactile sensors are electrically noisy, nonlinear, and, above all, strongly hysterethical. Moreover, they have poor dynamic response, quite low sensitivity, large drift and fatigue problems.

Two conductive elastomer sensors, incorporating proprietary elastomer material, and comprising, respectively, an array of 8 x 16 sensing elements over an active area of about 10 mm by 20 mm and an array of 16 x 16 sensing elements over an active area of about 40 mm by 40 mm, are commercially available (Barry Wright Co. 1984b).

An interesting extension of conventional conductive elastomer transducer technology has been proposed by Raibert and Tanner (1982), who developed a device that is at once a high resolution tactile array sensor and a special purpose parallel computer. The two investigators replaced the passive substrate of conventional tactile sensors with a custom-designed, large scale integration device that performs transduction (through a layer of conductive elastomer), tactile image processing and communication. A modification of the original concept (which included a 6 x 3 array sensor with 1 square millimeter cells) superimposes the conductive elastomer over a notched silicon substrate to make an all-digital VLSI tactile sensor (Raibert 1984).

One variation on conductive elastomer sensors is carbon fibers which, even though not elastomers, possess some attractive features such as high mechanical strength, low cost, flexibility and high resistance. A prototype carbon-filament tactile sensor designed at the University of Warwick in the United Kingdom demonstrated the extremely wide dynamic range (4 to 5 decades) obtainable with this technology (Larcombe 1981).

Although careful material selection can minimize hysteresis and long term creep, these effects will ultimately limit the performance of a conductive elastomer-based tactile sensor. Thus, some researchers have considered the more mature and proven technology of strain gauges, especially in their solid state version (Petersen et al. 1985, Wong and Van der Spiegel 1985). A family of tactile sensors incorporating solid state silicon strain gauges and capable of measuring contact forces perpendicular to their surface in the range of 0-9 N, with an accuracy of .04 N and hysteresis limited to .0009 N, are commercially available (Transensory Devices, Inc. 1984). Although solid state technology offers great promise for the design of future miniature, high resolution and "smart" tactile sensors, current silicon sensors are stiff and fragile. A novel approach investigated at Stanford University's Center for Integrated Systems may offer a solution to these problems. By exploiting a technology developed for fabricating flexible arrays of temperature sensitive pn diodes for biomedical applications (Barth et al. 1985), Stanford researchers have designed an array of silicon "islands", containing both piezoresistive pressure sensors and temperature sensors, held on a flexible polyimide substrate and interconnected by photolithografically defined gold leads.

3.2. Magnetic and electromagnetic sensors

The development of tactile sensor arrays based on the magnetoelastic properties of some different materials, including recently developed amorphous metals (metallic glasses), is mainly the result of continuous research efforts from investigators at the Naval Surface Weapons Center and at the National Bureau of Standards, U.S.A.. A material is magnetoelastic if there is a relationship between changes in its internal magnetic moment (and hence B field), changes in the mechanical forces applied to or by it, and changes in its physical length (Vranish et al. 1982). In particular, a magnetoelastic material exhibits the magnetostrictive effect when it shows a change either in its length due to changes in its internal B field, or in its internal B field induced by changes in its length.

A number of different sensor configurations have been explored by the N.B.S. investigators. Initially they proposed a relatively simple force/torque vector sensor incorporating a few magnetoelastic elements working as strain gauges (Vranish et al. 1982), and later more sophisticated magnetoresistive (Vranish 1984) and magnetoinductive (Vranish 1986) arrays.

Another configuration has been proposed for a magnetoresistive tactile array by Luo et al. (1984), who used amorphous ferromagnetic material for the fabrication of a 256-element array with a center-to-center distance of 2.5 mm. The authors claim this sensor to be particularly rugged and durable.

Outstanding sensitivity (potentially, orders of magnitude higher than that of standard strain gauges), linearity and dynamic range, no observable mechanical hysteresis, low cost and corrosion resistance are the properties that render magnetoelastic materials attractive. However, some aspects of these materials, such as the need for shielding and the brittlness of some of them, are troublesome. Others, such as temperature stability, require additional investigation.

An interesting concept has been proposed for a magnetic tactile sensor capable of detecting not only normal, but also torque and tangential forces (Hackwood et al. 1983). The sensor consists of one or more tiny magnetic dipoles embedded in a compliant medium on a substrate containing one or more magnetic sensors. When a force deforms the compliant medium, the embedded dipoles are displaced from their original positions: the magnetic sensors detect the change in magnetic field and produce electric signals which are processed externally to the array. The small size of the magnetic sensors (for instance magnetoresistors) helps to accommodate many sensors in each sensing element: with four sensors it is possible to detect four degrees of freedom of dipole motion. As suggested by the authors, this property may virtually enhance the pattern discrimination capacity of this tactile sensor, similar to that of the human skin, and allow the possibility to minutely and continuously shift and/or rotate the sensor array over a surface.

Finally, one should mention a new type of electromagnetic sensor capable, in principle, of simultaneous multifunction sensing, i.e. proximity, tactile, material properties diagnosis, flaw location and flaw sizing (Auld and Bahr 1986). An experimental, five-coil eddy current probe, consisting of a single drive coil with two pairs of smaller pickup coils, has been fabricated and preliminary tested. Multiple sensing functions are realized by reading the pickup coils in different combinations of operation. In particular, if a compliant dielectric layer is placed between the coils and an object, the sum signal of the pickup coils gives a measure of tactile pressure after contact is made with the object. Such tactile effects are actually observed in nondestructive evaluation (NDE) probes, despite efforts to avoid them by rigid probe construction. The authors claim that, by using and extending NDE technology, the reported difficulties of electromagnetic sensors (such as sensitivity to ambient electrical noise, uncalibrated for proximity measurement unless the material properties of the object are known, and inadequate spatial resolution) can be overcome.

3.3. Capacitive sensors

Proposed first by Boie (1984), capacitive tactile sensors have immediately attracted substantial research attention.

The feasibility of capacitive tactile sensing has been demonstrated in a prototype 8 x 8 element transducer, which implemented an idea originated by Nicol (1981) for the development of a sensorized platform for biomechanical studies. Boie's tactile sensor consists of a three layer sandwich structure: the top layer are columns of compliant metal strips over a central elastic dielectric sheet: the bottom layer is a flexible printed circuit board with rows of metal strips and multiplexing circuits. In this configuration, the sensor is a capacitor array formed by the row and column crossing, with the middle layer functioning as a dielectric spring. An improvement of this design has been proposed by Siegel et al. (1985), who devoted particular attention to optimize the dielectric layer by employing a dual layered dielectric to achieve both a large force range and sensitive operation, and by forming the material as a sheet with protruding tabs to improve the linearity of positional response to an applied force. Further work conducted to the fabrication of an integrated tactile sensor comprising both a 8 x 8 array of force sensing cells with a 1.8 mm center to center spacing, and a thermal sensor incorporating a 4 x 4 array of thermistors. This integrated sensor is intended to be eventually mounted on the UTAH/MIT dexterous hand (Siegel et al. 1986).

High performance, high resolution tactile sensors based on semiconductor capacitive structures have been also proposed. Capacitive pressure sensors are about an order of magnitude more sensitive, for a given device size, and more than an order of magnitude less sensitive to temperature than analogous piezoresistive sensors. A 8 x 8-element tactile array on 2-mm centers of capacitive cells has been obtained on a silicon wafer at the University of Michigan (Chun and Wise 1985). More recently, an integrated capacitive tactile sensor consisting of a conductive rubber layer placed on top of an isolating rubber layer, which is in turn attached to a silicon wafer on which aluminum rectangles define the sensing elements and in which also the addressing logic and the signal conditioning circuits are realized, has been proposed at Delft University of Technology in Delft, The Netherlands (Wolffenbuttel and Regtien 1986). The operation of the 9x9-element prototype sensor that has been fabricated, involves the capacitive coupling of a sinusoidal driving voltage from the selected aluminum rectangle, through the natural rubber layer to the conductive rubber layer: when the applied force deforms the rubber, a change in the capacitance is detected.

3.4. Electro-optic sensors

Electro optic sensors rely on the modulation of a light source by the mechanical deformation of a flexible material. Optical tactile sensors can be very sensitive, can offer the opportunity for versatile designs, and are immune to electromagnetic interference.

Several physical effects can be used to convert mechanical energy into optical signals.

An optical sensor developed at the Jet Propulsion Laboratory in Pasadena employs a flexible membrane with a reflecting surface, along with an infrared light source and 16 optical fibers and light detectors (Bejczy 1983). When a contact force deflects the membrane, the intensity of the light reflected back to the photodetector is modulated. Although the linear relationship between these two parameters is poor, substantial improvements of overall sensor performances are expected from a more favorable arrangement of light sources, carriers, and detectors (leading to more effective matrix scanning techniques), as well as, ultimately, from integrated optics.

Other optical sensors based on mechanically induced modulation of a reflected light beam have been proposed. At the MIT Man-Machine Systems Laboratory a tactile sensor has been designed in which light is trasmitted through a bundle of optical fibers to an elastic reflective surface and back through another bundle of optical fibers to a video camera (Schneiter 1982). In this prototype design, the spatial resolution is extremely high: about 330 sensitive spots per square centimeter. The most attractive feature of this sensor is that the distribution of forces over its sensitive area can be captured and processed in almost the same way and using the same techniques as visual images.

A different approach has been used by a Japanese group for the development of a 15x15-element array arranged on 2 mm centers. The device is comprised of: a) a light emitting diode matrix formed on a ceramic substrate, b) a photodetector matrix (fabricated using a newly developed planar type amorphous silicon photoreceiving film), and c) a pressure sensitive section composed of a transparent silicone rubber, with 1 mm diameter cylindrical studs, and a white rubber light reflecting sheet piled upon each other (Nomura et al. 1985).

A class of similar optical tactile sensors, all, in essence, evolving from an idea originally proposed for the development of a device for displaying pressure distribution under the foot of standing or walking patients, have recently been described by different groups. The measuring principle, common to all devices, consists of frustrating the total internal reflection of a light beam travelling in a light guide, by utilizing the properties of reflection between objects of a different refractive index. In the sensor configuration illustrated in Fig. 3 (Tanie et al. 1984) an elastic sheet is placed on an acrylic plate which conducts the light emitted by a halogen lamp by total internal reflection.

Pressure applied to the elastomer sheet causes a contact to occur between the sheet and the surface of the light guide, thus modifying the total internal reflection conditions (in fact the acrylic has a higher refractive index than both air and elastomer. The sheet scatters the light at the points where it contacts the acrylic surface; the light passes out through the lower surface of the acrylic plate and is collected by an array

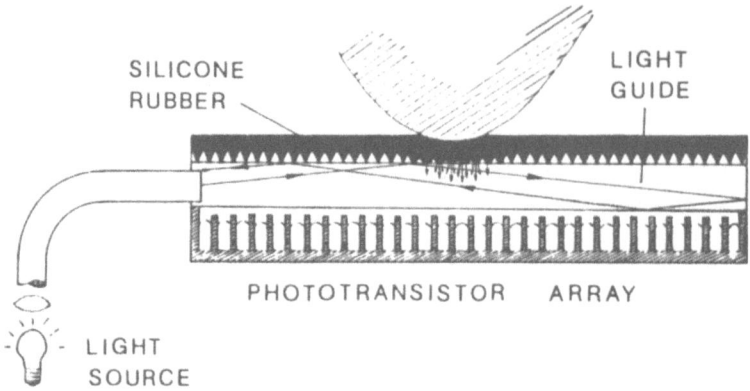

Fig. 3. Frustrated light tactile sensor

of phototransistors. The intensity of the light scattered at each contact point is approximately proportional to the magnitude of the pressure between object and plate.

A second, similar tactile sensor configuration has been proposed by Mott et al. (1984), whose sensor includes a compliant membrane stretched above, but not in close contact with, the acrylic plate. Light reflection frustration occurs when the membrane is brought into contact with the plate by pressure exerted on it. An interesting feature of this sensor, that has been designed and built to be so small so as to be actually incorporated in the finger of a robotic gripper, is the fact that a visual analog of the contact force distribution is obtained by focusing the scattered light on a charge-coupled device imager.

Another sensor, based on the same principle has been described by Begej (1984). In his optical tactile sensor, Begej utilized different textured transducer membranes in order to investigate the influence of such parameters as the thickness, the size and shape of the microtexture, the mechanical properties of the membrane on the depth and lateral spatial resolution, the force or strain response, the degree of hysteresis, and the time response characteristics of the tactile sensor. A sensor density of 54 elements per square centimeter was obtained over an active area measuring 22 mm x 25 mm.

Two interesting evolutions of the optical principle involving the frustration of total internal reflection have been proposed recently. King and White (1985), in a paper in which they presented a few stimulating ideas on the use of optical techniques for tactile sensing, described the concept of a shear-responsive rubbery skin incorporating elements that indicate the direction and magnitude of the shear stress by the displacement of the contact area from an equilibrium position, and the concept of sensors able to detect local torque by light polarization. Begej

(1986) has reported on a fingertip-shaped optical tactile sensor which includes 256 sensing sites distributed on the fingertip in two densities (1 mm spacing over a 13 mm x 13 mm central area and 3.2 mm spacing over the remaining area) and capable of detecting normal forces in the range 0-0.4 N per sensing site.

Some of the concepts described in this paragraph form the basis for commercial robotic tactile sensors (Tactile Robotic Systems 1985, British Robotic Sistems Ltd. 1984). Still another simple idea has been embodied in a commercial electro-optic tactile sensor for industrial applications (Lord Corporation 1985b), in which the deflection of elements protruding from a touch surface-an elastomer-reduces the light transmitted to detectors by physically blocking the light. This sensor comprises a 10x16 array of 160 sensitive sites over an area of 18 mm x 28 mm, and has the attractive feature of being designed for gripper mounting, with or without the addition of the force/torque vector sensor module already mentioned in paragraph 2.4.

3.5 Ferroelectric polymer sensors

Some synthetic polymers possess substantial piezoelectric and pyroelectric properties. The best known among such synthetic polymers is polyvinylidene fluoride (PVF2 or PVDF), a fluorinated polymer with a highly desirable combination of chemical and mechanical properties. PVF2 is available in thin flexible films that can be made strongly piezoelectric and pyroelectric. These polymer films have many features that render them attractive for the design of robotic transducers (Dario et al. 1983, Dario 1986), such as excellent linearity and dynamic response, high strain sensitivity, easy conformability to complex shapes, and low cost. However, since ferroelectric materials are intrinsically unable to detect truly static events and are simultaneously sensitive to force and temperature variations, they are not suitable for "general purpose" tactile sensors, and special attention must be used to effectively exploit their unique properties.

A PVF2-based tactile sensor was investigated at the University of Utah (Schoenberg et al. 1984), using ultrasonic (US) techniques to measure the amount of compression originated by contact force in a thin rubber layer. The same concept has been used by other investigators (Grahn and Astle 1984) in a closely packed array of US pulse-echo ranging PVF2 transducers capable of measuring contact pressure over a 2000:1 range, and with a response time for sensor elements of approximately 6 microseconds.

At the University of Florida, a corrugated PVF2 transducer has been proposed as tactile sensor (Patterson and Nevill 1986). The sensor, whose functioning is based entirely on dynamic motion, incorporates two perpendicular PVF2 elements covered with a ridged silicone rubber pad simulating the ridges of the human fingertip. When the sensor pad is moved across a surface and the signals originated by the vibration pattern is recorded and analyzed, it is possible to recognize different sample objects, to read the Braille alphabet, and to distinguish between different grades of sandpaper.

Many other groups are investigating possible applications of ferroelectric polymer technology for robotic tactile sensing (Dario 1986).

In recent years, our laboratory has been addressing the problem of designing a tactile sensor intended to be the core of a tactile sensing system aimed at investigating and, hopefully, reproducing most of the sensory capabilities of the human tactile system. The design features of our PVF2-based tactile sensor will be discussed in the next section, along with the features of the robotic system of which the sensor is a critical component. Here we shall only mention a PVF2 tactile sensor configuration we have developed for mounting in an industrial robot gripper (Fiorillo et al. 1987). This gripper sensor comprises an array of 128 elements over an area of 25 mm x 50 mm, and is associated with a computerized unit for matrix scanning and signal reconstruction. The flat tactile sensor also incorporates a covering rubber pad designed to mechanically decouple the sensing sites (thus reducing both cross-talk effects and artifacts due to shear stresses in the PVF2 film) and to shield the PVF2 sensors from sudden temperature variations, as required for practical applications in industrial environments.

4. AN ARTIFICIAL TACTILE SENSING SYSTEM

As an example of integration of sensory, motor, and control functions aimed at obtaining the dexterous behavior required for robot haptic exploration, in this section we shall briefly describe the design features of a simple robotic system we have implemented in our laboratory. The hardware part of the system comprises a single anthropomorphic finger equipped with joint torque and position sensors and incorporating a skin-like tactile sensor in its fingertip. It also comprises electronic units that perform some of the low level control functions and that interface the finger sensors and actuators with the computer which controls higher level functions. The main objective pursued was to set up a system to investigate basic issues in artificial tactile perception. A description of the finger system's component has been given in previous papers (Dario et al. 1985, Bicchi et al. 1985, Dario et al. 1986). Here we shall summarize the features of the finger and the tactile sensor, and emphasize the importance of integrating sensory information with the motor commands necessary to elicit them at various control levels. In particular the fundamental importance of a system approach to the design of a sophisticated tactile sensing system will be pointed out.

4.1 Anthropomorphic robot finger

The finger we have designed and fabricated has an anthropomorphic configuration, and is composed of four rigid links connected by hinge joints providing a total of four degrees of freedom. The two-degree-of-freedom articulation of the proximal phalanx of the human fingers is reproduced by two separate joints with perpendicular axes. The finger is presently mounted on a rigid fixture, as shown in Fig. 4, that quite severely limits its exploratory capabilities. The intention is to eventually connect it to a multi-degree-of-freedom manipulator making it capable of following complex object surfaces.

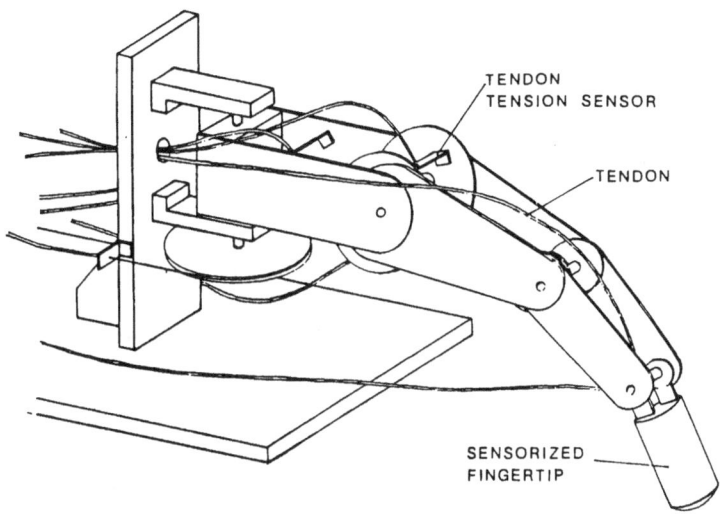

Fig. 4. Tendon actuated 4 d.o.f. sensorized finger

Despite this limitation, finger dexterity is sufficient to investigate fundamental problems associated with the simple tactile exploration procedures we intend to replicate.

Each articulation is driven via plastic coated stainless steel tendons, routed through flexible and incompressible sheaths, and actuated by remotely located dc servomotors. We have reduced friction effects associated with this configuration by including these effects in the force control loop, and by superimposing a "dither" vibration to the motor signal. One dc motor actuates each finger joint through a pair of opposed tendons, which are pretensioned to half the maximum required tension to avoid slackering of one of the tendons during high force exertion. The mechanical proportioning of the finger was based on the assumption that it should be capable of exerting a maximum contact force of about 10 N and moving at a maximum speed of .1 m/s.

Joint position and velocity are detected by incremental encoders located coaxially with the driving motors; joint torques are measured by tendon tension sensors consisting of strain gauge instrumented cantilevers located at the outlet of the conduits guiding the drive cables. Furthermore, a PVF2-based, composite tactile sensor is located on the distal phalanx of the finger.

The different force sensors of the finger have essentially separate functions. In fact, the strain gauge-based devices measure tendon tensions (and, consequently joint torques), while the force sensitive elements incorporated in the fingertip skin-like sensor are aimed at detecting the local distribution of contact forces in order to reconstruct object features (at high control level). A version of the finger being currently developed in our laboratory will also incorporate a contact resolving fingertip sensor, conceptually similar to the one proposed by Brock and Chiu (1985). When this vector sensor is available, it will be primarily dedicated to low level compliant motion control.

4.2 Skin-like tactile sensor.

According to the approach proposed in this section, the main goal of the fingertip tactile sensor is to provide the upper level hierarchies of the control system with "exteroceptive" data useful in describing, as thoroughly as possible, the physical (and even chemical) properties of the explored object. The low level control of compliant motion, instead, would be carried out using the force feedback signal detected by a fingertip force sensor. Presently, however, the fingertip does not incorporate this type of sensor. Some of the information provided by the tactile sensor is exploited for force control.

The tactile sensor we have designed is based on the technology of the ferroelectric polymer PVF2, and is capable of extracting, when appropriate motor acts are commanded to the supporting articulated finger, information about the explored object. The sensor, intended to mimic, at least functionally, even if not morphologically, the human finger pad skin, includes two sensing layers (a deep, "dermal" layer and a superficial, "epidermal" layer separated by a compliant rubber layer (Dario et al. 1984). The dermal layer is primarily sensitive to the local, contact-induced normal stress. It has relatively high spatial resolution (128 circular sensing sites, diameter 1.5 mm, center to center spacing 2.5 mm, as depicted in Fig. 5), and it is capable of providing a quasi-static response to force signals. Thus, the role of the dermal sensing elements can be conceptually associated to that of the slowly adapting (SA) receptors of the human skin, which are sensitive to the tiny spatial features of the indenting object (Phillips and Johnson 1982).

The epidermal layer includes only a few sensing sites (7 circular elements, diameter 1.5 mm, center to center spacing 2.5 mm) concentrated in a small area (the tactile "fovea") of the fingertip and particularly sensitive, like the quickly adapting (QA) skin receptors, to dynamic contact stimuli. Being backed by the rubber compliant layer (which has the twofold role of providing compliance to the fingertip sensor and to shield the dermal sensing layer from sudden temperature variations), the epidermal sensor is primarily sensitive to membrane strains such as those originated when the fingertip is rubbed along a grooved pattern.

The epidermal sensor comprises also a graphite ink resistive layer that, when connected to a power supply, heats the epidermal sensing elements and mantains a temperature gradient between the fingertip and the

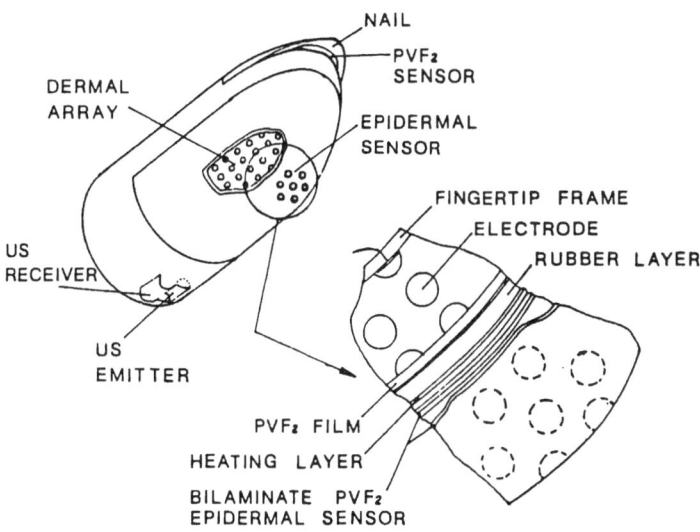

Fig. 5. PVF2-based sensorized fingertip

explored object (usually at room temperature). This provides a means of characterizing the thermal properties of the object's material composition. In general, in fact, the role of the tactile fovea (which is constituted by the epidermal sensor and by those dermal sensing sites immediately underneath) is to detect signals that the high level controller can process and interpret as intrinsic or estrinsic object features (such as object temperature, material hardness and thermal properties, surface roughness).

A distinctive feature of this sensorized fingertip is the ability (deriving from its finger-like curved shape) to explore details of complex shapes and cavities (Buttazzo et al. 1986). The accurate reconstruction of these object features could be facilitated by the possibility of detecting, through the slowly adapting, high spatial resolution dermal sensors, quasi static variations of contact pressure. Because of this possibility, a method has been devised to process the piezoelectric signal in order to reconstruct the force signal in each sensing element of the dermal matrix with an acceptable bandwidth (.02-100 Hz)(Fiorillo et al. 1987).

Two additional sensing capabilities have also been incorporated in the fingertip. The first is a nail-shaped element that includes a PVF2 film sensor capable of detecting the vibration of the nail when it is rubbed along tiny, even irregular, grooves (Buttazzo et al. 1986). The PVF2 nail sensor works like the pick-up of a grammophone, and the resulting signal can be processed in the same way as in (Patterson and Nevill 1986) to identify specific surface features. A second addition to the basic sensing capabilities of the fingertip sensor consists of an array of curved US transducers, also made out of thin PVF2 film. This array, presently comprising only two contiguous elements (a concave focussing emitter and a convex, wide acceptance angle receiver) will ultimately constitute the same

epidermal sensor, and will be used either as a proximity imager or as a
true tactile sensor (Dario et al. 1987).

Finally, the proposed fingertip sensor is expected to include, in the
near future, additional sensing elements, capable of measuring other
important physical and chemical parameters related to sensor-object tactile
interaction, such as local shear contact forces, humidity, and pH.

The overall functioning of both "internal" (joint torque and position)
and "external" (tactile) sensors can be better understood if considered in
the context of the tactile system of which they are important components.
This aspect is illustrated in the next paragraph, with reference to the
case of a multi-level control architecture intended for testing the
feasibility of a few elementar sensory-motor sequences ("tactile
subroutines") which form the basis for more complex exploratory haptic
procedures.

4.3 System control architecture

In order to approach the problem of replicating some of the exploratory
processes typical of human haptic perception, we have considered the
simplified, three-level control architecture depicted in Fig. 6 (Dario and
Buttazzo 1987).

At the bottom level of the control hierarchy, servo loops control
finger joint positions and torques through feedback signals from internal
sensors. This control level is designed to execute the hybrid
(position-force) commands sent by the middle level controller. A block
diagram of the low level control system is given in Fig. 7.

The hybrid commands Q, T and C are, respectively, the nominal joint
position vector (the desired finger position), the nominal joint torque
vector (corresponding to the desired contact force), and a compliance
matrix (which controls the stiffness of the movements and acts as a balance
factor between the torque command and the position command). The hybrid
command U, resulting from mixing, via software, the position and torque
commands (the upper control levels can decide the stiffness of the finger
during exploration) are sent to an analog control unit, which drives the
finger actuation system. However, a fundamental concept of the proposed
control system is that, while the force feedback signal is used only at the
bottom control level, the position signal (the actual joint position vector
A) is processed both at the bottom and at the upper control levels.

This is based on the consideration that the actual position of finger
joints is useful not only for controlling finger motion, but can be use-
fully processed in order to obtain exteroceptive information. For instance,
the absolute position of finger joints, associated with data on the contact
between the finger and the explored object, allows the high level controller to infer the shape of the explored object (one should also observe
that in case the force feedback signal were provided by a contact resolving
fingertip sensor, the force signal could also be processed at high level,
in order to enhance system perceptual capabilities (Salisbury 1984).

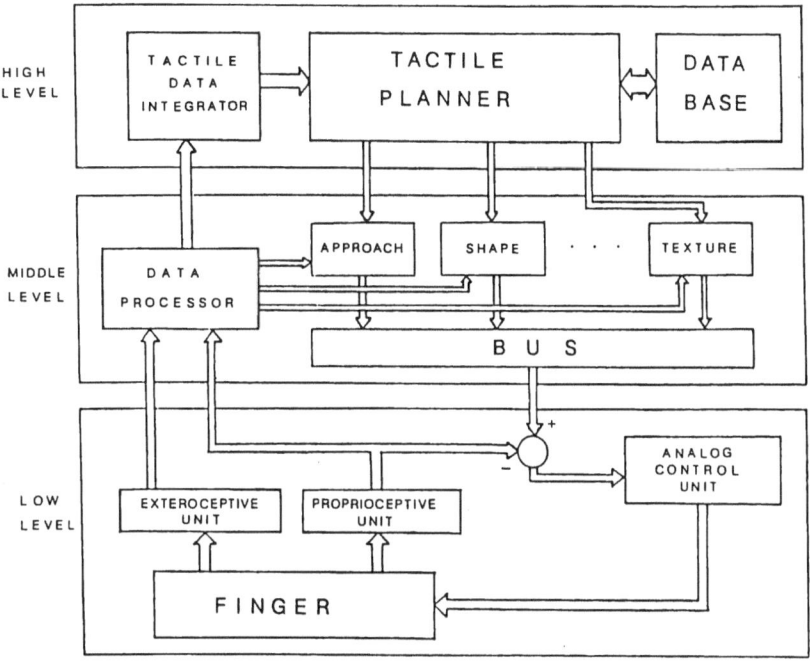

Fig. 6. Three-level control architecture for the study of robot haptics

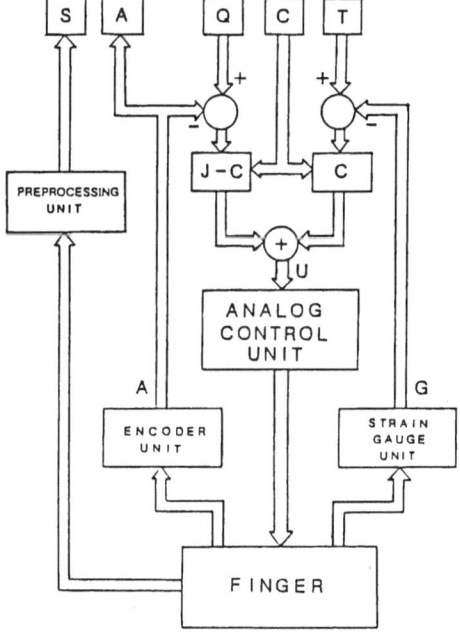

Fig. 7. The low control level

The PVF2 tactile sensor signals are processed at different levels. The signals detected by the dermal sensors are used to calculate a (somewhat approximated) contact point between the fingertip sensor and the explored object. As the geometry of the fingertip is known, this information allows for the estimation of the normal local contact surface, which represents the direction along with the desired contact force to be exerted, and the derivation of the common tangential plane in which a step of the exploratory motion can be commanded. The same dermal sensor signals, however, are also sent to the high level controller, because they contain useful information for perceptual purposes (e.g. for reconstructing object shape from surface normals, for extracting geometrical object primitives, or for calculating material hardness (Dario and Buttazzo 1987)). On the other hand, epidermal sensor signals are used only at high control levels, because they contain true tactile data, used only for perceptual purposes.

Actually, this multilevel signal processing and control architecture clearly illustrates the meaning of the distinction we have introduced between force and true tactile sensing.

The intermediate control level has the function of managing autonomously the execution of tactile subroutines, i.e. sequences of sensory-motor acts each aimed at extracting a specific object feature, upon selection by the high level controller. We have investigated the feasibility of some different tactile exploratory subroutines, namely: APPROACH (it has the function of controlling the motion of the finger until it touches an object based on proximity sensing; it can also deal with potentially dangerous situations, such as the contact with hot or sharp objects), SHAPE (it allows the finger to follow object surface and the high level controller to infer the 3-D shape of the object as the locus of the contact points), TEXTURE (it accesses the finish of parts of the object surface), HARDNESS (it provides data usable by the high level controller to infer the elastic properties of object material; in the present dermal/epidermal sensor configuration, a parameter closely related to the Young's modulus of the object material is calculated from the ratio between the epidermal and dermal signals elicited when the tactile fovea is pressed on the object and then released), and THERMAL (it recognizes different materials based on their thermal properties (Dario and Buttazzo 1987).

Consider for instance, a procedure for estimating the texture of a surface, consisting of rubbing the surface while the finger exerts a predetermined adjustable force on the rubbed surface. A figure of merit of surface roughness is obtained by analyzing the signal elicited during this procedure in the epidermal sensor (or in the nail sensor). Conceptually, this procedure can be decomposed in a) a low level motor act (the "rubbing" motion), and in b) a high level perceptual act (the extraction and analysis of the "vibration" signal). As in humans, the "rubbing" motion should be carried on almost automatically, while the full attention of the brain should be involved in the analysis of the vibration pattern.

In order to understand in detail how an exploratory procedure involving similar complex interactions between sensory and motor functions, and between different control levels, can be managed by the proposed artificial exploratory system, we illustrate here the execution of the subroutine THERMAL. This subroutine, illustrated in Fig. 8, aims at correlating the temperature variation induced in the epidermal tactile sensors by the heat flowing from a source within the sensor (the resistive ink layer) towards the object under examination, with the thermal diffusivity (during the thermal transient) or conductivity (at equilibrium) of the object's material.

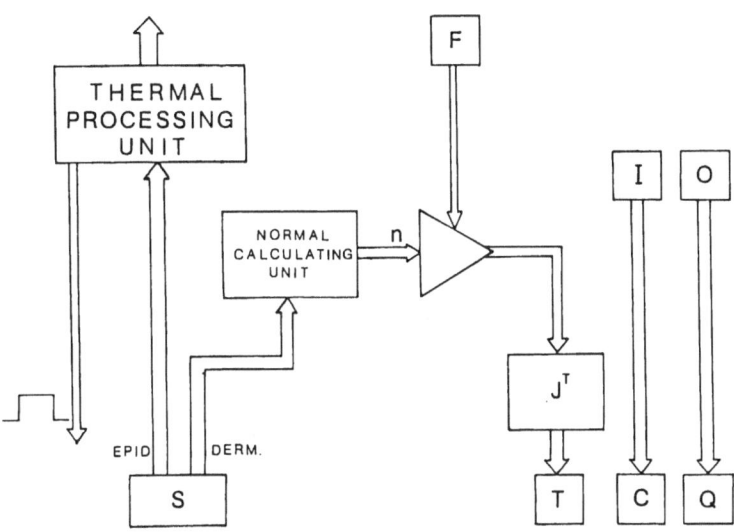

Fig. 8. Block diagram of the subroutine THERMAL

The first exploratory step involves positioning the fingertip in a convenient position relative to object surface. This can be done by locating the tactile fovea on a flat portion of the surface. Then the tactile fovea is pressed slightly on the object and the contact is maintained for a few seconds.

As shown in Fig. 8, the high level control commands the motion (Q = 0), the degree of compliance (C = 1), and the contact force F (that is converted into the joint torque vector T by applying the Jacobian transpose matrix). The signal measured by the dermal sensors (that indicates the direction of the normal unit vector to the contact point (Bicchi et al. 1985) allows for calculating the direction along with the contact force is to be applied. This signal, however, does not require any "conscious" attention from the high level controller. On the contrary, the signal generated by the epidermal sensor, when a power supply unit is commanded to heat the graphite layer, is continuously recorded by the high level controller, deriving useful data characterizing material thermal properties. We have tested THERMAL in an experiment during which the fingertip pressed a sample object with a force of 5N (the higher the contact force, the better the thermal contact between the sensor and the object). The resistive layer heated the epidermal sensors at a temperature of 40°C. The signals detected by an epidermal sensor element when the fingertip touched objects made of different materials (but with the same shape) on a flat portion of their surface, allowed clear distinction of not only metals from plastics or wood, but even, within each material class, different metals and plastics.

Ideally, the top control level should reproduce superior brain functions, such as the capability of deciding tactile exploratory strategies (and of commanding their execution by calling appropriate tactile subroutines), and of processing sensory information in order to construct a model of the explored object. Furthermore, it should be able to coordinate various sensing modalities and to "fuse" together different sensory data. Investigating the features of such upper control level is an open research field for artificial intelligence.

5. CONCLUSION

In this paper we have analyzed the state of the art of robotic contact sensing and have proposed some hopefully useful considerations for the design of artificial tactile systems. The number of robotic applications which would require some form of contact sensing and control is extremely wide, ranging from manufacturing operations to artificial perception. This fact, associated with the relative novelty of the study of tactile sensing in robotics, makes it difficult to provide specific, experimentally verified criteria for the choice of the hardware and software components of the tactile system most suitable for each application.

In general, however, one could say that for relatively simple assembly operations carried out using conventional grippers, the availability of appropriate control techniques, rather than the need for adequate force sensors, is the true limiting factor. This is not the case for more complex applications. The achievement of a high degree of dexterity, for instance, is conditioned both by the inclusion of dedicated sensors in the end effector, and by the appropriate use of the resulting sensory data in a sophisticated control architecture.

The attempt to investigate artificial tactile perception obviously poses the most challenging requirements to a tactile system. We have illustrated the complexity of these requirements by referring to a simple robotic prototype developed in our laboratory.

In this context, the information and the considerations proposed in this paper, although by necessity not exhaustive, could not only be useful as a guidance for the design of a tactile sensing system, but also stimulate further research efforts in the field of robotic active perception.

ACKNOWLEDGMENTS

We thank Antonio Bicchi and Giorgio Buttazzo for their contributions to the work on tactile perception described in this paper. We also wish to thank Riccardo Di Leonardo, Raffaello Francesconi and Fabrizio Vivaldi for their technical assistance, and Daniela Fantozzi and Iolanda Giusti for their dedication in typewriting the manuscript.

The work described in this paper was financially supported, in part, by the Italian Government (MPI 40%), by NATO Scientific Affairs Division (Grant C.R.G. 85/224) and by Grants from Axis S.p.A., IBM Italia S.p.A. and Polysens S.p.A. Support for M. Bergamasco was provided by a fellowship from Polysens S.p.A..

REFERENCES

Allan R (1985) Nonvision sensors. Electronic Design, June 27: 103-115

Auld BA, Bahr AJ (1986) A novel multifunction robot sensor. Proc. IEEE Int. Conf. on Robotics and Automation: 1791-1797, San Francisco

Bajcsy R (1984) What can we learn from one finger experiments? In: Brady M, Paul R (Eds) Robotics Research, Mit Press: 509-527, Cambridge, MA

Barry Wright Corporation (1984a), Watertown, MA, ASTEK Model FS6-120A

Barry Wright Corporation (1984b), Watertown, MA, Sensoflex Tactile System

Barth PW, Bernard SL, Angell JB (1985) Flexible circuit and sensor arrays fabricated by monolythic silicon technology. IEEE Trans. on Electron. Devices, ED-32(7):1202-1205

Begey S (1984) An optical tactile array sensor. Proc. SPIE Conf. on Intelligent Robots and Computer Vision, Vol. 521:271-280, Cambridge, MA

Begey Corporation (1985) Littleton, CO, TSA-32-V4 Optical Tactile Sensor Array, Technical Bulletin No. 1

Begey Corporation (1986) Littleton, CO, FTS-2 Fingertip-Shaped Tactile Sensor, Technical Bulletin No. 2

Bejczy AK (1977) Effect of hand-based sensors on manipulator control performance. Mechanism and Machine Theory 12:547-567

Bejczy AK (1983) "Smart hand"-Manipulator control through sensory feedback, January 15, JPL D-107 Report

Bicchi A, Dario P, Pinotti PC (1985) On the control of a sensorized artificial finger for tactile exploration of objects. Proc. '85 IFAC Symp. on Robot Control, Barcelona, Spain

Boie RA (1984) Capacitive impedance readout tactile image sensor. Proc. IEEE Int. Conf. on Robotics: 370-378, Atlanta

British Robotic Systems Ltd. (1984), London, UK

Brock D, Chiu S (1985) Environment perception of an articulated robot hand using contact sensors. Proc. ASME Winter Annual Meeting, Miami

Brown MK (1986) The extraction of curved surface features with generic range sensors. Int. J. Robotics Res., 5(1):3-18

Buttazzo G, Dario P, Bajcsy R (1986) Finger based explorations. Proc. SPIE Conf. on Intelligent Robots and Computer Vision, Cambridge, MA

Chun K, Wise KD (1985) A high-performance silicon tactile imager based on a capacitive cell. IEEE Trans. Electron Devices, ED-32(7):1196-1201

Coiffet P (1983) Robot Technology. Interaction with the Environment(2), Prentice-Hall Inc., Englewood Cliffs, NJ

Craig JJ (1986) Introduction to robotics. Chapter 9: Force control of manipulators. Addison-Wesley Publishing Company, Reading, MA

Cutkosky MR, Wright PK (1984) Active control of a compliant wrist in manufacturing tasks. Proc. 14th ISIR: 517-528, Gothenburg, Sweden

Cutkosky MR (1985) Robotic grasping and fine manipulation. Kluwer Academic Publishers, Boston, MA

Dario P, Domenici C, Bardelli R, De Rossi D, Pinotti PC (1983) Piezoelectric polymers: New sensor materials for robotic applications. Proc. 13th ISIR/Robots 7, Paper MS83-393, Chicago

Dario P, De Rossi D, Giannotti C, Vivaldi F, Pinotti PC (1984) Ferroelectric polymer tactile sensors for prostheses. Ferroelectrics 60(1-4):199-214

Dario P, Bicchi A, Vivaldi F, Pinotti PC (1985) Tendon actuated exploratory finger with polymeric, skin-like tactile sensor. Proc. IEEE Int. Conf. on Robotics and Automation: 701-706, St. Louis

Dario P, De Rossi D (1985) Tactile sensors and the gripping challenge. IEEE Spectrum 22(8):46-52

Dario P (1986) Ferroelectric polymer transducers for advanced robots. To appear in: Herbert JM, Wang TT, Glass AM (Eds), Application of Ferroelectric Polymers, Blackie and Son Ltd, Glasgow, UK

Dario P, Bicchi A, Fiorillo A, Buttazzo G, Francesconi R (1986) A sensorized scenario for basic investigation on active touch. Pugh A (Ed) Robot Sensors. Tactile & Non-Vision, 2, IFS (Publications) Ltd, Bedford, UK, and Springer-Verlag: 237-245, Berlin Heidelberg New York Tokyo

Dario P, Bergamasco M, Femi D, Fiorillo A, Vaccarelli A (1987) Proc. IEEE Int. Conf. on Robotics and Automation, Raleigh, NC

Dario P, Buttazzo G (1987) An anthropomorphic robot finger for investigating artificial tactile perception. Int. J. Robotics Res. (in press)

De Fazio TL, Seltzer DS, Whitney DE (1984) The IRCC instrumented remote center compliance. The Industrial Robot 11(4):238-242

Fearing RS, Hollerbach JM (1985) Basic solid mechanics for tactile sensing. Int. J. Robotics Res. 4(3):40-54

Fiorillo A, Dario P, Bergamasco M (1987) A sensorized robot gripper. Robotics (submitted)

Grahn AR, Astle L (1984) Robotic ultrasonic force sensor arrays. Proc. Robots 8: 21/1-17, Detroit

Grimson EW, Lozano-Pérez T (1984) Model-based recognition and localization from sparse range or tactile data. Int. J. Robotics Res., 3(3):3-35

Hackwood S, Beni G, Hornak LA, Wolfe R, Nelson TJ (1983) A torque-sensitive tactile array for robotics. Int. J. Robotics Res., 2(2):46-50

Harmon LD (1982) Automated tactile sensing. Int. J. Robotics Res., 1(2):3-32

Harmon LD (1984) Tactile sensing for robots. In: Brady M, Gerhardt LA, Davidson HF (Eds) Robotics and Artificial Intelligence NATO ASI Series, Springer-Verlag: 109-157, Berlin Heidelberg New York Tokyo

Hillis WD (1981) Active touch sensing. MIT AI Memo AIM-629, Cambridge, MA

Ishikawa M, Shimojo M (1982) A tactile sensor using pressure-conductive rubber. Proc. 2nd Sensor Symp, IEE of Japan: 189-192, Tsukuba, Japan

Jacobsen S, Wood J, Knutti DF, Biggers KB (1984) The UTAH/MIT dexterous hand: Work in progress. Int. J. Robotics Res. 3(4):21-50

Johansson RS (1979) Tactile afferent units with small and well demarcated receptive fields in the glabrous skin area of the human hand. In:Kenshalo DR (Ed) Sensory functions of the skin of humans, Plenum Press: 129-145, New York

Kasai M, Takeyasu K, Uno M, Murakaoka K (1981) Trainable assembly system with an active sensory table possessing six axes. Proc. 11th ISIR: 7-9, Tokyo, Japan

King AA, White RM (1985) Tactile sensing array based on forming and detecting an optical image. Sensors and Actuators, (8):49-63

Larcombe MHE (1981) Carbon fibre tactile sensors. Proc. RoViSec 1: 273-276, Stratford, UK

Lestelle D (1985) Gripper with finger built-in force/torque sensor. RoViSeC 5:69-78, Amsterdam, The Netherlands

Lord Corporation (1985b), Cary, NC, Tactile Sensors Series 200

Lord Corporation (1985a) Cary, NC, Lord Instrumented Gripper

Luo RC, Wang F, Lin Y (1984) An imaging tactile sensor with magnetoresistive transduction. Proc SPIE Conf. on Intelligent Robots and Computer Vision, Vol. 521:264-270, Cambridge, MA

Mason MT (1981) Compliance and force control for computer controlled manipulators. IEEE Trans. on Systems Man. and Cyber. SMC-11, 6:418-432

Mason MT (1982) Compliant motion. In: Brady M, Hollerbach JM, Johnson TL, Lozano-Pérez T, Mason MT (Eds) Robot Motion, Planning and Control, MIT Press: 305-322, Cambridge, MA

Morris KA (1985) Tactile sensing for automated assembly. Report LL-1201, Lord Corporation, Cary, NC

Mott DH, Lee MH, Nicholls HR (1984) An experimental very high resolution tactile sensor array. Proc. RoViSeC 4: 241-250, London, UK

Nicol K (1981) A new capacitive transducer system for measuring force distribution statically and dynamically. Transducer Tempcon '81, London, UK

Nomura A, Abiko I, Shibata J, Watanabe T, Nihei K (1985) Two-dimensional tactile sensor using optical method. IEEE Trans. Components, Hybrids, Manuf. Technol., CHMT-8(2):264-268

Ogorek M (1985) Tactile sensors, Manufacturing Engineering, 94(2): 69-77

Patterson RW, Nevill GE (1986) Performance of an induced vibration touch sensor. Pugh A (Ed) Robot Sensors. Tactile & Non-Vision, 2, IFS (Publications) Ltd, Bedford, UK and Springer-Verlag: 219-228, Berlin Heidelberg New York Tokyo

Paul RP (1981) Robot manipulators: Mathematics, programming and control. MIT Press, Cambridge, MA

Pennywitt KE (1986) Robotic tactile sensing. Byte January 1986: 177-200

Petersen K, Kowalski C, Brown J, Allen H, Knutti J (1985) A force sensing chip designed for robotic and manufacturing automation applications. Proc. IEEE Int. Conf. on Solid-State Sensors and Actuators: 30-32, Philadelphia

Phillips JR, Johnson KO (1981) Tactile spatial resolution II: Neural representation of bars, edges and gratings in monkey primary afferents. J. Neurophysiology 46(6):1192-1203

Pugh A (Ed) (1986) Robot Sensors. Tactile & Non-Vision, 2, IFS (Publications) Ltd, Bedford, UK

Purbrick JA (1981) A multi-axis force sensing finger. Proc. RoViSeC 1

Raibert MH, Tanner JE (1982) Design and implementation of a VLSI tactile sensing computer. Int. J. Robotics Res. 1(3):3-18

Raibert MH (1984) An all digital VLSI tactile array sensor. Proc. IEEE Int. Conf. on Robotics: 314-319, Atlanta

Salisbury JK, Craig JJ (1982) Articulated hands: Force control and kinematic issues. Int. J. Robotics Res. 1(1):4-17.

Salisbury JK (1984) Design and control of an articulated hand. Proc. Int. Symp. on Design and Synthesis, Tokyo, Japan

Salisbury JK (1985) Kinematic and force analysis of articulated hands. In: Mason MT and Salisbury JK Jr (Ed.) Robot hands and the mechanics of manipulation, MIT Press, Cambridge, MA

Schneiter JL (1982) An optical tactile sensor for robots. MIT Master Thesis of Science in Mechanical Engineering, MIT, Cambridge, MA

Shimano BE (1978) The kinematic design and force control of computer-controlled manipulators. Ph.D. Thesis, Stanford University Computer Science Department, Stanford Artificial Intelligence Laboratory AIM 313

Schoenberg AA, Sullivan DM, Baker CD, Booth HE, Galway C (1984) Ultrasound PVF2 transducers for sensing tactile force. Ferroelectrics, 60(1-4): 239-250

Siegel DM, Garabieta I, Hollerbach JM (1985) A capacitive based tactile sensor. Proc. SPIE Conf. on Intelligent Robots and Computer Vision, Cambridge, MA

Siegel DM, Garabieta I, Hollerbach JM (1986) An integrated tactile and thermal sensor. Proc. IEEE Int. Conf. on Robotics and Automation: 1286-1291, San Francisco

Snyder WE, St. Clair J (1978) Conductive elastomers as sensor for industrial parts handling equipment. IEEE Trans. on Instrum. and Measure, IM-27(1):94-99

Stokic D, Vukobratovic M, Hristic D (1986) Implementation of force feedback in manipulation robots. Int. J. Robotics Res. 5(1):66-76

Tactile Robotic Systems (1985), Sunnyvale, CA

Tanie K, Komoriya K, Kaneko M, Tachi S and Fujikawa A (1984) A high resolution tactile sensor. Proc. RoViSeC 4: 251-260, London, UK

Transensory Devices, Inc. (1984) Fremont, CA, Tactile perceptions

Van Brussel H, Simons J (1978) Automatic assembly by active force feedback accommodation. Proc. 8th ISIR, Stuttgart, West Germany

Van Brussel H, Belien H, Thielemans H (1985) Force sensing for advanced robot control. Proc. RoViSeC 5:59–68, Amsterdam, The Netherlands

Vranish JM, Mitchell EE, Demayer R (1982) Magnetoelastic force feedback sensors for robots and machine tools. Proc. SPIE Conf. on "Robotics and Industrial Inspection", Vol. 360:253–263, San Diego

Vranish JM (1984) Magnetoresistive skin for robots. Proc. "Robotics Research. The next five years and beyond" Paper MS84–506, Bethlehem, PA

Vranish JM (1986) Magnetoinductive skin for robots. Proc. 16th ISIR: 599–631, Bruxelles

Wang SSM, Will PM (1978) Sensors for computer controlled mechanical assembly. The Industrial Robot 5(1):9–18

Watson PC, Drake SH (1975) Pedestal and wrist force sensors for automatic assembly. Proc. 5th ISIR:283–291, Chicago

Whitney DE (1985) Historical perspective and state of the art in robot force control. Proc. IEEE Int. Conf. on Robotics and Automation: 262–268, St. Louis

Wolffenbuttel RF, Regtien PPL (1986) Integrated capacitive tactile imaging sensor. Proc. 16th ISIR: 633–641, Bruxelles

Wong K, Van der Spiegel J (1985) A shielded piezoresistive tactile sensor array. Proc. IEEE Int. Conf. on Solid–State Sensors and Actuators:26–29, Philadelphia

Wu C (1985) Compliance control of a robot manipulator based on joint torque servo. Int. J. Robotics Res. 4(3):55–71

Zhang H, Paul RP (1984) Hybrid control of robot manipulators. Report TR–EE 84–27, Purdue University, West Lafayette

Section 3

PHYSICS OF TRANSDUCTION AND TRANSDUCING TECHNIQUES

ANALOGS OF BIOLOGICAL TISSUES
FOR MECHANOELECTRICAL TRANSDUCTION:
TACTILE SENSORS AND MUSCLE-LIKE ACTUATORS

Danilo De Rossi, Claudio Domenici and Piero Chiarelli

Centro "E. Piaggio", Università di Pisa
Via Diotisalvi 2, 56100 Pisa, Italia
and
Istituto di Fisiologia Clinica del C.N.R.
Via Savi 8, 56100 Pisa, Italia

ABSTRACT

In this article the authors report their current attempts toward the development of new "skin-like" tactile sensors and "muscle-like" linear actuators potentially useful in the design of dexterous end effectors. The underlying design philosophy resides on mimicking electromechanical conversion properties of biological tissue making use of synthetic piezoelectric polymers or polyelectrolyte gels. A brief introduction is also given to the physical mechanisms which govern mechanical to electrical transduction in polymeric systems. It is a belief of the authors that substantial progress in the development of sophisticated tactile sensors and artificial muscles can be obtained by resorting to a "molecular bionics" approach.

1. INTRODUCTION

The development of dexterous robotic end effectors, capable of operating in an unstructured environment, and prosthetic systems intended to help in restoring lost manipulative functions in humans, both call for major breakthroughs in touch sensing and soft actuator technologies.

In recent times, increasing attention has been given to new research approaches devoted to understand and possibly mimick biological transduction and information processing mechanisms.

On the time course of evolution nature has developed unmatchable skillness in solving complex tasks, as those related to the interaction of living structures with the environment. Despite the enormous complexity inherent in closely mimicking biological structures and functions, a biomorphic approach to automated manipulation can offer suggestions for improved

NATO ASI Series, Vol. F43
Sensors and Sensory Systems
for Advanced Robots
Edited by P. Dario
© Springer-Verlag Berlin Heidelberg 1988

electromagnetic, weak and strong forces is generating a lot of excitement in the world of theoreticians. Some of these considerations can be used in the sensor field. In the first section of this paper the First and Second Law of Informatics will be introduced. Subsequently, in the second section the various kinds of measurands are related to six energy or signal domains. In the final section the main physical and chemical effects that can be used for the conversion of non-electrical signals and their place and time derivatives into electrical ones are enumerated. Moreover a three-dimensional diagram, the "Sensor Effect Cube", in which all these effects can be grouped in an elegant way, is presented.

First and Second Law of Informatics.

Information processing systems, such as measurement and control systems always consist of three units. In the first unit, labeled "input transducer", or more popularly, "sensor", the measurand is converted into a signal whose form is different from that of the measurement.

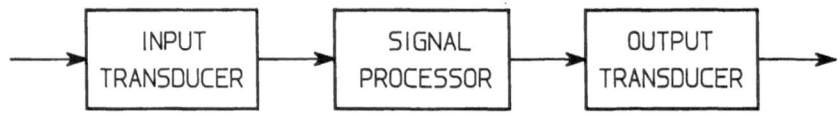

Fig. 1. Functional block diagram of information processing
        systems.

Because of the availability of inexpensive micro-electronic components, conversion into an electrical signal is the most obvious choice. The second unit is labeled "signal processor". In this unit the electrical signal is amplified or converted from an analog into a digital signal. The last unit is labeled "output transducer". In this device the electrical signal is again converted into a more usable form. We distinguish display, actuating, storage and transmitting devices. In a display the electrical signal is converted into a signal form which can be perceived by one of our senses.

A variety of meanings can be attached to the word "information". According to the I.E.E.E. Standard Dictionary of Electrical and Electronics Terms information is:

   "The meaning assigned to data by known conventions".

Though, this definition is rather perspicuous it is not very helpful in classifying transduction mechanisms and techniques. Bearing in mind that for information to be acquired it must be transported, we are more interested in knowing how information is carried from one place to another.

Therefore, we present a central hypothesis on the transport of information, implicitly already introduced by Wiener (2) and Brillouin (3), which states that information is something that must be carried by mass or energy. Strictly formulated, this hypothesis, which we call the First Law of Informatics analogous to the First Law of Thermodynamics, states:

"No information can be carried from a source (object of measurement) to a receiver (sensor) without mass or energy transport between the source and the receiver".

There are some interesting consequences to our hypothesis. One of the first is that, if telepathy exists, there has to be an unknown energy or mass transport between telepathic persons. If we assume that such unknown mass or energy does not exist, then telepathy must be impossible. A second consequence appears when we combine our hypothesis with the Second Law of Thermodynamics, which states that within a closed system in which energy is transported the total amount of entropy has to increase or remain constant. This leads to the Second Law of Informatics, in accordance with the Second Law of Thermodynamics, which states that:

"Whenever information is transported the total entropy of the system is increasing or remains constant".

Both Laws of Informatics indicate that energy and mass play a crucial role in information transport. Therefore, classification of transduction mechanisms and techniques should also be based on the classification of the different energies and possibly mass. This will be followed through in the next section.

Energy or signal domains

In modern information-processing systems mass as a carrier no longer plays an important role. Today energy, and especially electrical energy, is the dominant carrier. This has much to do with the rapid advances in the field of micro-electronics. From a physical point of view, we can roughly distinguish the following forms of energy:

1. RADIANT energy (or electromagnetic energy) is related to radio waves, microwaves, infrared, visible, and ultraviolet light, X-rays and gamma-rays;

2. GRAVITATIONAL energy concerns the gravitational attraction between masses;

3. MECHANICAL energy pertains to mechanical forces, pressure, displacements, etc.;

4. THERMAL energy has to do with the kinetic energy of molecules of which the temperature is a measure;

5. ELECTRICAL energy deals with electrical fields, electrical currents and magnetic fields;

6. CHEMICAL energy has to do with the electromagnetic forces with which the atoms are held together in a molecule;

7. NUCLEAR energy concerns the binding energies which keep the nuclei together;

8. MASS energy which is described by the well-known Einstein relation;

For instrumentation purposes it is not necessary to consider all eight of these energy forms. It is evident that measurement systems based on the use of nuclear or mass energies will not attract many potential users. The grouping of energies and signals elaborated by Lion (4) is commonly used. He distinguishes six groups, which are shown in Table 1 with some ways these energies can be employed to carry information.

Table 1. Six groups of signals with examples.

RADIANT signals	intensity, wavelength, polarization plane and phase
MECHANICAL signals	force, pressure, torque, flow, volume tilt, acoustic wavelength, amplitude and phase
THERMAL signals	temperature, temperature differences and temperature changes
ELECTRICAL signals	voltage, current, charge, resistance, inductance, capacitance, pulse duration, frequency and dielectric constant
MAGNETIC signals	field intensity and direction, magnetization and permeability
CHEMICAL signals	composition, toxicity, oxidation- reduction potential and pH

Each type of signal can be applied to the input of a transducer or may appear at the output. When the input and output signals of one transducer have the same form, we use the term signal processor. A gearbox is a mechanical, a heat exchanger a thermal signal processor. Thanks to the silicon planar technology, we have at our disposal a huge number of very sophisticated components such as microprocessors, memories,

A/D converters and operational amplifiers. Because the price/performance ratio of these components decreased dramatically, in the long run all signal processors in instrumentation systems will certainly be replaced by electronic signal processors.

When the input and output signals of one transducer do not have the same form, we use the term transducer. The word "transducer" is derived from the latin verb "traduco" and usually means a device that transfers energy from one system to another thus in accordance with the Second Law of Informatics. In the input transducer a non-electrical signal will most often be converted into an electrical signal. This signal then can be processed by the electrical signal processor. At the output transducer the electrical signal is again converted into a non-electrical signal. In case of a display the output signal has such a form that it can be perceived by one of our senses. The term display is most often used, when vision is involved. However, it is not unreasonable to speak of an acoustical display, in case hearing is involved or to speak of a chemical display, in case smell or taste is meant. In generalized form, a display is defined as an output transducer that converts electrical information in such a way, that one of the human senses can detect this information. Output transducers are also used to cause action. When the action is of a mechanical nature, the term "actuator" is often used. However, when an electrical signal is converted into one of the other domains, e.g., a solid-state laser generating radiant energy, no general terms for these devices are available. Therefore, it may be useful to apply the label "actuators" to all output transducers initiating some radiant, mechanical, thermal, magnetic or chemical action.

Several diagrams can be used to indicate the many possibilities that exist to convert and process information. For instance in a 6 x 6 matrix all transducers and signal processors can be placed in one of 36 the elements. Figure 2 shows another helpful diagram indicating the five signal conversions that are possible in input transducers. Each sphere represents a signal domain and in each transducer, indicated by an arrow, conversion occurs in the direction of the electrical signal domain. By employing the Seebeck effect, for instance, a transducer can be constructed which converts a temperature difference into an electrical signal. The Seebeck effect is then represented by the arrow between the thermal and electrical sphere. The piezoelectric effect is on the arrow between the mechanical and electrical domain, because with the piezoelectric effect a mechanical signal can be converted in an electrical signal and vice versa.

Many transducers can be based on the use of solitary effects like the Seebeck or piezoelectric effect. However, in studying the transducer field one also rather frequently finds sensors in which not one, but two or more solitary effects are used. For instance, in a certain flow transducer the mechanical measurand: flow, is at first converted into a temperature difference which is in turn converted into an electrical signal. In the diagram of Fig. 2 these conversions are indicated by the dashed arrows. In such a sensor two effects are apparently used in tandem, so that the term "tandem transducer" is appropriate.

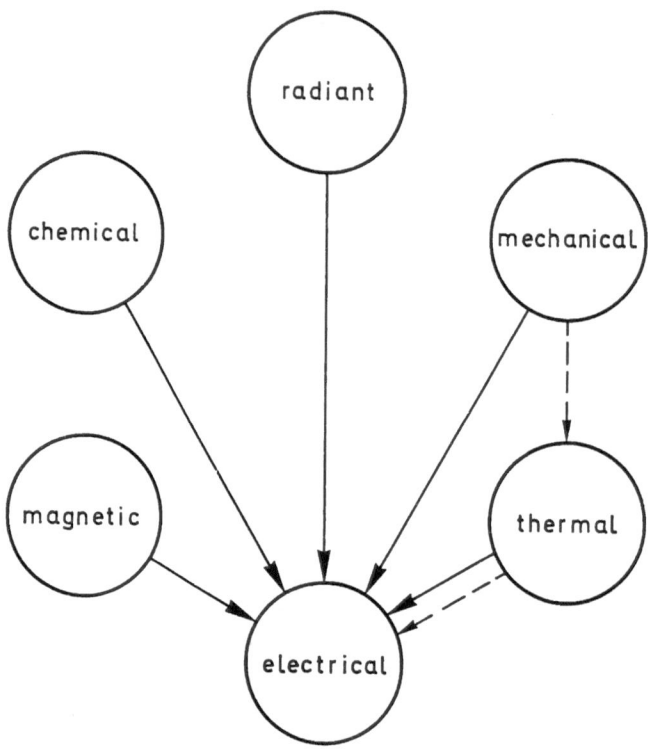

Fig. 2. Diagram indicating the five possible signal conversions in input transducers. A tandem transducer is indicated by dashed lines.

A very large number of physical and chemical effects can be used for signal conversion in transducers. In the next section the most important effects will be presented.

Physical and chemical effects for transducers.

From the moment that physics and chemistry started to attract the attention of scientists, they were intrigued by the many existing relations between the different physical quantities they could measure with their primitive measurement devices.

For instance, in 1821 Seebeck discovered the thermo-electric

effect, more commonly called the Seebeck effect. This effect ,
describes the phenomenon in which in a closed electrical
circuit composed of two dissimilar metals an electric current
flows when the junctions are kept at different temperatures.
Seebeck erroneously believed that the deviation of a small
magnet close to the circuit was caused by the forming of
north-poles at the cold end and south-poles at the warm
junction. Later on, scientists were able to give a good
phenomenological description of the effect and to list all
possible combinations of metals with the appropriate Seebeck
voltages. At present also semiconductors are studied and it
appears that these materials have very large Seebeck
coefficients and, therefore, can be used in many transducer
applications.

Not until quantum and statistical physics became common
knowledge did a correct explanation of the Seebeck effect
became available. Because today the interest in sensors or
transducers for above mentioned reasons is very strong, also
the study of suitable physical and chemical effects and their
suitability for the construction of sensors is intensified all
over the world (5,6).

Because of the advances in the micro-electronics field, signal
processors will always be based on modern electronics requiring
input transducers that convert from a non-electrical domain to
the electrical domain and output transducers that convert from
the electrical to a non-electrical domain. In principle only
effects that allow these conversions are of interest. The most
important effects for both conversions are here presented.

RADIANT to ELECTRICAL

> Photovoltaic effect, photoelectric effect,
> photoconductivity, photodielectric effect,
> photomagnetoelectric effect, photoelectric
> emission, thermal photoelectric effect,
> Becquerel effect.

MECHANICAL to ELECTRICAL

> Piezoelectric effect, piezoresistance, lateral
> photovoltaic effect, acoustoelectric effect,
> thermoelastic effect, Dember effect, Matteuci
> effect, triboelectrification, Elster and Geitel
> effect.

THERMAL to ELECTRICAL

> Thermoelectric effect, thermoresistive effect,
> thermodielectric effect, Nernst effect, thermionic
> emission, phonon drag, thermal photoelectric
> effect, thermoelastic effect, superconductivity

MAGNETIC to ELECTRICAL

> Faraday-Henry law, Hall effect, magnetoresistance,

Shubnikov de Haas effect, Matteuci effect, Suhl
effect, superconductivity, photomagnetoelectric
effect, Nernst effect.

CHEMICAL to ELECTRICAL

Volta effect, galvanoelectric effect, Donnan
potential, chemical electric effect, chemical
dielectric effect, concentration dependence of
conductivity and potential, secondary emission,
plasma resonance.

For output transducers the following effects are of interest:

ELECTRICAL to RADIANT

Electroluminescence, P-N junction luminescence,
Franz Keldysh effect, Kerr electro-optic effect,
Pockel's effect.

ELECTRICAL to MECHANICAL

Electrostriction, piezo-electric effect, Quincke
effect, Johnsen Rahbeck effect

ELECTRICAL to THERMAL

Electrothermal effect, Peltier effect, Ettinghausen
effect, Thomson effect

ELECTRICAL to MAGNETIC

Ampere's law, electromagnetic effect

ELECTRICAL to CHEMICAL

Electrochemical effect

As already indicated useful sensors can also be constructed
employing two transducers in-tandem. In such a case one of the
transducers will have an electrical input or output, whereas
the other half of the sensor will consist of an element that
converts a non-electrical signal into another non-electrical
signal belonging to a different signal domain. It is evident
that the possibility of tandem-transducers leads to many more
sensors than one would expect allowing only single signal
conversions. The effects which can be used in such tandem
transducers are enumerated elsewhere (7).

The "Sensor Effect Cube"

Another way of classifying physical and chemical effects is to

divide them into self-generating and modulating effects. In
self-generating effects the energy of the output signal is
directly converted from the energy of the input signal. In
modulating effects the main energy source of the output signal
is not the input signal, but an auxiliary energy source. The
energy of this source can belong to any of the signal domains.
This division has some practical consequences. When one wants
to measure a very weak signal, one's most obvious choice is a
modulating effect. On the other hand, if the signal is strong
and the use of a battery poses serious problems or if a sensor
without offset is required a self-generating effect is
preferable.

This division offers us a practical way to classify all
physical effects. This classification describes a physical
effect in terms of three parameters, indicating the domains of
the input signal, the output signal and the auxiliary energy
source. The input signal is the quantity to be measured, while
the output signal is the signal which is processed further.

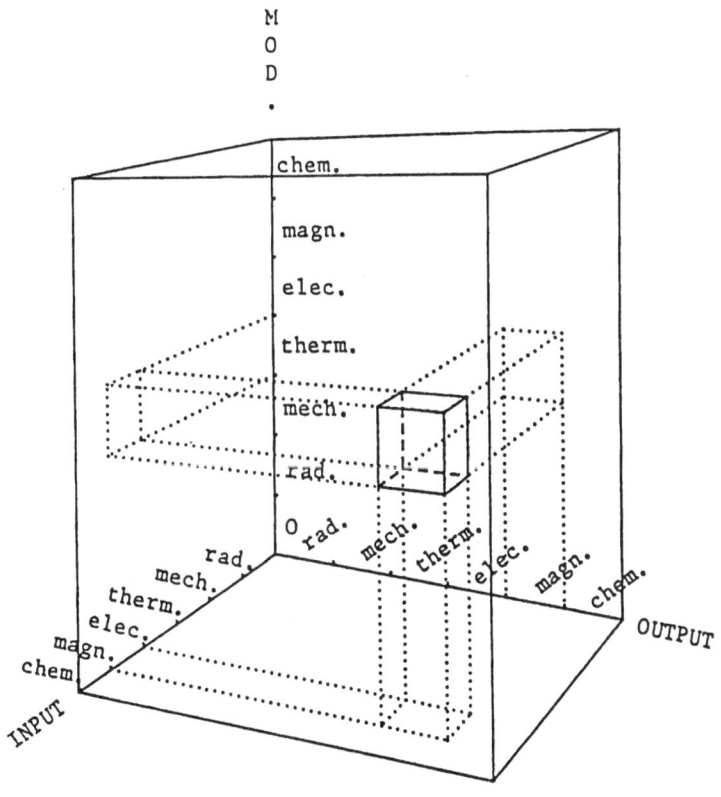

Fig.3. The "Sensor Effect Cube", a three-dimensional diagram
       containing all physical and chemical effects.

This classification can be represented in a three-dimensional
diagram in which each of the axes is a parameter and every
effect can be placed in at least one point of the diagram,
depending on its parameters (8).
The self-generating effects needing no auxiliary energy source
are placed in the base plane. This plane is denoted by the zero
along the Z-axis. In case of modulating transducers the vectors
along the X- and Y-axes indicate the form of the input and the
output energy, respectively, while the vector along the Z-axis
indicates the form of the input signal. We have found that
every effect described in the literature can be placed in at
least one cube. It appears that the "Sensor Effect Cube" is the
complete set of effects physically possible. Most cubes will
contain more than one effect. In a way, the cube is of more
interest to the research scientist because of the empty spaces
than the filled ones. For example the place representing a
single effect that allows the transduction of the time
deritative of the temperature into an electrical signal is
glaringly empty.

Conclusions

The ever-growing demand on the part of the industry for more
and better sensors offers a continuing challenge to scientists
working in this field. In recent years we have indeed seen many
and rapid innovations and the end is not in sight by far.
However, in spite of the many practical applications that have
appeared, the fundamentals of the field have still not been
thoroughly investigated. In the paper is shown, that based on
more practical than fundamental reasoning it is possible to
represent all transduction mechanisms and techniques in a three
dimensional diagram the so-called: "Sensor Effect Cube".

References

1. S. Middelhoek and A.C. Hoogerwerf, Smart sensors: when
   and where?, Sensors and Actuators, 8 (1985) 39-48.
2. N. Wiener, Cybernetics, The M.I.T. Press and John Wiley
   & Sons, New York, 2nd edn., 1961.
3. L. Brillouin, Science and information theory, Academic
   Press Inc., New York, 1956.
4. K.S. Lion, Transducers: Problems and prospects, IEEE Trans
   Industr. Electron. Contr. Instrum., IECI-16 (1969) 2-5.
5. D.W.G. Ballantyne and D.R. Lovett, A dictionary of named
   effects and laws in chemistry, physics and mathematics,
   Chapman and Hall, London, 1972.
6. J. Schubert, Physikalische Effekte, Anwendungen,
   Beschreibungen, Tabellen, Physik Verlag, Weinheim, 1982.
7. S. Middelhoek and D.J.W. Noorlag, Signal conversion in
   solid-state transducers, Sensors and Actuators, 2 (1982)
   211-228.

8. S. Middelhoek and D.J.W. Noorlag, Three-dimensional
   representation of input and output transducers, Sensors
   and Actuators, 2 (1981/82) 29-41.

ANALOGS OF BIOLOGICAL TISSUES
FOR MECHANOELECTRICAL TRANSDUCTION:
TACTILE SENSORS AND MUSCLE-LIKE ACTUATORS

Danilo De Rossi, Claudio Domenici and Piero Chiarelli

Centro "E. Piaggio", Università di Pisa
Via Diotisalvi 2, 56100 Pisa, Italia
and
Istituto di Fisiologia Clinica del C.N.R.
Via Savi 8, 56100 Pisa, Italia

ABSTRACT

In this article the authors report their current attempts toward the development of new "skin-like" tactile sensors and "muscle-like" linear actuators potentially useful in the design of dexterous end effectors. The underlying design philosophy resides on mimicking electromechanical conversion properties of biological tissue making use of synthetic piezoelectric polymers or polyelectrolyte gels. A brief introduction is also given to the physical mechanisms which govern mechanical to electrical transduction in polymeric systems. It is a belief of the authors that substantial progress in the development of sophisticated tactile sensors and artificial muscles can be obtained by resorting to a "molecular bionics" approach.

1. INTRODUCTION

The development of dexterous robotic end effectors, capable of operating in an unstructured environment, and prosthetic systems intended to help in restoring lost manipulative functions in humans, both call for major breakthroughs in touch sensing and soft actuator technologies.

In recent times, increasing attention has been given to new research approaches devoted to understand and possibly mimick biological transduction and information processing mechanisms.

On the time course of evolution nature has developed unmatchable skillness in solving complex tasks, as those related to the interaction of living structures with the environment. Despite the enormous complexity inherent in closely mimicking biological structures and functions, a biomorphic approach to automated manipulation can offer suggestions for improved

NATO ASI Series, Vol. F43
Sensors and Sensory Systems
for Advanced Robots
Edited by P. Dario
© Springer-Verlag Berlin Heidelberg 1988

sensors and actuators design and it could provide indications for implementing more efficient strategies of sensorimotor control.

In biological systems, basic to the implementation of sophisticated mechanical interactions with the environment, is the ability of specialized subcomponents to perform efficient electromechanochemical (EMC) energy conversion. Beside muscular contraction, the most thoroughly studied phenomenon, numerous other biological EMC processes exist related to cell motility, mechanoreception and connective tissue energy transduction. Differently from muscle mechanics and mechanoreception, only in the last few decades electromechanical (EM) and EMC phenomena in biological tissues and their synthetic analogs have become an area of active research (1,2).

The recent interest in EM transduction proper of synthetic polymers originates from its relevance in devising relatively simple models to fit more complex energy transduction mechanisms in biological tissues and because of the unique applications which have been developed based on these materials (3).

In this article, an introduction is given to the most relevant mechanisms which govern mechanical to electrical transduction in macromolecular systems by discussing their dipolar (piezoelectric) or ionic (streaming potential) origin; general concepts of chemical to mechanical transduction, through electric excitation, in charged polymer networks are also discussed to provide the basis for the illustration of synthetic analogs of biological contractile elements.

Two different tactile sensing structures are then presented, based on an analogy with EM conversion properties of human skin constituent biopolymers. It is also indicated how a stress components selective tactile sensor may prove to be useful in solving the most basic mechanistic problem of object recognition from tactual information, i.e. the inverse elastic field contact problem.

Finally, contractile mechanisms instrumental for animal motility are briefly introduced and the point is made that, since no principle in their functioning is involved beyond the most general precepts of macromolecular science, synthetic analogs of biological contractile machines can be devised (4). The mechanisms of electrically driven polymer contractility are then outlined and isometric force generation capabilities and isotonic force-velocity curves of contractile polymer elements are reported, in connection to their possible use as linear actuators.

2. POLYMER ELECTROMECHANOCHEMISTRY

Macromolecular structure confers to a substance a unique combination of properties which does not belong to other types of molecular organization.

More specifically, because molecular structure and macroscopic dimensions of a polymeric substance frequently may be interrelated, microscopic dimensional changes deriving from conformational variations of macromolecules induced by chemical or electrical perturbation can manifest themselves at a macroscopic level in some supermolecular polymeric systems of both natural and synthetic origin (5).

A comprehensive and systematic analysis of the mechanisms governing electrical to mechanical energy conversion in polymers is not available. However, guidance for undertaking this analysis can be obtained from the study of piezoelectric organic materials (6) and mechanochemical and electrochemical processes in charged polymer membranes (7).

One class of macromolecules that is particularly important, both because of the variety of physicochemical phenomena which interplay in its structural and conformational organization and because of its widespread occurrence in biological systems, is the class of polyelectrolytes consisting of macromolecules containing ionizable groups chemically attached in a repetitive manner to the chain backbone (8).

Polyelectrolytes are fundamental components of biological systems where, for instance, gels are formed by water and proteins, water and nucleic acids or by water and polyelectrolytes of different nature leading to the formation of more complex structures such as chromosomes, cytoplasm, muscles, etc.

Fibrous proteins and polypeptides, which are synthetic analogs of proteins, show the ability to perform electromechanical signal transduction through various mechanisms both of dipolar (originating piezoelectric behavior) and of ionic nature (leading to streaming potentials and electromechanochemical behavior) which originate from their elemental peptidic bond and their ionizable aminoacid content. The basic concepts of polymer piezoelectricity, streaming potentials and electrically induced contractility are discussed in the following three subsections.

2.1 Polymer piezoelectricity

A material is said to have piezoelectric properties when it displays the ability to convert mechanical energy into electrical energy, or vice-versa. In physical terms, piezoelectricity has been defined as "the electrical polarization produced by a mechanical strain in crystals belonging to certain classes, polarization being proportional to strain and changing sign with it" (9). The inverse is also true, i.e., a change in electrical polarization (measured as the moment of dipole per unit volume) induces a proportional mechanical strain in piezoelectric materials.

A phenomenological model of piezoelectricity, as of pyroelectricity, is generally derived using concepts of equilibrium thermodynamics (9). A free energy function is developed to describe the elastic, electrical, and thermal variables of a crystalline material. The extensive variable electric displacement is written in the form of an exact differential as function of the intensive variables, elastic stress (σ), electric field (E), and temperature (T). Maxwell relations are derived between the various differential coefficients to give the following equations (Einstein summation convention is used):

$$ dD_i = d^T_{ijk} \, d\sigma_{jk} + \varepsilon^{\sigma,T}_{ij} dE_j + p^\sigma_i \, dT \qquad [1] $$

d_{ijk} , ε_{ij} and p_i respectively indicate the piezoelectric strain constants, the permittivities and the pyroelectric coefficients. Since it links a 2nd rank tensor to a vector, d_{ijk} is a third rank tensor and, as it holds for σ_{jk} , d_{ijk} is a symmetric tensor and it has 18 components. It is so possible to use the more concise matrix notation (10) where the piezoelectric coefficients are expressed as d_{ij} (i=1,2,3; j=1,2,......6).

Three major mechanisms have been identified (11) to be responsible for piezoelectric properties in polymers:

i.   piezoelectricity caused by strain at a molecular level;
ii. piezoelectricty caused by macroscopic strain of the material;
iii. piezoelectricity which originates from spatial nonuniformity of material constants or trapped space charge.

Only the first two mechanisms are piezoelectric in a narrow sense, since they imply the existence of molecular structures which are organized and ordered at a crystalline or paracrystalline level, while the third mechanisms can also be observed in amorphous materials. We limit our discussion here to the first two classes.

In a semicrystalline piezoelectric polymer some of the elements of the piezoelectric matrix can be null, depending on the orientation and dipolar structure of its macromolecules and consequently its symmetry.

Piezoelectricity in polymers is usually observed in oriented films and four different piezoelectric matrices have been postulated, and their structure has been experimentally verified (12). The four symmetry classes, their piezoelectric matrices and the most representative materials of each class are reported in Table 1.

The structural symmetry of uniaxially oriented biopolymers can be approximated by D_∞ or C_∞, depending respectively on the absence or presence of polarity in the orientation axis (10). It is noteworthy to mention that in complex and etherogeneous structures such as biological tissues, the reported matrix forms may not fully describe the piezoelectric behavior and hence, in general, all the coefficients have a finite value.

CLASS	PIEZOELECTRIC COEFFICIENT MATRIX						REPRESENTATIVE MATERIALS
$D_\infty (\infty 2)$	0	0	0	d_{14}	0	0	Uniaxially oriented cellulose films, uniaxially oriented synthetic polypeptides
	0	0	0	0	$-d_{14}$	0	
	0	0	0	0	0	0	
$C_\infty (\infty)$	0	0	0	d_{14}	d_{15}	0	Bone, tendon, wool, keratin, DNA
	0	0	0	d_{15}	$-d_{14}$	0	
	d_{31}	d_{31}	d_{33}	0	0	0	
$C_{\infty v} (\infty m)$	0	0	0	0	d_{15}	0	Biaxially oriented polyvinylidene fluoride, ceramic-polymer composites
	0	0	0	d_{15}	0	0	
	d_{31}	d_{31}	d_{33}	0	0	0	
$C_{2v} (2mm)$	0	0	0	0	d_{15}	0	Uniaxially oriented polyvinylidene fluoride
	0	0	0	d_{24}	0	0	
	d_{31}	d_{32}	d_{33}	0	0	0	

T A B L E   1

Synthetic piezoelectric polymers presenting a particular scientific or practical interest are uniaxially oriented synthetic polypeptides, such as poly-γ-benzyl-glutamate (PBG), which shows a $D_\infty (\infty 2)$ symmetry and polyvinylidene fluoride (PVDF) and some of its derivatives which, after drawing and poling, have a $C_{2v}(2mm)$ symmetry.

2.2 Streaming potentials in ionized macromolecular aggregates

An electric potential difference can be measured across a slab of a water-permeated ionized polymer network upon application of a pressure gradient generating a flow of liquid across the sample. The appropriate discipline for the analysis of these phenomena is the thermodynamics of irreversible processes (7) which establishes, in its classical form, linear coupled relationships between flows and forces operating into the system.

For the simple case of, one-dimensional flow, pressure and potential gradients within the polymer sample, when ionic concentration gradients are negligible, the coupled relations are written in the form:

$$\begin{pmatrix} U \\ J \end{pmatrix} = \begin{pmatrix} -L_{11} & L_{12} \\ L_{21} & -L_{22} \end{pmatrix} \begin{pmatrix} \Delta P \\ \Delta V \end{pmatrix} \qquad [2]$$

where U is the average fluid velocity relative to the solid matrix, J is the current density, ΔP is the pressure drop across the sample and ΔV is the electrical potential difference across the sample. The phenomenological Onsager coefficients L_{ij} depend on the electrical, mechanical and chemical properties of the macromolecular matrix and of the interpenetrating liquid.

Equations [2] provide a phenomenological description of a set of electrokinetic coupled phenomena which include streaming current ($-L_{21}/L_{11}$) and streaming potential ($-L_{21}/L_{22}$) (7). The streaming potential mechanism can be described using a continuum approach (13) by referring to Figure 1.

The ionized macromolecular network holds fixed negatively charged groups and the interpenetrating liquid contains mobile positive counterions in the proper amount to maintain electroneutrality. When a solid reference electrode compresses the sample against a porous counterelectrode (tipically Ag/AgCl reversible electrodes are used in the experiments) a fluid flow is generated. Fluid flow entrains mobile counterions, producing a spatial disomogeneity of them, leading to incomplete neutralization of the fixed charges on the solid matrix. This charge separation is the origin of a streaming potential field.

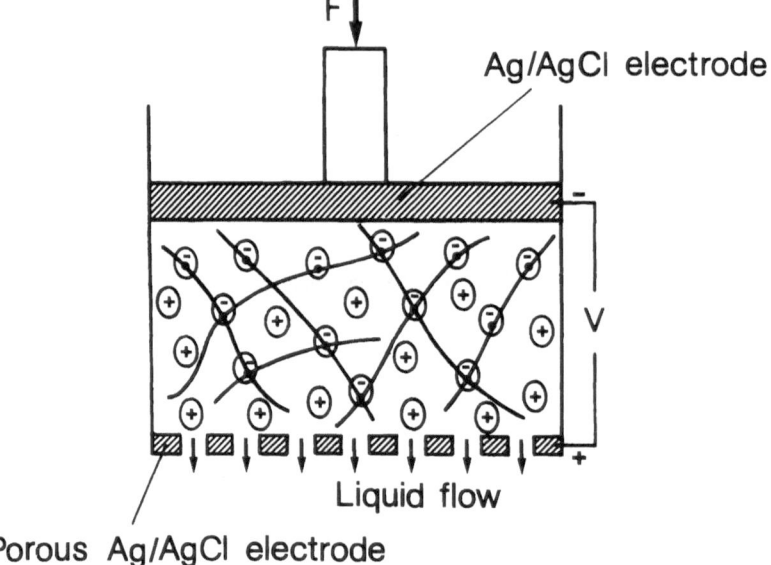

FIGURE 1. Schematic representation of the generation of streaming potentials. The sample, made of a negatively charged polymer network permeated by a liquid containing the positive counterions, is compressed between an impermeable and a porous electrode. The generated liquid flow creates a charge displacement by transporting a tiny amount of mobile positive charges. This charge separation is the origin of the streaming potential field.

Stress generated potentials, whose origin has been clearly ascribed to streaming potentials effects, have been observed in various forms of connective tissue (13), such as bone, tendon, cartilage and dermis and they appear to be a general property of the extracellular tissue matrix having a possible function in tissue growth, remodeling and repair (14).

2.3 Electrically induced contractility in polyelectrolyte gels

The first observations of large, reversible dimensional changes in crosslinked polyelectrolyte gels by chemically induced variations in the ionization state of the macromolecules date back to the early fifties (15). Mechanochemistry, the discipline devoted to the study of systems capable of converting chemical energy into mechanical work, has provided an adequate understanding of the mechanisms of contractile action induced in polyelectrolyte networks by thermal or mechanical stimuli (16).

Mechanochemical conversion phenomena can occur in both amorphous and semicrystalline polymers, although the physicochemical mechanisms which govern their behaviour may be different for these two class of materials. While crosslinked, water swollen, amorphous gels are typically isotropic on a macroscopic scale and their dimensional variations are mostly determined by osmotic forces which tend to reestablish a perturbed equilibrium through water transport, semicrystalline polymers are often highly anisotropic and mechanochemical conversion within them is typically associated with chemically mediated, crystalline-to-amorphous phase transitions (chemical melting) (17).

More recent discoveries (18) have shown that volume phase transitions in crosslinked polyelectrolyte gels can also be induced by electric fields. However, differently from mechanochemistry, little attention has been devoted to the study of the mechanisms governing electrical to mechanical energy conversion in polyelectrolyte gels and related experimental observations are scarce (19).

Different physical mechanisms can be invoked, in principle, to induce reversible volume changes in ionizable polymer gels with electric fields. If the effects of electric field on atomic and electronic polarizations are neglected, the primary electric effects on polyelectrolytes can possibly be grouped into five general classes: a) orientation of dipolar species, b) deformation of polarizable species and subsequent orientation of induced dipoles, c) influence of external fields on the dissociation constants of weak acids and bases and promotion of the separation of ion pairs into corresponding ions (second Wien effect), d) redistribution or motion of ionic species with subsequent changes in the network electrostatic free energy or leading to electrokinetic coupled phenomena, e) electrochemical reactions at the interfaces.

These primary electrical or electrochemical events may then be coupled to various conformational or structural rearrangements at a molecular level with can lead to macroscopic mechanical events when the material possesses a suitable degree of intermolecular bridges (covalent crosslinks or physical bonds).

3. MIMICKING SKIN TISSUES ELECTROMECHANICS

New artificial tactile sensing possibilities are opening up as more informations on human skin rheology and sensory modalities become available.

For a long time, neurophysiologists have investigated the role of skin receptors in the generation of electric signals, which are then transmitted to the central nervous system for further processing. Most investigators agree that even though some skin receptors can detect static or slowly varying parameters like pressure and temperature, most human tactile receptors can only sense rapidly varying parameters. Furthermore, there is evidence that static signals are virtually ignored by the brain. Phenomenologically, hence, the human skin behaves like a piezoelectric material.

Indeed, stress generated potentials (SGP) have been observed in human skin, both in dead dry preparations and in the living state (20, 21). It has also been suggested that piezo- and pyroelectric properties of human epidermis may play a fundamental transduction role for skin sense of touch and warmth (21). A detailed study performed on dead human skin (22, 23) either in the dry and moist state has shown that both dermis and epidermis act as mechanoelectrical energy converters. SGP in the dry state have been found to be of dipolar nature through piezoelectric effects, while in the moist state the contribution of electrokinetic phenomena (streaming potentials) to SGP appears to be preponderant.

Disregarding the actual origin of the physiological transduction mechanisms of the skin and the soundness of the hypothesis affirming that epidermal SGP can be perceived through the intraepidermal and the superficial dermal nervous network, it has been proved that mimicking the dipolar structure of skin can provide useful indications for the design of tactile sensors (24).

Here, two different approaches are reported to the possible development of tactile sensor materials and structures which mimic moist and dry skin mechanoelectrical conversion phenomena. A prototype tactile sensor which replicates the EM transduction phenomena operating in moist skin has been realized and it is schematically shown in Figure 2.

An electromechanical model of the extracellular matrix is simply made of a highly swollen, negatively charged rubber-like gel which is composed of a thermally crosslinked mixture of polyacrylic acid (20% by weight) and polyvinylalcohol (80% by weight).

The streaming potential mechanism operates in a slab of such a material when pressed in the configuration shown in Figure 2 and contact forces can be transduced in electric signals as shown in Figure 3.

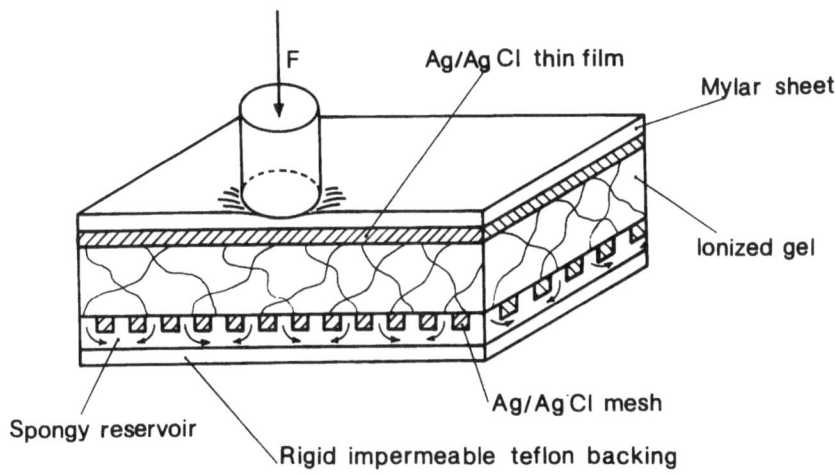

FIGURE 2. Skin analog implementing the streaming potential mechanoelectrical conversion.

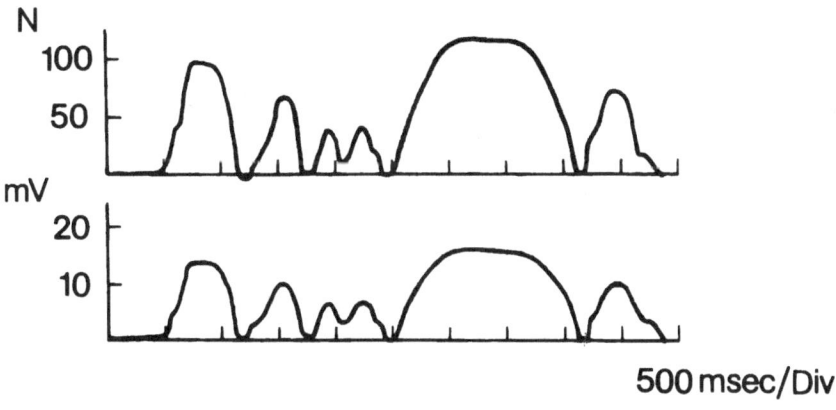

FIGURE 3. Simultaneous recording of the electric output (lower trace) of the skin analog sensor depicted in Figure 2 and the electric output (upper trace) of a Kistler piezoelectric cell mechanically in series to the sensor. The force was applied by a manually operated pusher plate.

Since the research on this aspect of material science for tactile sensing is in its very early stage, it is premature to claim its usefulness in tactile sensor technology; however, the easy conformability to complex shapes, softness, low cost and relatively high transduction sensitivity of this material make it a good candidate for skin-like touch sensing structures.

As already mentioned, very sophisticated tactile sensors can be devised by mimicking the piezoelectric properties of dry human skin.

From a structural model of human skin which is schematically drawn in Figure 4, it appears that the organization and orientation of fibrous proteins (keratin-like α-helical tonofibrils in the epidermis basal cell layer and collagen in the dermis) would confer to the skin portion around the epidermal-dermal junction (basal lamina) a very peculiar piezoelectric texture (22).

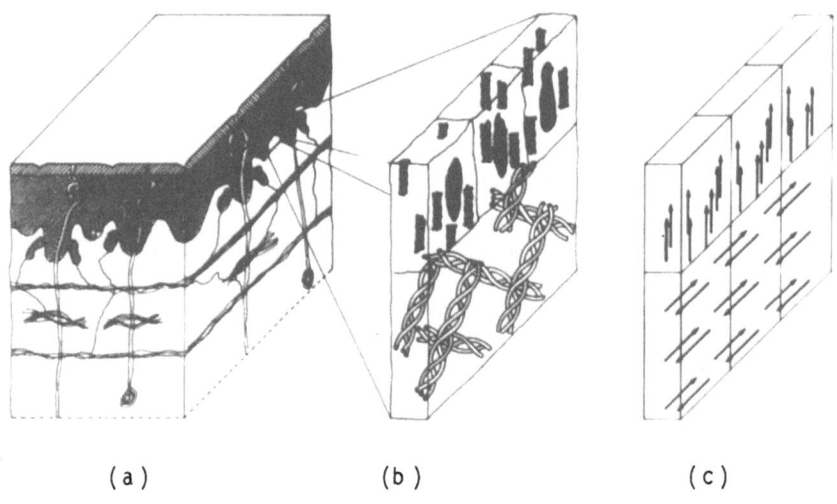

(a)   (b)   (c)

FIGURE 4. Structural model of the organization of human skin constituent biopolymers near the basal lamina. 4a): A cross section of human skin is depicted and the separation between epidermis and dermis (basal lamina) is indicated. 4b): The almost uniaxial orientation of keratin-like tonofibrils in the epidermis basal cell layer is indicated, perpendicular to the basal lamina; in dermis the collagen fibres are almost randomly arranged with a slight in-plane preferential orientation. 4c): The dipolar nature of skin is indicated by an uniaxial, exagonal symmetry structure in the epidermis and a planar, antipolar orientation in dermis.

The uniaxially oriented tonofibrils in the epidermal basal cell layer call for a C_{∞} (∞) symmetry class, while the collagen dermal networks has been shown to support only shear piezoelectricity as dictated by a D_∞ ($\infty 2$) symmetry class (see Table 1).

At the site of the basal lamina, hence, the piezoelectric coefficient matrix can be assumed to be the combination of the dermal and basal layer coefficient matrices:

$$d \;=\; \begin{pmatrix} 0 & 0 & 0 & d_{14}^{e} & d_{15}^{e} & 0 \\ 0 & 0 & 0 & d_{24}^{e} & (d_{25}^{e} + d_{25}^{d}) & 0 \\ d_{31}^{e} & d_{32}^{e} & d_{33}^{e} & 0 & 0 & d_{36}^{d} \end{pmatrix} \qquad [3]$$

where the coefficients all refer to a right-handed cartesian coordinate system having the z axis perpendicular to the epidermal-dermal junction and the superscripts e and d respectively indicate the epidermal and dermal components. This particular structure of the piezoelectric coefficient matrix confers to the material the intrinsic ability to electrically respond to the elastic stress field in all its six tensor components.

The capability of resolving the stress field in an elastic medium in its tensor components can be reproduced by means of properly arranged multielement piezoelectric polymer sensors.

The development of a compliant, stress-component selective sensor in an array is instrumental for approaching the most basic mechanistic problem in tactile sensing. The problem consists in an elastic field inverse contact problem, where the force distribution acting on a boundary of the sensor should be inferred from the spatially discrete knowledge of the elastic stress field over a surface inside the sensor. This problem, however, belongs to the class of ill-posed problems (25) where the existence, uniqueness and stability of the solution is not guaranteed and only through suitable constraints, formulated on the basis of physical considerations or complementary information (visual, kinestctic, proprioceptive, cognitive), solutions of the shape-from-touch problem may eventually be found.

The structure of the proposed multielement, multicomponent stress sensor is depicted in Figure 5.

Under two sets of assumptions, a transfer function (relating the voltage output of each sensor to the stress components) has been formulated and transducer dimensioning (interelement distance and depth of the elements into the embedding medium) has been accomplished (De Rossi D. et al., in preparation).

The following set of assumptions permits to calculate the piezoelectric response (transfer function) of the sensor:

I - the stress field is considered uniform over the interelement distance;

II - all the elements have the same thickness;

III - the electric measurements are performed by means of an ideal charge amplifier, measuring the charge which flows in an external circuit to neutralize the stress-induced charge appearing on the sensor electrodes.

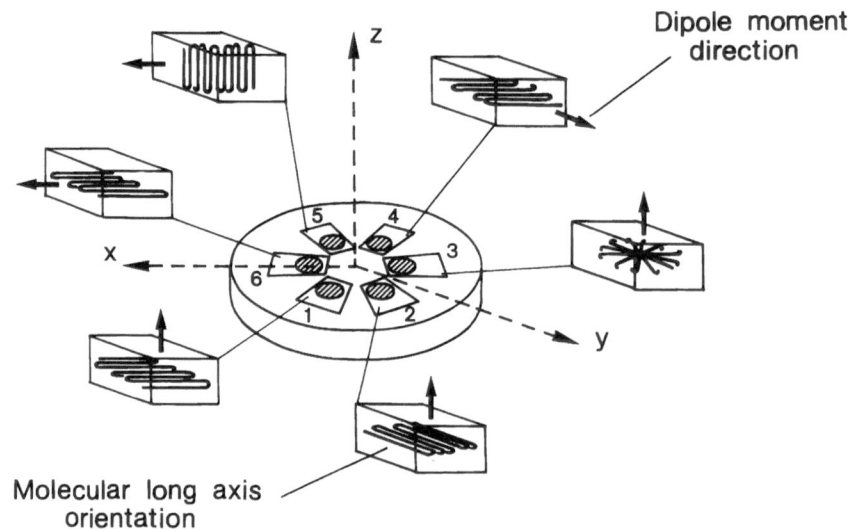

FIGURE 5. The proposed multielement, stress-component selective tactile sensor is embedded in a compliant, elastic medium. The six piezoelectric polymer elements, which have electrodes on their main surfaces, are made respectively of: 1) uniaxially oriented PVDF; 2) uniaxially oriented PVDF (the draw direction is rotated 90° in the plane in respect to element 1); 3) biaxially oriented PVDF; 4) PVDF sample obtained by cutting a thick uniaxially oriented sheet along it thickness, parallel to the draw direction; 5) PVDF sample obtained by cutting a thick uniaxially oriented sheet along it thickness, perpendicular to the draw direction; 6) uniaxially oriented PBG.

The calculated piezoelectric transfer function in the sensor coordinate system shown in Figure 5 is:

$$\varepsilon_3^i V_3^i = - t \left[D_{3j}^i \right] \sigma_j \qquad (j = 1,6) \qquad [4]$$

where V_3^i and ε_3^i are respectively the voltage appearing on each of the element electrodes and the element dielectric constants, the superscript i (i=1,6) indicates the various sensor elements, t is the element thickness and:

$$
\begin{bmatrix} D^i_{3j} \end{bmatrix} = \begin{pmatrix}
d^1_{31} & d^1_{32} & d^1_{33} & 0 & 0 & 0 \\
d^2_{31} & d^2_{32} & d^2_{33} & 0 & 0 & 0 \\
d^3_{31} & d^3_{32} & d^3_{33} & 0 & 0 & 0 \\
0 & 0 & 0 & d^4_{34} & 0 & 0 \\
0 & 0 & 0 & 0 & d^5_{35} & 0 \\
0 & 0 & 0 & 0 & 0 & d^6_{36}
\end{pmatrix} \qquad [5]
$$

Under the following set of assumptions:
IV - the thin sensor elements match the elastic properties and do not perturb the elastic field of the medium in which they are embedded;
V - the embedding medium is elastic, isotropic and seminfinite in extension;
VI - the sensor is loaded by a punctual vertical force directed across its center and the plane stress problem of solid mechanics can be applied; dimensioning of the sensor can be accomplished, determining the interelement distance and the sensor depth in order to obtain "stress aliasing" over the elements (in conformity to assumption I) and still retain sufficient stress sensitivity.

The development of a sensor capable to measure the 6 components of the stress tensor in a medium and of appropriate electronic means to read the electric signal from a piezoelectric polymer sensor array (26) are fundamental steps to approach the inverse field problem of touch sensing.

4. MUSCLE-LIKE ACTUATORS

Vertebrate muscles are very efficient devices for converting chemical energy into mechanical work, via electric excitation.
However, the intricacy of their structure and the highly elaborate biochemical "fueling process" strongly discourage any attempt to closely mimic their structure and operations by artificial means.
Nonetheless, the primary transducing element in muscle turns out to be remarkably simple. It consist of macromolecules arranged in ordered (helical) configurations which, under stimulus, are transformed to random coils. The random coil has a shorter statistical length than does the helical segment from which it is formed; hence, it delivers a force that

brings about contraction. This force, in fact, is calculable from the same concepts and theory that apply to polymer networks (27).

Other EMC mechanisms operate in aquatic unicellular organisms which appear to use more simple biochemical pathways for their motility and different organization and performance of their contractile machine.

Perhaps the best-known examples are spasmonemes (28), the contractile elements of some protozoa. Spasmonemes can be made to contract by simply adding calcium salts to the solution in which they are held. The size of the structures found in nature is very small. This may be so that a short reaction time can be achieved, since one may expect that the process involved is diffusion-limited. The rate of shortening can be 100 lenghts per second. However, the relaxation process takes several seconds and the stresses developed by the spasmonemes are small compared with those developed by muscle (about 15%). Their increase in volume is achieved by taking in water, possibly under osmotic driving forces.

Similarly, the known artificial contractile polymeric systems driven by chemical stimuli can also be divided into two classes: the molecular-conformational structures, i.e., those in which helix-coil transitions govern the contractile response and the elastoosmotic systems, i.e., those in which the deformation is caused mainly by the difference between the osmotic pressures in the liquid phase and the gel phase (the swelling pressure), .

Recently it has been reported that contractile phenomena of the elastoosmotic type (where the increase in volume is achieved by taking in water) can be induced in polyelectrolyte gels by electric fields to produce mechanical work (29).

In a study aimed to investigate the origin of this contractile behavior and to characterize the electromechanical response (30), it was concluded that the effect is primarily due to spatio-temporal pH gradients generated by electrode reactions which propagate inside the gel by electrodiffusion, causing mechanical rearrangement of the gel by changes in the statistical length of polymer chains.

As already demonstrated by the early studies of mechanochemistry, chain distension or contraction are caused by increase or decrease in electrostatic repulsion forces among electric charges on the polymer backbone. The amount of net charges fixed to the polymer chains and their electrostatic repulsion forces are determined by proton and salt concentrations in the interpenetrating solution whose changes affect the dissociation equilibrium of the weak polyacid component of the gel and perturbe the electrostatic interactions.

A preliminary characterization of gel actuators was also provided (30) focused on the isometric force generation capabilities at different sample lengths and on the force-velocity relationships under various afterloads.

For an artificial amorphous gel, made of a thermally crosslinked polyacrylic acid-polyvinyl alcohol (PAA-PVA) it was found that the maximum isometric active tension is generated by a sample when at its rest length (see Figure 6).

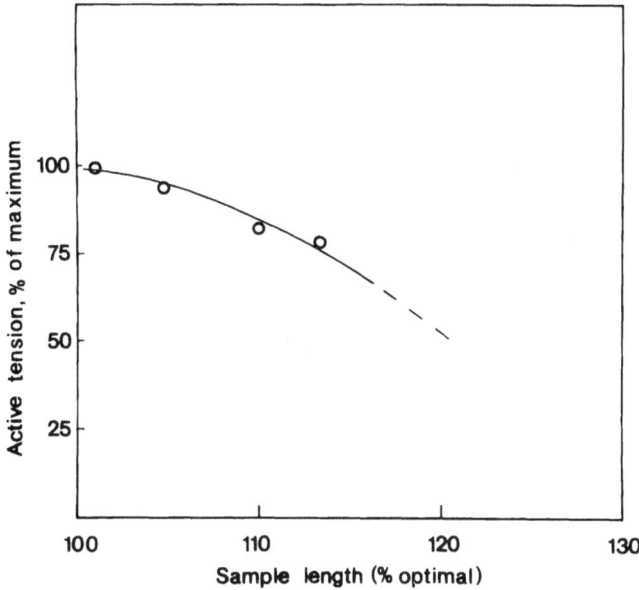

FIGURE 6. Isometric tension (% of the maximum observed value) generated by a PAA–PVA strip under EMC excitation (2.2 V across platinum electrodes), at various sample lengths (expressed as % of optimum length).

Contractile force densities (force by unit resisting cross-sectional area) of the order of 1-3 Kg/cm^2 are obtained under a 2.5 V potential difference at the electrodes, under isometric conditions.

The kinetics of gel contraction was also investigated either theoretically and experimentally and isotonic afterload contraction velocity curves are reported in Fig. 7.

The kinetics of contraction was found to be governed by two coupled phenomena i.e. protons diffusion inside the gel and mechanical readjustment through gel deswelling. For thermally crosslinked PAA-PVA gels, protons diffusion time constant has been found to be much longer and to rate limit the contractile phenomena kinetics.

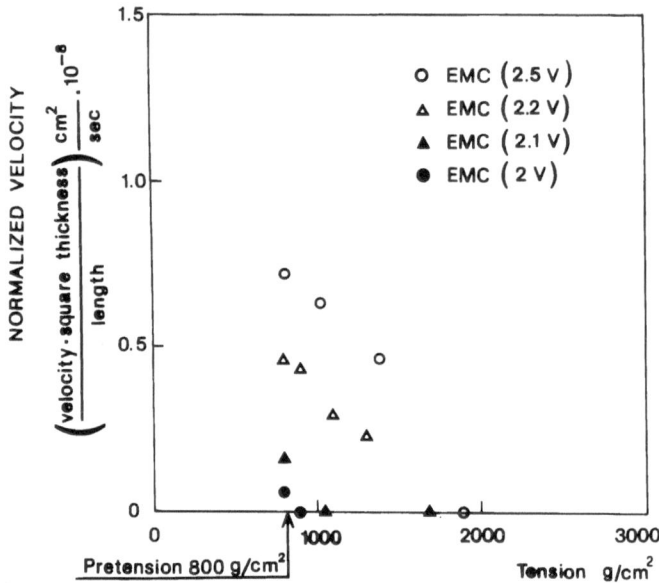

FIGURE 7. The contractile tension vs normalized contraction velocity of PAA-PVA samples when prestretched prior to isotonic contraction. Normalized velocity has been used to compare samples having different dimension. Voltage difference between platinum electrodes has been varied from 2 to 2.5 V.

By an extrapolation of the experimental data on gel contraction kinetics it can been inferred that a gel fiber 1 μm thick, 1 cm long would contract 15% of its original length in approximately 10 msec. However, 10^8 fibers in parallel would be required to exert an isometric contractile force of 3 kg. Either contraction force density and velocity of shortening come very close to those typical of animal muscles.

Such results encourage further studies on this class of materials and related technology to implement practically usefull devices. The choice of more appropriate electrodes, as liquid electrodes (31), and the availability of very thin, fatigue resistant hollow fibers are still pending problems in this respect.

REFERENCES

1) Sessler GM (ed)(1980) Electrects. Springer Verlag, Berlin.

2) Guzelsu N (1982) Mechanoelectrical effects in biological systems. In "Electronic conduction and mechanoelectrical transduction in biological materials", Lipinski B (ed). Marcel Dekker Inc., New York.

3) Galletti PM, Broadhurst MG, De Rossi D (Guest eds)(1984) Proceedings of the 1st Int. Symp. on Piezoelectricity in Biomaterials and Biomedical Devices, Ferroelectrics, 60 (N. 1/2/3/4).

4) Various authors (1982), Outlook for science and technology-the next five years, pp. 412-416. W.H. Freeman and Company, San Francisco.

5) Mandelkern L (1964) Contractile processes in fibrous macromolecules, Ann. Rev. Phys. Chem., 15: 421-448.

6) Fukada E (1974) Piezoelectric properties of organic polymers, Ann. New York Acad. of Sci., 38: 7-27.

7) Katchalsky A, Curran PF (1974) Nonequilibrium thermodynamics in biophysics. Harvard University Press, Cambridge.

8) Oosawa F (1971) Polyelectrolytes, Marcel Dekker Inc., New York.

9) Cady WG (1966) Piezoelectricity. McGraw Hill, New York.

10) Nye JF (1957) Physical properties of crystals. Oxford University Press, Oxford.

11) Hayakawa R, Wada Y (1973) Piezoelectricity and related properties of polymer films. Advances in Polymer Science, XI: 1-55, Springer Verlag, Berlin.

12) Fukada E (1974) Piezoelectric properties of biological macromolecules. Advan. in Biophys., 6: 121-155.

13) Grodzinsky AJ (1983) Electromechanical and physicochemical properties of connective tissue. CRC Critical Reviews in Biomedical Engineering, 9(2): 133-199.

14) Basset CAL (1971) Biophysical principals affecting bone structure. In "The biochemistry and physiology of bone" Vol. III, Bourne GH (ed), Academic Press, New York.

15) Kuhn W, Hargitay B, Katchalsky A, Eisemberg H (1950) Reversible dilatation and contraction by changing the state of ionization of high-polymer networks. Nature, 165: 514-516.

16) Katchalsky A, Lifson S, Michaeli T, Zwick H (1960) Elementary mechanochemical processes. In "Size and shape changes of contractile polymers", Wassermann A (ed), Pergamon Press, New York.

17) Flory PJ (1956) Role of crystalization in polymers and proteins. Science, 124: 53-60.

18) Tanaka T, Nishio I, Shao-Tang S, Ueno-Nishio S (1982) Collapse of gels in an electric field. Science, 218: 467-470.

19) Grodzinsky AJ, Shoenfeld NA (1977) Tensile forces induced in collagen by means of electromechanochemical transductive coupling. Polymer, 18: 435-443.

20) Shamos MH, Lavine LS (1967) Piezoelectricity as a fundamental property of biological tissues. Nature, 213: 2677-2679.

21) Athenstaedt H, Claussen H, Shaper D (1982) Epidermis of human skin - a piezoelectric and pyroelectric sensor layer. Science, 216: 1018-1021.

22) De Rossi D, Domenici C, Pastacaldi P (1986) Piezoelectric properties of dry human skin. IEEE Trans. on Electr. Insul., EI-21: 511-517.

23) De Rossi D, Domenici C, Pastacaldi P (1986) Human skin electromechanics: on the origin of stress generated potentials (Abstr.). Vth Meeting of the European Society of Biomechanics, Berlin.

24) Dario P, De Rossi D (1985) Tactile sensors and the gripping challenge. IEEE Spectrum, 22(8): 46-52.

25) Tikhonov AN, Arsenin VY (1977) Solutions of ill-posed problems. Winston/Wiley, Washington.

26) Dario P, De Rossi D, Domenici C, Francesconi R (1984) Ferroelectric polymer tactile sensors with anthropomorphic features. Proc. 1st IEEE Int. Conf. on Robotics, pp. 332-340, IEEE Computer Society Press, Atlanta.

27) Volkenstein MV (1983) Biophysics, pp. 423-431. MIR Publisher, Moscow.

28) Brown CH (1975) Structural materials in animals, pp. 76-77. A Halsted Press Book, J Wiley & Sons, New York.

29) De Rossi D, Parrini P, Chiarelli P, Buzzigoli G (1985) Electrically induced contractile phenomena in charged polymer networks: preliminary study on the feasibility of muscle-like structures. Trans. Am. Soc. Artif. Intern. Organs, XXXI: 60-65.

30) De Rossi D, Chiarelli P, Buzzigoli G, Domenici C, Lazzeri L (1986) Contractile behavior of electrically activated mechanochemical polymer actuators. Trans. Am. Soc. Artif. Intern. Organs, XXXII: 157-162.

31) Sandblom J (1972) Liquid membranes as electrodes and biological models. In "Membranes" Vol. 1, Eisenman G (ed), Marcel Dekker Inc., New York.

SOLID STATE TRANSDUCERS

Wen H. Ko
Electrical Engineering & Applied Physics Department
and
Electronics Design Center
Case Western Reserve University

ABSTRACT

This paper reviews the development of solid state sensors and
actuators during the last decade. Both solid state physical and
chemical sensors suitable for use in the measurement and control of
manufacturing processes will be discussed. The principle, the device
structure and the performance of physical transducers for temperature,
pressure, displacement, acceleration, flow, display, fluid injection
and controlled fluid flow valves is summarized. Chemical sensors for
humidity, combustive gases, ionic concentration in electrolytes and
large molecule concentration are also outlined for their potential
applications in automated manufacturing.

The technology of micromachining, one of the bases for solid state
transducers, is discussed with examples given to illustrate its capabilities and remaining problems.

The possibility of integrating signal processing circuits on or near
the transducer chip to fabricate "intelligent sensors" is presented
along with the functions that can be incorporated on the transducers.

.Future trends in solid state transducer research and the possibility
of beneficial collaboration between users and designers, as well as
between material scientists, technologists, device designers and
packaging engineers, is discussed.

INTRODUCTION

In the development of advanced robotic systems, one of the limiting
factors is the transducers (sensors and actuators) that interface the
robot with the changing environment. During the last decade much
attention has been focused on the development of input sensors and
output actuators to provide transducers for instrumentation used in
various areas including robotics, industrial, aerospace and biomedical
measurements. Both physical and chemical transducers have been
developed and many more are expected to be designed [1-7]. This
article will review state-of-the art solid state transducers, discuss
their performances and outline transducer technology, intelligent
transducers and future trends in transducer research. A limited
reference list (less than 5% of the vast amount of literature that has
been published), mostly in physical transducers, was selected subjectively by the author either because the paper describes a principle
discussed herein or because it represents the advanced technology on a

particular subject. Many of the references describe work in which the author himself participated.

Common electronic measurements and the transducers presently used are listed in Table I. Present transducers cannot satisfy the increasing demand for high performance including: sensitivity, stability, size and weight and ease of operation. Furthermore, with the advances in both artificial intelligence and electronic technology during the last decade, new and better control systems for robots are being designed using reliable, low cost, sophisticated VLSI circuits. These VLSI components and microprocessors make possible reliable, high performance systems to carry out complex control and signal manipulation functions. However, all these robotic systems demand new or better transducers.

With solid state electronic and micromachining technology, transducers can be designed and mass produced which offer: (1) uniformity of device performance, (2) higher reliability than individually assembled devices, (3) low cost when mass produced, (4) integration with electronic circuits that interface with other signal processing or computing circuits, and (5) better performance.

The application of silicon integrated circuits and micromachining technology to the development of high performance, long-term stable, computer compatible microsensors for control applications has been the focus of some efforts and has shown great potential and promise. Table II lists some of the solid state transducers reported in the literature that can be integrated with electronic circuits on the chip and that may have significance in Advanced Robotic Systems [8-81]. Several transducers are discussed below as selected examples. Emphasis is placed on physical transducers, because chemical transducers are discussed by other authors in this work [82, 83].

An important step has been initiated; the promising results excite many speculations. However, in order to realize these potentials, a more concentrated effort will be required to overcome the technological problems so that new generations of transducers can be produced that will satisfy various application demands. Research in solid state transducers will be of primary importance in the development of advanced robotic systems.

PHYSICAL TRANSDUCERS

As summarized in Table II, many solid-state transducers have been reported in the literature and developed to various degrees of completion. A few examples are selected and discussed.

1. Temperature

Machine and environmental temperatures are measured with thermal resistive devices--thermistor; thermoelectrical devices--thermocouple; p-n junction diodes; temperature sensitive resonant circuits, infrared radiation and chemical devices -- liquid crystals and others. Most of these principles can be applied to solid state sensors fabricated on semiconductor substrates.

P-n junction temperature sensors are now commercially available. For normal junction diodes with a constant current flowing through, the p-n junction voltage decreases about 2 to 3 mV/$^{\circ}$C as the temperature rises. Junction sensors with an integrated circuit interface can give direct voltage readings corresponding to temperature in $^{\circ}$F or $^{\circ}$C at the output [10]. The sensitivity of special junction devices is generally -1.2 V/$^{\circ}$C or -10%/$^{\circ}$C which is comparable to thermistors in sensitivity [11]. Many of these devices are commercially available from National, Motorola, PM1 Intensil (AD-590) and Analog Devices. Microstructured thermopile infrared detectors have been fabricated on silicon substrates for noncontact temperature measurement or IFR detection with a responsiveness of 6V/W and a time constant of 15 ms [12].

A p-n junction array was made to measure the temperature profile in the body [55]. The change of the energy band gap with temperature in semiconductor materials can be measured by the Optical Absorption Spectrum Edge. When incorporated with fiberoptics it provides a means for measuring body temperature with non-conductive devices. Liquid crystals or other materials that change their absorption characteristics with temperature can also be used with optical fibers to provide a nonconductive method for determining localized temperature.

2. Position and Motion

Position and displacement are measured with potentiometric devices, strain gauges, linear voltage differential transformers, capacitive displacement devices and ultrasonic devices. Velocity is measured with ultrasonic or optical Doppler effects, electromagnetic devices and integration of acceleration or differentiation of displacement. Acceleration is measured by a mass and its acceleration force. A two-dimensional position detector in the form of a six to eight millimeter rectangular chip of silicon crystal that can determine the two-dimensional coordinates of a light spot projected on the surface of the device with good linearity was reported [21,22]. Figure 2 illustrates that the displacement or position of a refracting surface can be accurately measured by optical reflection. An analog signal can be obtained at the output with an accuracy of a few microns [56]. The device was used to measure the motion of a tympanic membrane. It can be employed for measuring small deformation of stressed elements used in load cells and others. A silicon chip mass and spring accelerometer, 2 x 3 x 0.6 mm in size, weighing less than 0.02 gms was reported with a frequency response of 1000 Hz [30,31]. Several designs using piezoelectric films were reported to sense the inertia force [6,7]. These devices can be used in biological and physiological prostheses and sensory aids.

Figure 3 shows the principle and structure of a capacitive sensor used to measure the knee angle (the knee is a complex joint and it is difficult to measure the angle with conventional sensors). The sensor can provide needed feedback signals for computer controlled walking

[57] and can be integrated onto a silicon substrate. These devices are being used in research projects to provide feedback information for computer controlled prostheses.

3. Pressure

Methods for pressure transduction include measuring mechanical elastic properties of tubes and diaphragms, and using piezoresistive, capacitive and optical methods to convert mechanical deformation of metal, glass and silicon diaphragms into electrical signals representing pressure. Most of the present solid state sensors use piezoresistivity of silicon material to fabricate miniature force and pressure transducers. Figure 4 depicts a silicon diaphragm piezoresistive device for both gauge and absolute pressure. It measures 1 x 3 x 0.5 mm in size. Long-term baseline stability of 0.1% to 0.3% per month at 300 Torr Full Scale in body environment was observed [58]. The problem of packaging this device for industrial and control applications still needs to be solved. Several miniature silicon capacitive pressure transducers with some signal processing circuits integrated on the chip were reported [25-28]. A design with a bipolar oscillator, capacitance bridge and follower circuits integrated on the capacitor chip is shown in Figure 5 [24,60]. The optical reflection technique described in the paragraph "position and motion" has also been used to convert diaphragm deflection caused by pressure change into an electrical signal.

4. Flow

Fluid flow has been measured with electromagnetic devices, ultrasonic and optical Doppler devices, fiberoptics, pressure gradient and thermal dilution methods [61]. There are many review articles and textbooks discussing the established ultrasonic, electromagnetic and dilution techniques for flow measurement. Therefore, they are not discussed herein. Figure 6 illustrates a thermal transport device fabricated on a silicon chip that measures the temperature differences at the up and down streams of a heat source, referring back to the flow velocity [18]. Integrated circuits can be fabricated on the flow sensor chip to perform simple processing functions [19].

5. Actuators

Solid state microstucture research is a relatively new area. Existing actuator designs include cantilevel beams, ink jets [35], and display [36]. These principles and techniques can be used to design drug pumps and flow control devices.

Other types of actuators can be controlled by low power electrical signals to regulate micro-flow or position at the sub-micrometer level. Work on miniature drug pumps has been reported in the literature [62].

CHEMICAL TRANSDUCERS

Solid state chemical sensors have been developed in the following areas.

1. Humidity Sensor

Humidity can be measured by the change of (1) impedance of plastic material absorbing water vapor, (2) the surface impedance (resistance and capacitance) of insulators, such as Al_2O_3, with closely spaced electrodes and (3) the dew point when the sensor temperature is lowered. One solid state humidity sensor used a silicon chip on top of the Peltier cooling element. The chip contained two interdigitated electrode structures forming a capacitor. As the temperature is lowered, water vapor starts to condense on the structure, drastically changing its capacitance due to the large dielectric constant of water [39]. Other designs utilized porous oxide layers sandwiched between a base metal and a top electrode [5]. The charge flow transistor oscillator also offers an interesting alternative as a humidity sensor [38]. Humidity sensors are discussed in other parts of this book [82].

2. Gas Sensors

H_2 gas can be monitored with Pd or Pt gate MOSFET [40, 41]. The hydrogen permeates through the metal gate into the metal insulator interface producing a change in the threshold voltage of the FET. At $150^\circ C$, 10 ppm of H_2 can be detected. Other reducing and oxidizing gases can be monitored by the impedance of SnO_2, ZnO, TiO_2 and other oxide surfaces [42-44]. The combustive gas monitor used in gas heated homes and various laboratories already has a sizable commercial market. These gas sensing oxides can be deposited on silicon substrates and then modified with various catalytic materials deposited on the surface to sense various gases with improved sensitivity and selectivity, as shown in Figure 7. A gas chromatographic air analyzer was fabricated on a silicon wafer using solid state electronic techniques [42]. Similar techniques can be used to design and fabricate gas sensors that measure the change of mass and thermodynamic properties when gas molecules are deposited on the sensor surface.

3. Ion Sensor

Ion Sensitive Field Effect Transistors, ISFET, is an M-O-S FET with the gate electrode removed to the reference electrode and with the insulator exposed to the electrolyte. It can be used to measure ionic concentration in body electrolytes. Concentrations such as pH, K^+, Na^+, Ca^{++}, Ma^{++}, Cl^-, F^-, etc. can be measured with various modified insulators or by adding a membrane to the gate insulator [45-50]. Thin and thick film sensors for PO_2, PCO_2 and other gas and ionic species in solution are also being developed using silicon substrates and photolithography. Multiple sensors and reference electrodes can be integrated on the chip with the exposed area on the order of 10 um^2, thus alleviating many problems due to flow and protein deposition. Fiberoptics have been used for oxygen saturation measurement in blood. Improvement in signal analysis and sensor design should lead to the development of on-line catheter-type monitoring devices. Chemical sensors in liquids are discussed in other parts of this book [82].

4. Enzyme and Molecular Sensors

By enclosing, trapping or absorbing enzymes on substrate materials on
the surface of chemical sensors, the product of the enzymatic reaction
can be sensed as a means of detecting the presence of enzymes on the
substrate [50-52]. Similar techniques can be used for body protein
detection and the measurement of other molecular substances. The
possibility for on-line determination of proteins and moleculars will
greatly improve environmental health and some phases of industrial and
manufacturing processes. Stability and packaging are the major
current problems encountered with these types of sensors.

Combining solid state electronic technology and micromachining methods
to produce new and improved chemical sensors opens a vast new area
limited only by the creative approaches taken and the manner in which
innovative concepts are implemented.

NEW TECHNOLOGIES - MICROMACHINING

In order to design and develop new or improved transducers some new
technology will be needed. The field of micromachining where silicon
or other materials in micrometer dimensions can be machined or fabricated has been growing for the last decade [34, 63]. Although the
basic skill was developed through integrated circuit technology,
micromachining is particularly important for transducers and is being
developed by transducer research groups.

The major micromachining functions are:

1. Etching and Etch Stop. Wet etching, using chemical and electrochemical techniques, has been developed for silicon, metal and other
materials [64,65]. Dry etching techniques using plasma, ion beam and
spark erosion have been reported and can be used to machine three-
dimensional structures from a semiconductor or other substrate
[66,67]. Silicon diaphrams thinner than 1 micron with an area greater
than 1 $(mm)^2$, micro-cavities, beams, and bridges in micron scales,
have been fabricated in laboratories.

2. Bonding. Three-dimensional structures for transducers can also be
fabricated by bonding different layers together. For example, silicon
or GaAs can be electrostatically bonded to pyrex glass (#7740) to form
a hermetically sealed unit with a junction flatness below a micrometer
[68]. Other bonding techniques use sputtered glass, both 7740 and low
melting temperature glasses, as a sealing layer to bond silicon to
silicon substrates or silicon to metal [63,69,70]. Metal compounds
have also been used as a brazing material to seal silicon to metal,
ceramics, etc.

3. Selective Deposition. Many techniques have been developed that
can selectively deposit layers of conductive, semiconductive and
insulating materials of various properties on the substrate. These
include: evaporation, sputtering, ion beam sputtering, plasma and
chemical vapor deposition (CVD) [71, 72, 73]. Single crystal silicon,

polysilicon, SiO_2, Si_3N_4, Al_2O_3 and organic films have been deposited on silicon and other substrates to form three-dimensional micro-structures [74]. Single chip multiple sensors can also be fabricated with these techniques.

4. Feedthrough, Holes and Packaging. In transducer design there is need for insulated electrical connections between bonded layers which require holes and micro-chambers fabricated on the substrate and particularly require a method to package those transducers for implant or indwelling applications where the devices have to be in communication with the biological system. At the same time, leads and signal processing circuits have to be protected from the corrosive and highly conductive fluids in the body. The literature reported includes the following methods [1,2].

1. Diffused Al column with thermal gradient to form conductive paths across a silicon wafer [75]
2. Anisotropic etched back contact such that lead connection to outside is not exposed to sensor environment [76]
3. Plasma etched holes for feedthrough [77]
4. Laser drilled holes [78]
5. Spark erosion holes [79]
6. Other techniques such as ion beam milling [80], centrifugal etching [81] etc.

INTELLIGENT TRANSDUCERS

The concept of intelligent or integrated transducers is to fabricate part of the signal processing circuitry on the same chip with the transducer. This approach will greatly enhance the performance of the transducer and also make it compatible with computing and other high level signal processing circuitries without additional interfaces. VLSI technology and solid state transducers have made this possible. When a transducer is made on a semiconductor or other solid state material, electronic circuits can be fabricated on the same substrate without significantly reducing yield and reliability or increasing the processing steps. The increased performance is greater than the additional design effort. Such an approach is being made practical and many integrated sensors are being designed.

The functions of integrated transducers, besides transduction, are:

1. Impedance transformation and amplification
2. Compensation and error correction
3. Coding and modulation to provide digital output
4. Signal averaging and redundancy
5. Reliability checking fault detection and alarm
6. Remote sensing and telemetry
7. Power reduction scheme and time sharing
8. System parameter integration

The development of implantable pressure transducers at Case Western Reserve University can be used to illustrate the evolution of integrated sensors [24,25]. Pressure is a common biomedical parameter that is important to cardiovascular, respiratory and locomotive systems. The same sensor with small modifications, can be used to measure force, stress, strain, displacement, acceleration, flow, etc. However, long term implantable pressure sensors with time stability of 1% per year and suitable size and weight for small animals are still to be developed. In the past decade significant progress has been made in characterizing the device, understanding the stability problem and developing design techniques. As a result of this progress, many high performance industrial pressure transducers are now available. With careful design, implantable biomedical sensors are within reach.

Figure 4 illustrates a piezoresistive pressure transducer, with a built in reference chamber designed to measure absolute pressure [59]. This device has reached 3%/year stability after 3 months aging, at 300 mm Hg full scale range and is being used to measure body fluid pressure with additional packaging. The disadvantages are: (1) limited sensitivity 10-50 uV/mm Hg - Volt, (2) sensitive to differential stress, therefore, sensitive to side-way forces or touch, and (3) stability needs to be improved.

To overcome these limitations capacitive pressure transducers using microelectronic techniques were studied [24]. The performance is better than similar piezoresistive devices. However, they need a 0.5 MHz power supply and the output impedance is high - (200kHz).

Figure 5 shows a capacitive pressure transducer integrated with the 1/2 MHz oscillator and an emitter follower amplifier. This device, when finalized, will receive d.c., 3-5 volt, 5-10 ma and give d.c. output of + 200 mV corresponding to + 300 mm/Hg (or any other range designed to operate in) at a low output impedance of 50 - 200 ohms. Other integrated pressure sensors with analog or digital output have been reported [27, 28, 29].

This evolution in pressure sensor development is believed to be the general direction for other solid state microsensors [60]. The sensors being designed now will be able to go directly to the last stage after sufficient experience is accumulated.

FUTURE TRENDS

Besides utilizing new principles and materials, research work should be directed toward the packaging of biomedical transducers to meet the previously stated requirements. Many techniques and procedures developed for heart pacemakers can be adopted for this purpose. Methods need to be found for incorporating first stage signal processing circuitry into sensors so that the requirements for packaging can be reduced (such as reducing the impedance level of all external connections) and the signal transmission improved thereby producing a new family of integrated transducers.

Several transducers can be integrated on a single chip or in a single package, so that they can be correlated with computing devices to give better reliability and confidence or a more useful indication of the subject's health status. As an example, pressure, thermal conductivity and vibration may be measured simultaneously to provide a touch sensation for the robot finger. Similarly, various gas concentrations can be measured simultaneously to indicate the safety condition of the industrial environment.

The new generation of transducers should be computer compatible, highly reliable, and easily interfaced to other systems. The concept of integrating electronic signal processing circuits on the transducer chip to generate intelligent transducers is being pursued. Many new designs and developments are expected in the near future. The combination of integrated transducers (sensors and actuators) with microcomputers and VLSI chips is expected to change our concept and approach to measurement technology and control system design. Functional approaches to assess the status of the environment and form a basis for decision, instead of measuring individual parameters, is the trend of future research and development.

ACKNOWLEDGEMENTS

The assistance of my colleagues at the Electronics Design Center and its staff is genuinely appreciated.

This work was partially supported by NIH grants RR80057, RR02024 and NS-19174.

REFERENCE LIST

Journals, Special Issues and Proceedings

1. Sensors and Actuators, (ed. S. Middelhoek) 1981-85.

2. IEEE Trans. Electron Devices, Special Issues ED-26, 12; (Dec.
    1979) and ED-29, 1; (Jan. 1982) (ed. K. Wise).

3. A series of proceedings of the Workshop on Biomedical Sensors
    organized at Case Institute of Technology, Case Western Reserve
    University, Cleveland, Ohio, USA, published by CRC Press, inclu-
    ding Indwelling and Implantable Pressure Transducers (eds. D.
    Fleming, W. Ko and M. Neuman), 1972, and The Theory, Design and
    Biomedical Applications of Solid State Chemical Sensors (eds. P.
    Cheung, et al.) 1982.

4. Proceedings of the 1st and 2nd Sensor Symposium, IEE Society,
    Japan, (ed. S. Kataoka) 1981 and 1982.

5. Proceedings of the International Meeting on Chemical Sensors,
    Kodansha, Japan, (ed. T. Seiyama) 1983.

6. Abstracts of the 2nd Int. Conf. Sol. St. Sensors and Actuators,
    Delft, The Netherlands, May 31-June 3, 1983. Full papers pub-
    lished in Sensors and Actuators, Vol. 4. 1983/84.

7. Digest of 1985 International Solid State Sensors and Acutators
    Conference, June 11-15, 1985, Philadelphia, Pennsylvania, U.S.A.
    Published by the International Coordinational Committe on Sensors
    and Actuators, Electronics Design Center, CWRU, Cleveland, Ohio
    44106

Sampled Articles

8. Ko, W. and Hynecek, J. Dry electrodes and electrode amplifiers.
    Chapter in Biomedical Electrode Technology (eds. H. Miller and D.
    Harrison), New York, Academic Press, 1974, p. 169.

9. Wise, K., Angell, J. and Starr, A. An integrated-circuit
    approach to extracellular microelectrodes. IEEE Trans. Biomed.
    Eng., BME-17:238, 1970.

10. Meijer, G. An IC temperature transducer with an intrinsic ref-
    erence. IEEE Trans. Sol. St. Cir., SC-15:370, 1980.

11. Schaffer, H. and Koeder, O. A sensitive all silicon temperature
    transducer. Sensors and Actuators 4:661, 1983.

12. Lahiji, G. and Wise, K. A batch-fabricated silicon thermopile
    infrared detector. IEEE Trans. Electron Devices, ED-29(1):14,
    1982.

13. Royer, M. et al. ZnO on Si integrated acoustic sensor. Sensors
    and Actuators 4:357, 1983.

14. Swartz, R. and Plummer, J. Integrated silicon-PVF$_2$ acoustic transducer arrays. IEEE Trans. Electron Devices, ED-26(12):1921, 1979.

15. Yeh, Y., Muller, R. and Kwan, S. Detection of acoustic waves with a PI DMOS transducer. Japan J. App. Phys., 16-1 suppl.:517, 1977.

16. Zieren, V. and Duyndam, D. Magnetic-field-sensitive multicollector n-p-n transistors. IEEE Trans. Electron Devices, ED-29(1):83, 1982.

17. Popovic, R. and Balter, H. Dual-collector magnetotransistor optimized with respect to injection modulation. Sensors and Actuators 4:155, 1983.

18. Huijsing, J., Schuddemat, J. and Verhoef, W. Monolithic integrated direction-sensitive flow sensor. IEEE Trans. Electron Devices, ED-29(1):133, 1982.

19. Van Putten, A. An integrated silicon double bridge anemometer. Sensors and Actuators 4:387, 1983.

20. Rahnamai, H. and Zemel, J. Pyroelectric anemometer: Preparation and flow velocity measurements. Sensors and Actuators 2:3, 1981.

21. Noorlag, D. and Middelhoek, S. Two dimensional position sensitive photodetector with high linearity made with standard IC technology. IEEE J. Sol. St. Elec. Dev., 3:75, 1979.

22. Petersson, G. and Lindholm, L. Position sensitive light detectors with high linearity. IEEE J. Sol. St. Circ., SC-13:392, 1978.

23. Lubke, K., Rieder, G. and Thim, H. A high-speed high-resolution two-dimensional position-sensitive GaAs Schottky photodetector. Sensors and Actuators 4:317, 1983.

24. Ko, W., Bao, M. and Hong, Y. A high-sensitivity integrated-circuit capacitive pressure transducer. IEEE Trans. Electron Devices, ED-29 (1):48, 1982.

25. Ko, W. et al. Capacitive pressure transducers with integrated circuits. Sensors and Actuators 4:403, 1983.

26. Borky, J. and Wise, K. Integrated signal conditioning for silicon pressure sensors. IEEE Trans. Electron Devices, ED-26 (12):1906, 1979.

27. Sugiyama, S., Takigawa, M. and Igarashi, I. Integrated piezoresistive pressure sensor with both voltage and frequency output. Sensors and Actuators 4:113, 1983.

28. Yamada, K., Nishihara, M. and Kanzawa, R. A piezoresistive integrated pressure sensor. Sensors and Actuators 4:63, 1983.

29. Smits, J. et al. Resonant diaphragm pressure measurement system with ZnO on Si excitation. Sensors and Actuators 4:565, 1983.

30. Roylance, L. and Angell, J. A batch-fabricated silicon accelerometer. IEEE Trans. Electron Devices ED-26(12):1911, 1979.

31. Petersen, K., Shartel, A. and Raley, N. Micromechanical accelerometer integrated with MOS detection cicuitry. IEEE Trans. Electron Devices ED-29(1):23, 1982.

32. Chen, P. et al. Integrated silicon microbeam PI-FET accelerometer. IEEE Trans. Electron Devices ED-29(1):27, 1982.

33. Hok, B., Ovren, C. and Gustafsson, E. Batch fabrication of micromechanical elements in GaAs-Al $xGA_{1-x}As$. Sensors and Actuators 4:341, 1983.

34. Sansen, W., Vandeloo, P. and Puers, B. A force transducer based on stress effects in bipolar transistors, Sensors and Actuators 3:343, 1982.

35. Petersen, K.E. Silicon as a mechanical material. Proc. IEEE 70:420, May, 1979.

36. Cadman, M. et al. New micromechanical display using thin metallic films. IEEE Elect. Lett. EDL-4:3, 1983.

37. Bel, N. Integrated capacitive imaging display for the blind. Proc. 4th European Conf. Electronics, Germany, (eds. W. Kaiser and W. Proebster) Netherlands, North Holland Pub. Co., 1980, p. 549.

38. Senturia, S., Garverick, S. and Togashi, K. Monolithic integrated circuit implementations of the charge flow transistor oscillator moisture sensor. Sensors and Actuators 2:59-72, 1981.

39. Regtien, P. Solid state humidity sensors. Sensors and Actuators 2:85-95, 1981.

40. Lundstrom, K., Shivaman, M. and Svensson, C. A hydrogen sensitive Pd-gate MOS transistor. J. App. Phys. 46:3876, 1975.

41. Poteat, T. and Lalevic, B. Transition metal-gate MOS gaseous detectors, IEEE Trans. Electron Devices ED-29:123, 1982.

42. Terry, S., Jerman, J. and Angell, J. A gas chromatographic air analyzer fabricated on a silicon wafer. IEEE Trans. Electron Devices ED-26:1880, 1979.

43. Fouletier, J. Gas analysis with potentiometric sensors. A review. Sensors and Actuators 3:295, 1982.

44. Croset, M., Schnell, P., Velasco, G. and Sielka, J. Study of calcia-stabilized zirconia thin film sensors, J. Vac. Sci. Technol. 14:777, 1977.

45. Wen, C., Chen, T. and Zemel, J. Gate controlled diodes for ionic concentration measurement. IEEE Trans. Electron Devices ED-26:1945, 1979.

4 . Bergveld, P. Development, operation and application of the ion sensitive field effect transistor as a tool for electrophysiology. IEEE Trans. Biomed. Eng. BME-19:342, 1972.

47. Cheung, P., Ko, W., Fung, C. and Wong, A. Theory, fabrication, testing and clinical response of ion selective field effect transistor devices. in Theory, Design and Biomedical Applications of Solid State Chemical Sensors (eds. P Cheung et al.) Boca Raton, Florida, CRC Press, 1978, pp 91-118.

48. Siu, W. and Cobbold, R. Basic properties of the electrolyte-SiO_2-Si system: physical and theoretical aspects. IEEE Trans. Electron Devices ED-26:1805-1815, 1979.

49. Matsuo, T. and Wise, K. An integrated field effect electrode for biopotential recording. IEEE Trans. Biomed. Eng. BME-21:485, 1974.

50. Janata, J. and Huber, R. Chemically selective field effect transistors in ion-selective electrode. in Analytical Chemistry (ed. H Freiser) New York, Plenum Press, 1980, Vol. 2 pp. 31-79.

51. Lubbers, D. and Opitz, N. New fluorescence photometrical techniques for simultaneous and continuous measurements of ionic strength and hydrogen ion activities. Sensors and Actuators 4:473, 1983.

52. Peterson, J., Fitzgerald, R. and Buckhold, D. Fiber optic probe for in vivo measurement of PO_2. Analyt. Chem. 56:62, 1984.

53. Ko, W., Bergmann, B. and Plonsey, R. Data acquisition system for body surface potential mapping. J. of Bioengineering 2:38-46, 1977.

54. Prohaska, O. et. al. A 16-fold semi-microelectrode for intracortical recording of field potentials. Electro-enceph. Clinical Neurophysiology 47:629, 1979.

55. Barth, P. and Angell, J. Thin linear thermometer arrays for use in localized cancer hyperthermia. IEEE Trans. Electron Devices ED-29:144, 1982.

56. Green, L. and Ko, W. Optical displacement measurement device. EDC Design Memo 297-A, Case Western Reserve University, 1980.

57. Ko, W., Wang, S. and Marsolais, E. Altitude sensor for angle measurement in neural prosthesis. Proc. 37th ACEMB, Los Angeles, Sept. 1984, p. 106.

58. Leung, A., Ko, W., Spear, T. and Bettice, J. Intracranial pressure telemetry system using semicustom integrated circuits. Part I. Overall Development. submitted to IEEE Trans. BME. 1985.

59. Ko, W., Hynecek, J. and Boettcher, S. Development of a miniature pressure transducer for biomedical applications. IEEE Trans. Electron Devices ED-26:1986, 1979.

60. Ko, W. and Fung, C. VLSI and intelligent transducers. Sensors and Actuators 2:239, 1982.

61. Cobbold, R. Transducers for Biomedical Measuarements, New York, John Wiley, 1974.

62. Spencer, W. A review of programmed insulin delivery systems. IEEE Trans. BME. MBE-28:3, 1981.

63. Micromachining and Micropackaging of Transducers, Eds. Fung, C.D., Cheung, P.W., Ko, W.H. and Fleming, D.G., Elsevier Scientific Publishing Co., Amsterdam, 1985.

64. Theunissen, M. et. al. Application of preferential electrochemical etching of silicon to semiconductor device Technology. J. Electrochem. Society, Vol. 117(7):959, 1970.

65. Bassous, E. Fabrication of Nevel 3-D microstructure by anisotropic etching of (100) and (110) Silicon. IEEE Trans. Electron Devices ED-25:1178, 1978.

66. Coburn, J. Plasma assisted etching. Plasma Chemistry and Plasma Processing, 2(1):1 1981.

67. Flamm, D. and Donnelly, V. The design of plasma etchants. Plasma Chemistry and Plasma Processing, 1(4):317 1981.

68. Eugelkrout, D. et. al. Current research in adhesiveless bonding of cover glass to solar cells. 16th IEEE Photovoltaic Spec. Conf. 1982, p. 108.

69. Brooks, A. and Donovan, R. Low temeprature electrostatic silicon to silicon seals using sputtered borosilicate glass. J. Electrochem. Soc. 119:545, 1972.

70. Ko, W., Suminto, J. and Yeh, G. Bonding techniques for microsensors. In Micromachining and Micropackaging of Transducers, Elsevier Scientific Publishing Co., Amsterdam, 1985.

71. Herring, R. Advances in reduced pressure silicon epitaxy. Solid State Technol. 22:75, 1979.

72. Kern, W. and Bau, V. Chemical vapor deposition of inorganic thin films in Thin Film Processes, J. Vossen and W. Kern, Eds., New York, Academic Press, 1978.

73. Bean, J. Silicon molecular beam epitaxy as a VLSI processing technique. 1981 Technical Digest, IEEE-IEDM Proc. 1981, p. 6.

74. Guckel, H. and Burns, D. Planar processed polysilicon sealed cavities for pressure transducer arrays. 1984 Technical Digest, IEEE-IEDM 1984, p. 223.

75. Anthony, T. and Cline, H. Migration of fine molten wires in thin silicon wafer. J. Appl. Phys. 49:2777, 1978.

76. Huang, J. and Wise, K. A monolithic pressure-pH sensor for esophageal studies. 1982 Technical Digest, IEEE-IEDM, 1982. p. 316.

77. Tung, C. Plasma etching of silicon with D.C. bias. M.S. thesis, Case Western Reserve University, Cleveland, Ohio, May 1984.

78. Ehrlich, D. et. al. Fabrication of through-wafer via conductors in Si by laser photochemical processing. IEEE Trans. on Comp. Hyb. and Manuf. Technol. CHMT-5(4):520, 1982.

79. Van Osenbrugge, C. High preceission spark machining. Philip Tech. Rev. 30:195, 1969.

80. Bollinger, D. and Fink, R. A new production technique:ion milling. Solid State Technol. 25:79, 1980.

81. Kuiken, H. and Trjburg, R. Centrifugal etching: a promising new tool to achieve deep etching results. J. Electrochem. Soc. 130(8):1722, 1983.

82. Regtien, P. "Development and application of Humidity sensors", chapter   of this book.

83. Bergveld, P. "Development of and application of chemical sensors in liquids", chapter   of this book.

TABLE I.

TRANSDUCERS USED IN ROBOTIC MEASUREMENTS

A. Physical Parameters

Measured Quantity	Principle or Devices Used
Temperature	Thermistor, P-N Junction, Infrared
Sound & Vibration	Microphones, Variable Z
Light	Transmission, refraction, absorption, LED, Detectors
Magnetic	Permeability, field
Force and Pressure	Piezoresistance, capacitance, deformation
Displacement, Velocity, Acceleration	Capacitive, magnetic, laser, ultrasonic, optical, ultrasound, acceleration force
Flow - (fluid, air)	Electromagnetic, optical & ultrasonic Doppler, pressure drop, streaming potential
Volume	Displaced volume, dye dilation
Imaging - (shape, surface, 3-D structure)	X-ray, ultrasound, microwave, nuclear isotope
Time	Delay, response time

B. Chemical Parameters

Humidity	Dew point temperature, resistance and capacitance change due to condensation or absorption
Gas concentration	Gas sensitive electrodes, chromatography, chemical analysis
Electrolyte Concentration	ion selective electrodes, chemical analysis
Moleculars	Enzyme and chemical analysis

TABLE II

SOLID STATE TRANSDUCERS

Parameters Measured	Principles Used	Reference
Electrical potential and impedance	Active electrodes with amplifier	8
	Micro-electrode	9
Temperature and infra red radiation	Junction diode with IC	10
	Bulk barrier diode	11
	Thermo-piles on Si	12
Sound and Ultrasound	ZnO on MOSFET amplifier	13
	PVF_2 with FET amplifier	14
	ZnO integrated into gate of MOSFET	15
Light	LED and diode lasers	
	CCD and photodetectors	
Magnetic field	Multicollector transistor	16
	Carrier domain movement	17
	Magnetoresistance	4
Fluid flow	"Hot wire anemometer" with IC	18
	double bridge detector	19
	pyroelectric detector	20
Two dimensional position	Light sport position sensitive photodetectors on Si or on GaAs	21-23
Pressure	Deformation of Si diaphragm by capacitance	24,25
	or piezoresistance changes with IC	26-27
	Resonant diaphragm with ZnO on Si in the feedback loop	29
Acceleration	Inertia force on micromachined cantilevel beams, on Si with IC, or on GaAs	30
		31,33
Force	P-N junction stress effects	34
Actuators	Piezoelectric wafer operated ink jets	35
	Electrostatic shutter display	36
	Electrostatic force on skin	37
Moisture, Humidity	Charge flow transistor	38
	Oxide films	5,39
Gas concentration	MOSFET with Pd or Pt gate	5,40,41
Chemical species in air	Si wafer gas chromatograph	42,4,5
	Oxide films, potentiometric	6,43,44
Ionic concentration in electrolytes	Ion sensitive diode	45,46
	Ion sensitive FET	47,48,49
Chemical ions and molecules, solution	Chemically sensitive FET	4,5,50
	Optical absorbing of fluorescent indicators	51,52

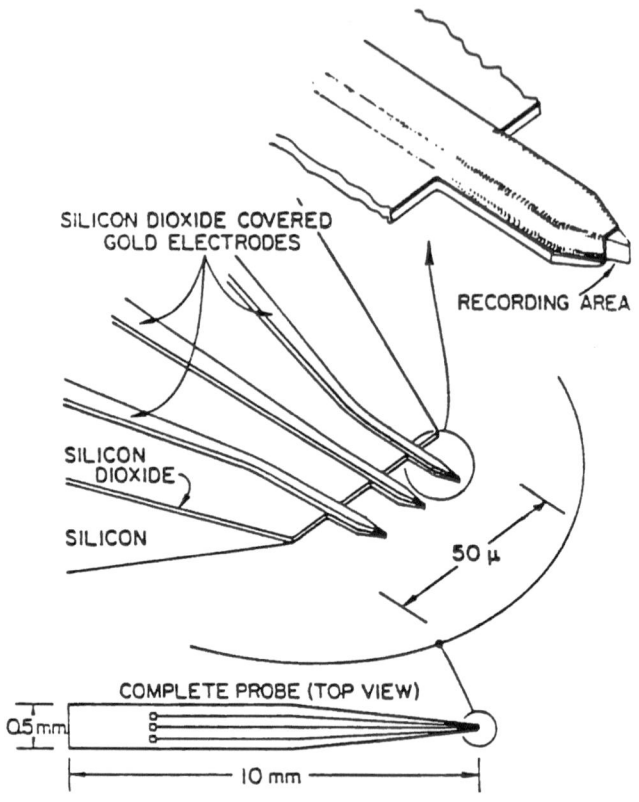

Figure 1.   A multiprobe microelectrode for

biopotential recording [9].

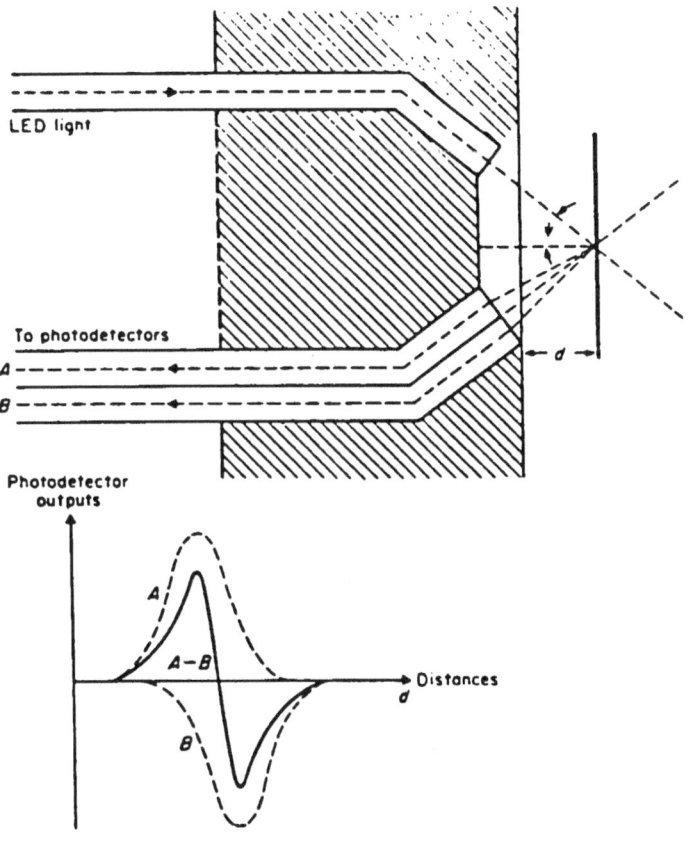

Figure 2.   Principle of optical displacement
            detector [56].

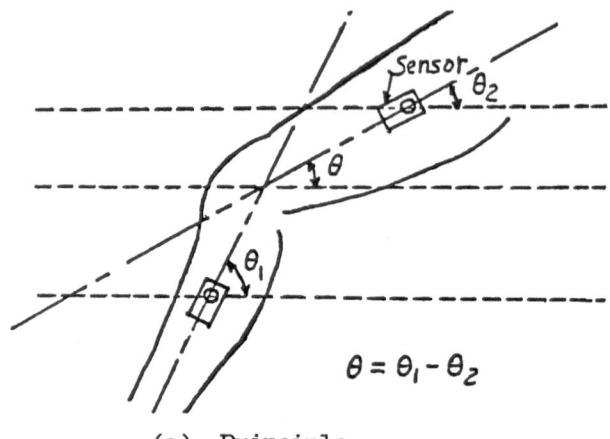

(a) Principle

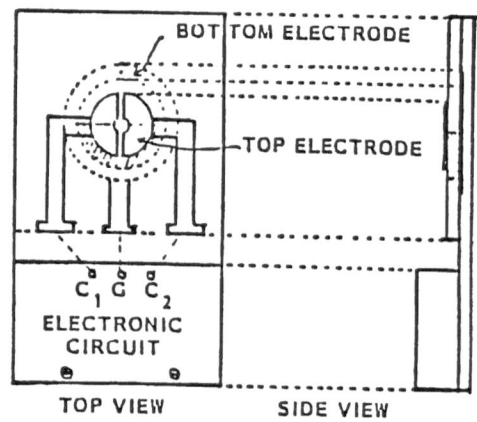

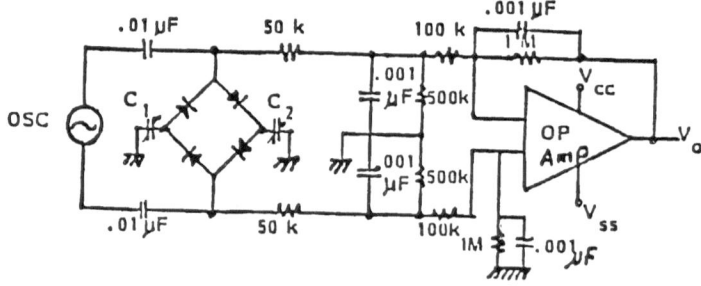

(b) Structure and Circuit Diagram

Figure 3.   Principle and structure of the
knee angle sensor [57].

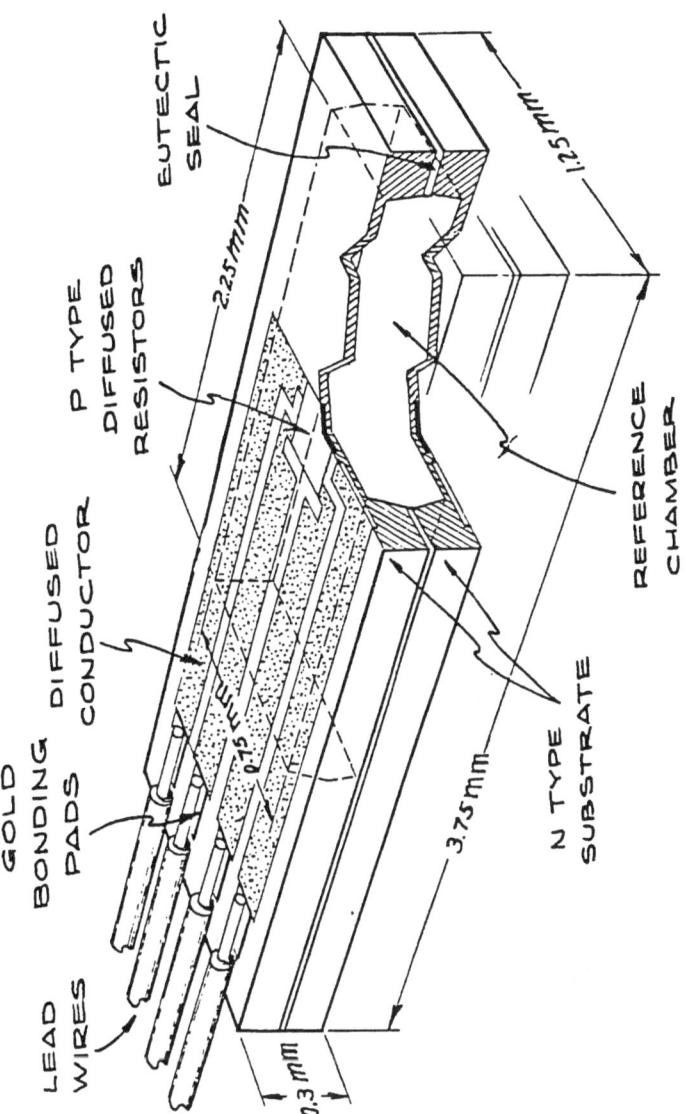

Figure 4.   Silicon diaphragm piezoresistive
pressure sensor for absolute
pressure measurement [24].

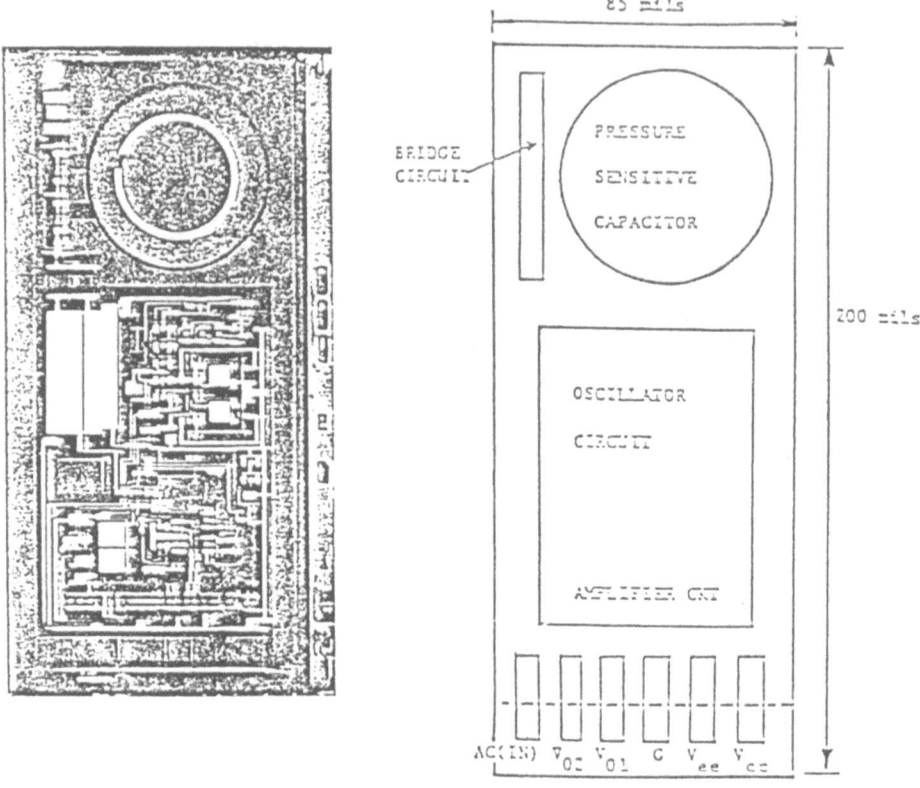

Figure 5 A silicon capacitive pressure transducer with bipolar signal processing circuit on the chip

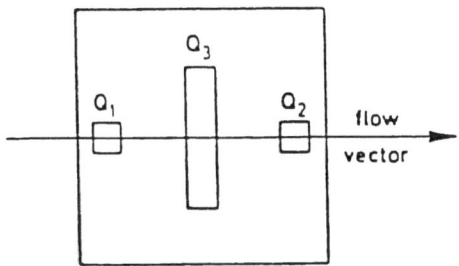

Basic chip layout.

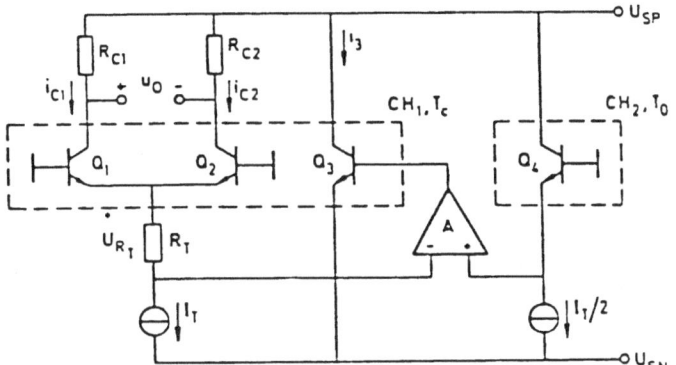

Circuit of the direction-sensitive flow transducer.

Figure 6.   Silicon integrated flow sensor [18]

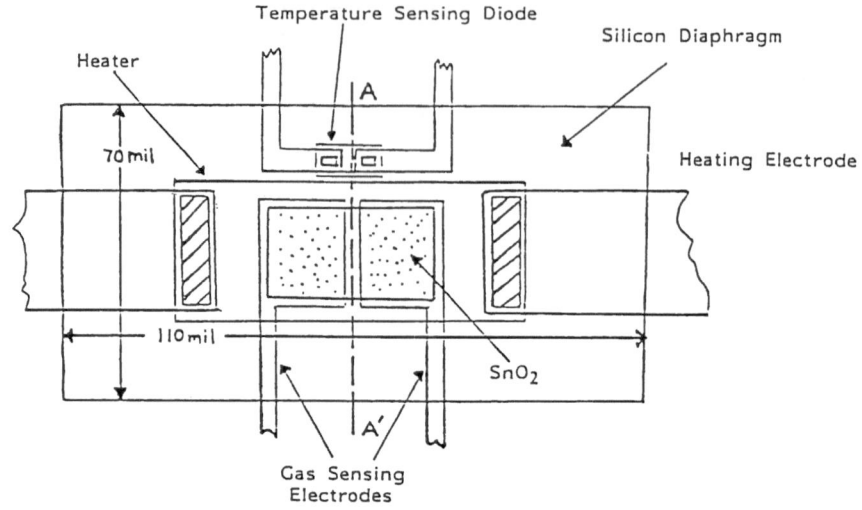

Layout of the Gas Sensor

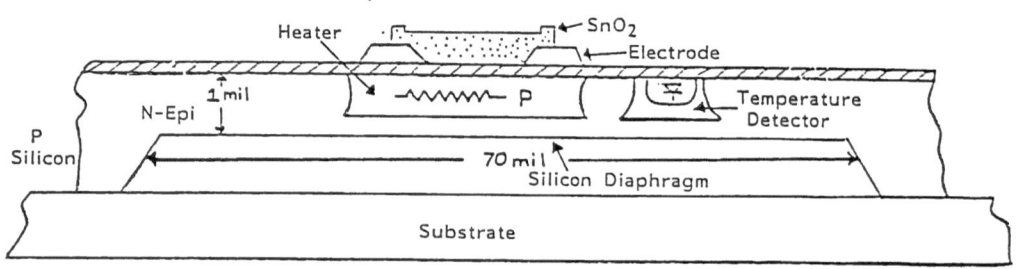

Cross-section of the Gas Sensor   (A – A′)

Figure   7   Structure of a silicon substrate gas sensor

First Results for Computer Vision Range Measurements

Ruzena Bajcsy
Eric Krotkov
Max Mintz

GRASP Laboratory
Department of Computer and Information Science
University of Pennsylvania
Philadelphia PA   19104

Abstract

This paper is a first step toward integrating sensor measurements of distance. Its major contribution is to identify and present qualitative models for the errors and mistakes introduced in three particular computer vision distance measurements: range from focus, range from point-based stereo, and range from line-based stereo.

These range measurement techniques are presented as computations, and their dominant sources of error are analyzed qualitatively. We propose to quantify the underlying models for these three range estimation techniques by deriving approximate confidence procedures for the intrinsic parameters and functions which characterize each technique.

1. Introduction and Motivation

When one deals with measurements of any kind, it is always the case that this measurement is accompanied by some error. What is an error? According to the methodology prevalent in physics, chemistry, and other sciences, it is a deviation from some ideal, or some standard. What is an ideal or a standard? The ideal comes from a theoretical description of the phenomenon which invariably is put into some mathematical language. This mathematical representation is also called a model. The word "model" is chosen not accidentally. Indeed it means that it is an idealization of the physical reality under precise conditions. A standard, on the other hand, is usually a very precise physical realization of the ideal with very small error. A standard is the best in some sense that we can get in a physical realization of the model. Error is different from discrepancy. The latter is the difference between two measured values of a quantity, for example obtained by two different measuring devices. There is still a further classification of errors: systematic error and random or stochastic error. Both these errors if well understood can be and should be included into the model, naturally using different mathematical tools for the systematic and stochastic errors. These tools are partly the subject of this paper. Both these errors can be evaluated by calibration of the instruments against standards. While the small random error is related to precision, the small systematic error corresponds to high accuracy.

Now that we have explained what we mean by error, we need to clarify what is a mistake? In our view a mistake is not a large error but a failure of a device or an identifiable component of a device. We realize that there is a very tenuous division line between large errors and mistakes. This is analogous to the problem in pattern recognition of variation

NATO ASI Series, Vol. F43
Sensors and Sensory Systems
for Advanced Robots
Edited by P. Dario
© Springer-Verlag Berlin Heidelberg 1988

2. BASIC PRINCIPLES OF FIBRE OPTIC SENSORS

Fibre optic sensors may be loosely classified as intrinsic devices (Figure 2) and extrinsic devices (Figure 3). In the former, light guided in an optical fibre is modulated within the guiding path whilst in the latter the light is allowed to emerge from the fibre, is modulated and then returned either via the same or another fibre. Therefore, intrinsic sensors are inevitably mechanically simpler, though suffer from the obvious disadvantage that they respond to environmental stimuli throughout their length unless considerable care is taken in the design of the interaction processes. Intrinsic sensors are extremely sensitive devices. On the other hand extrinsic sensors are more complex mechanically but are relatively simple to arrange so that most of the optical modulation (though not all) occurs in the region of interest.

The detection of the modulated radiation is, in principle, a straightforward process though it should be emphasised that optical detectors are capable of simply responding to the overall intensity of the light instant upon them. Most, if not all, fibre optic sensors impose analogue modulation on the optical carrier so that the ability to resolve the various possible values of the measurand depends not only upon the modulation process itself but more fundamentally, upon the charateristics of the optical carrier. Light – at least in most forms of photodetection – manifests itself as a quantum phenomena so that the detection process obeys particle counting statistics[9].

The most important single implication of this is that the optical carrier itself – even when closely controlled in intensity – exhibits power fluctuations determined by the Poissson statistics of the photons constituting the light beam. If the light beam consists of N photons per second then clearly there will be an intrinsic noise level of N photons per second imposed by this process. This in turn limits the resolution with which optical detection may be performed.

One very attractive feature of optical fibre sensors is the range of optical parameters which may be modulated by the measurand of interest. In this respect fibre optics differs from virtually every other transduction medium e.g. it is only possible to encode electrical signals in terms of the number of electrons passing through a detector (per unit) time. Optical systems on the other hand may be implemented using modulation of intensity (for instance via a shutter), state of polarisation (for instance using environmentally modulated birefringent crystals), optical phase (by causing the measurand to interact with one arm of a fibre optic interferomenter), optical frequency (through the Doppler effect), wavelength distribution (for instance using rotating prisms or evironmentally sensitive filters) and modulation waveforms (e.g. involving transient decay measurements or vibrational measurements). These are all illustrated schematically in Figure 4.

These modulation mechanisms all interact differently but it is useful to restate that finally optical intensity is detected. In most optical fibre sensors approximately 1mW of optical power is available and typically this is equivalent to a few times 10^{15} photons per second. Therefore, observing the Poisson statistical relationship mentioned previously we find that this corresponds to a possible intensity resolution of around 1 part in 10^7. Even though there are many exceptions it is a useful guideline to think that a fibre optic sensor system will have potentially this resolution available to be divided within the transduction elements as seems fit. This does, of

course, indicate a very wide dynamic range possibility and again this is one
of the attractive features of fibre optic sensors when compared to other
transduction processes. It will, of course, be apparent that considerable
care is needed in the sensor design to realise this dynamic range and it
will also probably be apparent that this dynamic range could either be
exploited in a single sensing device or divided across the elements of an
array. It may well be that this latter configuration is the most attractive
for many applications in robotics.

3. DEMONSTRATIONS

Examination of the literature[10] soon demonstrates that optical fibre
techniques have been used to monitor the vast majority of useful measurands.
Many of the instruments described in the literature are still laboratory
demonstrations, though a few have gone forward as the basis of a commercial
product. It is, however, probably more appropriate to describe in this
paper a range of sensor systems which may be useful in robotics. The
descriptions given here are brief but the references given will provide
further insight into their properties and applications.

3.1 The Fibre Optic Gyroscope

The fibre optic gyroscope exploits the Sagnac effect in the measurement
of inertial rotation[12]. A Schematic diagram of the gyroscope is shown
in Figure 5. In simplistic terms its principle of operation may be
viewed from the point of view of an imaginary observer sitting on the
beam splitter (or fibre optic directional coupler) input into the loop.
Clearly light injected into the loop will emerge from the counter
propagating direction sooner than the light propagating with rotation.
This time difference is measured as an optical phase difference and is
linearly dependent upon the rotation rate.

A considerable research effort has been expended on the gyroscope[13]
and even though the gyroscope is perhaps the most physically complex of
fibre optic sensors, it is also the case that most information is known
about this instrument. Much of this stems from work fuelled by the
potential of the gyroscope as a low cost acceleration insensitive
rotational measuring device which may operate in high vibration
environments and may be configured in a wide range of shapes and sizes.

The sensitivity levels which have been recorded and are in the region of
$10^{-2\circ}$ per hour with a long term drift of the order of $10^{-1} - 10^{-2\circ}$ per
hour (in an integration time of 1 second) in a 1 hour observation
period. In robotics, the fibre optic gyroscope has been considered as
an instrument which may be suitable for co-ordinate calibration.
Perhaps its most useful contribution could be as an angular transducer
for monitoring the relative position of limbs in robot arms. In this
application sketched in Figure 6, the relative angular positions of the
limbs may be monitored to about 0.1° over operational periods of several
tens of minutes prior to re-calibration. This would achieve accuracy
comparable to a precision shaft encoder without the necessity for a
complex mechanical assembly and could be commercially attractive.

3.2 Temperature Sensing Using Fibre Optics

Temperature measurement is often the most straightforward to implement regardless of the transduction technology. Fibre optics is no exception to this general rule and the majority of current commercial fibre optic sensors are temperature probes. Most of these probes are based upon changes in the chromatic properties of materials induced by changes in temperature. Examples include the exploitation of rare earth fluorescence spectra[14] and the changes in the bandgap of semiconductor material in turn inducing changes in either the fluorescence or absorption spectrum. The temperature probes, therefore, consist essentially of a short (relatively) wavelength optical source exciting a suitable material into fluorescence followed by observation and measurement of the flourescence spectrum using a suitable spectrometric system. Figure 7 shows the principles of one such device based upon the variations in semiconductor fluorescence.

This particular device, which is commercially available[15] , has the advantage that the excitation and fluorescence spectra are all in the near infra-red and so may be transmitted without substantial attenuation over very long lengths of optical fibre. The bandgap of a semi-conductor varies by of the order of 2mV per degree centrigade temperature change so that in the range -50°C to $+240^{\circ}$C a change in bandgap of the order of a 100mV would be expected. The shift in fluorescence wavelength is, therefore, in the order of 10% over the temperature range of interest. The returned signals, are, however, quite small so that long integration times (here of the order of 1 second) are required to perform the measurement with sufficient accuracy. The major advantages of this system are, of course, very small size and total immunity to electrically generated interference. The device has, therefore, found applications in high voltage measurement, microwave heating, diathermy and related applications.

At present this instrument is expensive but in the longer term one could envisage an array of minute optically read transducers performing thermal imaging or thermal tactile monitoring, perhaps as one component of an artificial finger tip.

3.3 Limit Stops

Perhaps the simplest fibre optic sensor is the limit switch which functions in an analogous way to an electrical micro switch. A number of such devices are currently commercially available usually based on modifications to conventional electrical switches. The general principle is shown in Figure 8. The apparent simplicity of these devices is deceptive since there remain a number of engineering problems associated with their construction, especially concerning optical loss through the gap. However, the modifications required to improve this are relatively few and high quality products should be available in the very near future.

3.4 Silicon Microtransducers and Fibre Optics

Silicon micromachining is the subject of another chapter in this volume[7] and a comprehensive account of the basic silicon micromachining technology is given in reference 8. In the context of fibre optic systems, silicon micromachining offers the potential of extremely precise mechanical fabrication in very tiny structures. It, therefore, becomes quite feasible to consider a range of transducers which may be attached directly to the end of an optical fibre which is in itself of the order 100 microns in diameter.

One such example is shown in Figure 9 which illustrates a silica bridge spanning an anisotropically etched hole in a silicon substrate. This bridge has a characteristic self-resonant frequency at a fundamental frequency of about 270kHz with observable overtones up to 1MHz (these resonances may be excited directly using conversion of optical energy into mechanical energy via heating on a metallic layer evaporated on to the silica substrate.) The resonant frequency is in turn a function of temperature (by about 0.1% per degree C) 50 that a compact and precise temperature measuring probe may be configured. The optical power levels returned from the sensor are comparable with the incident power levels (in the micro Watt region) so that integration times are much shorter and the thermal response may be limited by the thermal time constant of the sensing chip. This in turn is of the order of microseconds so that extremely rapid thermal changes may be monitored[18].

This technology is currently in its infancy, though the considerable interest which it has aroused indicates that it may well find a wide range of applications both for point and array transduction. The compatibility of the fabrication process with silicon integrated circuit manufacture indicates that an appropriate combination of high performance and economic costing will become available.

3.5 Proximity Sensing

The photonic sensor described in reference 1 has been a commercial product for approaching 20 years and is a simple distance monitoring device. It is designed for stable measurement of the location of an arbitary sample from the end of a fibre. Proximity sensing in robots presents a different type of problem and in many cases all that is required is an alarm signal or a rate of change of position indication. In these cases the requirements for accurate system referencing may be considerably eased.

One example of a rate of change sensor is the Doppler probe shown in Figure 11[17]. This probe is primarily designed to monitor the motility of microscopic bodies in fluids. Many variations on this theme have described[18,19] for use in various forms of flow measurement and anenometry, though in all cases the principle remains much the same. In the context of robot systems, such detectors could be useful to determine the rate of approach of a hand to a target object. The technique is, of course, entirely analogous to its ultrasonic equivalent though the frequency shifts involved are much greater and so very slow approaches may be monitored with a high degree of confidence. Velocities ranging from micrometers to meters per second may be monitored with ease.

The basic intensity modulated position probe concepts may also be modified with much higher depth resolution with that achieved with the photonic device[20] and may in principle approach sub nanomemter levels. For tactile purposes, surface imaging and related tasks a depth resolution of perhaps 1 micron from an array would appear to be more than adequate. Such an array could be configured in the form shown in Figure 12. In principle something of the order of 10 - 100 tactile "pixels" per square mm could be feasible, though this device has yet to be demonstrated in practice.

4. DISCUSSION

The mainstream of fibre optic sensor research and development has a ten year history though longer lived exceptions like the photonic device do exist. The technology is consequently still in its youth and relatively little commercial exploitation has succeeded to date though there is considerable and increasing activity in the market place. The applications area for fibre optic sensors are now well identified and centre around traditional sensing functions involving high radiation levels or intrinsic safety. However, the potential of sensors and sensor arrays[23] is only just becoming recognised and in these areas features such as the low overlap between sensing points in an array and the potential very high packing density offer the possibilities of performing functions which have not hitherto been possible. This, of course, also relies upon supporting technological advances at the array fabrication level and here it seems likely that some variation on photolithographically defined anisotropic micromachining will have a considerable part to play. In the relatively near future it is likely that significant developments will be made in these areas and the potential contributions to advancing the science of robotics promise to be substantial.

REFERENCES

1. C Menadier, C Kissinger and H Adkins, "The Fotonic Sensor", Instruments and Control Systems, 40, 1967 p114.

2. B Culshaw, "Optical fibre and signal processing", Peter Perignus, Stevenage 1984

3. Proceedings of NATO Advanced Study Institute International School of Quantum Electronics - Optical Fibre Sensors, Erice, Sicily, May 1986

4. T G Giallorenzi, J A Bucaro, A Dandridge, G H Sigel, J H Cole, S C Rashleigh and R G Priest, "Optical fibre sensor technology" IEEE J Quantum Electronics QE-18, 4, pp626-655 April 1982

5. A Harmer-Batelle Geneva Market Survey and Technical Report on Optical Fibre Sensors

6. ERA Technology, Leatherhead, Market Survey an Optical Fibre Sensor Technology (A MacGregor)

7. W Ko, This volume

8. K E Peterson, "Silicon as a mechanical material", Proc IEEE, $\underline{70}$ p420-455 1982

9. See, for instance A Yariv, Quantum Electronics, Wiley 1975

10. International Conference Series on Optical Fibre Sensors (London 1983, Stuttgart 1984, San Diego 1985, Tokyo 1986) Proceedings available via National Learned Societies (IEE, VDE, IEEE etc.)

11 SPIE Meetings on Fibre Sensor Technology available from SPIE, Bellingham Washington, USA

12. B Culshaw and I P Giles, "The fibre optic gyroscopes", J Phys E, January 1983

13. Fibre Optic Gyroscopes - The First Ten Years. Conference to be held Cambridge Mass September 1986

14. Luxton - Mountain View Ca. Product Literature

15. ASEA Innovation- Vesheros, Sweden Product Literature

16. S Venkatesh and B Culshaw, "Optically activated vibrations in a micromachined silica structure", Electronics Letters, $\underline{21}$, p315, 1985

17. R B Dyott, "The fibre optic Doppler anenometer", IEE Journal MOA, Jan 1978, $\underline{2}$, p13.

18. DISA GmbH Product Literature Fibre Optic Doppler Velocimeter

19. J Erdmann and D C Soreide, "Fibre optic laser transit velocimeter". Applied Optics, $\underline{21}$, 11, 1976 (June 1982)

20. R O Cook and C W Hamm, "Fibre optic lever displacement transducer", Applied Optics $\underline{18}$, 19, October 1979, p3230

21. B Culshaw, "Distributed and multiplexed fibre optic sensor system", in reference 3.

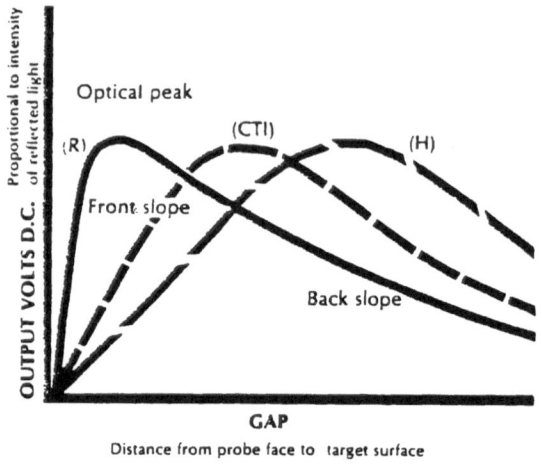

FIGURE 1 :   Fotonic Sensor Response for Various Bundle Structures

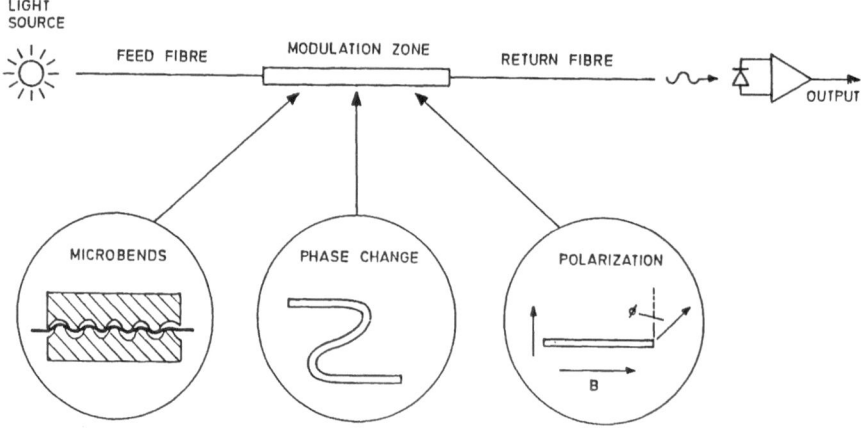

FIGURE 2 : Intrinsic Sensors

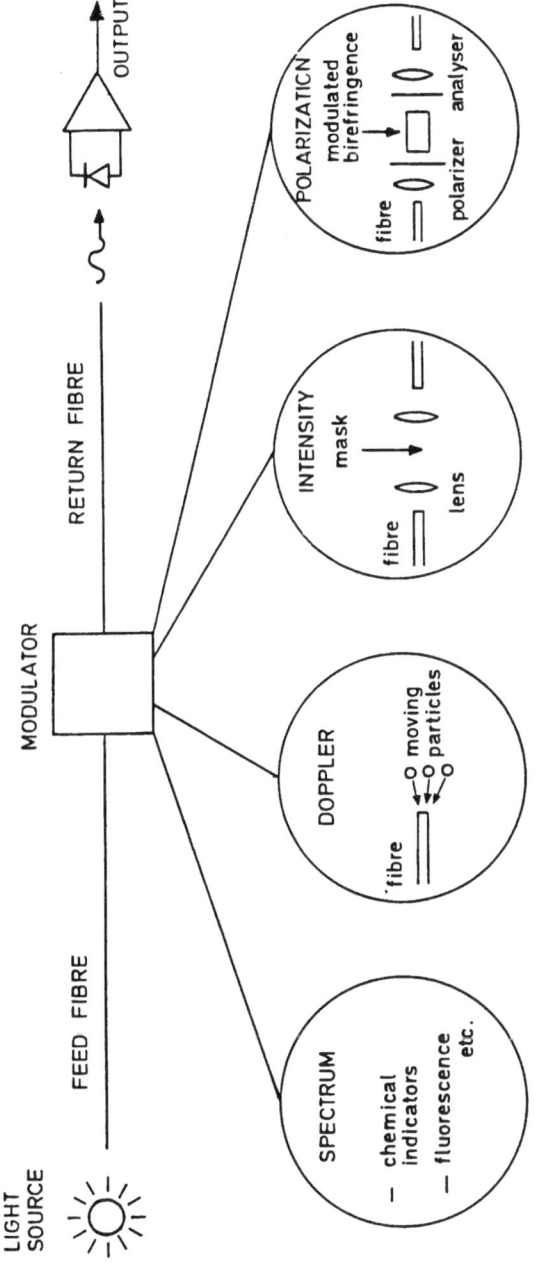

FIGURE 3

EXTRINSIC SENSORS

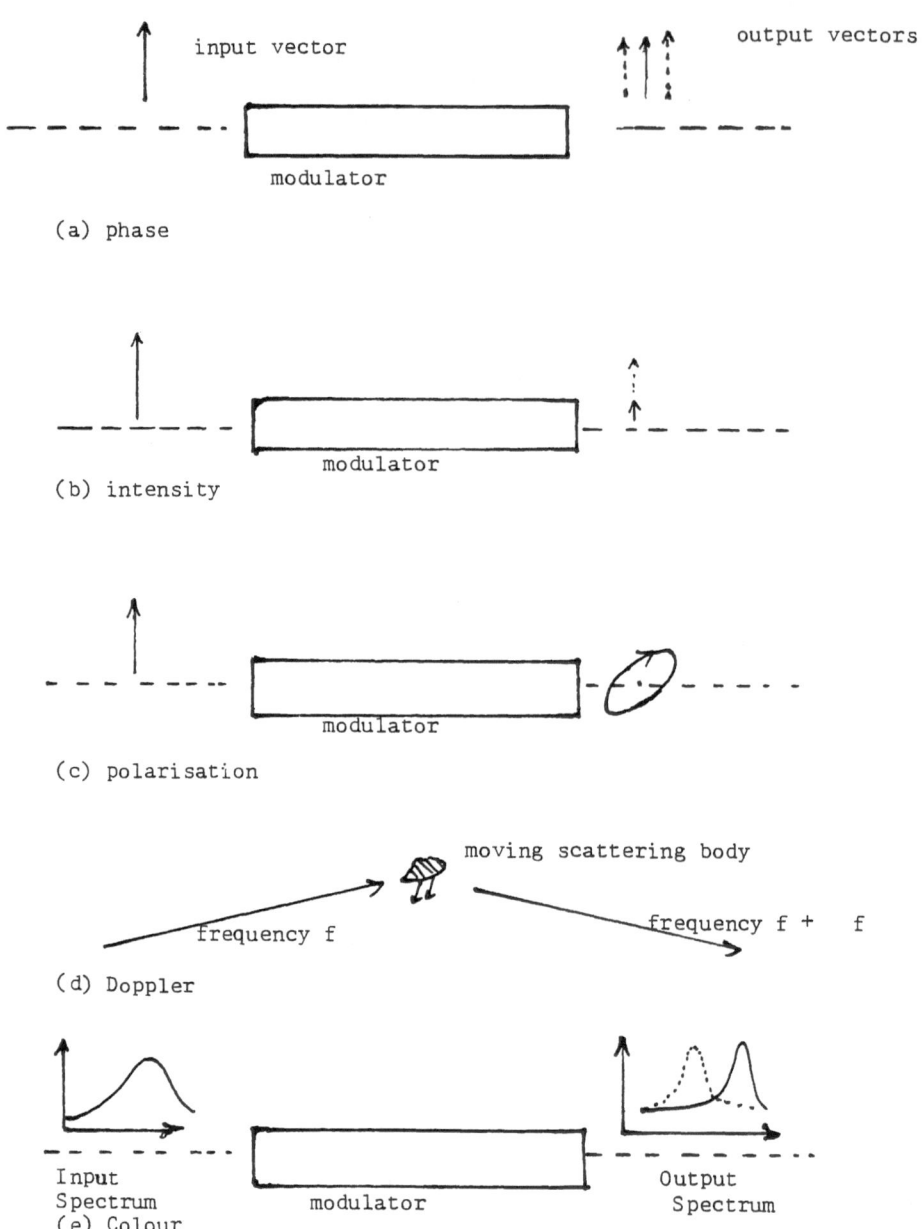

FIGURE 4: Schematic of basic optical modulation techniques

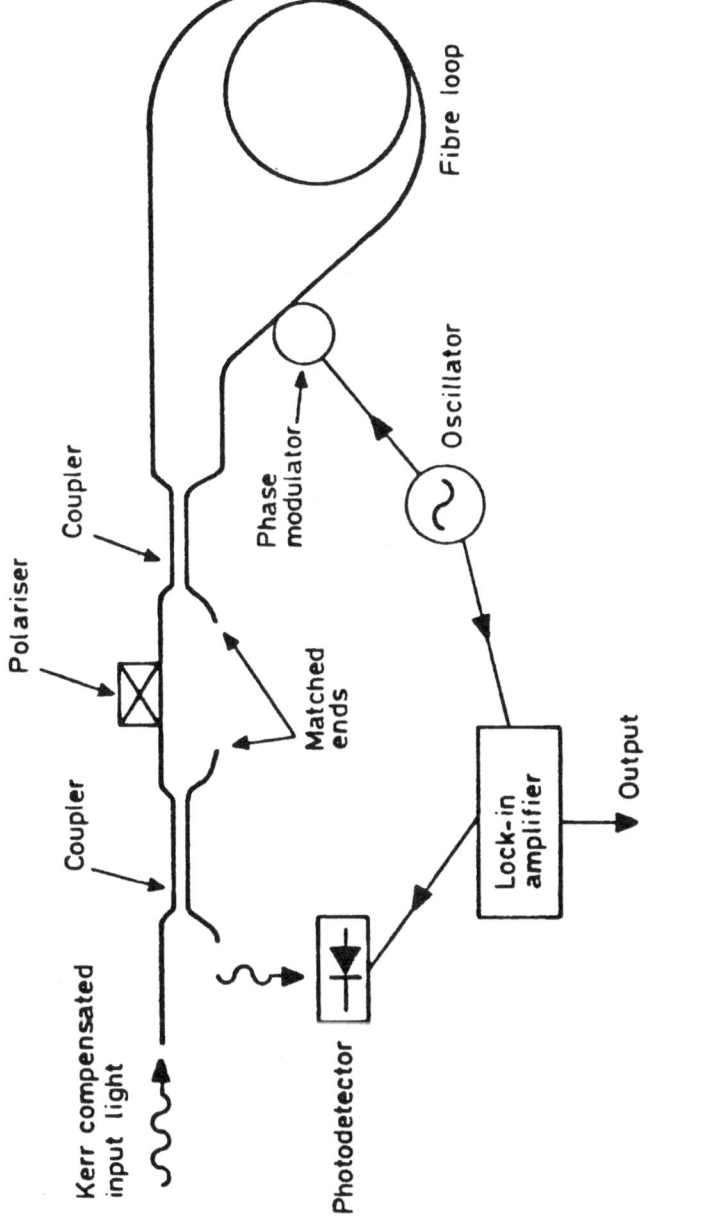

FIGURE 5: THE SIMPLEST LOW NOISE GYROSCOPE SHOWING THE USE OF LOCK-IN AMPLIFIER DETECTION

254

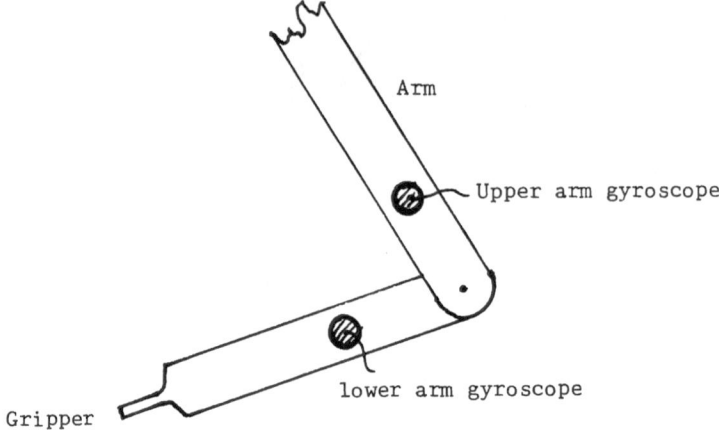

Figure 6: Possible use of fibre gyroscopes in robot arms

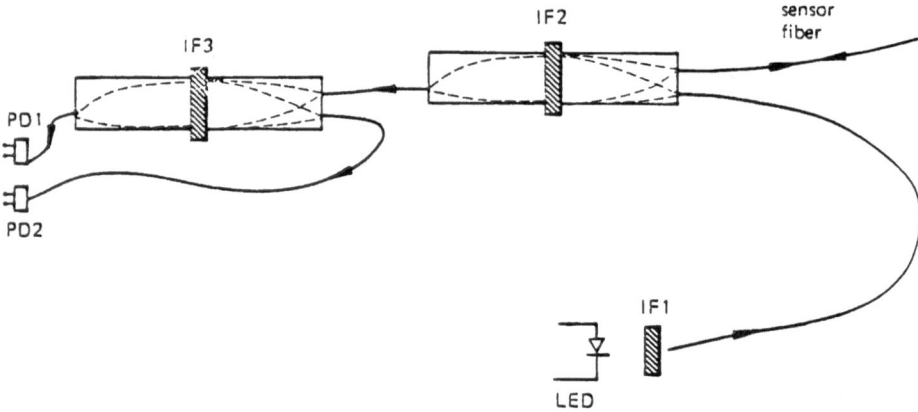

Figure 7: Optoelectronic system of the fiber-optic
          temperature sensor. The system is constructed
          using graded index rod lenses (GRIN) and
          interference filters (IF). A standard light-
          emitting diode (LED) and photodiodes are used.

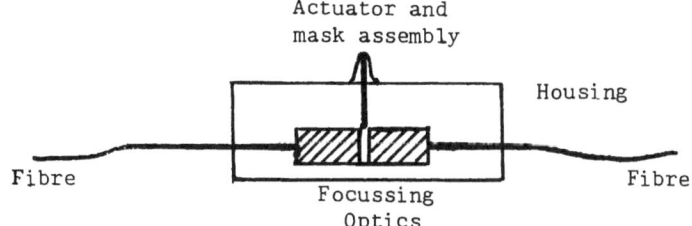

Figure 8: Scehmatic of fibre optic limit switch

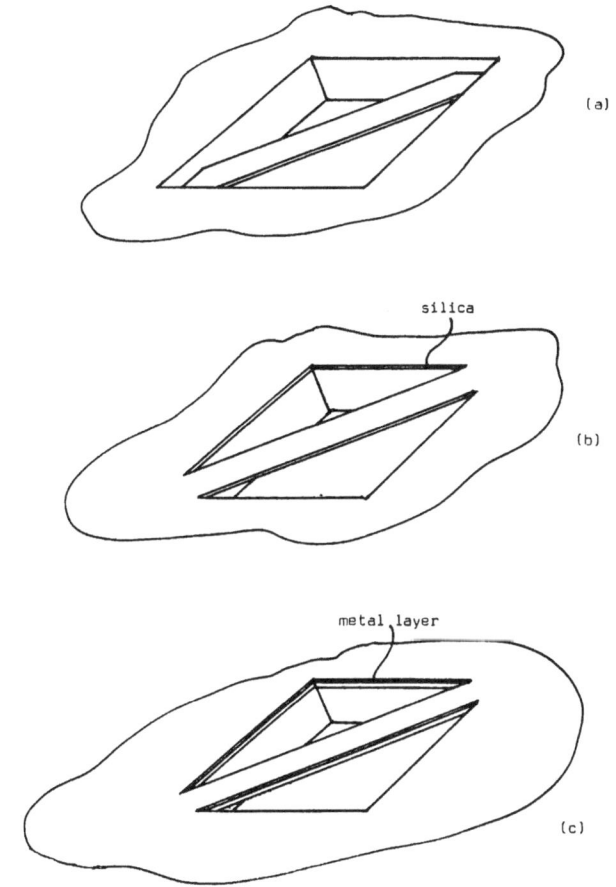

Figure 9:   Mechanical bridge structures fabricated (a) from silicon using a buried boron diffusion; (b) in silicon dioxide or nitride over an etched well; (c) metal coated

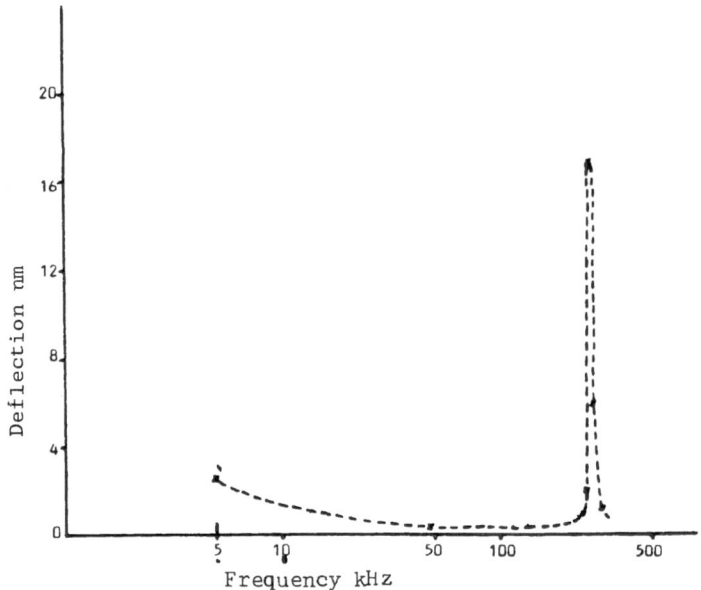

Figure 10:Frequency response of typical structure in figure 9

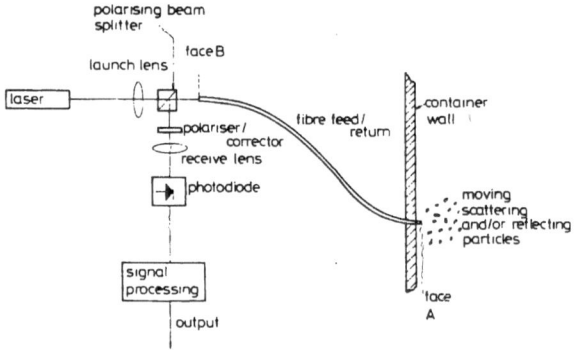

Figure 11: Schematic diagram of a fibre optic
Doppler anenometer.

Sensing surface

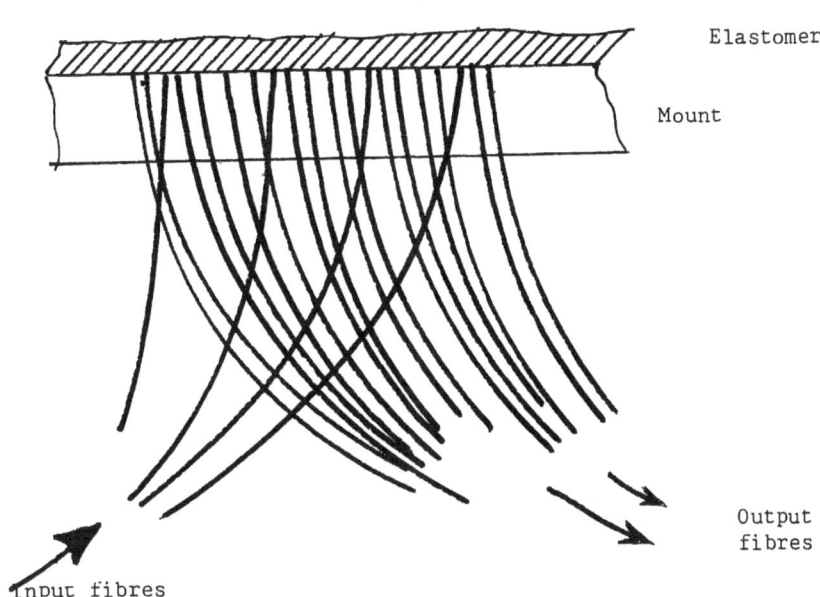

Elastomer

Mount

Output
fibres

Input fibres

Figure 12 Schematic diagram of fibre optic tactile sensor

Section 4

Applications

Models of Errors and Mistakes in Machine Perception:

First Results for Computer Vision Range Measurements

Ruzena Bajcsy
Eric Krotkov
Max Mintz

GRASP Laboratory
Department of Computer and Information Science
University of Pennsylvania
Philadelphia PA   19104

Abstract

This paper is a first step toward integrating sensor measurements of distance. Its major contribution is to identify and present qualitative models for the errors and mistakes introduced in three particular computer vision distance measurements: range from focus, range from point-based stereo, and range from line-based stereo.

These range measurement techniques are presented as computations, and their dominant sources of error are analyzed qualitatively. We propose to quantify the underlying models for these three range estimation techniques by deriving approximate confidence procedures for the intrinsic parameters and functions which characterize each technique.

1. Introduction and Motivation

When one deals with measurements of any kind, it is always the case that this measurement is accompanied by some error. What is an error? According to the methodology prevalent in physics, chemistry, and other sciences, it is a deviation from some ideal, or some standard. What is an ideal or a standard? The ideal comes from a theoretical description of the phenomenon which invariably is put into some mathematical language. This mathematical representation is also called a model. The word "model" is chosen not accidentally. Indeed it means that it is an idealization of the physical reality under precise conditions. A standard, on the other hand, is usually a very precise physical realization of the ideal with very small error. A standard is the best in some sense that we can get in a physical realization of the model. Error is different from discrepancy. The latter is the difference between two measured values of a quantity, for example obtained by two different measuring devices. There is still a further classification of errors: systematic error and random or stochastic error. Both these errors if well understood can be and should be included into the model, naturally using different mathematical tools for the systematic and stochastic errors. These tools are partly the subject of this paper. Both these errors can be evaluated by calibration of the instruments against standards. While the small random error is related to precision, the small systematic error corresponds to high accuracy.

Now that we have explained what we mean by error, we need to clarify what is a mistake? In our view a mistake is not a large error but a failure of a device or an identifiable component of a device. We realize that there is a very tenuous division line between large errors and mistakes. This is analogous to the problem in pattern recognition of variation

NATO ASI Series, Vol. F43
Sensors and Sensory Systems
for Advanced Robots
Edited by P. Dario
© Springer-Verlag Berlin Heidelberg 1988

within a category (errors) and the difference between categories (mistakes). Taking this analogy in to heart, mistakes may be modeled in the decision theoretic framework or represented in some inheritance graph. We shall explore these possibilities later in the paper.

We have been perplexed why these questions of errors and mistakes were not raised before in the vision community. We think that one reason is that in the past most of the results of various image processing and vision algorithms has been creating new images that are graphically displayed and evaluated by human observers. This kind of evaluation is inadequate from many real applications. For example, in industrial inspection, where the visual sensor acts as a measuring device or a probe and the results are used for making decisions on the quality control of the product, the accuracy, precision and error of these measurements become an important issue. Another domain where visual measurements must be used in a feedback is in robotic applications, both in manipulation and mobile robots. Simply, in all cases where a decision is based on visual measurement, it is obvious that the goodness and reliability of these measurements becomes the central issue.

Assumptions and the domain

We shall explore our ideas on errors and mistakes in the domain of computer vision with one particular task: Measurement of the three-dimensional distance. The assumption is that we do not know *a priori* the distance. We also assume that we have a pair of cameras observing one static, indoor, well-illuminated scene. We shall consider three different approaches of obtaining the distance through vision, as illustrated in Figure 1.

1. Distance from focus;
2. Distance from stereo, using a point matcher;
3. Distance from stereo, using a line matcher.

There are two reasons why we have chosen to discuss these three particular methods for measuring distance. First, we have implemented and are trying to evaluate them, so we are beginning to understand their theoretical and practical capabilities and limitations. Second, they can be viewed as processes which provide complementary and redundant measurements. Two measurements are *complementary* (independent) if they measure the same physical quantity with different processs; here, focus and stereo provide complementary measurements of distance. Two measurements are *redundant* if they measure the same physical quantity with the same processs; here point-based and line-based stereo provide redundant measurements of distance, because they use the same process (triangulation) with different features (points and lines). This is important because if these measurements can be integrated into one best estimate of distance, we will have a basis for integrating measurements from any processs, whether complementary or redundant.

This paper is a first step toward a methodology for integrating sensor measurements using specific sensor models. In our view, its major contribution is to identify and present preliminary models for the errors and mistakes introduced in three particular distance measurements, and to begin thinking about how to combine them. Future work will address quantitative models of errors and mistakes in computer vision distance measurements, and tactile sensing.

This paper consists of eight sections. Section 2 discusses the errors and mistakes introduced in the image formation process. Section 3 describes and analyzes the range from focus technique. Section 4 discusses the computation of distance from stereo disparities, and some errors that apply to all such computations. Section 5 presents and analyzes the

Figure 1. Architecture of range computations.

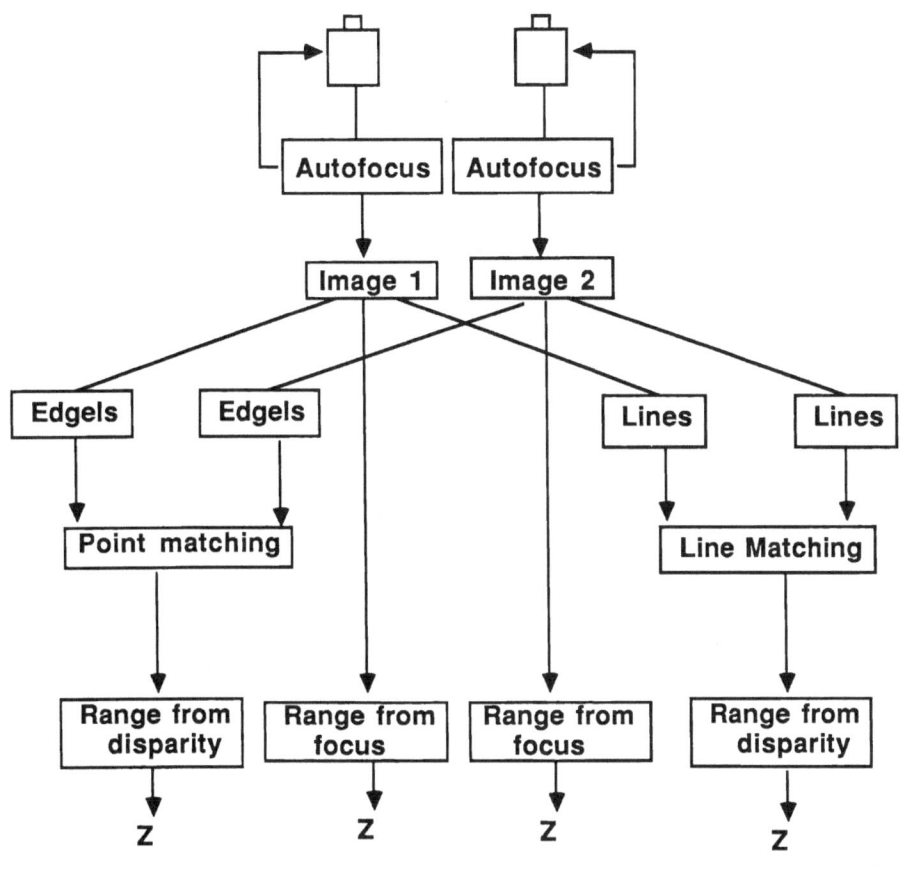

point-based stereo technique. Section 6 presents and analyzes the computation of stereo disparities based on finding and matching lines. Section 7 presents a framework for error analysis using confidence procedures. Section 8 ends the paper with general remarks, final conclusions, and the work to be done in the future.

2. Image formation

To the extent that all the distance computations compute local first derivatives of the intensity function, any noise in the image will propagate and be amplified. Hence it is important to understand and model the noise in the digitized values. A good review of the noise characteristics of CCD transducers can be found in the article by Purll [10]. This section will discuss our first results in modeling the noise in our camera system. A more detailed analysis is being prepared by Krotkov, McKendall and Mintz [8]. A crude model of the image formation process is illustrated in Figure 2, which lists some salient features of each component.

Figure 2. Image formation.

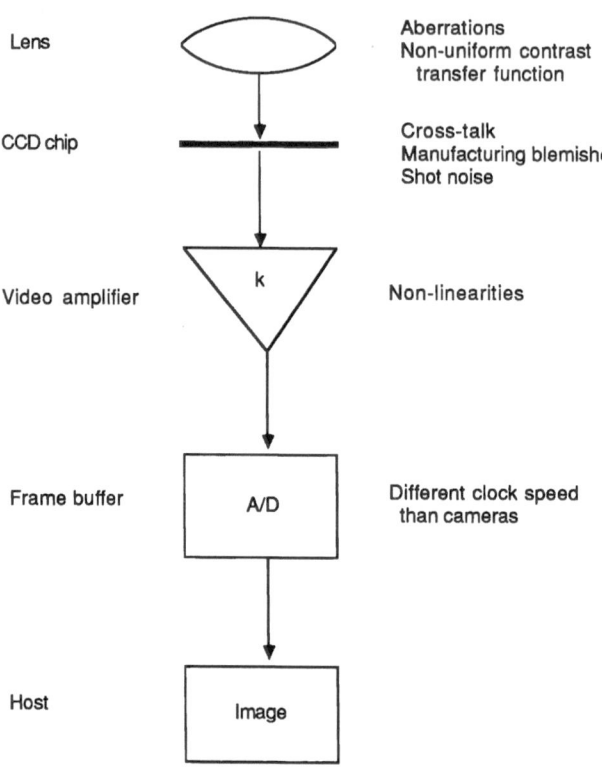

Lens		Aberrations Non-uniform contrast transfer function
CCD chip		Cross-talk Manufacturing blemishes Shot noise
Video amplifier	k	Non-linearities
Frame buffer	A/D	Different clock speed than cameras
Host	Image	

Notation

We will denote the image intensity function as the three-dimensional function I, with spatial arguments u and v, and temporal argument t. Since the image intensity is represented as 8 bits, $0 \leq I(u,v,t) \leq 255$.

Much of the analysis involves taking a time series of images. Let $\bar{I}(u,v)$ denote the sample mean of the image intensities over N time samples:

$$\bar{I}(u,v) = \frac{1}{N}\sum_{i=1}^{N} I(u,v,t) .\tag{1}$$

The spatial variance in a 5x5 neighborhood of the means is computed by:

$$s^2(u,v) = \sum_{i=-2}^{2} \sum_{j=-2}^{2} (\bar{I}(u+i,v+j) - \bar{I}(u,v))^2 .\tag{2}$$

2.1. Spatial noise analysis

Experiments were conducted to determine the *spatial* characteristics of the digitization process. The experiments reveal the "signature" of the digitizer with no illumination, with constant illumination, and with illumination from a room scene.

Dark signature

To identify the dark signature of each camera a sequence of $N=100$ images is taken with the lens cap on. The sample mean over time $\bar{I}(u,v)$ of each pixel $0 \le u,v \le 511$ is computed using Equation (1).

For each camera, there are a small number of pixels with non-zero mean and non-zero variance. Repetition of the experiment reveals that the non-zero values occur at the same locations in the respective images. These non-zero values are probably caused by the fact that the CCD elements being digitized into those locations are "blemished". The manufacturer allows a certain number of blemishes on each CCD chip. An element is considered blemished if it exhibits a spurious output (in comparison to its nearest neighbors) of more than 10% of the saturation voltage [6, p. 72]. The blemishes are caused by material variations across the chip, variations in element area, and processing variations in manufacturing (the way the silicon is grown). The dark signal, electric current formed by thermal leakage which is indistinguishable (by the video amplifiers) from photocurrent, also carries element-to-element non-uniformities but these are of small magnitude and are effectively "averaged out" by computing $\bar{I}$ [10].

Uniform illumination signature

Next a sequence of images is taken with a uniform illumination, accomplished by placing a nylon diffuser directly over the lens, essentially employing a translucent lens cap. The mean and variance are computed according to Equations (1) and (2), respectively, with $N=100$.

The first result from this experiment is that the intensities recorded along a 20-pixel band along the exterior border of the image have a significantly lower mean and higher variance than the intensities recorded in the interior. This is an artifact of (1) digitizing a 380x488 array of detectors into a 512x512 array of values, and (2) the fact that the digitizer and the cameras operate at different sample rates, 4.77 MHz and 7.16 MHz, respectively. This is clearly a mistake in the design of our system, and rather than model the behavior of the border values, we choose to ignore them, and use pixels only from the interior.

The second result is that there are some "stuck" pixels. These have large variances, and are located in the same places that were marked as "blemished" above.

The third result from this experiment is that the intensities form an annular pattern, with pixels at a given distance from the center (lying on an ellipse) having equal intensities, and with pixels at greater distances from the center having lesser intensities. This suggests that on-axis light rays are attenuated less than the off-axis rays, i.e., the contrast transfer function of the lens is maximum at its center and diminishes with distance from the center. The intensities seem to vary smoothly inside this annular distribution, and it appears that they vary linearly with field angle (distance to the center of the lens).

2.2. Temporal noise analysis

A number of experiments are being conducted to identify and model the temporal noise characteristics of the digitization process [8]. It is clear that the intensity samples are not temporally independent, i.e., the noise process is characterized by time dependency. This is a significant finding, which profoundly affects the filtering necessary to reduce the noise. We are still investigating the quantitative characteristics of the noise process, and its physical origins.

2.3. Summary

A model of the image formation is a necessity for any further consideration of errors and mistakes, since its parameters come to play in the algorithms for focus, stereo and all other vision and image processing algorithms. Our approach is to treat the whole camera system as a black box and make hundreds of input/output measurements and develop a stochastic model of its behavior. This methodology is especially appropriate when the system is too complicated or not well-enough understood to model its physical behavior explicitly. We feel however that predictive modeling must be used whenever it is possible, since this is the only way to obtain absolute benchmarks, and hence assert absolute mistakes. Having the black box approach only one can speak only about mistakes in a statistical sense.

The conclusions we can draw about the camera system so far are: (1) There are blemished pixels along the border and scattered around the interior of the image; (2) the lens attenuates the intensity of off-axis rays; (3) the intensity samples are not temporally independent. It remains to develop quantitative models of these phenomena.

3. Range from Focus

This section will describe a method for computing range from focus, as well as its inherent and practical limitations. The ideas and results presented here are reported in considerably greater detail by Krotkov [9].

3.1. Method

Our method for computing range from focus involves two steps. First, given the projection $P'=(u,v)$ onto the image plane of an object point $P=(X,Y,Z)$ (Z unknown), we find the lens focal length f which brings P' into sharpest focus. Second, given f from step one, we compute the Z-component of P using the thick lens law of first-order optics.

The first problem, how to best determine the focal length providing the sharpest focus on an object point at an unknown distance, is decomposed into two parts: (i) how to measure the sharpness of focus with a criterion function, and (ii) how to optimally locate the mode of the criterion function. The criterion function, which approximates the magnitude of the intensity gradient, can be stated as:

$$\sum_{x,y} S(x,y) \tag{3}$$

where

$$S(x,y) = \sqrt{(i_x * I(x,y))^2 + (i_y * I(x,y))^2} \; ;$$

$$i_x = \begin{bmatrix} -1 & 0 & 1 \\ -2 & 0 & 2 \\ -1 & 0 & 1 \end{bmatrix} \; ; \qquad i_y = \begin{bmatrix} 1 & 2 & 1 \\ 0 & 0 & 0 \\ -1 & -2 & -1 \end{bmatrix} \; ;$$

* denotes convolution; and x and y range over a (small) neighborhood around P'. In practice this criterion function takes its maximum value when the image is sharply focused, proves to be unimodal, varies monotonically with focal length on either side of the mode, and is relatively easily computed. However it is fairly sensitive to noise in the digitized samples, as illustrated in Figure 3. This problem is solved presently by averaging over many samples.

Figure 3. Criterion function temporal variations.

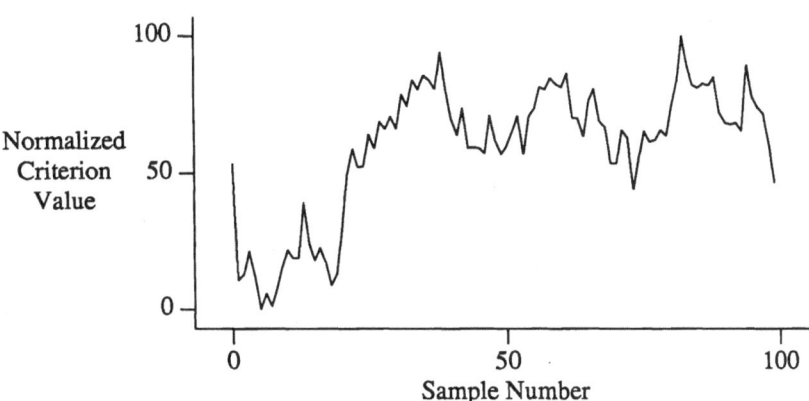

The intensities in a 20x20 window were sampled at 30 Hz, with no filtering. The criterion function was then evaluated for each sample.

The Fibonacci search technique is employed to locate the mode of the criterion function, since it is the optimal method for finding the extrema in a unimodal function (the focus criterion function) of one variable (focal length). In general, the Fibonacci search successively narrows the search interval until its size is a given fraction of the initial search region. In practice the Fibonacci search strategy performs extremely well, providing images that appear sharply focused to observers after approximately 11 iterations.

The second problem, how to compute the distance to an object point given the focal length of sharpest focus, is solved by application of the Gaussian thick lens law. The object distance Z is computed as

$$Z = \frac{k(f+a)}{k-f-a} - b \tag{4}$$

where f is the focal length providing sharpest focus on P', and k, a, b are empirically determined constants.

Using Fibonacci search for the peak of the criterion function and Equation (4), the distance of objects between 1 and 4 meters can be computed with an error less than 5 percent of the object distance. The results for objects at greater distances are not yet available.

3.2. Error analysis

There are a number of limitations of this method. Perhaps the most significant is that it applies to only one point at a time. Thus with a single camera, range from focus can not be computed in parallel. Other limitations include the precision to which d_{in} and f can be measured, the performance of the criterion function, and the nonlinearity of Equation (4).

Spatial quantization is one limitation of this method which can not be circumvented by more precise measurements or better equipment or algorithms. Because the photoreceptors have finite area, an object point may lie at a number of different distances and still be imaged sharply on the same receptor. The distance in object space between the nearest plane and the farthest plane at which satisfactory definition is obtained is the *depth of field*

$$\frac{2Zafc(Z-f)}{a^2f^2 - c^2(Z-f)^2} \tag{5}$$

where f is the focal length, c is the largest dimension of the photoreceptor cell, a is aperture diameter, and Z is the object distance. From Equation (5), it is clear that as c increases, so does the depth of field. The accuracy of the range computation is actually less than the depth of field of the lens, i.e., the computation is as accurate as physically possible. This shows that the dominant source or error in range from focus is the dimension of the photoreceptor cell.

It is a mistake to try to focus on an inappropriate window, one which contains no features or the projection of a depth discontinuity or an occluding edge.

4. Computing range from stereo disparities

There are some errors encountered by any computation of range based on finding disparities. First we will present a method for computing range, and then discuss the important error parameters.

4.1. Method

This section describes the transform from relative disparities (distances in image space) into absolute distances (distances in E^3), based on the geometry illustrated in Figure 4 (loosely based on the analysis by Torre *et al* [13]). We do not assume that the cameras are parallel.

If (u_l, v_l) are the coordinates in the left image of the perspective projection of an object point $P = (X, Y, Z)$, then by similar triangles:

$$u_l = f \frac{x_l}{z_l + f} \quad \text{and} \quad v_l = f \frac{y_l}{z_l + f} \tag{6}$$

Similarly for the right image:

$$u_r = f \frac{x_r}{z_r + f} \quad \text{and} \quad v_r = f \frac{y_r}{z_r + f} \tag{7}$$

The two image coordinate systems are related by a translation $\underline{D} = [a \ b \ c]^T$ and a 3x3 rotation matrix $[R]$ (determined by the three Euler angles Φ, θ, Ψ under the z, x', y'' convention [4, p. 108]) such that

$$\begin{bmatrix} x_r \\ y_r \\ z_r \end{bmatrix} = \begin{bmatrix} R \end{bmatrix} \begin{bmatrix} x_l - a \\ y_l - b \\ z_l - c \end{bmatrix} \tag{8}$$

Figure 4. Stereo geometry.

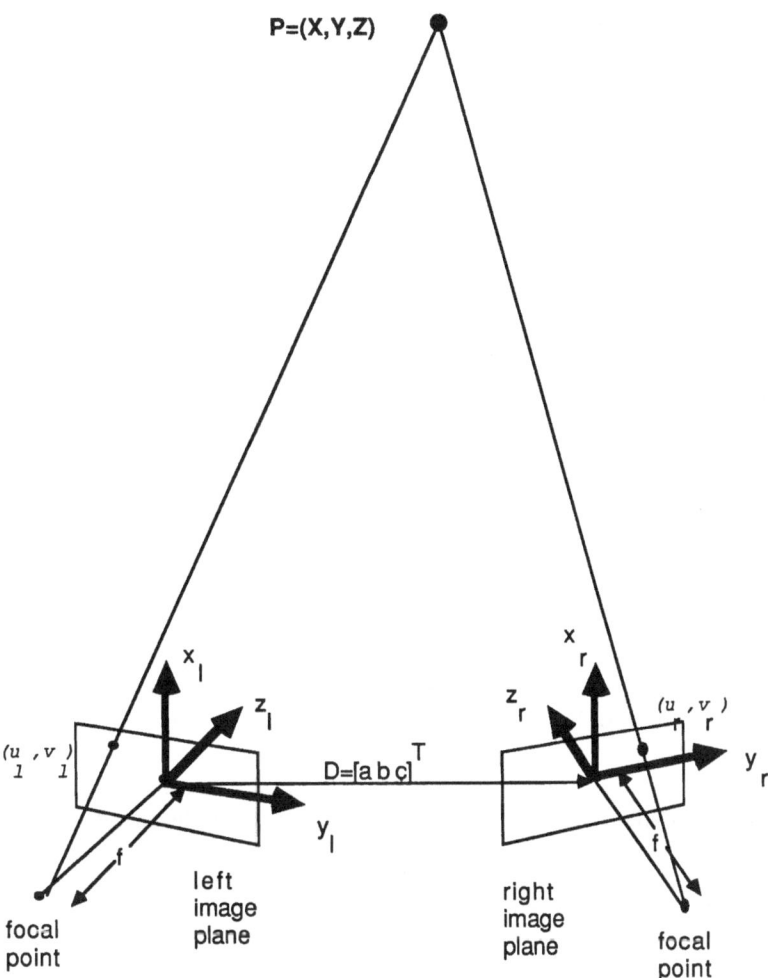

The geometry underlying the transformation from relative disparities to absolute distances.

Assuming that the scan lines are registered, $a=0$. Assuming that the optic axes are coplanar (i.e., the cameras are not tilted with respect to each other), $\Psi=0$. Since our cameras can not roll (rotate around the z-axis), $\Phi=0$.

Under these assumptions Equation (8) is given by

$$\begin{bmatrix} x_r \\ y_r \\ z_r \end{bmatrix} = \begin{bmatrix} 1 & 0 & 0 \\ 0 & \cos\theta & \sin\theta \\ 0 & -\sin\theta & \cos\theta \end{bmatrix} \begin{bmatrix} x_l-a \\ y_l-b \\ z_l-c \end{bmatrix} \tag{9}$$

Since the matching procedures produce horizontal disparities, we want to express z_l in terms of v_l and v_r. Using the equations in v_l and v_r derived by similar triangles, and solving algebraically for the object distance $Z \equiv z_l$ yields

$$Z = \frac{f}{\xi}\left[(b-v_l)(v_r \sin\theta+f\cos\theta)+fv_r+c\,(f\sin\theta-v_r\cos\theta)\,; \right] \tag{10}$$

where f is the focal length of the lens, θ is the convergence angle, and $\xi = \sin\theta(v_l v_r+f^2)+f\cos\theta(v_l-v_r)$. For $c=\theta=0$ this reduces to the familiar

$$Z = f\frac{b-v_l}{v_l-v_r}\,. \tag{11}$$

4.2. Error analysis

The accuracy of the matcher and the registration procedure limit the accuracy of the computed disparities; this will be discussed in the sections on the matcher error analysis. For given disparities, the range computation is limited by the precision in measuring b, θ, c, and f. Torre et. al. [13] analyzed range errors due to mechanical positioning errors of the cameras, the focal length of lenses, as well as the errors due to lens distortion, in matching and quantization in the picture.

Another error in the distance measurement is due to the fact that an area is projected into one point. This area naturally is bigger with the increased distance from the camera. This last error has been studied by Solina [12], who considered the worst case, and showed that the relative range error ΔZ for focal length f is:

$$\Delta Z = \frac{aZ}{(bf-aZ)} \tag{12}$$

where a is the pixel size, Z is the object distance, and b is the distance between the focal points (the baseline). From this analysis it follows that a small pixel size is most desirable. This however is quite costly. Another two parameters can be controlled, that is to use longer focal length and/or increase the baseline. Unfortunately both of these parameters if not chosen carefully have bad consequences: too long a focal length decreases the sharpness of focus and too long a baseline makes the matching problem more difficult if not impossible.

5. Point-based stereo

This section will describe a method for computing stereo disparities by matching edge points, as well as its inherent and practical limitations. The ideas and results presented here are drawn from Smitley [11]. The matcher is based on the following assumptions: (1) The two images are properly aligned so that the scan lines correspond to the epipolar lines and hence one can reduce the stereo matching problem into a one-dimensional problem. (2) Matching occurs only there where edge point features exist. (3) Edges occurring on horizontal lines cannot be matched, they are ambiguous. (4) There is an *a priori* range limit on disparities. All disparities outside of this range are rejected.

5.1. Method

First a registration technique developed by Izaguirre [7] brings the epipolar and scan lines into coincidence. The matching algorithm begins by locating *edgels* in each image I at multiple resolutions. For a Gaussian window G with a standard deviation σ, let the edge detection filter $f(x,y) = \nabla G(x,y)$. Let $P(x,y)$ denote the output of the edge detection filter at (x,y). $P(x,y)$ is defined by the convolution of $f(x,y)$ with $I(x,y)$, i.e., $P(x,y) = f*I(x,y)$, which has magnitude $||P(x,y)|| = \sqrt{P_x^2 + P_y^2}$ and direction $\theta = \tan^{-1}\left[\dfrac{P_y}{P_x}\right]$. *Edgels* are local maxima in $||P||$, computed using Canny's [3] method of non-maximum suppression.

Matching proceeds from coarse (blurred by G with larger σ) to fine resolution. The search window is determined by the size of the filter applied before the edge detection process begins. A left edgel at $L=(u_l,v_l)$ matches a right edgel at $R=(u_r,v_r)$ provided that their gradient vectors are similar ($||P_L|| \approx ||P_R||$ and $\theta_L \approx \theta_R$) and that the intensity distributions around L and R are correlated. Matches from coarser resolutions are used to guide the search for matches at finer resolutions. The disparity of a match (L,R) is the horizontal pixel distance $|v_l - v_r|$.

5.2. Error analysis

The matcher has been tested on diverse images, and generally matches between 75 and 90 percent of the vertically oriented edgels. It has proven to be robust in the presence of noise and poor illumination. However, the matcher commits both errors and mistakes.

Matching errors, i.e., errors in the disparity values, are caused by inadequate localization of edge points. In the absence of noise the localization uncertainty is equal to σ. Hence the smaller the σ, the more sensitive is the edge detector. To model the localization error in the presence of noise, we extend the analysis presented by Bajcsy, Liebman and Mintz [1] to two dimensions, and express the response of the edge filter to an intensity function $I(x,y)$ which is the sum of a noise-free image $I_0(x,y)$ and a zero mean white noise process $v(x,y)$ with constant spectral density r. The response of the edge filter to the noise process $v(x,y)$ is again a second-order wide-sense stationary stochastic process with zero mean and variance ρ, where

$$\rho = \int_{-\infty}^{\infty} \int_{-\infty}^{\infty} r\, f^2(x,y)\,dxdy = \frac{\sqrt{2}r}{8\pi\sigma^4} \tag{13}$$

The effect of noise on estimating the location of (x^*,y^*) -- the maximum of $P_0(x,y)$, i.e., the maximum response of the edge filter to $I_0(x,y)$ -- can be determined by considering the effective signal-to-noise ratio SNR at point (x^*,y^*):

$$SNR(x^*,y^*) = \frac{P(x^*,y^*)}{\sqrt{\rho}} \tag{14}$$

For a unit ramp intensity function in one dimension, $SNR(x^*)$ is proportional to $\sqrt{\sigma}/r$. This result is not significantly different in two dimensions, and the important point is that the signal to noise ratio depends directly upon σ.

From the analysis above, it follows that the disparity error due to point-matching of an edge element pair is a function of σ. Without noise, the disparity error is $\pm\sigma$. With noise the disparity error is proportional to $\sqrt{\sigma}/r$. In summary, the error of the whole process is the sum of all the errors from the cameras and point localization.

Since the disparity errors are a function of point localization, we propose that the point matching process itself results not in errors but in mistakes only. A mistake, i.e. a false match, occurs only when one of the four stated assumptions does not hold. This kind of mistake is analagous to the failure of a physical device.

6. Line-based stereo

This section will describe a method for computing stereo disparities based on matching lines, as well as its inherent and practical limitations. The ideas and results presented here are drawn from the forthcoming thesis by Henriksen [5].

6.1. Method

There are two basic steps in this stereo approach, and hence two different error sources: line extraction and line matching to compute stereo disparities.

The line extraction procedure goes as follows:
1. Compute the gradient of the grey-value function at every pixel in the picture.
2. Group pixels into edge-support regions based on similarity of gradient orientation.
3. Approximate the grey-value function in an edge support region by plane.
4. Compute a weighted average of the grey-values in the edge-support region, and use it to determine a horizontal plane.
5. Intersect the two planes computed in step 3 and step 4, giving an infinitely long straight line.
6. Project the line computed in step 5 onto the picture plane, and intersect it with the boundaries of the edge-support region, giving a line segment.

This algorithm produces a polygonal description of a picture, i.e., the picture is described by a set of line segments.

Line matching goes as follows:

1. Construct an Adjacency Graph such that: (a) for a given line it can tell which are neighbors to the given line, and (b) for a given point it can tell which lines are in the neighborhood of the given point.

2. Construct a Disparity Graph providing similar information as the Adjacency Graph but in a smaller neighborhood.

3. Generate Hypotheses, in four steps:
   1. Select a line segment L (let us say in the left) image.
   2. Compute the midpoint of the line segment.
   3. Compute the epipolar line in the right image corresponding to the midpoint of L.
   4. Find all line segments R in the right image which satisfy the following constraints:
      a) R intersects the epipolar line corresponding to the midpoint of L
      b) $|\operatorname{disp}(L,R)| < \operatorname{disp}[\max]$
      c) $length(L) \approx length(R)$

d) $arg\,(\Delta L\,){\approx}arg\,(\Delta R\,)$

e) $||L\,||{\approx}||R\,||$

The set of line pairs {L,R,disp} found under step 4 is a set of possible matches, and the elements in this set are called the hypotheses derived from L.

4. Verify hypotheses by testing whether the hypothesis satisfies two constraints: (i) Uniqueness - a line segment in one image can only have one match in the other image; (ii) Continuity - the disparity of neighboring line segments must have similar disparities.

6.2. Error analysis

As in the case of the point matcher, the line matcher commits both errors and mistakes. Because the line matcher uses more local information than the point matcher, its error parameters are more complicated.

In the line extraction process the computation of the gradient causes error σ, which is exactly the same as the point localization error in the disparities computed by the point matcher. In addition to the point localization error, the grouping of edges into regions is a source of error which has several parameters. The first parameter can be quantified by the size of the edge-support neighborhood. The larger the neighborhood, the larger the danger of including irrelevant information. The second parameter can be quantified by the least-squares residual for fitting the grouped points to a plane. The larger the residual, the worse is the fit to a plane.

In the stereo matching procedure errors, may occur during the hypothesis generation part, where the properties of matched lines must agree up to a certain ε measuring the tolerance with which two angles or two lengths must agree. This is not so large. Mistakes occur during the verification part where the uniqueness constraint (that is a line segment in one image can have only one match in the other image) prevents matches of similar lines in the neighborhood. This is illustrated in Figure 5. Figure 5a and 5b show the original left and right images; although the reproductions do not show it so clearly, there is a significant contrast difference in the two images. Figure 5c and 5d show the lines extracted from the original images. In the window marked, the left image contains two parallel lines, and the right image contains only one. This is due to (i) the contrast difference between the two images, and (ii) the fact that the gradient directions in the left image vary little and in the right image they vary greatly. As illustrated in Figure 5e, in this case the system understandably matches only one line instead of two. This is an example of an error (failure of line detection) compounded into a mistake (not matching a line).

It is an interesting consequence of this analysis that for more regular objects one needs bigger and bigger features for a unique match.

7. Error Analysis Via Confidence Procedures

We propose to quantify the underlying models for these three range estimation techniques by deriving approximate confidence procedures for the intrinsic parameters and functions which characterize each technique. The confidence measure will provide us with a methodology for: (i) quantifying the individual sources of error based on sensor measurements, (ii) deriving error measures for computed parameters, i.e., the effect of the propagation of the measurement errors, and (iii) obtaining sensitivity models for our measurement techniques and computational algorithms.

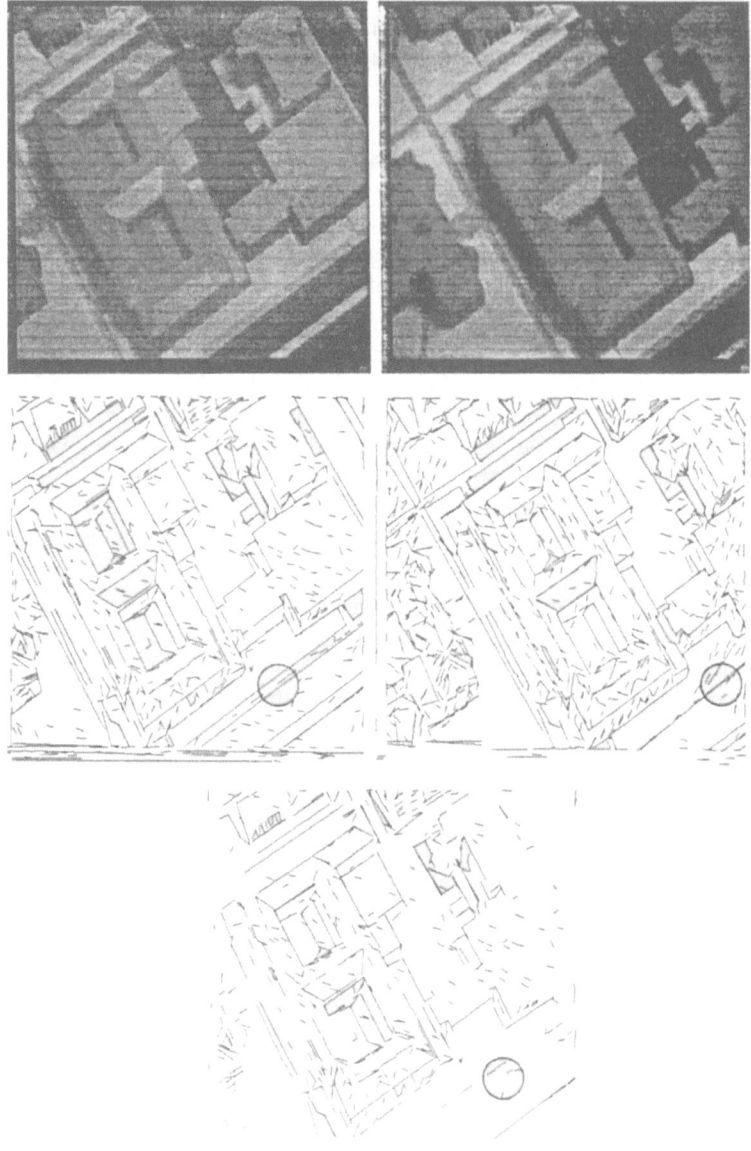

(a) Left image. (b) Right image. (c) Left lines. (d) Right lines. (e) Matches.

7.1. Concepts and Definitions

Confidence procedures can be viewed as set-valued estimates, as opposed to point estimates. In order to compare these concepts, we make use of the following specific example. Suppose $\underline{Z} = (Z_1, Z_2, \cdots, Z_n)^T$ denote a vector of n independent identically distributed (IID) observations which are normal with mean θ and variance σ^2 ($N(\theta,\sigma^2)$). Assume that σ^2 is known, and that θ is an unknown point in a known parameter set Ω. In this example we will assume that $\Omega = [-d, d]$. Then, based on this given measurement model, we find: (i) a point estimate for θ based on on $\underline{Z}$; and (ii) a fixed size confidence procedure for θ based on $\underline{Z}$.

7.2. Point Estimation

In order to obtain a point estimate for θ, we shall employ the method of maximum likelihood estimation (MLE) (Bickel and Doksum, [2]). The method of maximum likelihood dictates that we compute $\delta_M(\underline{Z})$, where:

$$\delta_M(\underline{Z}) = arg_\theta \max f(\underline{Z} \mid \theta) \tag{15}$$

and $f(\underline{Z} \mid \theta)$ denotes the conditional density of $\underline{Z}$ given θ.

The application of the MLE method in this example yields:

$$\delta_M(\underline{Z}) = d, \quad \overline{Z} < d; \tag{16}$$

$$\delta_M(\underline{Z}) = \overline{Z}, \quad |\overline{Z}| \le d; \tag{17}$$

$$\delta_M(\underline{Z}) = -d, \quad \overline{Z} < -d; \tag{18}$$

where $\overline{Z} = \frac{1}{n} \sum_{i=n}^{n} Z_i$.

7.3. Confidence Procedures

The decision rule $\delta^*(\underline{Z})$ defines an optimal fixed size confidence procedure of width $2e$, if for all $\delta(\underline{Z})$:

$$\inf_\theta P_\theta[\theta \varepsilon C^*(\underline{Z})] \ge \inf_\theta P_\theta[\theta \varepsilon C(\underline{Z})]; \tag{19}$$

where:

$$C^*(\underline{Z}) = [\delta^*(\underline{Z})-e, \delta^*(\underline{Z})+e]; \text{ and} \tag{20}$$

$$C(\underline{Z}) = [\delta(\underline{Z})-e, \delta(\underline{Z})+e]. \tag{21}$$

We illustrate the solution $\delta^*(\underline{Z})$ to this problem for the case where $d = 3e$:

$$\delta^*(\underline{Z}) = 2e, \quad a+2e \le \overline{Z}; \tag{22}$$

$$\delta^*(\underline{Z}) = \overline{Z}-a, \quad a \le \overline{Z} \le a+2e; \tag{23}$$

$$\delta^*(\underline{Z}) = 0, \quad -a \le \overline{Z} < a; \tag{24}$$

$$\delta^*(\underline{Z}) = \overline{Z}+a, \quad -a-2e \le \overline{Z} \le -a; \tag{25}$$

$$\delta^*(\underline{Z}) = -2e, \quad \overline{Z} < -a-2e; \tag{26}$$

where: the parameter a satisfies $2F(-a-e) = F(a-e)$; and F denotes the CDF of an $N(0,\sigma^2/n)$ variate.

The general solution to this problem appears in [14]. This solution can be extended to include a wide variety of non-Gaussian sampling distributions with uncertain scale factors, as well as ε-contaminated sampling distributions. The details of these extensions appear in [15].

8. Discussion

This paper has presented an analysis of the errors and mistakes made in computing the distance of objects using three particular range measurement techniques: focus, point-based stereo, and line-based stereo. First we looked at the noise in the image formation process. Then the range measurement techniques were presented in some detail, and their most important error parameters were identified. Finally confidence procedures were proposed to quantify those error parameters.

The noise in the image formation process propagates into all of our computations, so it is very important to model this noise. Our first qualitative findings about this model are: (1) there are blemished pixels along the border and scattered around the interior of the image; (2) the lens attenuates the intensity of off-axis rays; (3) the temporal noise is probably not normally distributed. It remains to develop quantitative models of these phenomena.

For the range from focus computation the errors include the precision with which the focal length can be measured, the temporal variations of the criterion function due to digitization noise, and the size of the photoreceptor cells. The latter is the dominant error for our present implementation. Mistakes are committed by trying to focus on a window with no features, or with features at different object distances.

The errors in the range from stereo disparity computation are due to the precision in measuring (i) the focal length of the lens, (ii) the displacement between the cameras, (iii) the convergence angle, and the errors in the disparities computed by each of the matching algorithms. The accuracy of the stereo disparities computed by point-matching is determined by the uncertainty of point localization, which is $\pm\sigma$ No errors are introduced by the matching, only mistakes. The accuracy of the stereo disparities computed by line-matching is determined by the uncertainty of point localization ($\pm\sigma$ and errors in line-finding which can be quantified by the size of the grouping region and the least-squares residual.

In general, the error of a matcher is disproportional to the complexity of the matched feature. So the edge-based matcher will be more error-prone than the line-based matcher. This confirms the intuitive notion that edges are less reliable features than let us say lines or other bigger features. In fact as we said before the matcher makes only mistakes. Hence our final important conclusion about computing range from stereo: One cannot accept the disparity values from stereo without some guidance which can come either from *a priori* expected knowledge about the distance or from other more direct measurement of the distance, like range from focus or vergence. Then one proceeds: if the range from stereo agrees with some other measurement of the same distance then neighboring distance values computed from stereo are accepted otherwise they are rejected.

In our view, the major contribution of this paper is to identify and present preliminary models for the errors and mistakes introduced in three particular distance measurements, and to begin thinking about how to combine them. Future work will explore the quantitative aspects of the error parameters that have been identified.

Acknowledgements: This work was supported in part by NSF/DCR 8410771, Airforce F49620-85-K-0018, DARPA/ONR, ARMY/DAAG-29-84-K-0061, NSF-CER DCR82-19196 A02, by DEC Corp., IBM Corp., and LORD Corp.

References

1. Bajcsy, R., M. Mintz, and E. Liebman, "A Common Framework for Edge Detection and Region Growing," *Submitted to the 9th Intl. Conf. Pattern Recognition*, Paris (October, 1986).

2. Bickel, P. J. and K. A. Doksum, *Mathematical Statistics,* Holden-Day (1977).

3. Canny, J. F., "Finding Edges and Lines in Images," *MIT AI-TR-720* (1984).

4. Goldstein, H., *Classical Mechanics,* Addison-Wesley (1966).

5. Henrikson, K., "Line-based Stereo Matching," *University of Pennsylvania TR* (1986).

6. Imaging, Fairchild CCD, *CCD: The Solid State Imaging Technology*, 1980.

7. Izaguirre, A., P. Pu, and J. Summers, "A New Development in Camera Calibration -- Calibrating a Pair of Mobile Cameras," *IEEE Conference on Robotics and Automation,* St. Louis, pp. 74-79 (March, 1985).

8. Krotkov, E. P., R. McKendall, and M. Mintz, "Statistical Models of Camera System Noise," *In preparation* (1986).

9. Krotkov, E. P., "Focusing," *University of Pennsylvania TR-CIS-86-22* (1986).

10. Purll, D. J., "Solid-state image sensors," in *Automated Visual Inspection,* ed. B. G. Batchelor, Elsevier Science Publishing Company, New York (1985).

11. Smitley, D. L., "The Design and Analysis of a Stereo Vision Algorithm," *University of Pennsylvania TR MS-CIS-85-27* (1985).

12. Solina, F., "Errors in stereo due to quantization," *University of Penn. T.R.* (December, 1985).

13. Torre, V., A. Verri, and A. Fiumicelli, "The Stereo Accuracy for Robotics," *International Symposium of Robotics Research*(3), pp. 89-93 (1985).

14. Zeytinoglu, M. and M. Mintz, "Optimal Fixed Size Confidence Procdures for a Restricted Parameter Space," *Ann. Statist.* 12, pp. 945-957 (1984).

15. Zeytinoglu, M. and M. Mintz, "Robust.Fixed Size Confidence Procedures for a Restricted Parameter Space," *Submitted for publication to Ann. Statist.* (1985).

W. Jüptner
BIAS - Bremer Institut für angewandte Strahltechnik
Ermlandstr. 59, D-2820 Bremen 71

1. Introduction

Robots are flexible handling systems. To use this flexibility best, the robots should get more intelligence, especially with recognizing sensors. One necessary form of recognition is the identification of objects in the three-dimensional space, or more precisely: there must be a projection of the object to be handled - or not - into the "brain" of the robot, fig. 1.

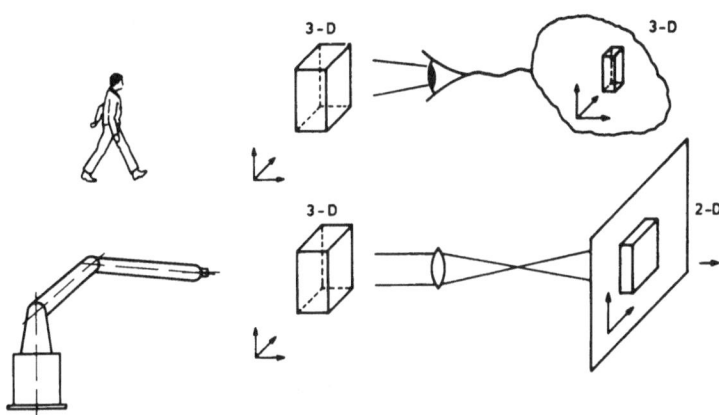

Fig. 1: Human and Robotics Imaging

Usually, TV-cameras are used as image detectors, which deliver a two-dimensional projection of the three-dimensional object in space. Even, if by optical methods like multidirectional viewing or light-intersection the third dimension is evaluated, a lot of frames have to be stored for one object. The number of information stored before the extraction very soon exceeds some million bytes. This seems to be very much compared to usual robot vision, but one should remember that there is an preprocessing by the object selection and transportation today /1/.

Holography might be a good tool to overcome the problems, fig. 2: This imaging technique is the two-dimensional coding of a three-dimensional information: By means of holography, the image of a three-dimensional object is stored with all the depth information in a two-dimensional plane. Therefore holography is an ideal 3D to 2D-converter. Its limits will be discussed as well as technical realizations. However, holography has more potentiality:

NATO ASI Series, Vol. F43
Sensors and Sensory Systems
for Advanced Robots
Edited by P. Dario
© Springer-Verlag Berlin Heidelberg 1988

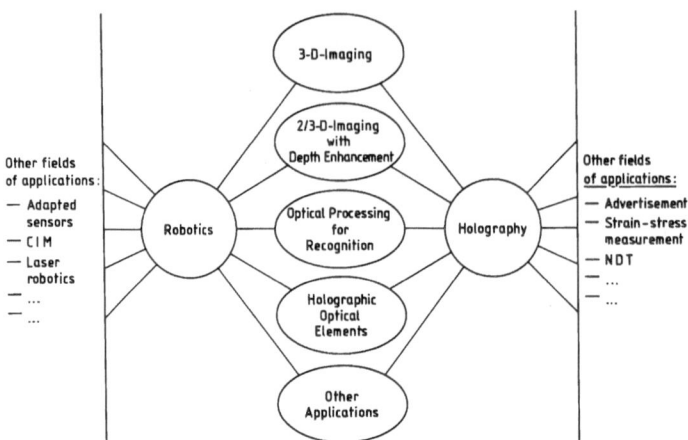

Fig. 2: Applications of Holography to Robot Vision

The hologram itself is useful in a special application as an optical computer. So it can be used for part recognition or part selection. The aspects for future developments will be outlined.

Some modern developments such as the "light-in-flight" method according to Abramson /2/ may become useful. This method works especially with light of short coherence length, which is more suitable to production-near applications. Furthermore, it might be a method of storing depth information in an easy way to be processed. About this method will be reported.

There are a lot of other possible applications of the holography in robot sensorics, which cannot be mentioned in this report at all. One example are the holographic optical elements (HOE) /3/. The most simple form of this is a lens, which can be adapted to different tasks. By expanding them to phase conjungated optics all errors of the optics and the influence of the environment can be corrected. Today the possible applications of these optics only can be estimated.

2. Basics of Holography

Holography is an optical method to store three-dimensional fields in two-dimensional memories /4/. For a better understanding of this technique, it shall be explained by an old but interesting experiment, which is theoretically described by Fresnel /5,6/: the Fresnel zone plate, fig. 3.

The radius of each ring is proportional to the square root of the ring number, fig. 4:

$$r_m \approx \sqrt{m}$$

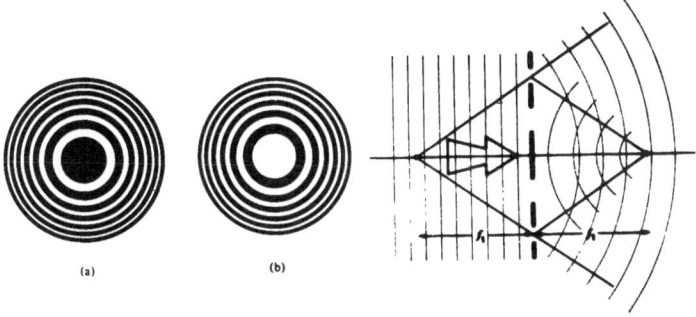

Fig. 3: Fresnel Zone Plate

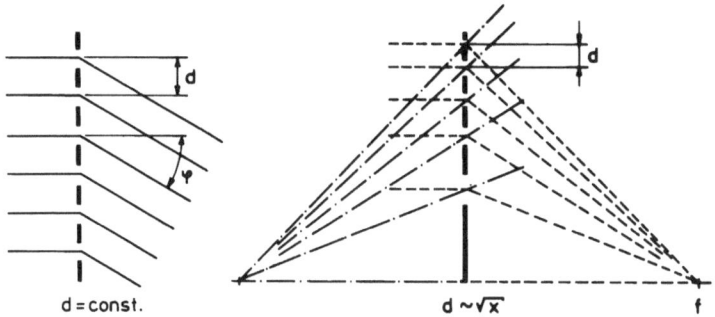

Fig. 4: Grating and Zone Plate Diffraction

One can derive by the laws of diffraction on a grating

$$\sin\varphi_n = n\,\lambda\,/\,d$$

with φ_n : diffraction angle for nth order
 n : order
 λ : wave length
 d : grating constant, difference between two slots

that a linear grating with slot distances proportional to the square root of the slot number diffracts a parallel light beam into three main parts: One part is only attenuated, one part is diffracted into a point with the distance f and one seems to come out of a point in -f. Of course, this is only valid for sinusoidal gratings, so the drawings might be thougt to be the binary image of these. An easy procedure to get a Fresnel zone plate was mentioned by Lord Raleygh. One has to take two beams, e.g. a parallel one and a point source beam on the axis of the parallel beams. In a certain distance F of the point source a photo plate has to be placed, in the most simple case perpendicular to beam axis. The illuminated plate can be developed. Then the amplitude distribution is similar like in a zone plate but sinsusoidal. By the laws of the Fresnel zone plate, one can prove that for this plate is valid

$$(1/a + 1/b) = m \lambda / r_m^2$$

with $a = \infty$: $b = f = F$ and $f = r_m^2 / m \lambda$

This leads directly to the principle of holography: The Fresnel zone plate produced with two beams is a hologram for one light point: If this light point is the object, it will be imaged by the second beam into the plate. After illuminating the processed plate with the parallel beam the virtual image and a real image of this source point is reconstructed in the real position. That's holography according to Gabor /7/. He suggested a comparable set-up to take a hologram from a slide with only a few informations in it, fig. 5.

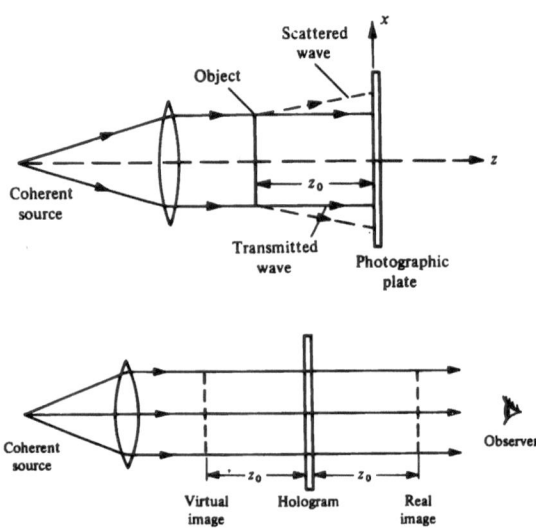

Fig. 5: In-line Holography (Gabor)

However, the disadvantage of the method is, that the reference beam source, the virtuell image and the real image are on one axis one behind the other. By this circumstance the observation of the virtuell image is difficult. Therefore, Leith and Upatnieks /8/ suggested an out of axis reference beam. Additionally, there is no must for having a parallel reference beam. This leads to the common form of set-ups used today, fig. 6.

In this case each surface point of the object can be thought to be a light point source as mentioned before. Of course, in practice there is not any longer such a simple zone plate, but a very complex distribution of black and white lines, fig. 7.

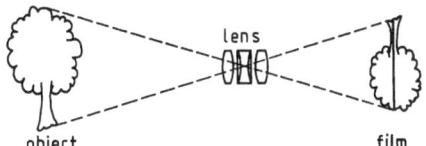

a) Photography: Imaging by lenses

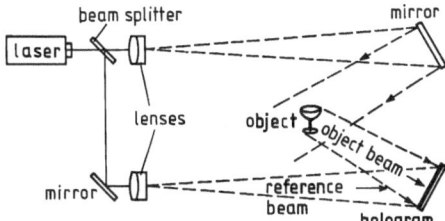

b$_I$) Holography: Storage of light waves

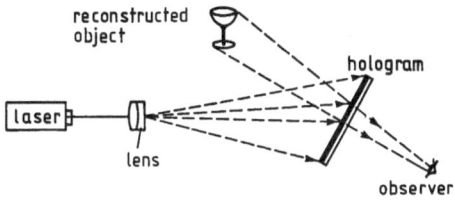

b$_{II}$) Observing a hologram

Fig. 6.: Common Set-up for Holography

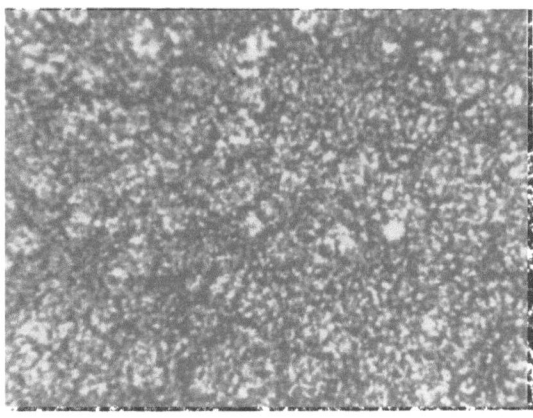

Fig. 7: Hologram Fringes

The mathematical description of holography is given as follows:
The object wave in the plate area may be expressed by

$$E_0 = E_{0o} * \exp / i (wt + \emptyset_0) / \tag{1}$$

with $\emptyset_0$ as a function of x and y. This object wave interferes with a reference wave in the form

$$E_R = E_{Ro} * \exp/i(wt + \emptyset_R)/ \qquad (2)$$

with $\emptyset_R$ as a (usually slightly varied) function of x and y. The plate stores the intensity, which is given by

$$I = (E_0 + E_R) * (E_0 + E_R)^{\otimes} \qquad (3)$$

$$= E_{Oo}^2 + E_{Ro}^2 + E_{Oo}*E_{Ro}*\exp/i(\emptyset_0-\emptyset_R)/ + E_{Oo}*E_{Ro}*\exp/-i(\emptyset_0-\emptyset_R)/$$

Equation (3) describes a time-constant intensity distribution, which carries the information of the object. This intensity distribution can be registered by any light sensitive medium with sufficient resolution, e.g. photographic plates. Normally the interference fringe spacing is in the order of some wavelength or less.

Now let's assume, that a photographic plate was exposed to the intensity according to equ.(3), developed, replaced and illuminated again with the reference beam. When the transmission is linear dependent on the intensity with a factor ß, then this results in:

$$T = T_0 + \beta*I$$

$$E_R*T = E_R*T_0 + E_R*\beta*E_{Oo}^2 + E_R*\beta*E_{Ro}^2 \qquad (4)$$

$$+ E_{Ro}*\beta*E_{Oo}*E_{Ro}*\exp/i(wt+\emptyset_0)/$$

$$+ E_{Ro}*\beta*E_{Oo}*E_{Ro}*\exp/i(wt-(\emptyset_0-2\emptyset_R))/$$

$$E_R*T = \beta * (E_{Oo}^2 + E_{Ro}^2) * E_R \qquad \text{reference wave}$$

$$+ \beta * E_{Ro}^2 * E_0 \qquad \text{object wave}$$

$$+ \beta * E_{Ro}^2 * E_{Oo}*\exp/i(wt-\emptyset_0)/*\exp/-2i\emptyset_R/ \qquad \text{real image wave}$$

This equation describes three waves behind the hologram: The diffraction order zero results in the incident reference wave. Additionally the old object wave appears, only changed by a temporally and spatially constant factor. The last term describes an object wave with negative phase distribution. This is a real image of the object, but its position is modified by the phase factor $2\emptyset_R$, so that the real and the virtual image are not in axis.

There are a lot of different possibilities to discuss the above mentioned equ.(4). But since it is not the aim of this report to discuss holography as a whole, more details might be studied e.g. in the book of Hariharan /9/ or others. In the following only the special points of interest are discussed, that are related to robot sensors.

3. Direct sensoring of objects

One point of interest is to image the object directly into the robotics memory, comparable to the TV-camera sensor. The equ.(3) and (4) mark the way to do it: Instead of recording the intensity with a photographic plate, it can be lead to an TV-camera target. Then this image of the interference pattern can be processed electronically. With the very well-known means of image digitizing the information gets into the memory of a computer. In a first step the non varying background I_G should be subtracted: The intensity distribution has the form:

$$I = I_G + I_O * \cos(\emptyset_k(x,y)) \tag{5}$$

from which I_G can be subtracted easily. So the distribution

$$I = a * \exp/i(\emptyset_O - \emptyset_R)/ + b * \exp/-i(\emptyset_O - \emptyset_R)/ \tag{6}$$

is stored in the memory. The "reconstruction" in the computer only needs the phase of the reference beam, so the resulting computation is

$$R * I = a * R_O * \exp/i\emptyset_O/ + b * R_O * \exp/-i(\emptyset_O - 2\emptyset_R)/ \tag{7}$$

which describes the whole field of waves and by this, of course the virtual and real image. The hologram intensity distribution and the amplitude and phase distribution of the reference beam in the same plan are known. So for each interesting plane the computer can analyse what a TV camera would see. With developed techniques for the image processing, it is possible to recognize the object. Unfortunately there are two disadvantages in the moment: the limited resolution of the camera target and the large number of computations needed for the image reconstruction.

The limited resolution of the camera target limits the maximum spatial frequency of the hologram interference fringes. This makes necessary a special arrangement for the object and the reference wave.

One equipment to match these conditions can be a set-up for Fourier transform holograms, fig. 8.

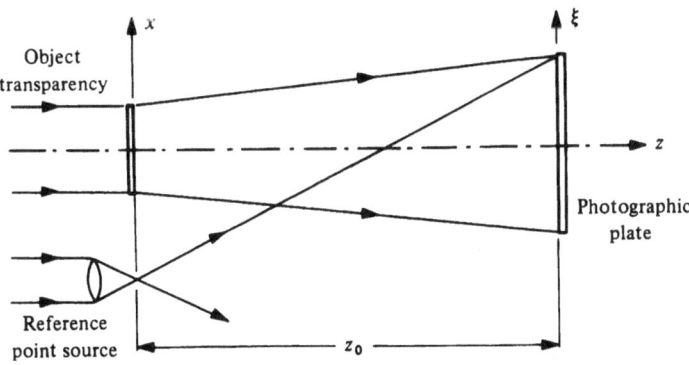

Fig. 8: Fourier Transform Hologram /9/

The object is illuminated with nearly parallel light. Additionally, the reference beam is focussed in a way, that it seems to come from a source in the object plane near the object. This results in low spatial interference fringe frequencies, suitable to camera targets. There is only one disadvantage: One has to know the objects position, but this is just the starting question in 3-D recognition. Furthermore, the condition of enough spacing of the interference fringes is only valid for small distances in depth. So, in common cases, this method is not suitable. Although there might be some applications where this method can do the work for recognition.

So another arrangement must be applied, which is comparable to the ESPI (Electronic Speckle Pattern Interferometry) set-up /10/, fig. 9.

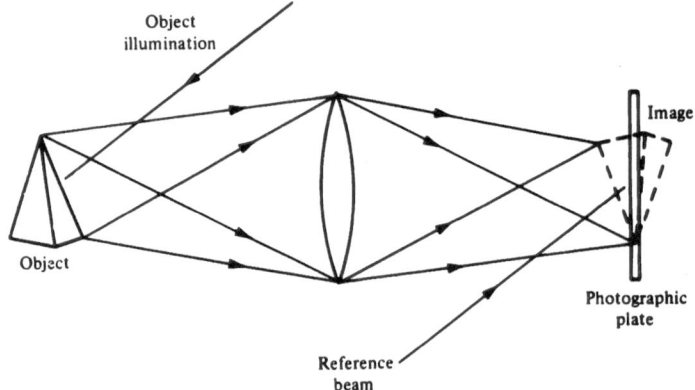

Fig. 9: Image Plane Hologram /9/

The object is illuminated with laser light and then projected onto the camera target, fig. 10. Additionally, a nearly parallel reference beam is given

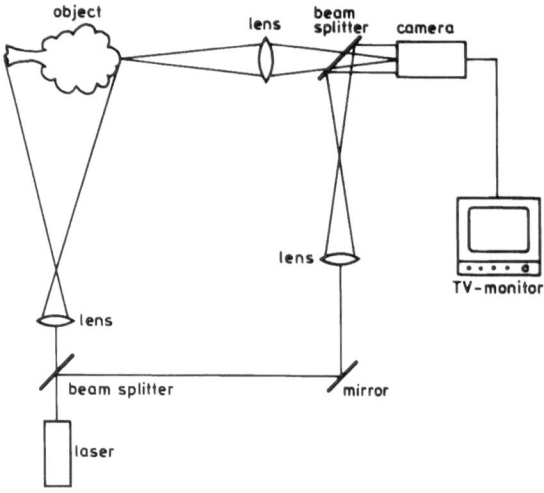

Fig. 10: TV Hologram Camera, nach /10/

parallel to the object beam onto the target and an aperture limits the spatial frequencies of the object image. So the interference fringe spacing is suitable to the resolution of the target. The advantage of the technique is the fast storage of the hologram and a high stability against disturbances. Unfortunately the noise increases with smaller apertures by creating larger speckles, fig. 11. So that at least, it is more a specklegram (ESPI set-up) than a hologram /11/. This method can be used for interferometric

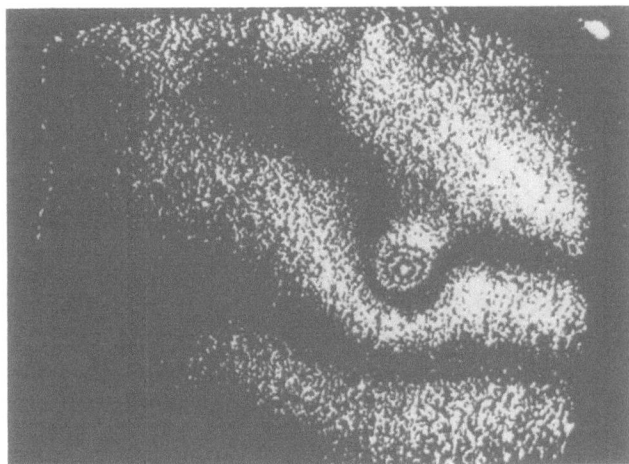

Fig. 11: TV Holographic Tire Testing /10/

measurements. It is mainly investigated and developed by the british group around Butters and Tyler.

Of course, the resolution of the targets limits the volume that really can be observed three-dimensionally. Assuming a common CCD-camera normally used in robot vision, the resolution is approximately 500x256 = 128.000 pixels. With this camera the surface of a cube with the sidelength a should be recognized really three-dimensional with a resolution of 1mm in each direction. Because five sides can be seen in maximum, this leads to

$$5 \, a^2 = 128.000 \text{ mm}^2 \qquad \text{and} \qquad a = 160 \text{ mm}$$

The easiest way out of this limiting restriction can be the development of easy to use high resolution cameras. Another way might be a step-wise recognition: First a large volume is observed with a low spatial resolution. Then, step by step, the resolution will be enhanced while the volume gets smaller.

Another disadvantage is the number of calculations and by this the large computation time, needed with todays computers for robots. But it will be a fact of time, that more powerful micro-processors are available: In the moment the 32-bit unit with 16 MHz cycle time gets available. This will not be the end of the capacity of microprocessors. Furthermore, microprocessor arrays will be developed. So, in near future the computation capability should be powerful enough to do the work of direct image processing out of an hologram.

4. Holographic contour mapping

Holographic contour mapping is one way for the evaluation of the depth information in an easy procedure, e.g. combined with the image plane holography. Herewith the object is imaged onto the target as described before. But the imaging is done with two different wavelength λ_1 and λ_2. Then the two images interfere and fringes appear which are a measure for the elevation. The holograms should be taken with a dye laser: The wavelength difference can be easily controlled in modern lasers with an accuracy down to a few nanometers and less. The difference in elevation dz from fringe to fringe is given by

$$dz = \lambda^2/2 \cdot d\lambda$$

for a given wavelength difference $d\lambda$.

For an example, the wavelength difference is assumed to be $d\lambda = 2$ nm with the basic wavelength of 630 nm. This leads to

$$dz = 100 \text{ um.}$$

The evaluation of the fringes can be done with the usual image analyzing techniques by detecting the lines of maximum or minimum intensities. More precisely and faster works a method newly developed in the BIAS /12/. The image is Fourier transformed twice with an unsymmetrical filtering in between. By this method the phase at each point in the fringe system can be calculated directly. Since the method needs two FFT, it requires faster computers for robot sensors than the common 16-bit systems of today.

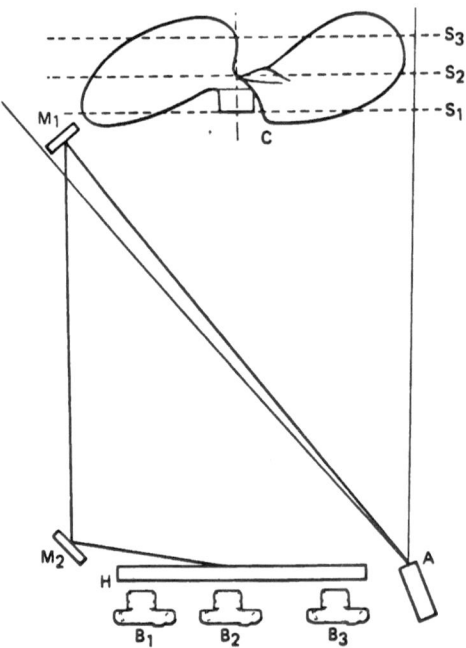

Fig. 12: Light-in-flight Holography /2/

A future aspect may be mentioned: The light-in-flight method according to Abramson /2/: For making the hologram by this technique laser light with small coherent length is used, e.g. given by very short laser pulses of pico- or femtoseconds, fig. 12.

If the reference beam travels nearly parallel to the photographic plate a hologram is recorded when the two short pulses - short in space: 0,3 mm for a one-picosecond-puls - from the object and reference beam meet each other. This condition is valid for different positions on the plate with respect to the depth of the object. In the set-up according to fig. 12 the depth of the object is monitored from the left to the right hand side of the plate.

This technique only needs the short coherence length not the short pulse. This means, that diode lasers should be applicable. Furthermore, the difference in time between the reflected object light from different parts of the object can be controlled by pulsing the laser diodes. This seems to be an interesting future research work.

5. Optical image processing, matched filters

Optical image processing can be a fast way of recognizing objects in robot vision. The standard optical processor is shown in fig. 13 /13/.

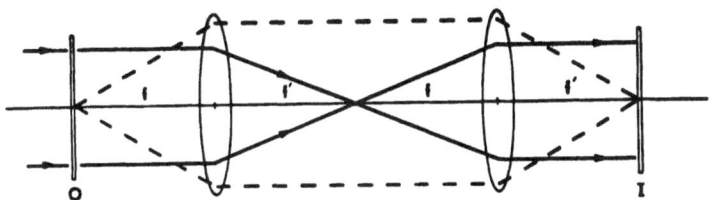

Fig. 13: Coherent Optical Processor /13/

The object is imaged with coherent light by a first lens into the focal plane which is the frequency plane by the theory of Abbe. This means, that in this area the frequency and phase content of the object is displayed.

An interesting excample is given by Dutton and coworkers /14/, fig. 14,15. The different methods of filtering and their results are demonstrated with a dusty mesh. So this seems to be a fast way for object recognition, the problem is to get the correct, the matched filter. This problem can be solved by holography. The filter can be produced as an hologram. Even more the filters might be stored in the memory of the robots intelligence. Then the incoming image is to be Fourier transformed, filtered and retransformed for object recognition.

A direct way to use optical processing with holography show equ.(1)-(4). As can be seen, both equations for the object and the reference wave are the same ones, only the indices are changed. So, if the reconstruction is done by the object wave, the reference wave is recovered and vice versa. Theoretically, this can be written as follows:

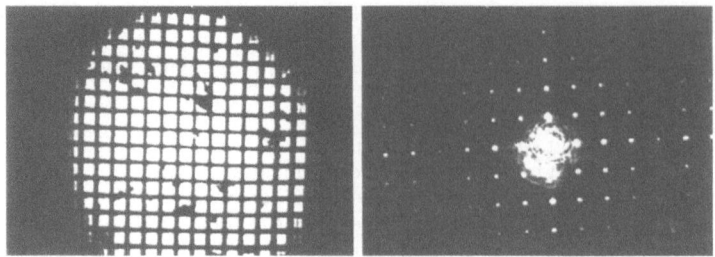

Fig. 14: Dusty Mesh and Frequency Transformation /14/

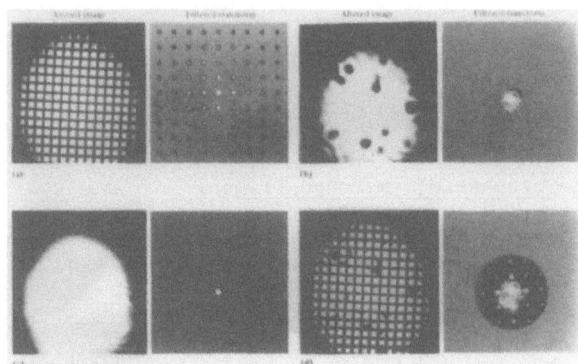

Fig. 15: Filtering Function and Result /14/

Assume the object wave from equ.(1), but with a changed amplitude and a little bit dislocated:

$$E'_0 = E'_{0o} * \exp/i(wt + \emptyset_0 + \Delta\emptyset)/ \tag{8}$$

with

E'_{0o} = new amplitude
$\Delta\emptyset$ = change in location

So the result is

$E'_0 * I$ = undiffracted parts

$$+ E_{0o} * E_{Ro} * E'_{0o} * \exp/i(wt + 2\emptyset_0 + \Delta\emptyset - \emptyset_R)/$$

$$+ E_{0o} * E_{Ro} * E'_{0o} * \exp/i(wt + \emptyset_R + \Delta\emptyset)/$$

The second term is something like the magnified image under an angle given by the original reference wave and dislocated by $\Delta\emptyset$. The third term reconstructs the reference wave with a certain additional angle given by $\Delta\emptyset$. If this wave is focussed into a detector plane, a light point directly indicates the appearence of the holographed object and its position.

One can describe this matched filter techniques in a more common way by introducing a linear function, which means to transform the object into a similar but not the same one. This can be done, e.g. by introducing the function h = h(x,y) in the way

$$\emptyset_0(x,y) \implies h * \emptyset_0(x,y)$$

In this case the third term of equ.(8) contains an additional factor in the following way:

$$\text{third term} = E_{0o} * C_{Ro} * E'_{0o} * \exp/i(wt + \emptyset_R + \Delta\emptyset)/ * \exp/i(h - 1)\emptyset_0/$$

This factor describes a small phase modulation of the reference beam and is a measure of the cross correlation of the two objects: The closer (h - 1) gets to zero the more the two objects are correlated, the less the reference wave is modulated. A more detailed information in this field is given e.g. by Goodman /15/. Only some remarks on additional potentialities in this application for robot vision should be given:

- The matched filter techniques can be applied in combination with the image plane holography. Then the requirements for spatial resolution drop down. The further processing can be done optically, additionally to a normal image processing. By this, both is done in parallel: vision and recognition.
- The matched filter technique can be applied in future as an additional coprocessor in the computer of the robot in the following way: the image of the object, taken by a normal TV-camera, is transfered to a liquid crystal display according to fig. 13 with matched filters, which can be liquid crystal displays, too. So it helps to recognize objects.

6. Conclusion

Holography and robot vision, this is a look into the future: Robots are well established and widely used tools. They need sensors for broader applications. The sensors must be fast enough, that means recognition frequencies should be expressed in terms of milliseconds. In this range applicable sensors are available. On the other hand: Holography is a fascinating technique for imaging with the whole three-dimensional information. Up today the real high resolution media for taking a hologram are photographic or thermographic ones with long exposure times even when they are processed in real-time. There are some steps to an electronic processing like the ESPI. But in the moment, the resolution is not high enough for a sensor suitable to robots. This will change in the future with High Definition TV's and following techniques. Furthermore, the images in ESPI are taken electronically but mainly for an TV output. Even when these data are processed interference patterns, filtering then is not sufficient in the moment for robot vision.

The next step to take the hologram electronically, process it and give it into an optical processor, e.g. for matched filtering, reaches the limits of the optoelectronic interfaces. Again this will be a question of time to solve the problems with the developments of better liquid crystal displays. The latter are available since some months for colour TV. Unfortunately the resolution is as low as about 50 thousand points. This is sufficient for leisure TV but not for optical processors. But again, this will change in the future.

Last not least, the direct processing of the holograms would be very interesting: All the mathematics are well known and easily to be described. On the other hand in terms of calculations it is a big task with millions of

FLOP's. So today the necessary computer of the robot would grow into the area of the CRAY computers. This is the state of the art with a standard of 16-bit-microprocessor-ICs and the 32-bit ones just starting their applications. Additionally, the normal computer structure is based on a single processor with one or more coprocessors for special tasks, e.g. number crunching. In both directions this is not the end of the developments: there will be 64-bit, 128-bit and more bit microprocessors in future, when new mask techniques allow the integration of millions of transistors. Furthermore, BIAS developed a multi-processor structure with only four different elements, microprocessor, bus, memory and knot point. The structure of the system could be chosen by the program: parallel or serial processing is possible. The system was tested in a Fast Fourier Transform Process with a drastical gain in processing speed. This idea can be realized in future with special IC's for the four elements and with higher integration, e.g. several microprocessors in one chip. The limit in the moment are adequate software strategies to choose the best structure and to modify the program structure respectively. So future developments will bring holography and robot vision closer to oneanother for better sensors in better robots.

7. References

/1/ NN.:   Tagungsband der 15. IPA-Jahrestagung
"Sensorsysteme zur Automatisierung der
Produktion", Stuttgart, 1982

/2/ Abramson, N. H.:   Holography Using Picosecond Lightpulses
Holographic Data Nondestructive Testing,
Proc. of SPIE, Vol. 370, p. 170-174

/3/ Rallison, R.:   Applications of Holographic Optical Elements,
Lasers & Applications, Dec. 1984, p. 61-68

/4/ Kiemle, H.;   Einführung in die Technik der Holografie
      Röss, D.:   Akad. Verlagsgesellschaft, Frankfurt/Main, 1969

/5/ Gerthsen, Chr.;   Physik
      Kneser, H. O.;   Springer-Verlag Berlin Heidelberg New York, 1977,
      Vogel, H.:   13. Auflage, S. 377-387

/6/ Hecht, E.;   Optics
      Zajac, A.:   Addison-Wesley Publ. Co., Philippines, 1974

/7/ Gabor, D.:   A New Microscopic Principle,
Nature 161, 1948, p. 777-778

/8/ Leith, E. N.;   Reconstructed Wavefronts and Communication Theory,
      Upatnieks, J.:   J. Opt. Soc. Am. 52, 1962, p. 1123-1130

/9/ Hariharan, P.:   Optical Holography,
Cambridge Monographs in Physics, Cambridge, 1984

/10/ Raimann, G.:   Eumig HT-10: Ein TV-Speckle Interferometer und
seine praktischen Anwendungen,
Herbstschule '77, Hannover 1977

/11/ Tyler, B.:   Electronic Speckle Pattern Interferometry
                             to be published in SPIE-Proc. 604

/12/ Kreis, Th.:   Digital Holographic Interference-Phase measurement
                             Using Fourier-transform Methods,
                             J. Opt. Soc. Am. A, Vol. 3, No. 6, 1986

/13/ Young, M.:   Optics and Lasers
                             Springer-Verlag Berlin Heidelberg New York, 1977

/14/ Dutton, D. et al.: Spectra Physics Laser Technical Bulletin No. 3

/15/ Goodman, J. W.:   Introduction to Fourier Optics,
                             McGraw-Hill, New York, 1968

ELECTRICAL PROXIMITY SENSORS

H.CLERGEOT*, D.PLACKO*, J.M.DETRICHE**

* Laboratoire d'Electrotechnique, SIgnaux et Robotique,
E.N.S 94230 CACHAN - FRANCE

** Unité de Génie Robotique Avancé,
91191 SACLAY - FRANCE

ABSTRACT

The subject of this paper is the use of electrical devices for proximity sensing. A general discussion is made about the concept of multisensory "perception" by the robot, and about the interest of proximity sensing for reducing the complexity of this problem. The main characteristics of sensors using the electrical field, the magnetic field, or electromagnetic waves are then discussed, from general physical arguments. Indications are given on actual sensor design and on available devices. An exemple of the characterisation of a multisensory system is given for an eddy current array developped in our laboratory.

I. INTRODUCTION

Measurement or perception ?

Sensors are used in robotics in a very different way than in metrology. Sensors for conventionnal measurement applications operate in a "one dimensional universe". All precautions needed (control of environment, compensations, ...) are used, so that the reading is a function only of one parameter . In actual operation of sensors for robotics, very often, these ideal conditions cannot be fulfilled and the output is a function of several features of its surroundings. In such a situation, the information of one isolated sensor may be very degraded. But, as will be discussed, a multisensor association can nevertheless restore a valuable image of the environment.

This recovery requires on one hand a more complete characterisation of the sensor, and on the other hand an appropriate model of the "universe" of the robot, resulting of some prior knowledge of the environment in which it will operate for a given task. Let us explain these concepts by two very simple exemples.

Suppose that the task is to clamp very carefully an iron sheet with a magnet fastened on the hand of the robot. The robot is programmed to bring the magnet at some distance of the metal plane. The final approach only is on the control of a sensory device placed on the hand. In a first eventuality it is known a priori that the angular position of the magnet with respect to the plane is correct : the only parameter to control is the distance d. In this "one dimensional universe", a conventional distance measuring sensor will do the job.

NATO ASI Series, Vol. F43
Sensors and Sensory Systems
for Advanced Robots
Edited by P. Dario
© Springer-Verlag Berlin Heidelberg 1988

In a second eventuality, it is known that there may be in addition a tilt angle θ along a given direction. This is again a very simple universe model, with only two parameters, d and θ. In order to control the tilt angle θ, we may introduce a second distance sensor. But the operation of such sensors is certainly affected by θ so that we need a more complete characterisation of the outputs, as a function of the distance and the tilt angle : $f_1(\theta,d)$, $f_2(\theta,d)$. This gives two equations to recover (θ,d) from the output signals of the two sensors.

From a conventionnal point of view, the tilt angle θ would be considered as a cause of systematic error in the measurement of d. In the present context, θ and d can be recovered exactly with two sensors, if the assumptions on the environment and the sensor response are correct.

Going a step further, from these exemples, sensors used in robotics snould be understood not as measuring devices in a conventional sense, but rather as "perception" devices in a more general acceptation. There, perception means fitting an internal representation of the universe by the robot to the true environment. This is done by trying to fit what the outputs of the sensors should be according to the model, and what they actually are.

This presentation gives a general frame necessary to understand the operation of robots under the control of a multisensory system, and emphasizes the importance of a complete characterisation of each sensor. The robot "brain" selects and scanns periodically a set of sensors to update or refine its representation of the universe (or some part of the universe).

Proximity sensors

Proximity sensors are short sighted sensors, or sensors with eye screens, like horses, so that they do not be frightened by a too complex universe. This is very important for short term industrial applications in robotics : there are not presently fast enough computers (nor probably smart enough strategies) to update in real time a complex representation of the full robot universe, as seen by a general view CCD camera for instance.

In the context of proximity sensing, the general lines of the robot hand motion are under operation of an upper control level (for instance a preregistred program). The robot then knows from a data base, at each step of the task, the local universe model for the robot hand : it will usually be very simple, with only a few parameters to identify. The proximity sensors are placed on the hand, with a perception field limited to the local environment. Their role is to provide a refined local representation for an accurate control of the hand. The more significant features of proximity sensors may be summarised as follows :

- simple local universe model
- small amount of information to process
- accurate, relative perception

Electrical proximity sensors

This communication is restricted to electrical proximity sensors. Only the determination of the geometrical parameters is considered. The measurement may rely on the electrical field, on the magnetic field or on electromagnetic waves, with in each case a limitation to a specific family of detectable targets.

II.GENERAL PHYSICAL BASIS

Field configurations

The ideal field configuration for different kinds of electrical sensors is represented in figure 1.

For capacitive sensors (fig. 1a) the field is basically unipolar and is affected by metallic or dielectric bodies.

Inductive sensors (fig. 1b) use a bipolar (or multipolar) field. The reluctance of the magnetic path is reduced when a magnetic material is close to the poles. Better results are for materials with high permeability, and low conductivity to avoid the effect of eddy currents. For the same reason inductive sensors operate at low frequency.

Eddy current sensors also use the magnetic field (fig. 1c), but with high conductivity metallic targets. The frequency used is high enough so that there is a very small penetration of the magnetic field in the metal (in the ordre of the skin depth, $\delta = \sqrt{2/\omega\sigma\mu}$). Then in this case there is an increase of the reluctance when the metallic target comes close to the sensor.

For electromagnetic wave sensors, there is a reflection of the emitted wave by a conductive, dielectric or metallic target. The reflected wave is collected by the emitter itself or an auxilliary receptor.

For each kind of sensors there is a family of targets for which the variation of the physical characteristics have small influence : high conductivity grounded targets for capacitive devices, high permeability materials for inductive units, high conductivity metals for eddy currents, high reflectivity for electromagnetic waves.

Images on a plane target used for field modelling

The concept of images, familiar in optics, finds an extension for the four kind of sensors, in the ideal situations just considered. This will be useful to make the connection between the field pattern of an isolated sensor and its characteristics in presence of a plane target.

The solution for the electromagnetic field must satisfy the Maxwell equations in the free space and the limit conditions on the surface of the target. Let us demonstrate that these limit conditions are satisfied for the field created, in the free space, by the sensor and its "image", symmetric of the sensor with respect to the plane surface of the target (fig. 1).

For the capacitive sensor and a conductive target (fig. 1a) the electric field mustbe orthogonal to the surface. This condition is satisfied for the field obtained by summing the effects of the sensor and its image, excited with opposite polarities.

For magnetic sensors, in the case of targets with high permeability (inductive sensors), again, the field must be orthogonal to the surface and an image with reversed polarity has to be used (fig. 1b). But for eddy currents (target with high conductivity, fig. 1c) the field is tangential, which condition is satisfied by the introduction of an image with the same polarity as the sensor.

In the case of wave propagation the field is the sum of the incident wave and the reflected wave emitted by the image, with a phase difference depending on the complex reflectivity of the target (fig. 1d).

Conclusion for the basic characteristics

The concept of image gives a better understanding of the operation of the sensor : detection of the target is through the measurement of the "reflected field" created by the image. Thus the sensitivity of the sensor for a target at distance d corresponds to the free space pattern of the sensor field at distance 2d, in direction of the target. This is a simple generalisation of ideas familiar for electromagnetic waves but with some specific features in each case.

For electromagnetic waves the field pattern can be made very directive when the wavelenght is small with respect to the dimension of the emitter. This is the case for microwaves (λ = 3cm for f = 10GHz) and naturally for optical waves. The amplitude varies as d^{-2} with the distance.

For electric sensors the field is basically unipolar, omnidirectionnal with a d^{-1} variation of the potential. The directivity can be improved by proper electric shields but remains poor. The sensor may be sensible to interfering electric fields.

Magnetic sensors use a dipolar (or multipolar) field. The consequence is a very sharper decay of the sensitivity at long ranges (d^{-3}) but a better immunity to interfering fields. Again, the directivity is poor and can be improved by the use of magnetic shields.

Use of a balanced configuration

At long range, the sensivity to motions of the target is very small and the drifts of the sensor become critical. The consequence is that, wherever this is possible, a differential design has to be used, to achieve a zero analog output for the isolated sensor, and an internal self compensation of the drifts.

An interpretation of the balanced configuration is that the excitation field is eliminated, so that the output is a function of the reflected field only.

III. DESIGN, FEATURES AND APPLICATIONS

Capacitive sensors

For the variation of the capacity C(d) of the probe electrode to the ground, two asymptotic forms can be derived, for small or large values of d.

At short range the formula for a plane capacitor can be used :

$$C = \varepsilon_0 \frac{S}{d} \qquad (1)$$

At long distance (fig. 2a), the potential V(d) may be calculated as the sum of the free space potential, Q/C_∞ and the potential V'(d) created by the image at distance 2d :

$$\begin{cases} V(d) = \dfrac{Q}{C(d)} = \dfrac{Q}{C_\infty} + V'(d) \\[2mm] V'(d) \simeq \dfrac{-Q}{4\pi\varepsilon_0 \cdot 2d} \end{cases} \qquad (2)$$

hence :

$$\frac{1}{C} = \frac{1}{C_\infty} - \frac{1}{8\pi\varepsilon_0 d} \qquad (3)$$

To improve the directivity of the sensor, a metallic shield can be added. Often, a grounded shield is used, but it introduces an important capacity to ground in parallel with the active one. Even a small relative drift of this capacity (due to dust, humidity, temperature, mechanical deformation) may give important measurement error. The right solution is to use a guarded shield.

The variation of the capacity may be measured by a bridge unbalance with synchronous detection. The capacity may also be incorporated in an oscillator, and cause a shift of the oscillation frequency /1/.

Capacitive sensors are available as non contact limit switches /2,3/ or for accurate distance measurement in the linear part of their characteristics /4/. Applications have been reported for measurements in very adverse temperature conditions /5/.

In robotic applications, array structures have been proposed for automatic seam tracking /6/. The principle of a guarded array is proposed in fig. 3a. The non directivity of capacitive sensors may be an advantage in the design of obstacle detection systems /7,8/

Inductive sensors

Asymptotic forms of the reluctance R (d) for long and short range can be derived. At short distance the law of magnetic circuits can be applied. S being the surface of the poles, the reluctance of the magnetic material is neglected. For the air gaps, equal to d, one finds (fig. 2b) :

$$R(d) = \frac{2d}{\mu_0 S} \qquad (4)$$

At long range, for a given flux $\emptyset$ in one pole, the circulation of H(d) is computed as the sum of the contribution of the excitation H_∞ of the isolated sensor, and the contribution of the excitation H'(d) created by the image (fig. 2b).

$$H'(d) \simeq \frac{\emptyset w}{4\pi\mu_0 (2d)^3} \qquad (5)$$

The circulation in the core being negligible

$$\int H\, d\ell = \mathcal{R}(d).\, \phi = \int H_\infty d\ell - w\, H'(d)$$

hence:

$$\mathcal{R}(d) \simeq \mathcal{R}_\infty - \frac{w^2}{4\pi \mu_0 (2d)^3} \qquad (6)$$

The directivity may be improved by the use of a screen made of high permeability material, but at the expanse of reduced sensitivity and important leakages.

The variations of the reluctance are measured by a bridge unbalance signal or the frequency shift of an oscillator. Direct detection of the pertubation of the magnetic field may be performed by a hall effect device, in particular if a DC excitation is used ; these devices will be described in a later paragraph.

Inductive sensors are used with low frequency excitations as non contact switches to detect the presence of magnetic bodies/2,3/.They can be used for precise distance measurement on limited range (for instance measurement of the thickness of a dielectric coating on a magnetic body).

For utilisation as proximity sensors, eddy currents seems to be more appropriate, as will be discussed now.

Eddy current sensors

At long distance an asymptotic form similar to the case of inductive sensors can be derived, but the polarity of the image is reversed so that the relation is now, instead of (5) :

$$\mathcal{R}(d) = \mathcal{R}_\infty + \frac{w^2}{4\pi\mu_0 (2d)^3} \qquad (7)$$

At short range the reluctance is limited only by the leakages and may become very high (figure 2c).

The effect of the physical properties is expressed by the equivalent penetration of the field :

$$\delta' = \mu_r \, \delta = \sqrt{2\mu_r / \mu_0 \omega \sigma} \qquad (8)$$

For conductrive materials, ω can be choosen high enough so that S' remains in order of, say, .1 mm and can be neglected or easily corrected for.

To improve the directivity and reduce the leakages, a high conductivity metallic shield may be used, with a far better efficiency than the magnetic shields used for low frequency inductive sensors (figure 4).

The possible measurement set ups are the same than for inductive sensors. Note that with magnetic sensors, a high precision differential configuration may be used (figure 4). The field is divided into a reference path and the measurement path. The signal is derived from a separate winding making the difference between the two fields. The structure is carefully balanced for the isolated sensor, so that the signal is a function of the reflected field only.

Eddy current sensors are available as non contact switches /2.3/ or as distance measuring devices / 9 /. They are appreciated for their ruggedness and relative insensitivity to electromagnetic interference. In robotic applications, they have been used for seam tracking / 10 /; more detail will be given in paragraph 4 about a special array for this purpose / 11 /.

Hall effect transducers

These devices offer the opportunity of detecting the magnetic field directly, instead of the f e m inducted in a winding by its variations. This makes possible the detection, for instance, of the constant field generated by magnets used as markers on a material. But even for AC modulated fields, Hall effect devices may be interesting because they are very small and easier to integrate than coils. They also have a very short response time.

Very efficient hall effect transducers are made possible by the use of high mobility semi conductors ($I_n A_s$, $I_n A_s P$; $I_n S_b$ has a very high mobility but is very sensible to temperature). The effect is used either in Hall fem generators, or in magnetoresistances.

Hall generators are linear and sensitive for low value of the field; an inconvenient is that their is no electrical isolation between the excitation and the measurement ports. Magnetoresistnaces are easier to handle (small, only two wires), but they are quadratic at low field : for better sensivity a constant polarisation field has to be used.

Note that integrated magnetoresistances are available encapsulated with a polarisation magnet, in differential configuration / 12 /. Typical characteristics are displayed in figure 5.

Non contact switches using hall effect sensors for detecting magnetic materials are also available / 2,3/.

Electromagnetic wave sensors

Typical designs are given on figure 6 for microwave (figure 6a) or optical frequencies (figure 6b).
At long ranges the reflected field varies as d^{-2}.

For small wavelenght it is possible to focuss the beam in a very small angle θ limited only by diffraction to $\theta \approx \lambda /D$, where D is the diameter of the sensor. The counter part is that a plane target can be detected only when it is perpendicular to the beam, to within $\pm \theta /2$. Nevertheless, for optical wavelenght, the mirror reflexion no longer holds due to scattering on the small irregularities of the target : for irregularities greater than the wavelenght the reflected power becomes closer and closer to a uniform repartition.

On the reflected beam the sensor may be sensible to :

a) the reflected power
b) the doppler shift of the frequency
c) the propagation time $\tau = 2\ d/c$

The first principle has been widely applied in robotics for very unexpensive optical devices / 13 /.

The doppler shift is used in microwave intrusion detectors / 14 /. The frequency of the wave reflected on a moving target is shifted by $\Delta f = 2v_r/\lambda$ were v_r is the radial velocity. This gives an additionnal opportunity to reject the excitation signal by frequency filtering, so that a great sensitivity can be achieved. The restriction is that the target must move. The minimum value of v_r, for 10 GHz detectors, is in the order of $\leqslant 1$ m/s.

To measure the propogation time, some kind of modulation has to be used. The more common are pulse modulation, sine wave modulation, or a linear frequency modulation. Pulse modulation is used in radar or laser telemetry but cannot be used for proximity applications.
(typical resolution could be $\sim$ 50 cm and minimum range $\sim$ 10 m). The sinewave modulation turns the lag τ in a phase shift $2\omega\tau$; using a high enough frequency F the method is very sensitive. We have developped in our laboratory a laser telemeter on this principle with F = 300 MHz and a resolution of . 1 mm / 15 /. When the frequency f is linearly swept, their is a difference $\Delta f = 2\tau (df/dt)$ between the excitation and the reflected wave. This principle is used in radio-altimeters, but it could be used used also for laser telemetry, where very large values of df/dt can be easily achieved, with laser diodes for instance.

IV. EXEMPLE OF MULTISENSORY ARRAY

A multisensor array, using a row of eddy current devices was developped in our laboratory for applications in automatic welding control. Besides the design of the sensor itself, we were faced, in a limited scale, to the problem of rebuilding a representation of the profile of the target from the sensor information, as considered in the introduction.

For this purpose one needs an a priori model for the profile, the characterisation of the output of any individual sensor of the array, and a criterion to fit the true output and the output computed from the representation.

The array, using elementary sensors similar to figure 4, is represented figure 3b. Each elementary sensor is calibrated to give the exact distanced of a metallic plane in front of it. If the profile presents small deviations $\delta(x)$ with respect to the average d, a linear relation was assumed between the deviation Δ of the reading of the sensor and $\delta(x)$, in the form

$$\Delta = \int h_d(x) \cdot \delta(x) \, dx \qquad (9)$$

where h_d (x) is an even fuction of the lateral distance x from the axis of the sensor, and decreases with increasing $|x|$ because the sensor is obviously less and less affected by deviations $\delta(x)$ farther from it. More precisely, from physical arguments, it can be deduced that h_d (x) is closely related to the repartition of the magnetic field on a plane at distance d. This gives the form /16/ :

$$h_d(x) = k \left(a^2.d^2 + x^2 \right)^{-3/2} \qquad (10)$$

where a was adjusted experimentally.

This gives the characterisation of the sensor. For the model of the profile, for seam tracking applications, a succession of line segment may be used, with the breakpoints as parameters.

A least squares criterionhas been used for fitting the sensor signals.

This sensor was experimented in actual arc welding operation / 11 /. The distance andlateral position of the welding torch were controlled by open loop actuators, the only feed back for positionning being given by the sensor array.

V.CONCLUSION

Some specific features of sensors usedfor robotics have been outlined. Very often they are used in conditions where the reading of one isolated sensor gives, at best, a degraded information. In a multisensor configuration the outputs of the sensors may be considered as elements of "perception" of the environnement. The recovery of a good representation of this environnement requires a more complete specification of the sensor and a model of the environnement, with the smallest possible number of unknow parameters.

The advantage of proximity sensor is that they have access only to a limited, relevant part of the robot "universe", liable to a simpler model. Proximity sensors may be considered under the aspects of

- directivity
- specificity to target
- range

Concerning the electromagnetic wave sensors, they can be made very directive, the counter part being that they may become very sensitive to the orientation of the target. This is less critical for very short wavelenght due to scattering by the surface irregularities. A great variety of targets can reflect an appreciable fraction of the incident power. Sensors measuring the reflected power are very sensitive to the target nature. The main advantage of time of flight measurement is that this dependance is eliminated. The operating range is connected to the directivity : using a collimated beam, all the emitted power can be concentrated on a remote target.

For the other kind of electrical sensors the directivity cannot be improved very much ($\Theta \simeq 45°$) the counter part being a good tolerance to the target orientation. An interest of capacitive and eddy current sensors is that their output is almost independant of the nature of the target for all good conductive material.

For these sensors the range may be considered as proportionnal to the width of the active front part of the sensor (say : $d_{max} \lesssim 5$ W).

REFERENCES

/1/ Bjoern E.Bjerede "On a class of capacitively tuned transducer oscillators", IEEE Trans. Inst. Meas.
Vol IM-18, N° 4, December 1969 pp 336-340.

/2/ Selection report "Proximity switches-detecting metal objects without touch" Modern Material Handling, April 1980 pp 98-101.

/3/ Non contact switch manufacturers: Control Engineering, Inc., Eaton Corp., Kaman Sciences Corp, Unimax Switch Corp, Veeder Root. Delavan Electronic Inc., Ivo. Industries, Electromagnetic components, Omion Electronics.

/4/ Distance capacitive measurement manufacturer, ex : Mechanical Technology Inc.

/5/ High temp. capacitive measurement manufacturer, ex : Hitec Corp.

/6/ F.B. Prinz, J.F. Hoburg "Sensors for seam charachterization in robotic arc welding" Carnegie Mellon University Pittsburg U.S.A.

/7/ "Protecting workers from robots" American machinist, 1984 pp 85-86 (General Motors Robotguard)

/8/ Obstruction avoidance system, AMI Proxaguard, manufactured by Assembly Machine Inc.

/9/ Distance eddy current measurement manufacturer, ex : Kaman Science Corp.

/10/ J.M. Detriché : "Detecteurs de joints pour robots de soudage à l'arc" AFCET Besançon p 99 Novembre 83.

/11/ D. Placko, H. Clergeot, F. Monteil "Seam tracking using a linear array of eddy current sensors" Proceed. ROVISEC 5 pp 557-568 Amsterdam, October 85.

/12/ Galvanomagnetic devices manufacturer, ex : Siemens.

/13/ B. Espiau : "Optical proximity sensors", NATO advanced research workshop on sensors and sensory systems for advanced robots, Mai 86, Maratea.

/14/ V.F. Fusco "A microstrif X-Band movement transducer" proceed. Sensor 84, pp 272-277, Paris June 1984.

/15/ H. Clergeot, D. Placko, S. Guillon "Laser range finding sensor for robotics", proceed Rovisec 6, Paris, June 1986.

/16/ D. Placko : "Dispositif d'analyse de profil utilisant des capteurs à courant de Foucault" Thèse 3ème cycle, Orsay Avril 84.

305

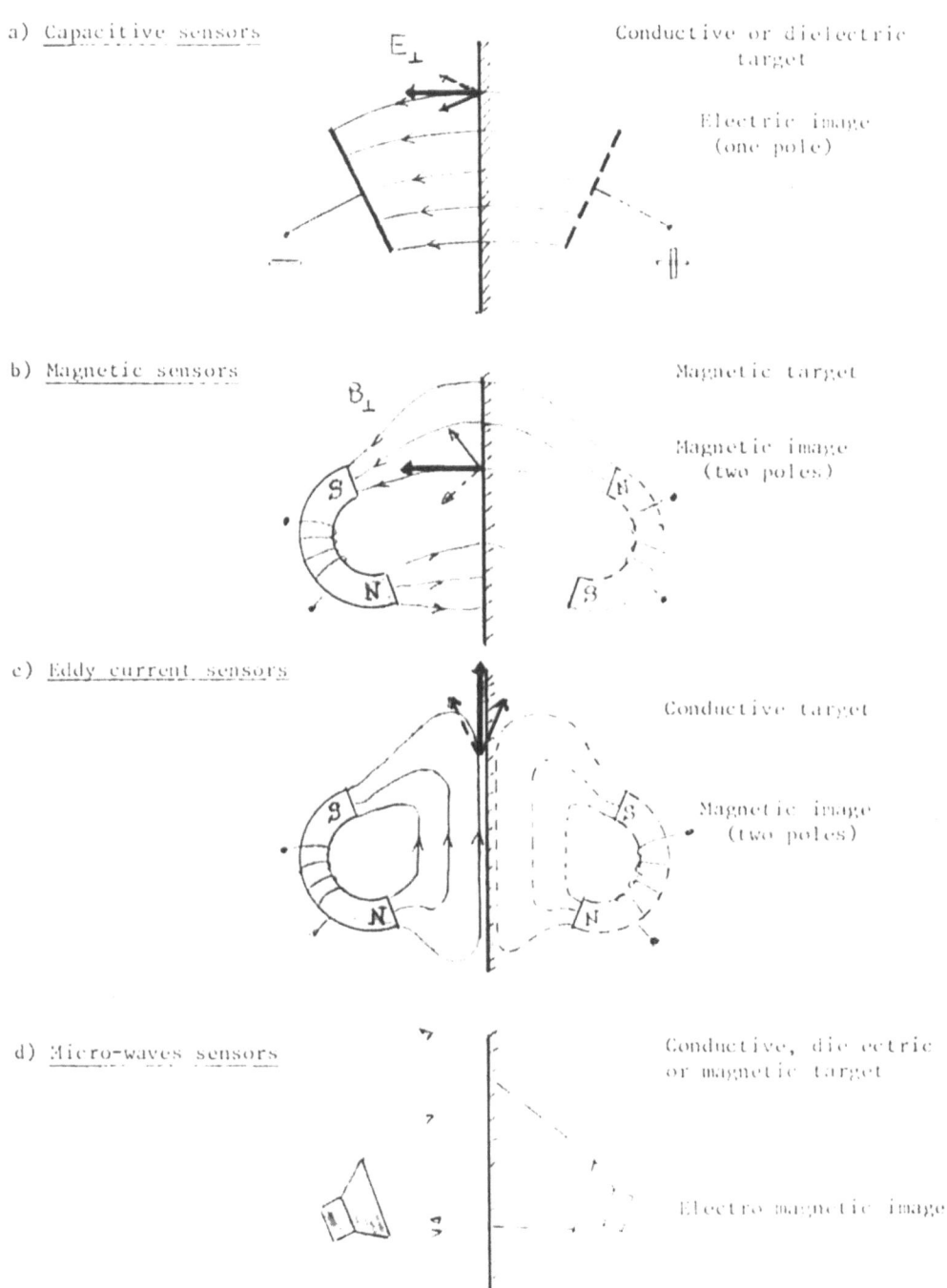

a) Capacitive sensors — Conductive or dielectric target / Electric image (one pole)

b) Magnetic sensors — Magnetic target / Magnetic image (two poles)

c) Eddy current sensors — Conductive target / Magnetic image (two poles)

d) Micro-waves sensors — Conductive, dielectric or magnetic target / Electro-magnetic image

Figure 1 : General presentation - Field configurations

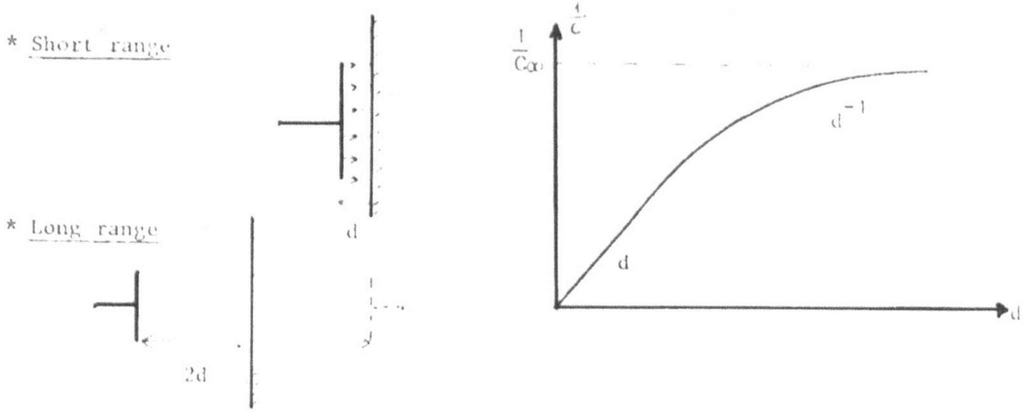

* Short range
* Long range
d
2d
1/C
1/C∞
d^-1
d
d

Figure 2a : capacitive sensors

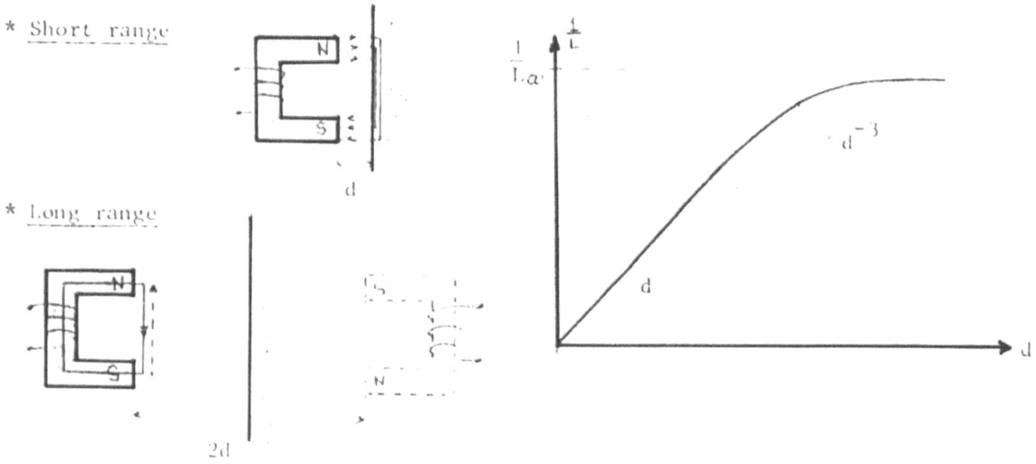

* Short range
* Long range
N
S
d
2d
1/L
1/L∞
d^-3
d
d

Figure 2b : inductive sensors

* Long range
N
S
2d
1/L
d^-3
1/L∞
d

Figure 2c : eddy current sensors

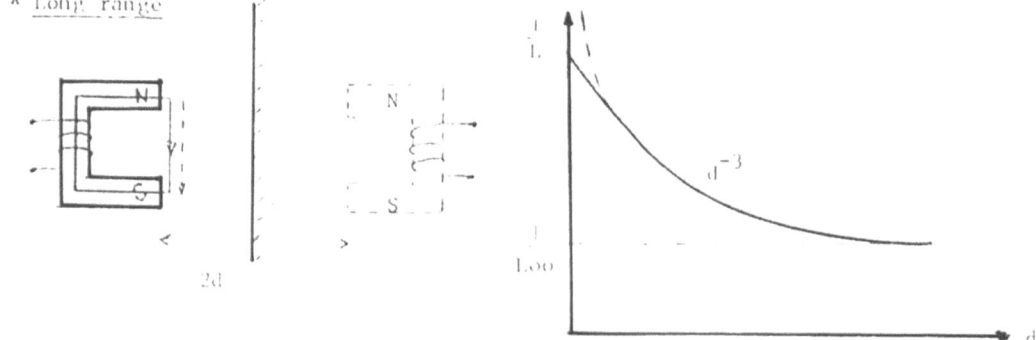

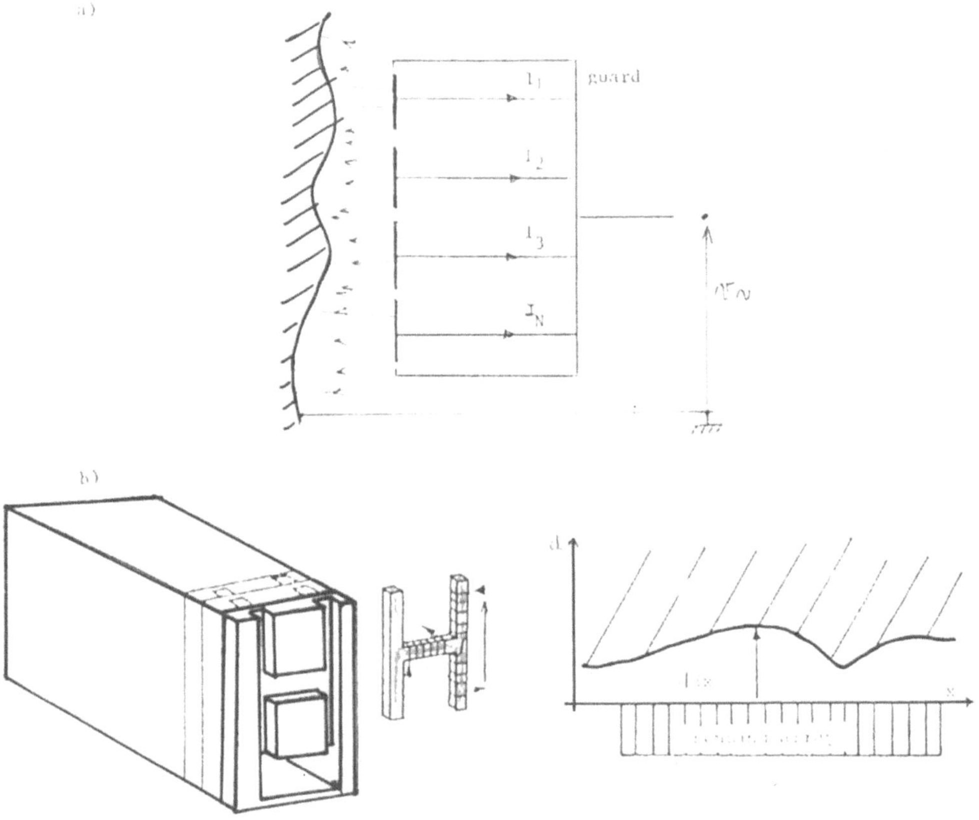

a)

guard

I_1

I_2

I_3

I_N

b)

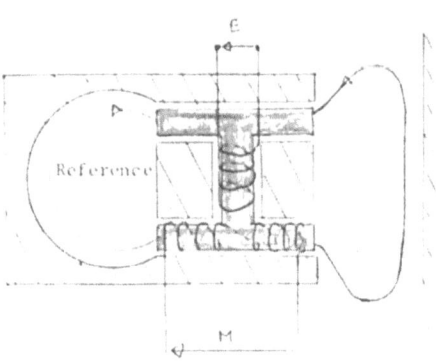

<u>Figure 3 : Array dispositions</u>

E

A

Reference

M

<u>Figure 4 : Balanced magnetic sensor</u>

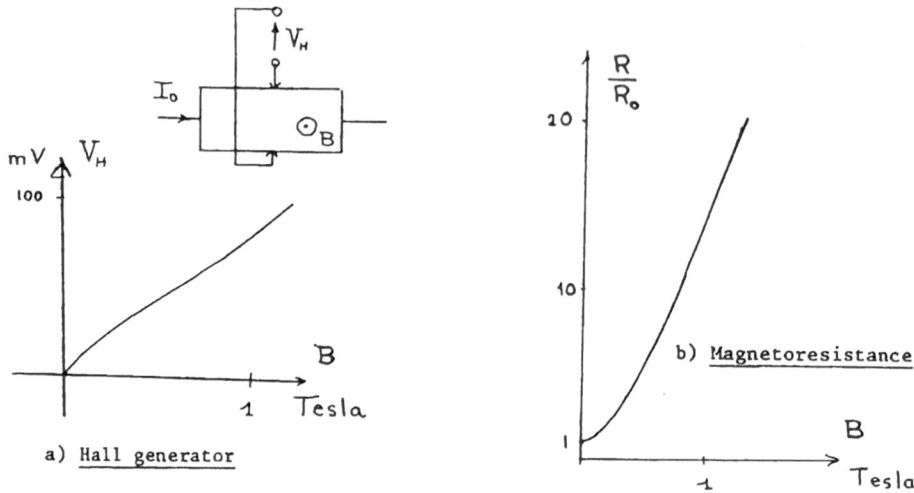

Figure 5 - Hall devices

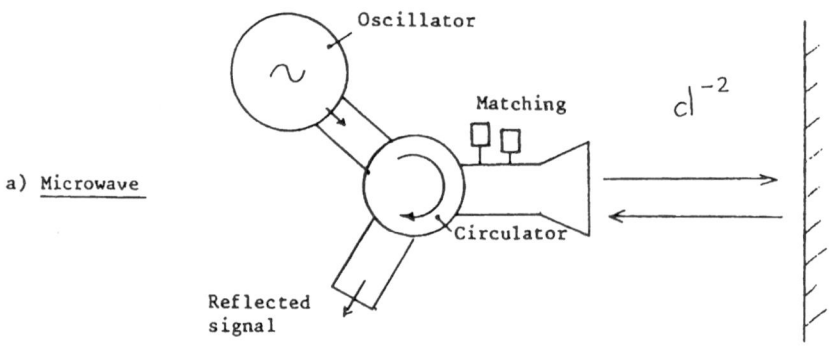

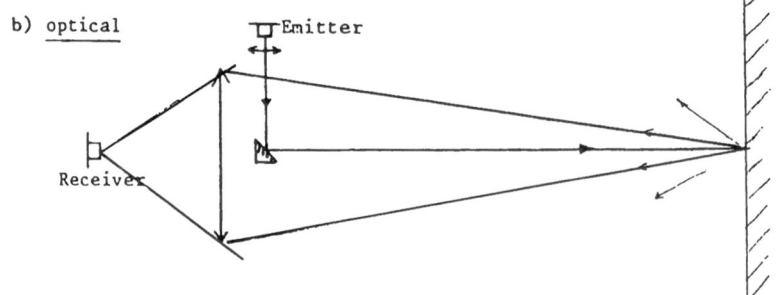

Figure 6 : E.M. Wave sensors

ULTRASONIC SENSORS

LEIF BJØRNØ

The Industrial Acoustics Laboratory
Technical University of Denmark
Building 352, DK-2800 Lyngby, Denmark

ABSTRACT

This paper reviews the field of ultrasonic sensors for use in robotics with emphasis on sensors based on piezoelectric materials for use in air and in water. A detailed discussion is given of demands to sensors used for range and proximity measurements as well as of factors influencing the accuracy of these measurements. The structure of the ultrasonic field of a piston source is emphasized and the influence of sensor backing and load as well as bond lines on characteristic features of the ultrasonic field is discussed. The use of focused sources and aperture shading is evaluated and ultrasonic sensor calibration methods are presented. Finally, several ultrasonic sensor types for use in air and in water are presented including bimorph sensors and their characteristics.

INTRODUCTION

Noncontact sensing of solid objects for robotics application is widely acknowledged as essential for instance for intelligent machine control. Ultrasonic sensing has been considered with particular interest due to the fact that it is rather inexpensive, it gives range information (time-of-flight) and potenti-

NATO ASI Series, Vol. F43
Sensors and Sensory Systems
for Advanced Robots
Edited by P. Dario
© Springer-Verlag Berlin Heidelberg 1988

ally, it may give 3-D surface information [1].

The more widespread industrial use of robots for instance in relation to flexible manufacturing systems and unmanned machine operations has emphasized the need for more reliable sensors - or transducers as they are most frequently called because a sensor is a transducer, but a transducer is not always a sensor - being able to operate in industrial environments [2,3,4].

Among other things due to the better acoustic coupling between transmitting and receiving elements - frequently made from piezo-electric materials - and liquids, ultrasonics has for many years been used for underwater purposes as for instance range measure-ments, and a considerable fund of knowledge has been established concerning the use in liquids of ultrasonic transducers. However, during recent years coupling procedures have been developed im-proving the energy transfer from a transmitter to air in a frequency range as high as 100 kHz to 1 MHz [5-11].

As robots are being used to a great extent in difficult acces-sible environments as for instance under water and a great depths this exposition of ultrasonic sensors will comprise sensors for use in water and in air without producing a sharp discrimination between the two applications. Due to the fact that ultrasonic sensors based upon piezoelectric materials are most frequently used, in particular for transmission and reception of high-frequency signals, this sensor type shall be treated in general in this paper, well knowing that ultrasonic sensors based on capacitive, inductive, electromagnetic principles etc. are being used in robotics. This choice is also influenced by the fact that silicon-based "smart sensors" frequently exploit the trans-duction effect found in piezoelectric materials.

Non-contact ultrasonic measurements may include determination of distance, thickness, position, perpendicularity, shape and orien-tation. In particular the distance determination using time-of-flight measurements which by an appropriate scale factor are converted into distance units is fundamental for the discussions in this paper. Resolutions by distance measurements down to 0.01 mm in air have been reported [8] and a resolution of 0.05 mm of distances up to 500 mm using frequencies up to 200 kHz in air is obtainable [9]. Qualities of ultrasonic sensors of vital importance to their use in robots are space resolution, sensitivity, dynamic range, linearity, hysteresis, time resolution and signal processing. Several of these qualities shall be treated in the subsequent sections.

Several factors affect the accuracy of ultrasonic distance measurements and have frequently led to a limited success of ultrasonic sensors simply because some fundamental rules for manufacturing and use of ultrasonic sensors have been violated. These rules, based upon knowledge created over years of ultrasonic sensor development and use, shall be treated in more detail in this paper. The accuracy of distance measurements is influenced by signal-to-noise ratios [12] where background noise frequently in a frequency range far above 100 kHz caused by aerodynamic noise sources will limit applications of ultrasound. Moreover, ultrasonic beam patterns, i.e. spotsize, depth of the field and alignment of sensor axis perpendicular to the reflecting surface [13] will influence the quality of the distance measurement. Furthermore, this quality is influenced by complicating factors like dimensional and material accuracy in the construction of the sensor together with the need for compensation for change in speed of sound in the air/liquid of the propagation path due to temperature changes, flow influences etc.

For instance the ultrasonic beam patterns or beam width may be controlled by variation of sensor diameter, variation of transmitted frequency, use of focused sources or by use of aperture shading. The construction and manufacturing phases are very essential for the applicability of an ultrasonic sensor and several defects to be observed too late during the application of the sensor are caused by mistakes made during these early phases of a sensors life. These defects can for instance comprise severe double pulsing, incorrect frequency, too long pulses, poor sensivity, subsidiary maxima, broken, damaged or wrong sized crystals, extra beams, very high internal echoes etc. all leading to reduced applicability or even rejection of the ultrasonic sensor for use in robots. Difficulties during the production phase may for instance lead to the strong variation in

pulse echo response of two commercial sensors purchased at the
same time and said to be nominally identical as shown in figure 1.

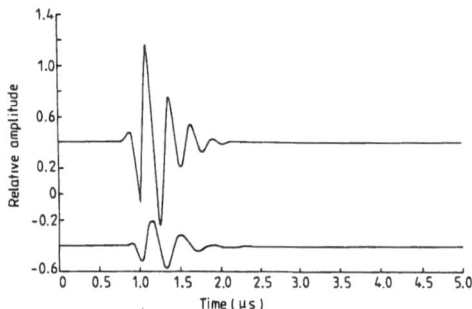

Figure 1. Pulse echo responses of two
nominally identical commercial
sensors. Reference [16].

It is evident that the efficiencies of the two sensors are very
different as frequency contents and pulse length vary from one
sensor to the other.
Several main factors have to be observed during sensor construction.
The five most essential sensor components, each of which being
dependent on the quality of several subcomponents, are sensing
element material, electrodes, backing and front impedance matching
materials including bond-line materials, electrical connections
and the housing. Several of these sensor components and their influence on the quality of the final sensor shall be discussed in
more detail in the subsequent sections.
For ultrasonic sensors based on piezoelectric materials the choice
of material has a strong influence on the sensor characteristics.
Commonly used sensor materials are for instance quartz,
$LiNbO_3$, Li_2SO_4 PZT in several modifications, barium titanate,
lead metaniobate, CdS, ZnO and PVDF. Factors to be considered, when
using one of these materials in ultrasonic sensors are for instance
maximum allowable temperature represented by their Curie point,
aging, piezoelectric constants, electromechanical coupling coefficients, dielectric constants, shape, acoustic impedance,

Q-values, cut/polarization, homogeneity etc. These factors of
which some are closely related to one another have a profound
effect on the qualities of the final sensor.

THEORETICAL AND EXPERIMENTAL BACKGROUND

a. THE PISTON SOURCE

Numerical data for the structure of the acoustical near- and farfield produced by a piston source have been available for many
years [14,15]. Simple analytical expressions have been developed
for the variation, in the farfield, of the sound pressure and soun
intensity as functions of radial source distance R and angle θ ιd
with the acoustic axis of the piston source, see figure 2.

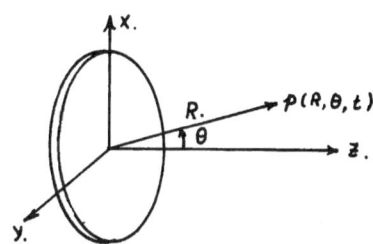

Figure 2. Piston source geometry used
in deriving the expressions
(1) and (2)

These farfield expressions are:

$$p(R,\theta,t) = j \; \frac{\rho_o c}{2} \; U_o \; \frac{a}{R} \; ka \; e^{j(\omega t - kR)} \left[\frac{2J_1(ka\sin\theta)}{ka\sin\theta} \right] \qquad (1)$$

for the sound pressure p, and

$$I = \frac{\rho_o c(ka)^2 U_o^2 a^2}{8r^2} \left[\frac{2J_1(ka\sin\theta)}{ka\sin\theta} \right]^2 \tag{2}$$

for the sound intensity I.

In the expressions (1) and (2) ρ_o and c denote the density and the velocity of sound in the medium in which the sound field is produced. U_o is the particle velocity amplitude at the source, while k denotes the wave number of the acoustic field. Both expressions show the influence of the dimensionless product ka on the amplitude as well as on the directivity of the acoustic field, the last mentioned being represented by the directivity function in the brackets of (1) and (2). Increasing frequency of the piston vibration, and thus increasing value of k, and increasing piston diameter will lead to a more narrow acoustic far-field beam and thus to an increased directivity index d expressed for the piston source by:

$$d = 10 \log \frac{(ka)^2}{1 - \frac{2J_1(2ka)}{2ka}} \tag{3}$$

Increasing values of d will, all other equal, lead to a reduced diameter of the spot insonified by the piston source.

The directivity of the piston source may also be expressed by their 3, 6 and 10 dB beamwidths which are the angles θ at which the intensity has dropped 3, 6 or 10 dB relative to the intensity on the acoustic axis. For a piston source emitting sound at a wavelength $\lambda = a/4$ these beamwidth values will be $\theta_3 = 7^\circ.4$, $\theta_6 = 10^\circ.1$ and $\theta_{10} = 12^\circ.9$, while the first hull of intensity will be reached of $\theta = 17^\circ.3$.

The directivety functions are dependent on the geometrical shape of the sound source and expressions to be used for their calculation may be found in many textbooks [15].

Figure 3 shows the farfield beam patterns for a circular piston source at various frequencies given as logarithmic plots (a) and (c) and as linear plots (b) and (d).

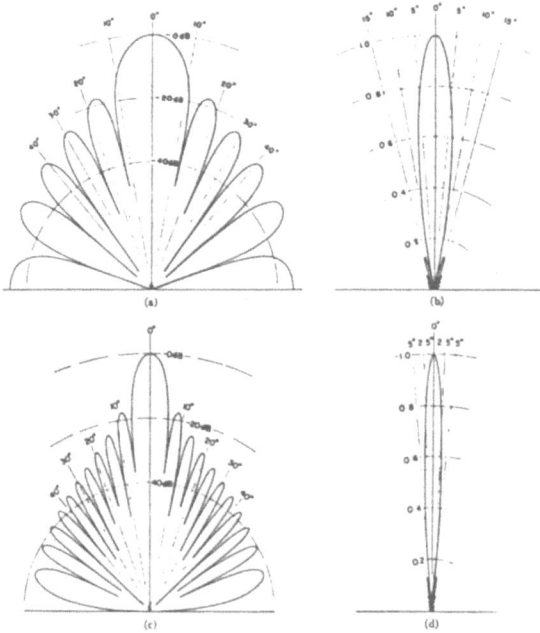

Figure 3. Farfield beam patterns of the
magnitude of sound pressure
produced by a circular piston
source at two frequencies
leading to a/λ = 2.78 for (a)
and (b) and a/λ = 5.62 for (c)
and (d). Reference [14]

The directivity of the sound field as well as the magnitude of the
sidelobes relative to the mainlobe are obvious on this figure.
While farfield calculation are relatively easy to perform ana-
lytically, nearfield calculations are mostly based on numerical
procedures. The magnitudes of the sound pressure in the nearfield of a circular piston source using a/λ = 2.5 and a/λ = 5.0,
respectively, are given in figure 4.

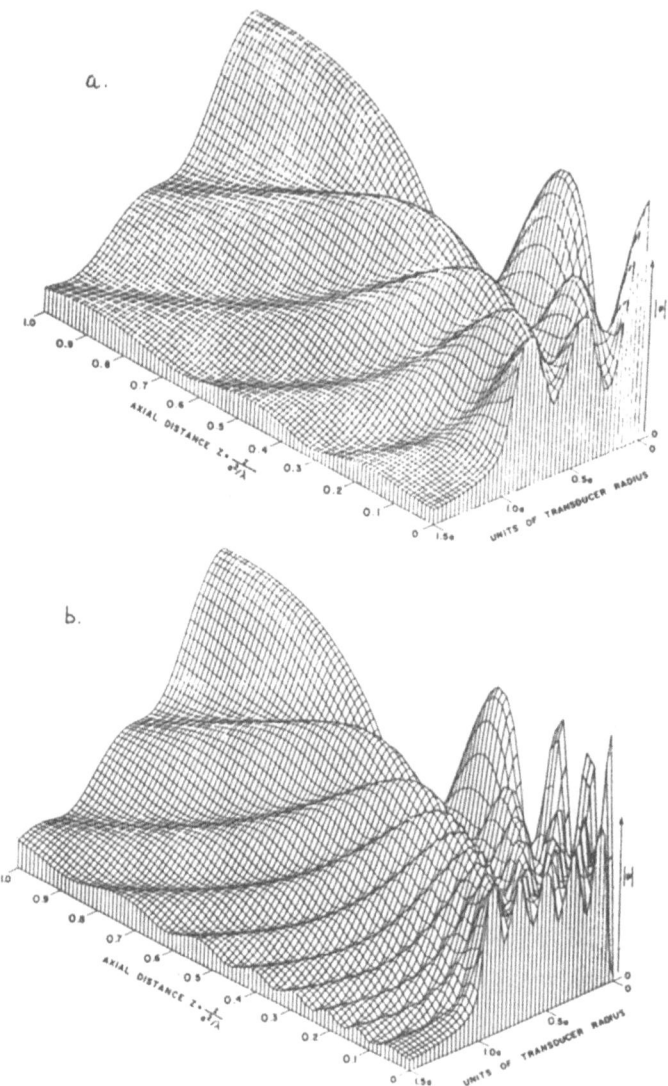

Figure 4. Magnitude of the sound pressure in
the nearfield of a cirkular piston
source at two frequencies leading
to (a): $a/\lambda = 2.5$ and (b):
$a/\lambda = 5.0$. Reference [14]

The fast variation of the sound pressure as a function of source
distance on and off the acoustic axis (the Z-axis) in particular
for $a/\lambda = 5.0$ may be seen from figure 4 and the increasing

complexity of the sound field with increasing a/λ is demonstrated in figure 4.

Plots of the 3 and 6 dB sound pressure contours also give information about the complexity of the nearfield compared with the
farfield. For a/λ = 1.0; 2.5 and 5.0 the computed sound pressure
contours in the near- and farfields of a circular piston source
are given on figure 5.

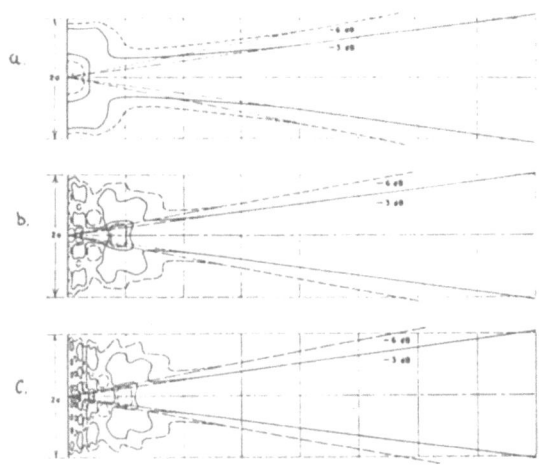

Figure 5. -3 dB and -6 dB sound pressure
contours in the near- and farfield
of a circular piston source at two
frequencies leading to (a): a/λ =
1.0, (b): a/λ = 2.5 and (c): a/λ =
5.0. Reference [14].

This figure shows that a focusing is taking place such that the
average sound intensity reaches a maximum within the minimum
spot size. Detailed nearfield sound pressure contour plots for
a/λ = 2.5 and 5.0 are given in figure 6 which emphasize the
strong variation in amplitude and phase found in the nearfield
of a piston source.

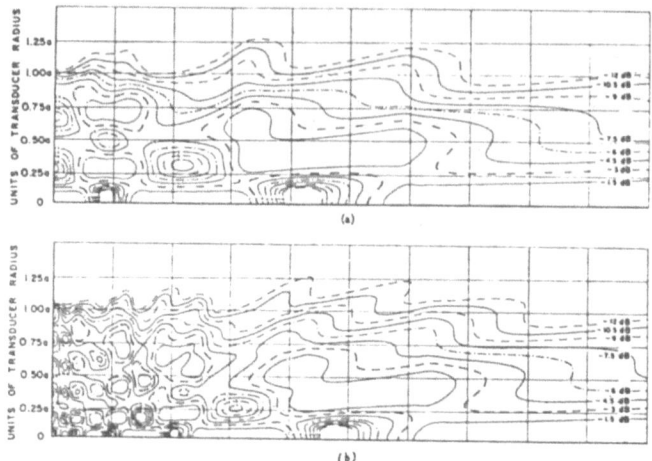

Figure 6. Nearfield sound pressure contours
for a circular piston source at
(a): $a/\lambda = 2.5$ and (b): $a/\lambda = 5.0$.
Reference [14]

By pulsed operation short pulses from a wide-band piezoelectric piston source used as transmitter and as receiver will
lead to an interaction between a direct wave and a diffracted
edge wave emitted in all directions from the edge of the
piston [17].While the direct wave pressure amplitude is propor-
tioned to U_o, see expression [1], the edge wave pulse on the
acoustic axis is an inverted replica of the direct wave. As also
the reception causes an edge wave influence, the transmission and
reception of a short pulse for instance when the piston source
is used for distance measurements, will lead to a train of pulses
making it difficult to select the right pulse to represent the
time-of-flight between the sensor and the object.
When the reflector is a small distance off the acoustic axis a
pulse widening has, moreover, to be taken into consideration.
This pulse widening, and thus the change in the spectral composition of the received pulse signal, increases with the diameter of the sensor and with the off-axis position of the object.
A further complicating factor on the shape of the acoustic beam
from a transmitter is the prospective existence of Lamb waves in
the transmitter. The Lamb waves may be considered as a parasitic
mode of vibration of the source where energy not only goes into

the thickness mode, but also into the radial mode of transmitter
vibration. Due to the fact that the group velocity of the Lamb
wave modes is normally greater than the velocity of ultrasound
in water, a substantial amount of the radial energy of vibration
enters into the fluid around the transmitter in a relatively
high-angled head wave, the angle of which is determined by the
ratio between the velocity of sound in the fluid and the Lamb
wave velocity. This head wave may at certain frequencies dominate
the transmitter beam pattern and will lead to difficulties in the
determination of the distance and direction to an object. The influence of Lamb waves may be reduced if the width of the transmitter element is reduced below the half wavelength of the Lamb
wave. The total width of the transmitter element may be retained
and thus the directivity characteristics of the transmitter if
the original element is devided into several minor elements forming an array. Figure 7 shows the dramatic effect of dividing the
transmitter element into 2 or 4 minor elements which together
lead to the original diameter of the transmitter element. From
figure 7 may be seen how the head wave dominates the beam patterns
when a single 4 mm in diameter element is used. The full line
represents the beam patterns influenced by the head wave while the
dotted line represents the beam patterns to be expected based
upon wavenumber and element diameter. The influence of the head
wave drops dramatically when the element is devided into 2 minor
elements and when 4 minor elements are used the beam patterns
come close to the theoretically predicted ones.
Modifications to the time course of a pulse emitted by a piezoelectric piston source may be obtained by varying the thickness
of the piston source as a function of the radial distance from
its center. A reduction in pulse length may be obtained by the
introduction of a tapered disc thickness, as the natural frequency
of the disc is determined by its thickness, and a broadening of the
bandwidth of the transmitter takes place when a tapering of the
disc thickness is introduced.

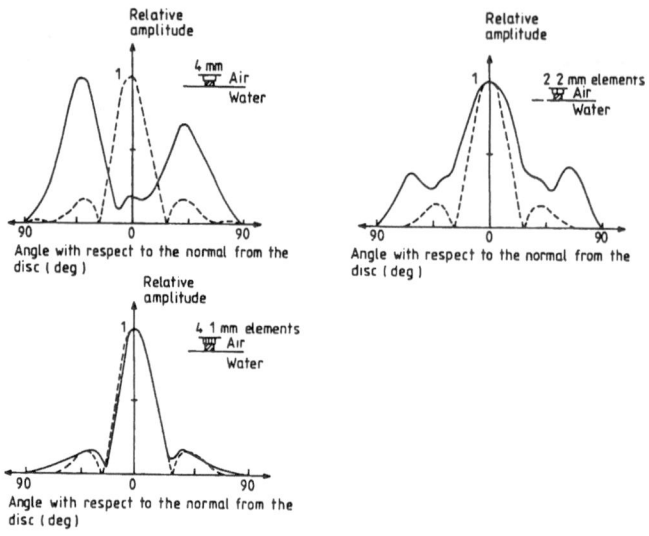

Figure 7. Lamb wave mode generated sidelobes, and
their reduction by increasing the number
of individual elements of the sensor, keeping
the sensor diameter constant. Full curve,
experimental results, broken curve; predicted
piston source beam patterns. Reference [16].

b. BACKING AND LOAD INFLUENCES

In order to improve the resolution, not only high frequencies, but
in particular short pulses are necessary. One period of a sine
wave will be desirable, but is frequently difficult to obtain.
A natural ringing of the transmitter will normally be present,
unless special countermeasures have been taken. One of these
countermeasures is to change the acoustic impedance of the backing of the piezoelectric disc material to match the impedance of
the disc material in order to suppress reflections at the interface between the disc and its backing. The influence of the
degree of acoustic impedance mismatch between the disc and its
backing on the pulse shape is given in figure 8.
In this figure the acoustic impedance, i.e. the product of the
density and the velocity of sound, of the disc material is
3×10^6 Nsm^{-3} while the acoustic impedance of the backing is
being varied from $Z_B = 3 \times 10^6$ Nsm^{-3} to $Z_B = 0.3 \times 10^6$ Nsm^{-3} leading
to a mismatch in acoustic impedance of 1:10 between the backing
and the disc. The ringing influence at increasing impedance mismatch is obvious.

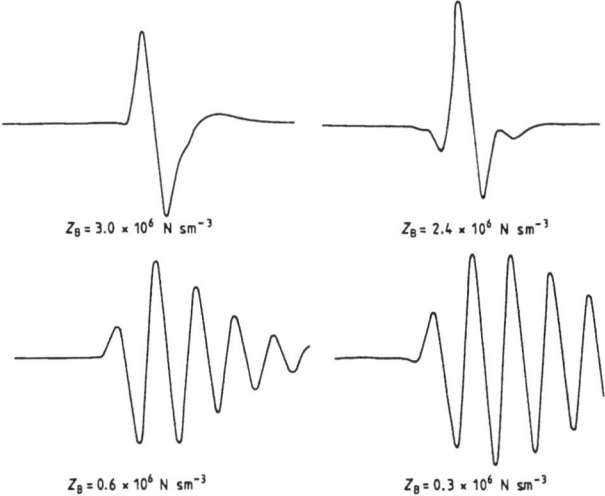

$Z_B = 3.0 \times 10^6$ N sm^{-3} $Z_B = 2.4 \times 10^6$ N sm^{-3}

$Z_B = 0.6 \times 10^6$ N sm^{-3} $Z_B = 0.3 \times 10^6$ N sm^{-3}

Figure 8. Pulse shape variation as a function of the acoustic impedance of the backing Z_B. Disc impedance is 3×10^6 Nsm^{-3}, everywhere. Reference [16].

As backing materials for piezoelectric disc are normally used a mixture of tungsten powder and epoxy or rubber. Such a mixture also possesses an acoustic attenuation effect which leads to a reduction of the amplitude of the ultrasound waves transmitted into the backing via the piezoelectric disc. By means of tungsten powder suspended in araldite it has been possible [16] to obtain values of backing density of $13 \cdot 10^3$ kg m^{-3} and sound velocity in the backing of 2300 ms^{-1} leading to an acoustical impedance of about $3 \cdot 10^7$ Nsm^{-3}, which is about 10% lower than the acoustical impedance of PZT. Lower impedance disc materials as for instance PVDF are more easy to match, but their piezoelectric constants are frequently not as high as the ones found in PZT. The use of a backing material reduces the Q-value, i.e. the quality factor, of the disc/backing system and leads to a lower emitted ultrasound amplitude. Thus, the price for short pulses is frequently a lower transduction efficiency.

In order to improve the transfer of ultrasonic energy from the vibrating disc to the load a front-face matching system is frequently introduced [25]. This system may consist of a single

layer, frequently of a quarter of a wavelength in thickness, of
a material having an acoustical impedance being the geometrical
mean of the disc and load impedances. However, more recent studies have shown that the use of two matching layers is more beneficial than the use of one and will lead to improved bandwidth
of the sensor, as well as a reduction in the pulse length may be
obtained. According to [18] in a two-layer sensor the intermediate
layers should have acoustic impedances Z_a and Z_b given by:

$$Z_a = Z_L^{3/7} \cdot Z_D^{4/7} \quad \text{and} \quad Z_b = Z_L^{6/7} \cdot Z_D^{1/7} \qquad (4)$$

where Z_L and Z_D denote the acoustic impedance of the load and the
disc, respectively.

Depending upon the application and available data processing
equipment, backing and/or front-face matching may be introduced
in the sensor construction and quite dramatic changes in the
sensor efficiency may be obtained.

c. BOND LINE INFLUENCE

In particular at higher frequencies the presence of a thin layer
of bond material may not have a negligible influence on the sensor
performance if the acoustic impedance of the layer is significantly lower than those of the materials on either side. The
thickness of the bond line as well as the bond material have to
be kept under control. Moreover, the quality of the bond may
have a considerable influence on the beam patterns of the sensor,
as poor or patchy bond lines may lead to an unpredictable sensor
performance. Thus the bond line is frequently one of the most
crucial factors in a sensor production.

Figure 9 shows the influence on the pulse shape arising from
various bond line thicknesses ranging from 0 to $\lambda/100$, where λ is
the wavelength in the bond material.

The bond lines in figure 9 are between a PZT disc of acoustic
impedance 33×10^6 Nsm^{-3} and a tungsten/araldite backing of acoustic impedance 27×10^6 Nsm^{-3}. The reflection influence of the thicke
bond line is obvious and a longer pulse is obtained. Therefore, st
for broad bandwidth applications a line thickness of not more

than $\lambda/200$ should be aimed at.

For other piezoelectric materials like quartz or PVDF which have a lower acoustic impedance than PZT, bond line thickness problems should not be found for frequencies up to, and for PVDF even above, 1 MHz.

The patchy structure of the bonds, where reductions in the bond quality or even small unbonded areas are randomly distributed across the bonded surface, will have an influence at all frequencies. Some reduction in the influence of the bond line may be obtained by the use of a bonding material having a somewhat higher acoustic impedance. Attempts to use lightly loaded araldite [19] as well low-melting solder materials for bondings between the disc and its backing have been made.

d. FOCUSED SOURCES

Apart from the self-focusing observed in the beam patterns on figure 5 no concentration of acoustic energy is found in the field produced by a piston source. However, focused transmitters have for a long time been used in NDT and in medicine in order to increase the lateral resolution in the focal region and in order to improve the signal-to-noise ratio. Focusing may be obtained by the use of a suitably curved transmitter or by the use of an acoustic lens in front of a plane piston source. Spherically as well as cylindrically focused transmitters are available.

For a spherical cap transmitter the acoustic pressure variation along the axis of the beam may be determined using expression (5) as [16]:

$$p(R) = p(0) \left(\frac{2D}{(1-R)} \right) \left[\sin \left\{ \frac{\pi}{\lambda} \left[(R-h)^2 + \frac{a^2}{4} \right)^{\frac{1}{2}} - R \right] \right\} \right] \qquad (5)$$

where R is the distance from the cap, D is the diameter of the cap, a is the radius of curvature of the spherical section, λ is the ultrasonic wavelength and h is a parameter given by:
$h = D - [D^2 - (a^2/4)]^{\frac{1}{2}}$.

As already showed by O'Neil [20] the lateral pressure variation in the focal plane is similar to the piston source farfield beam

pattern, but is strongly reduced in lateral dimensions and the
major beam is much more peaked. Moreover, the maximum sound pressure amplitude is obtained at a shorter source distance than predicted by geometrical considerations due to the influence of
diffraction effects.

The sound pressure variation in the lateral direction (x) may
be expressed by [16]:

$$p(x) = p_{max} \, 2J_1 \left(\frac{\pi Dx}{\lambda R_f} \right) (\pi Dx / \lambda R_f)^{-1} \tag{6}$$

where R_f and x denote the source distance at which the maximum
occurs and the lateral distance from the acoustic axis, respectively.
The use of focused sources in robotics may improve the distance
resolution and reduce the probability of false echoes in particular when used for alarm for the approach of a surface, where the
well-defined focal region can be of value to avoid ambiguity.
The use of several sensor elements in an array permits a variation
in the position of the focal point, i.e. dynamic focusing, to be
made using the same sensor, but varying the time of activation
between the elements. By means of such a multiple element array,
a so-called phased array, the direction of the acoustic axis may,
however, be varied by delaying the excitation time of each neigh-
bouring element along the array by the same time delay. The
phased array concept open up a flexibility in the tailoring of
ultrasonic beam patterns.

e. APERTURE SHADING

Aperture shading, where the main lobe of the beam patterns is
emphasized while the side lobes are suppressed, may lead to a
more well-defined beam and insonified region of an object.
Normally, a Gaussian beam shape in the acoustic pressure variation vertical to the acoustic axis is being aimed at.
Several methods for obtaining the Gaussian beam shape have been
developed over the years. Electrode configurations, including a
star-shaped electrode [21] and a reduced diameter of the back
electrode relative to the front electrode [22], have been tried
out. Moreover, a set of concentric annular electrodes have been
produced on a piezoelectric crystal using photo-etching techni-

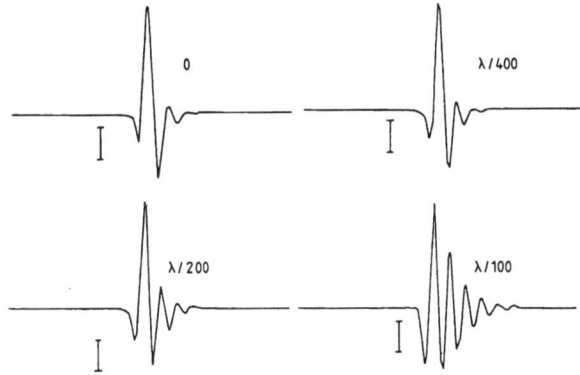

Figure 9. Variation in the pulse shape as a function of bond line thickness for a sensor having $Z_B = 24 \times 10^6 \mathrm{Nsm}^{-3}$. Reference [16].

Figure 10. Coordinates of electrodes for calculating the electric field. Reference [22].

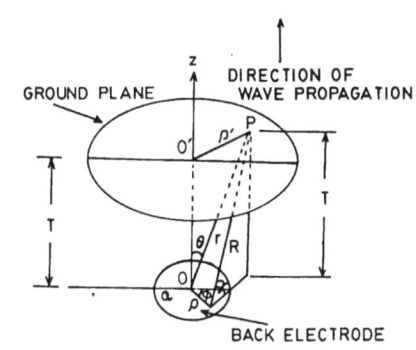

Figure 11. Electric field distribution for different ratios D/T. a. is the radius of the back electrode. Reference [22].

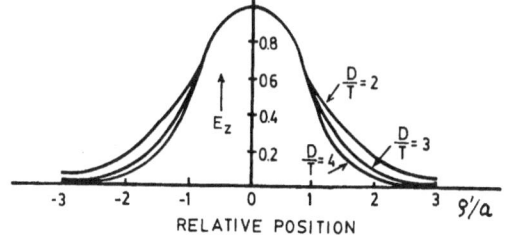

ques, and by varying the voltage supplied to each individual electrode a Gaussian radial distribution of the displacement
velocity has been obtained [23]. Also shading of a planar array
of sensor elements has been studied using the phased array concept together with a variation in the radial direction of the
voltage supplied to the elements.

The Gaussian field transmitter reported in [22] is shown in
figure 10.

The influence of the ratio of the diameter D (=2a) of the back
electrode and the thickness T of the piezoelectric crystal on
the electric field distribution on the crystal is shown in
figure 11 which shows the establishment of the Gaussian field
distribution across the crystal surface. The sound-pressure distribution in the horizontal and vertical directions across the
acoustic beam at the source distance Z = 10 cm for 6 MHz, and with
a comparison of the curves for 2 and 6 MHz in the vertical direction, are given in figure 12, which show the formation of a

Figure 12.
Sound pressure distribution
in the horizontal and
vertical direction across
the acoustic beam at the
source distance Z = 10 cm
for 6 MHz. 2 and 6 MHz
curves are given in the
vertical direction.
Reference [22].

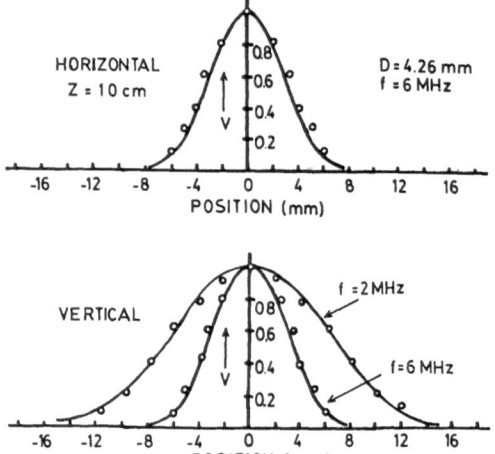

Gaussian shaped acoustic beam without the influence of side lobes
and with a good agreement between the experimental and theoretical data.

Another sensor with a reasonable directivity is the so-called
axicon sensor. This sensor consists in principle of an annular

element having a radial width less than one wavelength. It can
for instance be produced by masking the central region of a
conventional sensor by the use of an absorbing material. Each part
of the annulus will emit sound over a wide angular range, and
this sound will arrive in phase, only at points along the axis
of the annulus while sound arrivals outside the axis will take
place more or less out of phase, and thus not be able to reach
the amplitude obtained along the axis. A collimation of the
acoustic beam is thus produced which is of importance for the
use in robotics.

f. CALIBRATION

The calibration of ultrasonic sensors is mandatory and several
calibration procedures have been developed over the years. These
procedures, some of which have formed the basis of international
standards, are for instance: Reciprocity calibration which frequently is a time consuming, but a first order calibration method;
reflection by a reflector of known geometry, frequently a ball;
time delay spectrometry (TDS); Raman-Nath diffraction; schlieren
techniques; laser interferometry; the use of standard calibrated
sensors, i.e. a comparison method; calorimetric measurements;
radiation force measurements; etc. Also calibrations based on
measurements of the elements of an electric equivalent circuit
of a piezoelectric sensor, using admittance and impedance circle
plots, are useful procedures for determination of parallel and
series resonance frequencies as well as mechanical Q-values and
electromechanical coupling coefficients.
Sound power scattering from a sphere of radius r_o may be used
for calibration of ultrasonic sensors as the scattered power
as a function of kr_o where k is the wavenumber is a simple function of the distance r and the angle θ, given by:

$$I_{r\theta} \sim 0.11(kr_o)^4 \left(\frac{r_o}{r}\right)^2 (1-3\cos\theta)^2 \qquad \text{for } kr_o << 1 \qquad (7)$$

and

$$I_{r\theta} \sim \left(\frac{r_o}{2r}\right)^2 + \left(\frac{r_o}{2r}\right)^2 \cot^2\left(\frac{\theta}{2}\right) J_1^2 (kr_o\sin\theta) \text{ for } kr_o >> 1 \qquad (8)$$

Also the use of Raman-Nath diffraction, where a laser beam is diffracted by an ultrasonic wave, has proven to be useful for calibration purposes and for finding and classification of sensor defects [24]. These defects may arise from a too large spot of solder on a PZT disc, a patchy bond line or no bond at all over an essential part of the disc surface, cracks in the disc material etc. 3-dimensional plots of the ultrasonic field using Raman-Nath diffraction of laser light traversing the ultrasonic field in 4 different directions with a separation of 45° will give very detailed information of type and position of the defects [24].

ULTRASONIC SENSOR TYPES

a. BIMORPH SENSORS

The change in dimensions of a disc of piezoelectric material under voltages below the breakdown level are small, only a few microns for discs up to a centimeter in thickness. Such a disc also has considerable stiffness that it cannot respond to small forces asso ciated with large motional amplitudes. A bimorph structure can, however, be used to overcome these limitations to some extent.
The figure 13 is shown two bimorph structures.
The first structure (a) consists of two similar ceramic (PZT) discs attached to a central metallic disc. Both ceramic discs are electroded and they are poled in the same sense if the output terminals are to be the central metal disc and the outer, connected, electrodes. If the terminals, however, are the outer electrodes the ceramic disc are poled in the opposite sense. Structure (a) is referred to as a trilaminar composite. The biliminar composite, structure (b) on figure 13, is an alternative form of a bimorph consisting of a thin metal plate attached to a thin ceramic disc.
The central metal disc in (a) may be so thin as to be negligible from a mechanical point of view, structure (c) on figure 13, and it may even be replaced by a set of metallized holes running the length of the plate to be used during the poling of the ceramic.

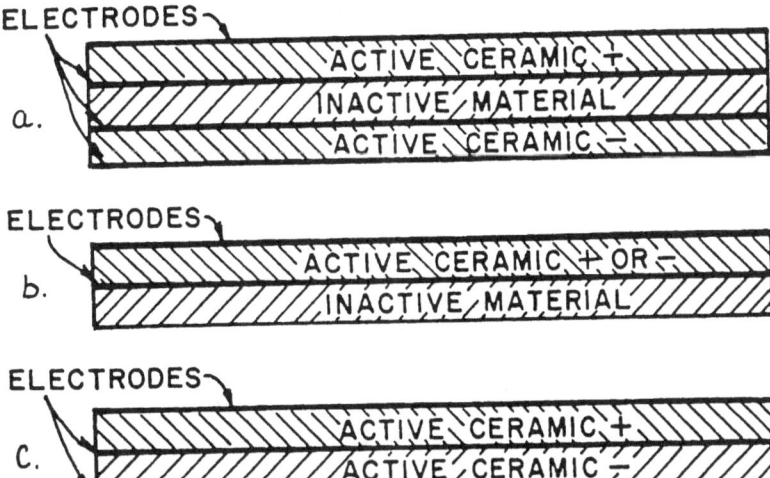

ELECTRODES

a. | ACTIVE CERAMIC + | INACTIVE MATERIAL | ACTIVE CERAMIC −

ELECTRODES

b. | ACTIVE CERAMIC + OR − | INACTIVE MATERIAL

ELECTRODES

c. | ACTIVE CERAMIC + | ACTIVE CERAMIC −

Figure 13. Bimorph composite structures for ultrasonic sensors. (a) trilaminar, (b) bilaminar and (c) bimorph.

Figure 14. Bimorph based microphone.

Figure 15. Microphone based on a bimorph element (1) connected with a membrane (2). (3) and (4) are a perforated lid and a porous material, respectively.

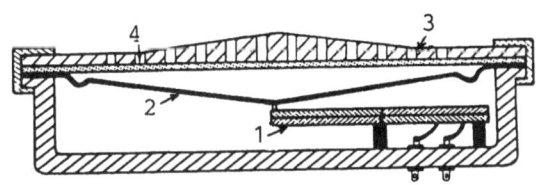

When an excitation voltage is applied to the electrodes of the
discs on figure 13, the piezoelectric ceramic composite flexes
in and out in a motion perpendicular to the disc plane.
Due to the magnitude of the flexural vibration amplitude considerable acoustic energy can be transferred to the fluid around
the composites. This energy transfer may take place either by
a direct coupling to the fluid medium or the composite may drive
a separate radiating structure, for instance a diaphragm,
directly or via a stiff connecting rod. An example of a microphone structure, using a bimorph element is given on figure 14.
In this simple structure the diaphragm is coupled to the centre
of a bimorph element which together with the cavity contributes
to the frequency response of the microphone. Another microphone
construction based on a bimorph element is given on figure 15.
In this microphone structure a cantilever shaped bimorph element
is supported in 3 corners and is free to move coupled to the
membrane in the 4th corner.
Some examples of characteristic transmitting and receiving
qualities as well as directivities measured by some standard
ultrasonic sensors for use in air are given in figure 16.

Figure 16. Frequency characteristics for re-
ception (a) and
transmitting (b)
of ultrasonic
signals by 3 ultra-
sonic sensors.
Their directivities
are given in figure
(c).

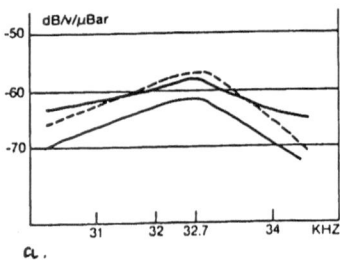

a.

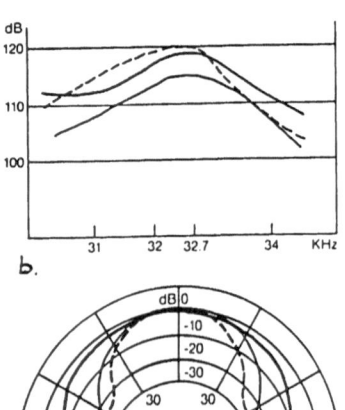

b.

c.

These ultrasonic sensors, produced by Projects Unlimited Inc.,
have a center frequency around 32.7 kHz and they have case diameters 16, 18 and 24 mm. Their receiving sensitivity range from
-63 to -67 dB re. 1 V/μ Pa, while their transmitted sound pressure
level range from 114 to 120 dB re. 20 μ Pa, all at the resonance
frequency 32.7 kHz. The directivity curves show, as emphasized
by the expression (1) and (2), that for constant frequency the
largest sensor diameter will lead to the most narrow beam.
The rise and decay time for the signal emitted by an ultrasonic
transmitter is dependent on the Q-value of the transmitter. The
convolution curve for an ultrasonic pulse having a carrier frequency of 40 kHz is shown in figure 17.

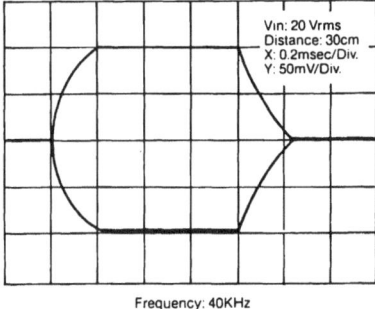

Figure 17.
Rise and decay course
by a 40 kHz ultrasonic
transmitter operating
in a tone-burst mode.

Frequency: 40KHz

With the rise and decay time being approximately 0.2 ms, the Q-
value of this transmitter is around 8, which is acceptable for
a transmitter to be used in air.
Within certain limits it is possible to increase the transmitted
sound pressure level by increasing the input voltage of the
transmitter in order to improve the signal-to-noise ratio, see
figure 18. There is, however, a limit for the applicability of
this procedure for signal-to-noise ratio improvements set by the
reverberation, where signals coming from other surfaces in the
sensor environment may confuse the distance measurement. Improved
applicability of this procedure may in particular be found by
sensors of high directivity.
Figure 18, moreover, shows the influence of temperature on the
frequency characteristic of a 40 kHz sensor. A temperature
decrease leads to an increasing resonance frequency, while the
transmitted sound pressure level is nearly unchanged.

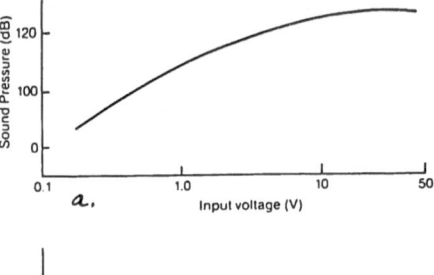

Figure 18.
Characteristic change
due to input voltage (a)
and temperature charac-
teristics (b) by a 40 kHz
sensor.

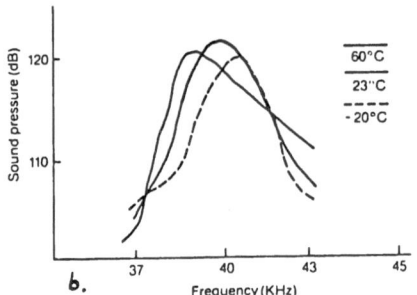

b. SOME SENSORS FOR USE IN AIR.

Sensors for use in air exist as closed as well as open constructions. As the sensor characteristics may change when the piezoelectric element is exposed to humidity, dust, temperature
changes, hostile atmospheric conditions, corrosion etc. it is
frequently necessary to use a closed construction in which the
sensing element very often is protected by a membrane made from
a metal or plastic material. 4 different types of ultrasonic
sensor constructions are given in figure 19 including open as
well as closed constructions.

On this figure (a) and (b) show open constructions, (a) having
displacement antinode at the center and (b) having displacement node at the center due to the coupling to the mass below.
(c) and (d) are closed constructions showing vibrations (c)
at the fundamental resonance and (d) at the first overtone.
The resonance tuning of the protecting membrane by varying its
thickness t and diameter D for a membrane material having a
velocity of sound c may be performed using the expression (9)
for the resonance frequencies f_n:

$$f_n = k_n tc/D^2 \quad \text{where } k_n = 1.88; \ 7.31; \ 16.4 \qquad (9)$$

where n is the number of the flexural modes, i.e. the number
of nodal circles (n = 0,1,2 ...).

While the diameter of the pieroelectric disc is not critical
for the types (a) and (b) on figure 19, the diameters for the
sensor types (c) and (d) should not exceed 0.35D and 0.25D,
respectively.
Also the use of piezoelectric plastic foils as for instance
PVDF has become very attractive for sensor constructions.
Due to the small displacements of the foil surface when operated in the thickness mode, the transmitters for use in air
and based on PVDF exploit the transverse coupling effect between
the thickness and the length of the foil, se figure 20.
In figure 20 a metallized piezoelectric high-polymer foil is
exposed to a varying electric field. By clamping two edges of
the foil the prolongation and shortening of the foil caused
by its thickness variations causes the emission of sound. In
order to stabilize the foil in neutral position it frequently
is provided with a sponge rubber backing.
For receiving purposes the piezoelectric plastic foil based
sensors compete with sensors based on electret foils.

c. SOME SENSORS FOR USE IN WATER

PVDF foils may also, due to their lower specific acoustic impedance than the one of ceramics, be used for transmitting and
receiving of signals in water. Some sensor constructions are
shown in figure 21 where a single-layer receiver backed by
unpoled PVDF, and its receiving response are given together with
a multi-layer (16 layer) PVDF transmitter. Also for the transmitter the piezoelectric element is surrounded by unpoled
PVDF in order to obtain the best impedance matching.

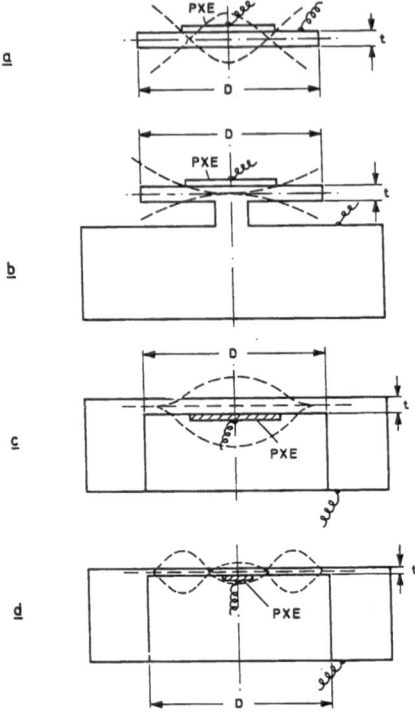

Figure 19. Four different types of rotational symmetric
ultrasonic sensors. Reference [26].

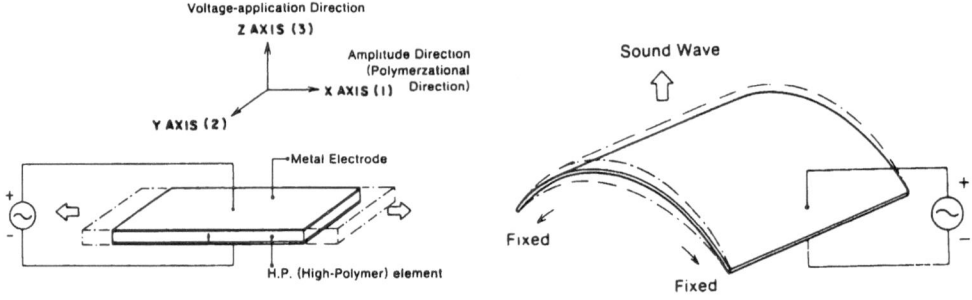

Figure 20. Piezoelectric high-polymer and its
transmitting mode.

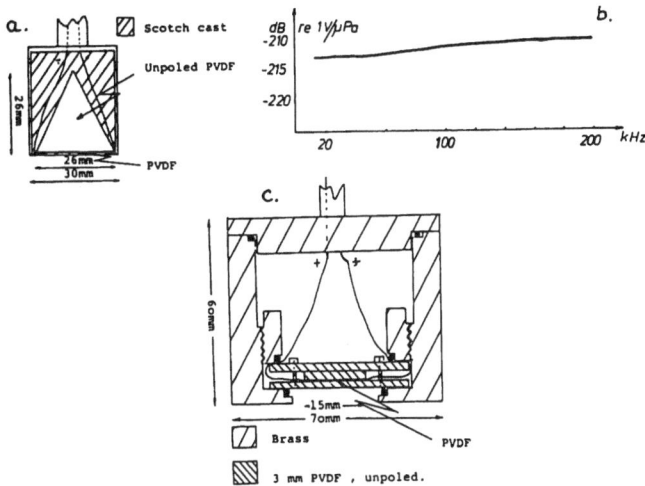

Figure 21. PVDF-based receiver (a) and its receiving
response (b). 16-layers PVDF transmitter (c).

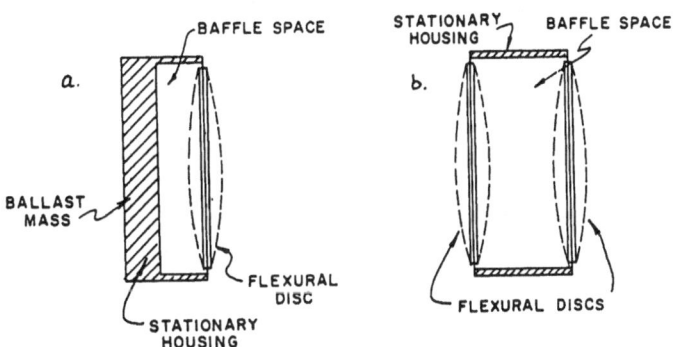

Figure 22. Flexural disc bimorph transmitters (a)
unidirectional and (b) bidirectional.

Som flexural disc transmitters based on bimorph structures
for use in water are shown in figure 22. Frequently the
baffle space is filled with oil instead of air, which leads to
different coupling factors and resonance frequencies. The bidirectional flexural disc transmitter may, moreover, have the
two discs driven out of phase. The efficiency of flexural disc
transmitters may be very high depending on the impedance match
with the medium surrounding the discs.

A broad variety of sensor types has over the years been constructed for underwater distance determination. The closed type
on figure 19 is frequently used, and for higher power transmission at lower frequencies piezoelectric elements clamped
between metal blocks forming head and tail parts of a sandwich
transmitter are being used. A more recent discussion of
ultrasonic sensors for underwater use may be found in [28].
In order to increase the bandwidth of a piezoelectric sandwich
transmitter a perforated structure of the head part has proven
to be useful [29].

CONCLUSION

There exist a hard-ware side of the ultrasonic sensor problem,
which due to the many factors influencing the final sensor
qualities frequently lead to the impression that sensor construction is more an art - or even black magic - than simpel
technology. This hard-ware side may lead to some difficulties
when sensors for robotics shall be developed, but fortunately
a considerable fund of knowledge, frequently earned "the hard
way", is available for the sensor engineer.

REFERENCES:

1. Prinz, F. B., The use of acoustic versus optic range
   sensors in manufacturing systems. J. Acount. Soc. Amer.,
   Suppl. 1, 79, 1986, S58.

2. Kleinschmidt P. and Magori, V., Ultrasonic robotic sensors
   for exact short range distance measurement and object
   identification. Proceeding of 1985 IEEE Ultrasonics
   Symposium, 1985, 457.

3. Tehon, S. W. and Roberts, C. R., An ultrasonic machine
   tool datum. Proceedings of 1985 IEEE Ultrasonics
   Symposium, 1985, 468.

4. Schoenwald, J. S., Strategies for robotic sensing using
   acoustics. Proceedings of 1985 IEEE Ultrasonics
   Symposium, 1985, 472.

5. Hickling, R. and Maria, S. P., Precision in ultrasonic
   gauging. J. Acoust. Soc. Amer., Suppl. 1, 79, 1986, S59.

6. Smith, R. W., Walters, R., Carlson, J. and Harris, R.,
   High-frequency acoustic systems for robotic applications,
   J. Acoust. Soc. Amer., Suppl. 1, 79, 1986, S60.

7. Billings, J. K., Noncontact ultrasonic gauging for
   industrial applications. J. Acoust. Soc. Amer., Suppl. 1,
   79, 1986, S60.

8. Meyer, P. A., Distance measurements using airborne sound.
   J. Acoust. Soc. Amer., Suppl. 1, 79, 1986, S60.

9. Knight, J. A. G., Pomeroy, S. C. and Beurle, R. L.,
   Ultrasonics in manufacturing. J. Acoust. Soc. Amer., Suppl. 1,
   Suppl. 1, 79, 1986, S59.

10. Cicco, G. D., Morten, B., Prudenziati, M., Taroni, A. and
    Canali, C., A 250 kHz piezoelectric transducer for
    operation in air: Application to distance and wind
    velocity measurements. Proceedings of 1982 IEEE Ultra-
    sonics Symposium, 1982, 321.

11. Schoenwald, J. S. and Smith, C. V., Two-tone CW acoustic
    ranging technique for robotic control. Proceedings of
    1984 IEEE Ultrasonics Symposium, 1984, 469.

12. Bass, H. E. and Bolen, L. N., Ultrasonic background noise
    in industrial environment. J. Acoust. Soc. Amer.,
    Suppl. 1, 79, 1986, S59.

13. Brown, M. K., Feature extraction techniques for recog-
    nizing solid objects with an ultrasonic range sensor.
    IEEE Journal of Robotics and Automation, RA-1, (4), 1985,
    191.

14. Zemanek, J., Beam behaviour within the nearfield of a
    vibrating piston. J. Acoust. Soc. Amer., 49, (1),
    1971, 181.

15. Kinsler, L. E., Frey, A. R., Coppens, A. B. and Sanders,
    J. V., *Fundamentals of Acoustics*. John Wiley & Sons, 1982.

16. Silk, M. G., Ultrasonic transducers for nondestructive
    testing. Adam Hilger Ltd., Bristol, U. K., 1984.

17. Hagman, A. J., Weight, J. P., Brown, A. F. and Quentin, G.,
    Diffraction effects with wide-band ultrasonic transducers.
    Proc. Ultrasonics International 1979, 447, Butterworth
    & Co. (Publishers) Ltd.

18. Desilets, C. S., Fraser, J. D. and Kino, G. S., The design of
    efficient broad-band piezoelectric transducers. IEEE
    Trans. Sonics & Ultrasonics, SU-25, 1978, 115.

19. Smith, W. M. R. and Awojobi, A. O., Factors in the design
    of ultrasonic probes. Ultrasonics, 17, 1979, 20.

20. O'Neil, H. T., Theory of focusing radiators. J. Acoust. Soc. Amer., 21, (5), 1949, 516.

21. Haselberg, K. von and Krautkrämer, J.: Ein ultraschall-strahler für die werkstoffprüfung mit verbesserten Nahfeld. Acustica, 9, 1959, 359.

22. Du, G., and Breazeale, M. A., The ultrasonic field of a Gaussian transducer. J. Acoust. Soc. Amer., 78, (6), 1985, 2083.

23. Zerwekh, P. S. and Claus, R. O., An ultrasonic transducer with Gaussian radial velocity distribution. Proc. 1981 IEEE Ultrasonics Symposium, 1981, 974.

24. Bjørnø, L. and Larsen, P. N., Transducer calibration using light diffraction. Submitted for publication in Ultrasonics.

25. Lakestani, F., Baboux, J. C. and Fleischmann, P., Broadening the bandwidth of piezoelectric transducers by means of transmission lines. Ultrasonics, 13, 1975, 176.

26. Philips Application Book, *Piezoelectric Ceramics,* J. von Randeraat & R. E. Setterington (Eds), second Ed. Jan. 1974.

27. Sasady, N. C., Hartig, A. and Bjørnø, L. Development of some transducers based on polyvinylidene flouride Proc. Ultrasonics International 1979, JPC Science and Technology Press, London 1979, 468.

28. Rijnja, H. A. J., Modern transducers. Theory and Practice. In L. Bjørnø (Ed.), *Underwater Acoustics and Signal Processing.* D. Reidel Publ. Coup., Holland, 1981, 225.

29. Lin, C. M., Hou, L. Q. and Ying, C. F., Analysis of broad-band piezoelectric sandwich transducer with perforated structure. Archives of Acoustics, 9, (3), 1984, 349.

ACOUSTIC IMAGING IN THREE DIMENSIONS

J. F. Martin, K. Marsh, J. M. Richardson, and G. Rivera

Rockwell International Science Center
1049 Camino Dos Rios
Thousand Oaks, CA 91360

ABSTRACT

An algorithm and method of implementation are described for obtaining high-resolution, three-dimensional acoustic images of simple objects in air using only a small number of transducer positions. Each transducer position provides a single large bandwidth pulse-echo waveform along a specified incident direction. Probabilistic estimation procedures using the full waveforms are used to reconstruct a three-dimensional image of the objects in the observed volume. Prior information about the physics of the scattering process, the statistical nature of the noise and the acoustic reflectivity of the target object are used to obtain resolution of the order of the smallest wavelength present in the acoustic pulses. Results of both synthetic and experimental tests of the algorithm are presented, as well as a brief description of the experimental apparatus and the algorithm used.

1.0 INTRODUCTION

Although considerable success has been obtained in imaging three-dimensional objects with optical techniques, there are many situations in which acoustic techniques might be a useful complement or replacement for optical machine vision. The principal advantage of acoustics is its ability to make range or image measurements through optically opaque media.

Of course, there are certain disadvantages relating to the low speed of sound, its high attenuation in air, and its long wavelength. Furthermore, the most limiting factor in practice is often the existence of a high level of ambient noise that is difficult or impossible to block.

NATO ASI Series, Vol. F43
Sensors and Sensory Systems
for Advanced Robots
Edited by P. Dario
© Springer-Verlag Berlin Heidelberg 1988

Despite these disadvantages, the value of acoustic imaging has certainly been demonstrated in medicine, nondestructive evaluation of materials, and exploration under the sea. As robots move into these applications, it will be necessary for them to include acoustic imaging and ranging systems in their battery of sensors. In addition to the traditional arenas of application, it is quite possible that acoustic imaging in air can provide estimates of object shape, texture, and density if practicable methods of making high-resolution measurements with acoustics can be found. Unfortunately, the usual methods and algorithms of acoustic imaging and estimation in air, water, or materials make it impossible to obtain resolutions better than several wavelengths and even worse, require hundreds or thousands of transducer positions for a three-dimensional image. Before proceeding to a description of the approach employed in this effort, a brief description of one class of the conventional methods of acoustic imaging will be presented.

2.0 CONVENTIONAL ACOUSTIC IMAGING

Figure 1 shows a simplified description of the radiation pattern emanating from a flat transducer excited to generate longitudinal waves. (For a detailed description of the characteristics of the radiation fields produced by transducers, see Ref. 1.) The key parameters are the radius of the transducer, denoted as a, and the operating wavelength λ. These determine the angular spread of the radiated beam, defined here as θ, and the transition point from "near-field" to "far-field", R_{FF}. For range values less than R_{FF}, the radiation field is complex and difficult to describe mathematically. For range values greater than RFF, the radiation field can be rather accurately described as a simple propagating wave.

In conventional practice, imaging applications involving a single transducer require physically scanning the transducer in two dimensions, in a sequence of rasters in either cartesian coordinates or spherical coordinates.[2,3] By any one of a

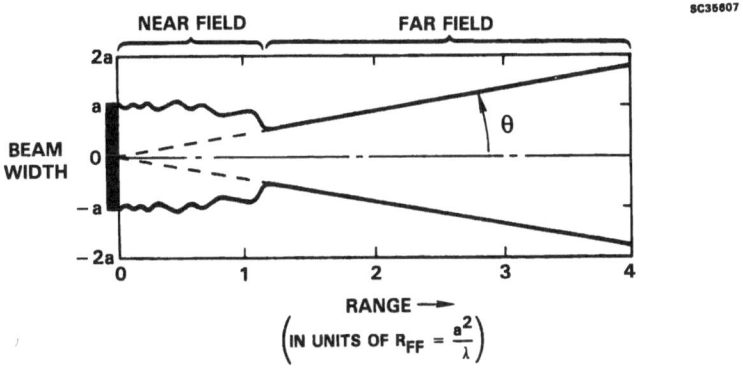

NEAR FIELD   FAR FIELD

SC35607

BEAM WIDTH

RANGE →

$$\left(\text{IN UNITS OF } R_{FF} = \frac{a^2}{\lambda}\right)$$

GENERALLY,
FOR R < R$_{FF}$, THE BEAM CAN BE FOCUSED BUT EXHIBITS COMPLEX INTERFERENCE
FOR R > R$_{FF}$, THE BEAM CANNOT BE FOCUSED AND SPREADS LINEARLY

Fig. 1 Simplified radiation pattern of a flat cylindrical transducer of radius a generating a longitudinal wave of wavelength λ. θ_6 represents the half-angle of the beam for an intensity 6 dB smaller than the on-axis intensity; it is approximately equal to 0.3 λ/a.

variety of methods, the time-of-flight is extracted from the earliest received pulse and used to estimate the range to the target scene at each point in the scan. The resulting image constitutes a range map of the scene, in which the range to the nearest object has been determined at hundreds or thousands of points.

The angular spread of the radiation pattern of the transducer or transducer array, combined with the range to the object, determines the lateral resolution possible with this approach. As shown in Fig. 1, the angular spread is a function of the wavelength and the transducer diameter. The range resolution is determined simply by the wavelength. As an example, consider an acoustic system operating at a frequency of 50 kHz; given a velocity of sound in air of 345 m/s, λ is 0.69 cm. A typical transducer is the one made by Polaroid, which has a radius of about 1.8 cm; the resulting half-angle at the 6 dB point is about 6 degrees. Figure 2 shows the expected

radiation pattern for this case. Note the presence of a main
lobe, with minor lobes occurring symetrically on each side.

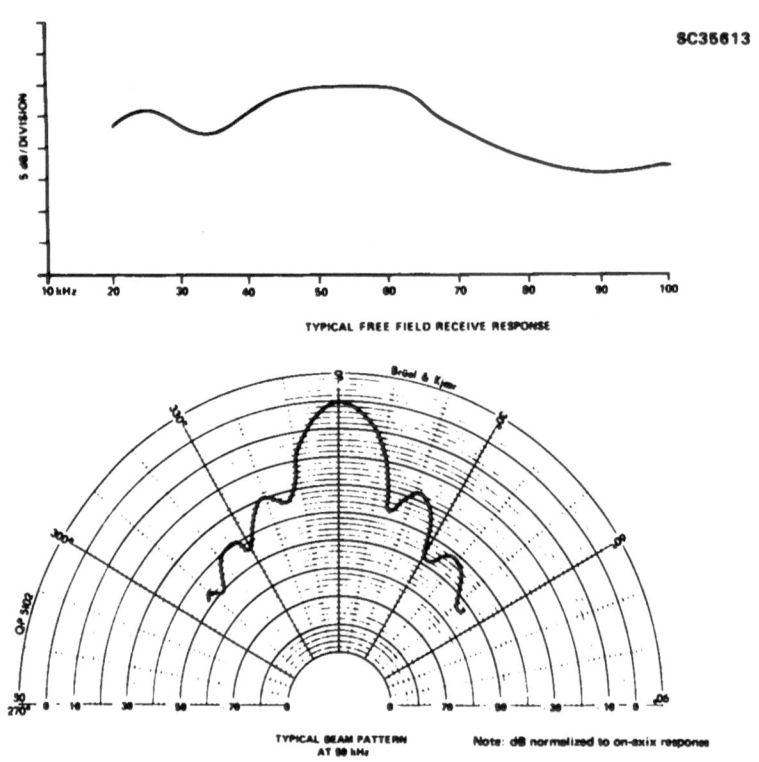

Fig. 2 Radiation pattern for an unfocused transducer of
        radius 1.8 cm operating at approximately 50 kHz.

The lateral resolution possible with this transducer is
approximately the size of the beam diameter, which is about
6 cm at a range of 25 cm. This means that objects or features
closer together than several wavelengths could not be resolved. The range resolution possible is a function of the
curvature of the object's surface: with a smooth flat surface
that fills the beam, a range resolution of 1/10 of a wavelength is easy to obtain. If there is substantial curvature
over the beam width, then the range resolution will be considerably worse because of the error in interpreting phase.

One standard method for obtaining higher lateral resolution is to focus the transducer by curving its face or by
inserting appropriate time delays in the transmitting and
receiving circuits of an array of small flat-faced transducers. The focal length is governed by the curvature of the
transducer face or the array time delays, but cannot in general be larger than R_{FF} as defined in Fig. 1. As with the
unfocused transducer, the resolution available with a focused
transducer is a linear function of the range to the target
objects; in this case, that is the focal length. The diameter
of the focal spot (Δy) for an intensity 6 dB down from the
maximum is approximately 0.6 F (λ/a), where F is the focal
length of the transducer or transducer array, and a is the
radius of the trandsucer.

For a focal length of 25 cm and a transducer radius of
4.5 cm, the value of Δy is 2.3 cm. This is smaller than the 6
cm obtained at 25 cm with an unfocused transducer, but is
still more than three times greater than the wavelength of the
acoustic energy used. Smaller values of Δy are possible only
if larger values of a are employed; such values may be impractical. Even if they are not impractical, it is still necessary to increase the amount of scanning required because the
spot size is now much smaller.

The second major problem with conventional acoustic imaging of objects in air is that the size of the returned echo is
a very unreliable indicator of size or shape: smooth, flat
surfaces are visible only over a small range of angles, convex
sharp corners are nearly invisible at any angle, concave corners can reflect as much sound as a smooth flat surface, and
so on. These effects can be quite serious, and require a substantial amount of modeling if a range map is to be correctly
interpreted.

In summary, the large beam width and the large number of
transducer positions required makes acoustic imaging in air
rather impractical by conventional techniques. It was the

objective of this research to demonstrate that it is possible to obtain a high resolution three-dimensional estimate of an object's shape in air with a small number of fixed transducer positions by employing a probabilistic approach. This approach is based on efficient use of a small amount of prior information and the availability of large bandwidth pulse-echo waveforms.

3.0 PROBABILISTIC ACOUSTIC IMAGING APPROACH

The present approach is a probabilistic one, based on measurement models using the Kirchhoff approximation for the scattering of acoustic waves. The measured data consist of pulse-echo waveforms taken in a small number of incident directions. The a priori information is in the form of a statistical distribution of possible acoustic impedance values. For imaging of solid objects in air, it is assumed that the acoustic impedance of the object is infinite (i.e., all acoustic energy is reflected), and hence everywhere in space the image can be represented by a three-dimensional characteristic function with only two possible values at each point, which can be defined as 0 (in air) and 1 (in the object). In many cases, it is not necessary to know this function everywhere, but rather it may suffice to know the elevation function of the visible surface, seen from some viewpoint.

In this paper we confine our attention to the case of a flat-bottomed object resting on a table where the upper surface of the object can be represented by a single-valued elevation function. An inversion technique for this case has been proposed by Richardson et al,[4,5] and will be discussed in detail below. It is an iterative method in which each step involves a linear Gaussian estimation problem.

3.1 Formulation of the Problem

We assume that an unknown solid object rests upon a rigid table within a known localization domain. A set of pulse-echo scattering measurements is made under conditions such that the localization domain and each transducer are in the far field of each other. We will define the incident wave to be the wave that would exist in the absence of the object and the scattered wave as the increment due to the presence of the object.

In formulating the measurment model, we limit our investigation to the case of objects described by single-valued elevation functions. We will use the Kirchhoff approximation for the scattering of acoustical waves under the assumption that the surfaces of the object and the table can be regarded as perfectly rigid. For the sake of simplicity, we will limit our discussion to situations in which acoustical shadows either cannot occur or can be neglected.

The appropriate measurement model is represented by the following expression

$$f(t,\vec{e}) = - \frac{\alpha c}{2\vec{e} \cdot \vec{e}_z} \sum_{\underline{r}} \delta \underline{r}$$
$$[p'(t - 2c^{-1}\vec{e}.\underline{r} - 2c^{-1}\vec{e}.\vec{e}_z \, Z(\underline{r})) \qquad (1)$$
$$-p'(t - 2c^{-1}\vec{e}.\underline{r})]. + \nu(t,\vec{e}) \quad .$$

The symbols in the above expression are defined as follows:

$f(t,\vec{e})$ = a possible measured scattered waveform at time t coming from a transducer having an incident wave direction e.

$\nu(t,\vec{e})$ = experimental error associated with f(t,e).

$p'(t)$ = time derivative of p(t), the measurement system response function. The latter is defined to be the waveform produced by the measurement system if a fictitious scatterer with an impulse response function R(t) = δ(t) is positioned at the origin.

$\underline{r}$ = two-dimensional vector $\vec{e}_x x + \vec{e}_y y$ giving positions in the xy plane (the plane of the table top). It takes values on a two-dimensional grid spanning the localization domain D_L. An elementary area of the grid is denoted by $\delta \underline{r}$.

$Z(\underline{r})$ = elevation function (i.e., the value of the vertical coordinate z on the top surface at a horizontal position $\underline{r}$).

$\vec{e}_z$ = unit vector pointing in the +z direction.

α = constant dependent upon the properties of air.

c = velocity of acoustic waves in air.

Equation (1) represents a summation of waves reflected to the receiver from each element of the object's surface using the Kirchoff approximation.

The time t is assumed to take a discrete set of values corresponding to an appropriate sampling rate over a specified observation interval. The localization domain D_L is defined by the inequalities: $- 1/2\, L \leqslant x < 1/2\, L$ and $-1/2\, L \leqslant y < 1/2\, L$.

To give a complete description of the measurement model, we must specify the a priori statistics of $Z(\underline{r})$ and $v(t,e)$. Here, we assume that both entities are Gaussian random vectors with the properties

$$EZ(\underline{r}) = 0 \quad , \tag{2}$$

$$EZ(\underline{r})Z(\underline{r}') = \delta_{\underline{r}\underline{r}'} \, \sigma_z^2 \quad , \tag{3}$$

$$Ev(t,\vec{e}) = 0 \quad , \tag{4}$$

$$Ev(t,\vec{e})v(t',\vec{e}') = \delta_{tt'} \, \delta_{\vec{e}\vec{e}'} \, \sigma_v^2 \quad , \tag{5}$$

$$EZ(\underline{r})v(t,\vec{e}) = 0 \quad . \tag{6}$$

In Eqs. (3) and (5) the Kronecker deltas, $\delta_{rr'}$, $\delta_{tt'}$, and $\delta_{\vec{e}\vec{e}'}$, are generalized in an obvious way to the case of non-integer and, in some cases, nonscalar subscripts.

Our problem is to determine the most probable elevation function given the results of scattering measurements. In more specific mathematical terms, our problem is to find the function Z that maximizes the a posteriori probability denoted by $P(Z|f)$. Here, the symbol Z represents the values of $Z(\underline{r})$ for all $\underline{r}$ and, similarly, f represents the values of $f(t,e)$ for all t and e. As is well known,

$$P(Z|f) = P(f|Z)P(Z)/P(f) \qquad (7)$$

where in the maximization process $P(f)$ may be regarded as constant. The factor $P(f|Z)$ is determined entirely by the model (1) and the a priori statistical properties of the measurement error $\nu(t,e)$. The factor $P(Z)$ is determined by the a priori statistical properties of the elevation function $Z(\underline{r})$.

3.2   Method of Solution

The determination of the most probable elevation function given the results of scattering measurements, i.e., the determination of $Z(\underline{r})$ that maximizes $P(Z|f)$, cannot be carried out by purely analytical means because of the nonlinear dependence upon $Z(\underline{r})$ in the measurement model (1). Several approaches have been tried in the past and are described in Ref. 6.

In the present paper we discuss a new method for dealing with the problems associated with the nonlinear dependence upon $Z(\underline{r})$ in the measurement model (2.1). The method involves an iterative procedure in which $f(t,e)$ and $p'(t)$ are initially subjected to a common smoothing operation that has the property that the smoothed version of $p'[t-2c^{-1}e\cdot\underline{r}-2c^{-1}e\cdot e_z\,Z(\underline{r})]$ can be linearized with respect to $Z(\underline{r})$ or the deviation of

$Z(\underline{r})$ from a first guess. Later stages of the procedure involve successive unsmoothing and linearizations with respect to incremental corrections to $Z(\underline{r})$. In more explicit mathematical detail, we assume that the smoothed versions of $f(t,\vec{e})$ and $p(t)$ are given by

$$f_m(t,\vec{e}) = H_m(t)*f(t,\vec{e}) \quad , \tag{8}$$

$$p_m(t) = H_m(t)*p(t) \quad , \tag{9}$$

where * denotes temporal convolution and $H_m(t)$ is the time-domain transfer function representing the low-pass filter associated with the mth stage. The exact model for the smoothed measurement process for the mth stage is clearly given by

$$f_m(t,\vec{e}) = - \frac{\alpha c}{2\vec{e} \cdot \vec{e}_z} \sum_{\underline{r}} \delta \underline{r} [p_m'(t - c^{-1}\vec{e} \cdot \underline{r}$$

$$- 2c^{-1}\vec{e} \cdot \vec{e}_z \ Z(\underline{r})) - p_m'(t - 2c^{-1}\vec{e} \cdot \underline{r}] + \nu(t,\vec{e}) \quad . \tag{10}$$

To present the general nature of the iterative method, we will introduce abbreviated notation. The above expression for the smoothed measurement process can be written in the compact form

$$f_m = G_m(Z) + \nu \tag{11}$$

in which the correspondences with the earlier, more explicit form are obvious. Linearization with respect to the deviation from the previous estimate gives

$$f_m = G_m(\hat{Z}_{m-1}) + A_m(Z-\hat{Z}_{m-1}) + \nu \quad , \tag{12}$$

where

$$A_m^T = [\frac{\partial^T}{\partial Z} G_m(Z)]_{Z=Z_{m-1}} \quad . \tag{13}$$

The best estimate of Z at the mth stage, i.e., $\hat{Z}_m$, is given by a straightforward application of linear estimation theory. A more detailed discussion is given in Refs. 4 and 5.

The next step in the iterative procedure is to select a time-domain transfer function $H_{m+1}(t)$ corresponding to a higher frequency roll-off in the low-pass filter [this represents an incremental unsmoothing of the previously smoothed $f(t,\vec{e})$ and $p(t)$]. We then use $\hat{Z}_m(\underline{r})$ as a new point in state space about which the measurement model is to be linearized. We then obtain eventually a new estimate $\hat{Z}_{m+1}(\underline{r})$ by the same procedure as before, except with the subscript m replaced by m+1.

The total recursion procedure is straightforward, at least in principle. We commence with $Z_0(\underline{r}) = 0$, or some other initial estimate, and a choice of $H_1(t)$ such that the characteristic wavelengths involved in the smoothed version of $f(t,\vec{e})$ and $p(t)$ are sufficiently long. The recursion process is carried on until the difference between successive approximations becomes sufficiently small according to a suitable criterion. At the present time, we have not established a procedure for selecting the sequence of transfer functions $H_m(t)$, and thus this selection remains a problem to be handled by computer experimentation.

4.0 COMPUTATIONAL EXAMPLE

In this section, we present an example of the above iterative approach using synthetic test data. The main purpose of this computation is to provide some insight into what imaging

performance is possible with a relatively sparse set of scattering measurements in the absence of scattering theory errors
and measurement error. The first type of error is avoided by
using the same Kirchhoff approximation in both the preparation
of synthetic data and the imaging procedure; however, a major
part of the Kirchhoff error is avoided by limiting our treatment to cases in which acoustical shadowing does not exist.
The second type of error, i.e., that due to imperfect measurement, is avoided by setting $\sigma_\nu = 0$ in the preparation of test
data; we must emphasize, however, that the in the imaging procedure σ_ν is assigned various positive values, a matter that
will be discussed in greater detail in a later paragraph.

In the production of synthetic test data we assume that
the object of interest is a squashed tetrahedron, i.e., a
regular tetrahedron with its vertical scale reduced by some
factor. In particular, we constructed three tetrahedra with
horizontal edges having a common length of 13.75 mm and with
vertical heights corresponding to scaling factors of 75%, 50%
and 25% relative to an ideal regular tetrahedron. In these
examples, we have assumed a set of five pulse echo measurements (one more than the hypothesized minimal number) each
with an incident direction

$$\vec{e} = - (\vec{e}_x \cos \phi \; k + \vec{e}_y \sin \phi) \sin \theta - \vec{e}_z \cos \theta \qquad (14)$$

given by the value of azimuthal and polar angles, ϕ and θ,
respectively, tabulated below

$$\theta = 0 \quad 54.7° \quad 54.7° \quad 54.7° \quad 54.7° \quad ,$$
$$\phi = - \quad 0° \quad 90° \quad 135° \quad 225° \quad .$$

The constant α in Eq. (1) was taken equal to 1 since its common value occurs in both the preparation of synthetic test
data and the imaging problem, and can be regarded as self-
cancelling. We used a signal-to-noise ratio η defined by

$$n^{-1} = \tilde{f}(t,\vec{e})_{max}/\sigma_\nu \qquad\qquad (15)$$

where $\tilde{f}(t,\vec{e})_{max}$ is the maximum (with respect to t and e) of the waveform corresponding to noiseless synthetic test data, and where σ_ν is the standard deviation of $\nu(t,e)$. The function p(t) corresponded in the frequency domain to a Hanning window between the frequency limits f_{min} and f_{max}.

In the iterations, f_{min} was fixed at 1 KHz, and various values of f_{max} assumed. Approximately 20 iterations were required, the first having used the parameter values f_{max} = 10 KHz, n^{-1} = 10, and the final iteration having used f_{max} = 40 KHz, n^{-1} = 33. The results are presented in Fig. 3. The first column in this figure presents the assumed tetrahedron used in the preparation of synthesis data. The second column presents the output of the iterative procedure, and the third column is a horizontal cross section at 30% of maximum. A significant amount of superresolution is present, and this is discussed in more detail by Richardson et al.[6]

5.0 EXPERIMENTAL RESULTS

We now describe an experiment in which the shape of a simple object was reconstructed successfully using a small number of pulse-echo measurements together with the algorithm described in Section 3.0. The experiment was conducted in an anechoic chamber of dimensions 6 ft × 4 ft × 4 ft. The object to be imaged was a pentahedron with a square base, sitting on a flat, circular rotatable table. The height of the pentahedron was 16.6 mm,, and the base width was 57.4 mm. Six incident directions were used, accomplished by the appropriate positioning of a single, closely spaced pair of transducers, one for transmitting and the other for receiving. The transmitter was a Dyneaudio D-21AF electromagnetic dome tweeter amplified by a NAD 2200 power amplifier. The receiver was a

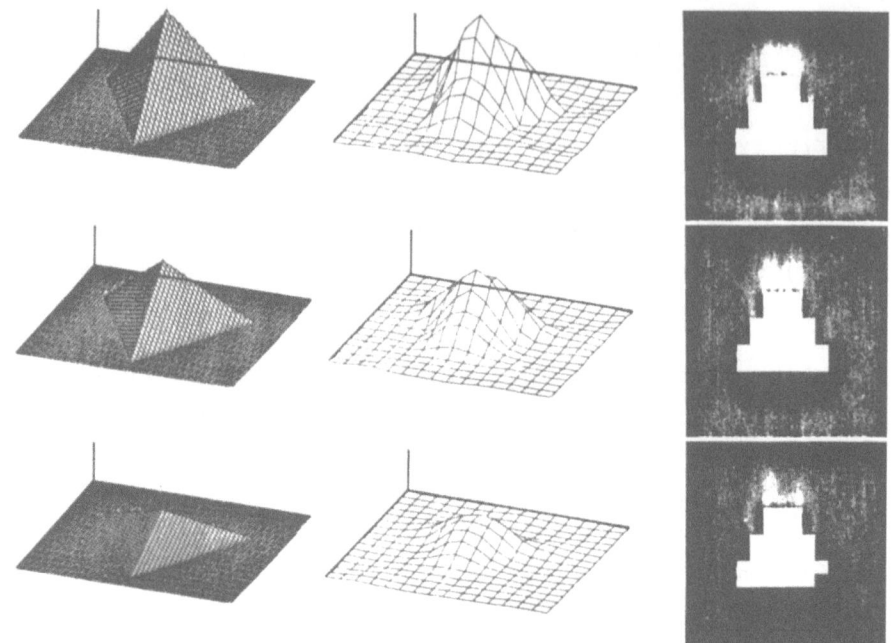

Fig. 3   Application of the iterative algorithm to squashed
         tetrahedra of three different assumed heights: (a)
         assumed object; (b) imaged object - hidden line
         representation; and (c) imaged object - horizontal
         cross section at 30% of maximum.

Brüel and Kjar type 4155 prepolarized condenser microphone
amplified by a B&K type 2230 precision integrating sound level
meter. The pulse-echo waveform and frequency response of the
combined system in pulse-echo mode, measured by reflecting an
incident pulse from a flat surface, is shown in Fig. 4. The
required polar angles θ were accomplished by mounting the
transducer pair at appropriate locations, and the various
azimuthal angles φ were produced by rotating the table. The
incident directions were as follows:

    θ =   0   45   45   45   45   45
    φ =   -   0   90   135   225   297

    Figure 5 shows the arrangement of the six different
locations of the transducers.

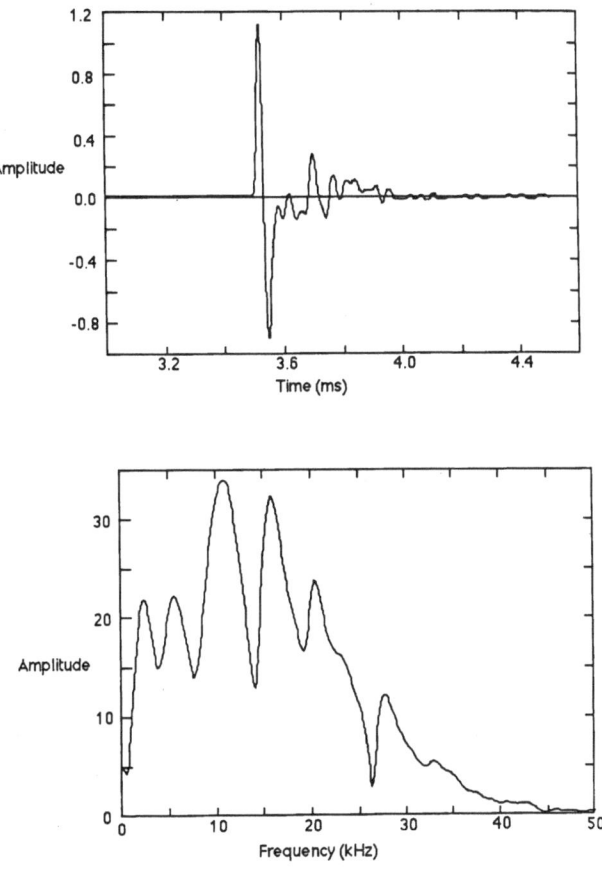

Fig. 4   Pulse-echo time waveform (a) for reflection from a
smooth, flat surface and its Fourier transform (b).

The waveforms were digitized using a Data Precision Data
6000, transferred to an Apple Macintosh for local display and
analysis and uploaded to a VAX 11/780 for image construction.
Figure 6 shows a block diagram of the apparatus. For each
incident direction, a set of waveforms was acquired with only
the table present, and then another set was acquired with the
pentahedron on the table. After averaging each set of waveforms (in order to reduce the background noise), the average
"table only" waveform was subtracted from the average "pentahedron + table" waveform to provide the time-domain measurement $f(t,\vec{e})$ required by our algorithm. The reference waveform
$p(t)$ was obtained by a pitch-catch experiment in which the

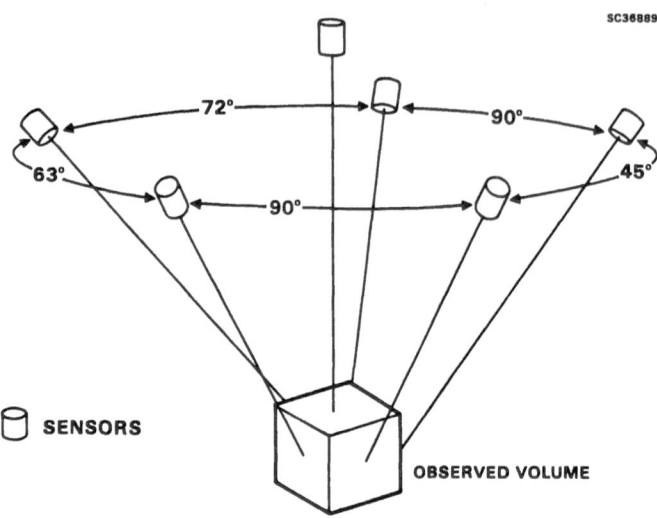

Fig. 5 Arrangement of transducer pair positions; all were
        1 m from the observed volume.

transmitter was aimed directly at the receiver. If we denote
the measured pitch-catch waveform by $V_p(t)$, then $p(t)$ is given
by

$$p(t) = (2/R)V_p(t - \tau) \tag{16}$$

where R is the distance between transmitter and receiver, in
this case 1 m. The quantity τ represents a time delay correction, to allow for the differing path lengths in the calibration experiment as compared to the pentahedron
measurements.

These data were used as input to the imaging algorithm
described in Section 3.0. The signal-to-noise ratio (as
defined by Eq. (15)) was 10. A total of 20 iterations was
required, each of which employed a Hanning window in the frequency domain. The upper cutoff of this window was gradually
increased in the same way as for the inversion using synthetic

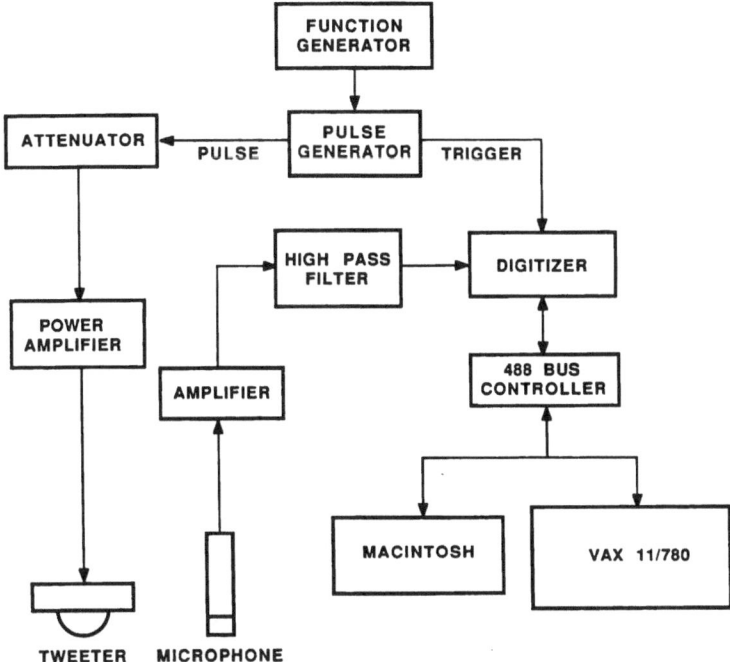

Fig. 6 Block diagram of the acoustic measurement system.

data as described in Section 4.0. The initial iteration used
an upper cutoff of 1 kHz, while the final iteration used
30 kHz.

The result of applying this sparse-data imaging algorithm
to the six waveforms is shown in Fig. 7, which represents the
most probable elevation function conditioned on the measurements. The elevation function is shown as a hidden line plot
from two different viewing directions. It can be seen that
the reconstruction is a reasonable approximation to the original pentahedron. There are, however, some problems with it,
the principle ones being:

1. The deduced height is too large by a factor of 1.7.
2. The baseline (representing the flat table) is rumpled.

VIEW ANGLE: 00.00 DEGREE ALTITUDE
45.00 DEGREE AZIMUTH

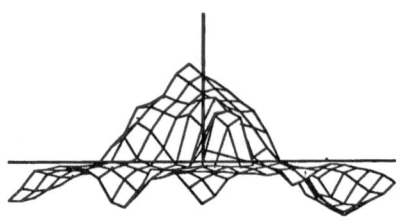

VIEW ANGLE: 60.00 DEGREE ALTITUDE
60.00 DEGREE AZIMUTH

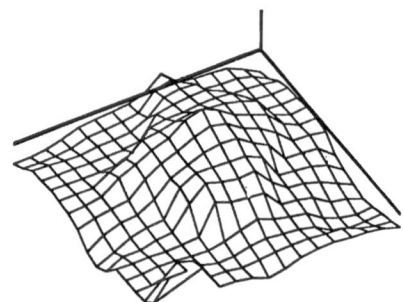

Fig. 7   Results of applying the new sparse-data imaging
algorithm to the experimental waveforms: (a) as
viewed from the side; (b) as viewed from 30° down
from directly overhead.

Both of these effects are consistent with an error in the
time delay calibration. We have confirmed this hypothesis by
a careful examination of the original data, and have arrived
at an improved scheme for performing the calibration. The experiment will be repeated, and the results presented in a
forthcoming paper. In any event, we are encouraged by the
present results, since they represent the first recognizable
reconstruction of an object other than a sphere which has been
made from sparse scattering data in air.

ACKNOWLEDGMENTS

This work was supported in part under DARPA contract number N00014-84-C-0085 and in part by Rockwell International Internal Research and Development funding.

REFERENCES

1. J. Krautkramer and H. Krautkramer, "Ultrasonic Testing of Materials," Springer-Verlag, 1977.

2. R.A. Jarvis, Computer Vol. 15, p. 8 (1982).

3. J.S. Schoenwald and J.F. Martin, "Acoustic Scanning for Robotic Range Sensing and Object Pattern Recognition," Proceedings of 1982 IEEE Ultrasonic Symposium.

4. J.M. Richardson, K.A. Marsh, J.S. Schoenwald and J.F. Martin, "Acoustic Imaging of Solid Objects in Air Using a Small Set of Transducers," Proc. of the 1984 IEEE Ultrasonics Symposium, p. 831.

5. J.M. Richardson, J.F. Martin, K.A. Marsh and J.S. Schoenwald, "Acoustic Imaging Systems (For Robotic Object Acquisition)," Final Report for DARPA Contract No. N00014-84-C-0085, 1985.

6. K.A. Marsh, J.M. Richardson, J.F. Martin and J.S. Schoenwald, "Acoustic Imaging in Robotics Using a Small Set of Transducers," Proc. of 4th Int. Conf. on Robot Vision and Sensory Controls, Oct. 1984.

Paolo Pelosi and Krishna Persaud

Istituto di Industrie Agrarie, Universita' degli Studi
via S.Michele, 4, Pisa, ITALY.
Dept. of Physiology and Biophysics, Medical College of Virginia
Box 551, MCV Station, Richmond, Virginia 23298, USA.

Abstract

The design of an artificial nose requires (a) a knowledge of the
"olfactory language", i.e. the molecular parameters of the odorants that
are measured by specific receptors in the nasal mucosa; (b) the
availability of gas sensors, specific, reversible and sensitive, able to
convert chemical informations into electric signals.

For what concerns the understanding of odour recognition at the
molecular level, it is known that specific receptor proteins, present in
the olfactory mucosa, discriminate the various molecular structures on the
basis of size, shape and position of functional groups: in this process the
hydrocarbon part of the molecule is often very critical.

Gas sensors, based on polymers of aromatic and heteroaromatic
molecules, that exhibit electrical conducting properties, can be used as
specific odour detectors. Their electrical resistance varies upon
interaction with several substances in the gas phase, with a specificity
related both to the presence of functional groups and to the shape of the
hydrocarbon part of the odorant molecule.

This characteristic, together with fast responses and high stability,
make such polymers suitable for designing an artificial nose. Moreover,
their organic nature would allow the synthesis of new polymers tailored for
detecting particular gases or for spscific needs.

Introduction

There is a rapidly increasing need for a reproducible and objective
method to evaluate odour quality and intensity. At present, such
measurements have to rely on subjective judgements of panels of trained
people: this is expensive, time consuming and not accurate. A device, which
mimics the olfactory discrimination mechanism would be of great interest to
several industries for quality control, such as detecting off odours in
canned food, or perfumery.

Such a device could be made from an array of transducers, which have
an overlapping specificity to most chemicals in the gas phase.

The construction of this apparatus would be the result of two lines of

NATO ASI Series, Vol. F43
Sensors and Sensory Systems
for Advanced Robots
Edited by P. Dario
© Springer-Verlag Berlin Heidelberg 1988

research:

(a) specific gas sensors would act as transducers for converting chemical information, present in the odorant molecules as functional groups, size and shape, into electric, optic, acoustic or any other kind of easily measurable signal;

(b) the elucidation of the biochemical mechanisms on which olfactory sensation is based, together with the relationships between molecular parameters and odours, would provide the necessary information coding system for making the gas sensors as similar as possible to natural receptors. Such similarity has to be searched for mainly in the type of information processiong to use in an artificial nose, rather than in the specificity or mechanism of action of the single gas sensors.

We are still far from understanding all the processes leading to olfactory perception and odour discrimination. However it has been established that an odour is related to the presence and position of functional groups, as well as to molecular size and shape. Therefore an artificial nose should ideally use gas sensors capable of discriminating the different compounds on the basis of such molecular parameters.

Concerning gas transducers, some types commercially available are made out of various metal oxides, that change their resistance on exposure to most chemicals in the gas phase. Such inorganic sensors have previously been used in a machine modelling the olfactory discrimination mechanism, but some characteristics, like slow response and recovery, easy poisoning and almost complete lack of specificity , greatly limit a prospective use for odour recognition and discrimination (46).

Organic conducting polymers are receiving increasing attention because of some unique characteristics and interesting applications in several fields of electronics. We have found that several of these polymers respond to gases with a reversible variation in their conductivity and that they could be used as sensors in an artificial nose. In fact, they are fast both in response and recovery, are not easily poisoned by odorants, and show selectivity towards different compounds. They can serve as models for olfactory receptors, in the sense that they can discriminate on the basis not only of functional groups, but also on the size and shape of the hydrocarbon part of the molecule. Finally, being organic polymers, they offer the possibility of being tailored for each particular need, by changing both the backbone of the polymer and the nature, size and shape of the side chains.

This paper reviews in the first part recent literature on these aromatic conducting polymers, with particular attention to their synthesis, structure determination and elucidation of the electrical conducting mechanism. Data are very scarce in this field and often not very clear, due to the difficulty in handling conducting polymers, which are usually completely insoluble, making most of the techniques for purification and structure determination not applicable to these compounds. Thus, a comprehension of the structure as well as of the conduction mechanism is essential for investigating the interactions between polymer sensors and odorants.

This work reports the synthesis of several new conducting polymers and the measurement of their responses to a series of odorants. These data, although still very preliminary, give evidence of some characteristics of these sensors that indicate useful possibilities for incorporation into an artificial nose. Although this task is still far from its accomplishment and will require greater understanding of the biological olfactory code, the sensors here described can find useful applications in many cases when odour measurements are needed, both for quality control and for detecting specific chemicals at low concentrations in the gas phase.

Olfactory language

The biological olfactory code is not yet known, but the basic criteria, used by the nose for odour recognition, are broadly defined. This kind of information is sufficient for designing a machine, which is able to discriminate odours.

If we take the visual system as an example and regard a camera as an "artificial eye", we realize that in both systems (the natural one and the artificial one) light informations are coded, using the same parameters, i.e. intensity and frequency. If, for instance we make a television camera . whose sensors respond only to intensity and degree of polarization of the light. but not to frequency. we would miss the colours and get images different from what we see; or, if the response to frequencies covered a different part of the spectrum from the visible one, again the pictures we get could look not very familiar. But it is not strictly necessary, for instance. that the colour vision be based on the same primary colours used by the human eye. For example. it is possible to make all colours and shades . that the human eye can discriminate, using yellow, red and blue as primary colours. as well as using red, green and blue.

In olfaction. the system is only much more complex, for two main reasons:
(a) instead of three basic colours there are probably some thirty basic odours. and therefore they are more difficult to identify;
(b) the relationship between chemical structure is not simple and linear, like the correspondence between colour and frequency; therefore, it is necessary to understand first these relationships and to define which molecular parameters are the most important in determining the odour sensation.

Although the mechanism of odour perception and codification is far from being fully understood, the criterion used by our olfactory system is known in its basic points, that enable us to design a machine for odour discrimination and recognition. which works with the same or a similar "language". These main points can be summarized as follows:
- the olfactory system can identify and recognize complex mixtures of odourants. without separating the components, but sensing the mixture as a single stimulus (11);
- the codification is accomplished. at the peripheral level, by a series of rather specialized receptors; the sensation is the combined response of all the receptors (11);
- the number of these receptors has been estimated. on the basis of the recognized specific anosmias. to around 30; each of them is tuned on a set of molecular parameters. that define a so called "primary odour" (2)(6);
- the molecular parameters for eight "primary odours" and the specificities of the corresponding receptors have been broadly defined, through the study

of the relative specific anosmias (3)(4)(5)(40)(41);
- proteins binding "urinous"(45), "camphoraceous"(20), and "green" smelling(42)(43)(44)(53) classes of compounds have been identified in the nasal mucosa of various animals and one of them has been recently purifie and crystallized (8).

All the results of olfactory thresholds measurements, studies of specific anosmias and biochemical characterization of odorant binding proteins indicate that:
- the peripheral receptors are not highly specific, but broadly tuned and partly overlapping, so that the olfactory system can respond, as a whole, to virtually any molecule, volatile enough to reach the nasal mucosa and of molecular weight not greater than 250 - 300;
- they recognize not only the presence of a functional group in the odorant molecule, but also the size and the shape of its hydrocarbon part, that is the "oriented profile" of the molecule: this is believed to occurr through a fitting mechanism similar to that well established for other ligand-receptor systems.

So, gas sensors to be used in an artificial nose do not need to be exact copies of the natural receptors, but should respond to odorants, according not only to the presence and position of functional groups, but also to size and shape of the molecules.

For instance, it is well known that large cyclic ketones (15-19 carbon atoms) can smell "musky" or "urinous", depending on the size and shape of the molecule, whereas cyclic ketones containing ten carbon atoms smell "camphor" or "minty", again depending on molecular shape(7). Gas sensors responding only to functional groups would classify the odours of all these molecules as very similar; other sensors, responding also to molecular size, but not to shape, would still fail to distinguish a "camphor" from a "minty" odorant, or a "musky" from a "urinous".

Synthesis and structure of conducting polymers

Conducting organic polymers are characterized by long chains of fully conjugated pi bonds, that are responsible for unique properties of these compounds, such as electrical conductivity, metallic appearance, black colour and almost complete insolubility in any solvent.

The simplest among these polymers, and the most extensively studied, is polyacetylene, constituted by linear chains of conjugated double bonds (-CH=CH-)n (14). Unfortunately, polyacetylene is very easily oxidable and has to be handled always under nitrogen. This fact, together with the absence of functional groups prevents its use for gas detecting purposes .

Recently great interest is arising around other conducting polymers, derived from aromatic or heteroaromatic compounds and costituted by monomer units connected one another through single bonds. This is a fast growing area of research: from 1977 to 1985 more than 2,000 papers appeared on the subject, the most interesting in the last two years. These polymers still contain a fully conjugated chain of π electrons, but, unlike polyacetylene, are very stable not only towards oxygen, but also towards most chemicals. Besides, the presence of an heteroatom in most of them provides interesting structural properties. For a prospective use as artificial olfactory receptors, chemical stability is essential, while the presence of an heteroatom would make the interaction with the odorants more specific, by establishing bonds in a preferred orientation.

By far the best studied among these polymers is polypyrrole, that was first prepared electrochemically in 1968 by Dall' Olio et al., and as a free standing film in 1979 by Diaz et al(18). Since then, many applications have been found for this polymer, ranging from its use as organic electrode to charge storing material in rechargeable batteries, as an ion gate membrane, or to protect semiconductor electrodes from photocorrosion (10)(20).

Polypyrrole still is the favorite among these conducting polymers, because of its ease of preparation, mechanical strength and good stability. However, other aromatic or heteroaromatic polymers have been prepared, either by electrochemical or chemical synthesis, particularly polythiophene, polyaniline and few others, as reported in Table I.

Table I. Conducting polymers based on aromatic and heteroaromatic monomers.

	References	
Monomer	synthesis and structure	electrical properties
Pyrrole	18 24 55	19 39 47
1-Methylpyrrole	30	30
3.4-Dimethylpyrrole	50 51	39 47 51
Thiophene	32 52 56	9 27 51
3-Methylthiophene	26	52
Aniline	12 34	29
Azulene	54	54
p-Phenylene benzoxazole	35	35
p-Phenylene	48	9 13
p-Phenylensulfide	25 49	49

The main questions around these polymers concern:
(a) determination of the chemical structure;
(b) elucidation of the conductivity mechanism;
(c) improving the mechanical properties.

All these problems are of great interest for a prospective use of the conducting polymers in gas detection and odour discrimination. In fact, understanding the chemical and electronic structure is necessary for studying the interactions with the odorants, as well as for designing new polymers with desired characteristics, while the availability of polymer films of good mechanical properties would allow the construction of sensors in a very reproducible way and make the measurements very accurate.

The structure of these polymers is in most cases still an open question, because, as already pointed out, their insolubility does not allow the use of most instrumental methods of analysis; besides, their black colour generally prevents the use of transmission spectroscopy. Techiques that have given information on the structure of polypyrrole, and to a minor extent of other conducting polymers, include: "magic angle" NMR for solid samples (50)(31). ESR (47)(23), ultraviolet photoemission (21), X-ray diffraction and X-ray photoemission spectroscopy (38).

X-ray and electron diffraction techniques have been of little use with polypyrrole, which has been found to be very poorly crystalline (50)(51). Both photoelectron spectroscopy and X-ray photoemission spectroscopy are in agreement with a structure of polypyrrole, mainly constituted by repeating units of pyrrole directly connected through 2 and 5 position (38) . Some degree of disorder is apparent from the spectra, probably due to the presence, at a minor extent, of bonds in the 3 and 4 position of the ring. That the polymerization occurs preferentially on the 2 and 5 position is also confirmed by the observation that 2,5-dimethylpyrrole does not polymerize (Pelosi and Persaud, unpublished), while the 3,4-dimethyl derivative very easily gives the polymer in the same conditions. This polymer looks much more regular from the XPS spectra (38). This type of structure is also in agreement with the signals obtained in magic angle spinning 13C NMR. The most favorite conformation for polypyrrole, as for other similar conducting polymers is planar, with the pyrrole units pointing alternatively in opposite directions. This situation affords a regular linear structure, with no steric hindrance, even in the case of the 3,4-dimethyl derivative. By contrast, a polypyrrole with all the rings pointing in the same direction would make a curved chain, with a higher degree of strain, while a 3,4-dimethyl derivative could not even be planar.

Another relevant element in the structure of polypyrrole and other conducting polymers is the presence of the so called "oxidized" and "reduced" forms: when prepared by electrochemical synthesis,it is the "oxidized" form that is obtained, which is conductive and contains a certain number of positive charges, balanced by anions of the electrolyte used in the polymerization. The ratio between monomer units and positve charges depends both on the monomer and the electrolyte. For instance, for polypyrrole perchlorate this ratio is 3, for polypyrrole tetrafluoroborate is 4, while in the case of polythiophene tetrafluoroborate is 3 and with polybithiophene perchlorate is 7 (38); when the anion is a straight chain alkyl sulfate or sulfonate, with a number of carbon atoms varying from 4 to

16. the electrochemically grown polypyrrole still contains one positive charge every 3-4 pyrrole units. These last polymers appear very attractive, because, probably due to the long chain anions, they show much more ordered structures than the conventional polypyrroles (55). The "oxidized" conducting form of polypyrrole can be switched to a neutral, yellow insulating form, by passing an electrical current of revered polarity.

All methods available for molecular weight determination of polymers or for their fractionation, according to molecular size, require that the sample be in solution and therefore cannot be used with conducting polymers. In the case of polyacetylene, the problem has been solved by using a radioactive catalyst, that remains attached to the polymer chain (14). With polymers made by chemical synthesis, like polythiophene, quantitative determination of the groups, that remain at both ends of the chain, can give a measure of the molecular weight, assuming that there are no branches on the polymer (32). For most of conductive polymers, that are synthesized electrochemically and therefore do not bear any end-group, these approaches cannot be used. In the case of polypyrrole, a monomer, substituted with tritium in the 2 and 5 positions has been polymerized, and then the residual radioactivity has been determined in the polymer, again assuming that the polymerization only affords linear chains with the pyrrole units connected exclusively through the 2 and 5 positions, and ignoring any exchange reaction between tritium and hydrogen that could take place during the polymerization.

Concerning the mechanism of electrical conduction, there are many papers in the literature, presenting theories more or less supported by experimental results, but the picture is still not clear and definite (9)(13)(15)(35)(39)(47)(52). However it is worth summarizing the main points, particularly those of relevance to the interaction of the polymer chain with the odorant:

- it is clear and well established that an alternating series of double bonds, or pi electrons, is necessary for the polymer to be conducting, as supported by the fact that all conducting polymers present such characteristic; however, this structural element alone is not enough for the conductivity: the electrons, present in valence orbitals, would need too high energy to be promoted in conduction bands; on the other hand, there is no complete conjugation, like in benzene, even in the chain of polyacetylene. The fact that a series of alternating electron is not an electrical conductor is well showed by polypyrrole, which in its neutral form is an insulator;

- a polymer of alternating electrons requires doping to became conductive. In polyacetylene, the dopants can be added to the polymer in different concentrations, affording materials with different conductivities, ranging from an insulator to an "organic metal"; in polypyrrole and the other aromatic and heteroaromatic polymers, when synthesized electrochemically, the dopants consist of positive charges and their counterions from the electrolyte;

- the polymers, as formed, contain a high concentration of unpaired electrons (polypyrrole has been used for many years as a reference sample in ESR spectroscopy) and, according to one of the theories, these radicals could be responsible for the electrical conduction. Recently this theory is being given less attention, because, at least in some cases is not in agreement with the experimental results (for instance, in polyacetylene, low concentrations of dopants reduce the number of unpaired electrons and increase the conductivity at the same time);

- the conductivity is now explained in polyacetylene with the presence of perturbations (solitons). that. in the presence of dopants. become charged and can thus carry electricity along the polymer backbone (14)(15); in polypyrrole and similar polymers the conductivity has been related to the presence of bipolarons. stabilised by the conjugated chain of double bonds, that can transfer positive charges from one macromolecule to the other (13)(47);

- finally it has been found that electricity is carried mostly on the surface of the polymer: this is of great practical interest both for the manufacturing and the use of conducting polymers as gas sensors: in fact, we have found that it is possible to coat a polypyrrole sensor (that presents good mechanical characteristics) with a layer of another polymer and to get a new sensor with the electrical properties of the coating material; besides. as the interaction between polymer and odorants occurr only on the surface of the sensor. the conductivity will be modified to a great extent. because there are no charges travelling in the bulk of the polymer.

By treatment with strong alkali (2 M NaOH). films of polypyrrole. prepared electrochemically. increase their electrical resistance 3-4 orders of magnitude (29). The polymer can be brought back nearly to its original value by treatment with strong acids (6M HCl) and the cycle repeated a number of times. with the resistance increasing slowly. but regularly. after each cycle. The phenomenon is explained. on the basis of optical spectra. recorded on very thin films. and on XPS analysis. as a chemical modification. The increase in resistance would accordingly be the result of a decrease in the number of positive charges and their counterions.

Many efforts have been aimed at the preparation of conducting polymers with good mechanical properties. Because these polymers cannot be melted or dissolved. or molded. they have to be prepared directly in the form they will be used. It is noteworthy the fact that. although polypyrrole has been known since 1968. it became an interesting subject of research only ten years later. when it was first prepared as a free standing film; again. one of the reasons why it is studied much more than other conducting polymers is due to its particularly good mechanical properties. However. films of polypyrrole are fragile and rather difficult to handle.

Therefore. pyrrole has been polymerized inside conventional polymers. such as polyvinyl alcohol. polyvinylchloride. polystyrene and several others (1)(17)(24)(33)(36)(37). The resulting materials retain most of the properties of the supporting polymer (can be bent. stretched and obtained in extremely thin films). with in addition a good conductivity. A particularly interesting case is the polymerization of pyrrole. and other aromatioc monomers. using polymeric electrolytes. such as sulphonated polystirene. sulphonated polyethylene. poilycarboxilate and others (16). The polymeric electrolyte at the same time supplies the dopant and improves. in some cases. the mechanical properties of the conducting films.

Most of the researches on conducting polymers. including polypyrrole and other polyaromatic compounds. are aimed at the synthesis of materials with very high conductivities. comparable to those of the metals. with the hope of succeding in making room temperature superconductors.

For our purposes. this is of little relevance; on the contrary. it is better to have a polymer with a not very high conductivity. in order to be able to measure variations in resistance in both directions.

There is no mention in the literature about a possible use of conducting polymers as odour detectors and discriminators. Only one patent deals with the use of gas sensors, based on polypyrrole and other polymeric and non polymeric substances for detecting combustion gases, in order to prevent fires (57). The method is based on a change in electrical resistance, in the presence of inorganic gases, that are commonly produced in fires, such as carbon oxides, nitrogen oxides, ammonia, sulphur dioxide and others.

Gas sensors based on conducting polymers

We present original results, concerning the synthesis of new conducting polymers, their response to a number of odorants and the use of dielectric constant measurement as a new tool for investigating the interaction between polymer sensors and odorants.

The use of conducting polymers as specific gas sensors has not yet been described in the literature, except for a french patent, dealing with the detection of some inorganic gases, that are commonly produced during combustion. No mention, however, is made to organic gases or odours, nor to the specificity of such polymers.

We attempted polymerization of a series of aromatic and heteroaromatic monomers, by anodic oxidation, in the presence of the electrolyte tetraethylammonium tetrafluoroborate, following the method described for the preparation of polypyrrole. The monomers, that gave a black coating of polymer on a stainless steel anode are listed in Table II. Other monomers, such as some methyl and dimethyl anilines, clearly underwent oxidation or oligomerization, as the solution darkened near the anode, but the compounds formed failed to adhere to the electrodes. It is not excluded that a polymer can be successfully obtained in such cases, by changing the solvent. In other cases, as with several derivatives of phenylidrazine, furane and imidazole, no reaction was apparent, under the conditions used, and the solution remained clear and colourless.

Table II. Monomers polymerized electrolytically on a stainless steel anode in 0.1 M tetraethylammonium tetrafluoborate, at 3 V for 10 min.

Monomer	Solvent	Conc.	Comments
1-methylpyrrole	99% MeCN	0.1	regular and compact
1-naphtylamine	80% EtOH	0.1	not compact, partly dissolves
m-phenylendiamine	70% EtOH	0.1	regular and compact
isovaleraldehyde anile	90% EtOH	0.025	regular and compact
3.5-dimethylaniline	40% EtOH	0.05	thin but regular
2.6-dimethylaniline	20% EtOH	0.1	brittle
4-methylaniline	50% EtOH	1.0	brittle, not regular, partly dissolves
N.N-dimethylaniline	80% EtOH	0.1	regular and compact
aniline	70% EtOH	0.1	regular but brittle
indole	80% EtOH	0.1	thick and regular
phenol	60% MeCN	0.2	regular and compact
2-chloroaniline	90% EtOH	0.1	thin but regular
3-chloroaniline	70% EtOH	0.08	partly dissolves
N-hexylaniline	70% EtOH	0.1	thin but compact
2-anilino-ethyl acetate	50% EtOH	0.1	brittle
2-anilino-2-methylbutyl acetate	70% EtOH	0.1	brittle
2-anilino-menthyl acetate	80% EtOH	0.1	thin and brittle
thiophene-2-aceto-nitrile	80% EtOH	0.1	thin and brittle
2-isobutylthiazole	99% MeCN	0.1	thin and compact

Of all the polymers synthesized. polypyrrole proved, as already known from the literature. to be the strongest and the easiest to handle, therefore. it was used for making sensors. Other sensors were prepared by polymerizing 2-chloroaniline. thiophene 2-acetonitrile, indole and 2-isobutylthiazole on the surface of polypyrrole sensors. Because conductivity in these polymers is a superficial phenomenon, the supporting polypyrrole did not interfere in the performances of these last four polymers. This was confirmed by the fact that. in some cases (2-chloroaniline and indole). the electrical resistance of the sensor increased. The data relative to the five sensors prepared are reported in Table III.

Table III. Electrical resistance of the polymers used as gas sensors. All polymers were grown on a polypyrrole anode, at a potential of 3V for 10 minutes.

Monomer	solution conductivity	sensor resistance
Pyrrole	$9.3 \times 10E-7$ mho	400-700 ohm (n=7)
2-Chloroaniline	$1.2 \times 10E-4$ mho	1-8 Mohm (n=4)
Thiophene-2-acetonitrile	$1.1 \times 10E-5$ mho	300-800 ohm (n=4)
Indole	$2.0 \times 10E-5$ mho	1-5 Kohm (n=4)
2-Isobutylthiazole	$6.0 \times 10E-7$ mho	150-600 ohm (n=4)

The responses of these five sensors. as changes in the electrical resistance. were measured with a series of 28 pure chemicals. most of them organic. in the vapour phase. at room temperature. The results are summarized in Table IV. The responses are given as increase or decrease in resistance. or no response. regardless to the intensity of the signal.

Table IV. Changes in electrical resistance in different sensors in response to various compounds.

Compound	Sensor				
	A	B	C	D	E
Acetic acid	−	−	−	−	−
Ammonia	+	−	+	−	+
Benzaldehyde	−	−	−	−	−
Benzene	+	−	−	0	+
Bromine	+	−	−	−	−
1-Butanol	−	0	0	−	0
Chloroform	+	−	+	−	+
Cycloheptanone	−	0	−	−	0
1.2-Diaminoethane	+	+	−	−	+
Dichloromethane	−	+	−	−	0
Diethyl ether	−	−	+	+	+
Diethyl ketone	−	−	+	−	−
Ethanol	−	−	−	−	−
Ethyl acetate	−	0	−	0	+
n-Heptane	−	0	0	0	0
1-Hexanol	−	+	−	−	0
beta-Ionone	−	−	−	+	−
2-Mercaptoethanol	−	0	−	−	−
Methanol	−	−	0	+	+
2-Methylpyrazine	−	−	−	−	+
1-Phenylethanol	−	0	−	−	0
alpha-Pinene	−	0	−	+	0
beta-Pinene	−	0	−	+	0
Piperidine	−	−	−	−	+
Pyridine	−	−	−	−	+
Pyrrolidine	−	−	−	−	+
Triethylamine	−	−	−	+	+

(+) increase in resistance. (−) decrease in resistance.
(0) negligible response.
A: polypyrrole. B: poly-2-chloroaniline,
C: polythiophene-2-acetonitrile,
D: polyindole. E: poly-2-isobutylthiazole.

A specificity of response clearly appears from the data, both with respect to sensors and to odorants: in fact, the same sensor responds differently to different odorants, and the same odorant produces different responses with the five sensors. The 28 odorants of Table IV gave 20 different sets of responses; with 14 of the odorants, discrimination could be achieved only on the basis of these qualitative measurements. By introducing the intensity of the signals, as an additional element, the discrimination becomes much easier, as the intensity of the responses can vary greatly from one odorant to another. Fig. 1 reports the actual variations of resistance, measured with two different sensors, in the presence of some odorants, to show how the responses differ in shape, direction and intensity. However, it is more convenient (see "Discussion") to discriminate only on the basis of a qualitative response and to use more sensors, when necessary.

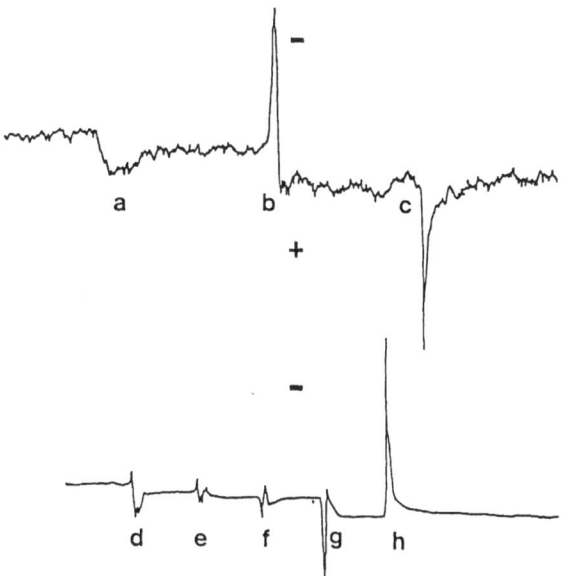

Fig. 1. Upper: responses of a polypyrrole sensor to (a) chloroform, (b) diethyl ether. (c) benzene. Lower: responses of a poly-2-isobutylthiazole sensor to (d) chloroform. (e) benzene, (f) diethyl ether, (g) 2-methylpyrazine, (h) ethanol.

A closer examination of the results of Table IV shows that the responses of the sensors depend not only on the functional groups on the molecules of the odorants, but also on other molecular parameters. It is difficult to make correlations on these few preliminary data, but it is interesting to point out some examples, that show how compounds of the same chemical class, differing only in the hydrocarbon part of the molecule, can be easily discriminated by these sensors. For instance, methanol and ethanol give opposite responses with sensors D and E; diethyl ketone and cycloheptanone give opposite responses with sensor C, while b-ionone can be discriminated in the same way from diethyl ketone, both by sensor C and D;

the five alcohols, methanol, ethanol, butanol, hexanol and phenylethanol can be discriminated using only sensors B and C.

All the sensors proved to be stable with time and give reproducible responses after at least six months at room temperature. When heated above 150 C all, except poly-2-chloroaniline, deteriorate very rapidly. Its stability at high temperatures allowed the use of a polychloroaniline sensor as specific detector for gas-liquid chromatography. The specificity of this sensor adds qualitative information to the gas chromatographic peaks. Fig. 2 shows a gas chromatogram of a mixture of isooctane and ethanol: the conventional FID detector responds to both compounds, while the polymer sensors only detects ethanol; in the cases of poorly resolved peaks, the polymer detector would allow an accurate determination of one of them.

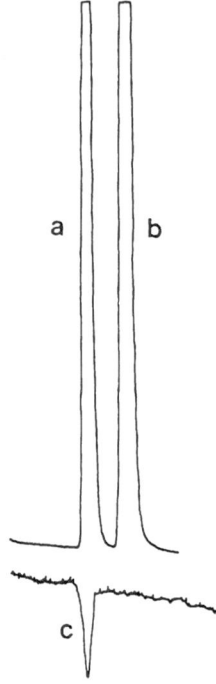

Fig. 2. A mixture of ethanol and isooctane was injected into a gas liquid chromatograph in which the gas outlet was switched between a flame ionisation detector and a poly-2-chloroaniline sensor. Detector temperature was 210 C and column temperature was 140 C, with nitrogen as carrier gas at 3 ml/min.
Upper trace: flame ionization detector response to (a) ethanol, (b) isooctane.
Lower trace: response of a poly-2-chloroaniline sensor to (c) ethanol. No response was recorded to isooctane.

The use of a gas chromatograph for delivering the odours allows more accurate determinations of the responses, because we know the amount of odorant used and also because the odorant reaches the sensor after having been purified through the column. By means of a gas chromatograph the stimulus-response curve has been measured, using ethanol as the odorant and polychloroaniline as detector. The curve, shown in Fig. 3, increases regularly and then shows saturation at high concentrations of ethanol.

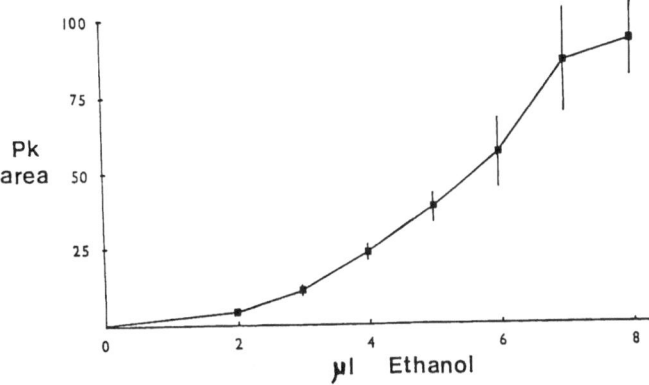

Fig. 3. The area of the response peak of a poly-2-chloroaniline sensor was plotted against the volume of ethanol injected into the gas liquid chromatograph. Detector temperature was 210 C and column temperature was 140 C. with nitrogen as carrier gas at 3 ml/min.

Discussion

The results indicate organic conducting polymers as a novel class of chemical transducers. with high selectivity, that can be used in all fields of gas analysis. particularly in the design of an artificial nose.

With reference to the last aspect, although the experiments with these polymers are still at an early stage, and the understanding on the olfactory system are far from clear and complete, the basic informations on both systems put into evidence some common aspects, that indicate organic conducting polymers as the most suitable transducers, now available, for the codification and discrimination of odours. These characteristics can be summarized as follows:

(a) their response to odours is fast and reproducible; the recovery is complete within seconds. even at room temperature; they are not poisoned by most odorants. including sulphur compounds: these characteristics make them far superior to commercial gas sensors, based on inorganic oxides;

(b) the response to odorants is selective and discrimination can be very clear. in some cases. with signals in opposite directions (increase or decrease of electrical resistance);

(c) the relationships between structure of the odorant and response include molecular parameters. such as size and shape of the molecule. that are very important in the codification of the odours by the olfactory system;

(d) the field of conducting polymers is extremely wide and a very great number of molecules can be polymerized. to give sensors with different

specificities; when required, sensors can be made very selective, by chemical modification of the polymer, or by synthesizing the appropriate substrate to polymerize, as shown by the results with asymmetric sensors;
(e) the electrochemical synthesis of conducting polymers makes possible the direct polymerization of several sensors on a single chip, using the technique of integrated circuits.

The main questions are centered around the interactions of the odorant molecules with the sensors:
- what kind of chemical bonds are established;
- what makes the interaction rather specific;
- in which way is the conductivity affected.

To answer these questions, we can follow two approaches:
(a) we can ignore the structure of the polymer in its detail and collect responses to a very great number of odorants; then, try to establish relationships between structure and response, with a sort of "pharmacological" approach, similar to that used in olfaction;
(b) we try to establish first the complete structure of the polymer sensors, and to clarify the mechanism of electrical conduction; then, using physico-chemical and chemical methods, we can look at the specific interactions with the odorants.

The second approach looks more attractive, but requires information, that, in some cases, are very difficult to get. For instance, the structure of polypyrrole has been more or less identified, but still there are doubts on its molecular weight, on a possible "secondary" or "tertiary" structure and how the macromolecules are packed together. Even less is known for the other polymers. Concerning the mechanism of electrical conduction, only theories are at present available. Therefore, it would be more practical to use both approaches at the same time. The collection of a large amount of data would give immediate application of these polymers in the analysis of gases and could suggest models for studying the mechanism of interactions with the odorants.

In the light of present knowledge, we can only say that the interaction of the odorant molecule with the polymer should involve the establishment of weak bonds, from hydrogen bonds to hydrophobic interactions: the conductivity of the polymer would thus be affected through an electron releasing or withdrawing effect; the interaction could also produce a deformation of the polymer backbone, twisting part of the chain out of the planar conformation and thus increasing the electrical resistance. It is not excluded that the variations in resistance can be the consequence of interaction of the odorants, not only with the polymer itself, but also with the dopant anions, whose presence and concentration is essential for the electrical conductivity.

Concerning the design of an artificial nose, using conducting polymer sensors, the results indicate that the basic requirements for synthetic odour receptors are met by such polymers and can be summarized in the following characteristics:
(a) by combining a sufficient number of different sensors, it has to be possible to get responses to practically all volatile molecules;
(b) the artificial nose, as a whole, and therefore at least some of the individual sensors, should be able to discriminate among molecules with the same functional group, but different in their hydrocarbon part: this can be easily checked, by testing series of homologues, and positional,

geometrical and optical isomers;

(c) the system should give a unique set of responses for each odour, a sort of a "spectrum" characteristic for every substance. Therefore, to have a number of combinations as large as possible, the number of different sensors cannot be too small. However, even if we use a code of three letters (increase in resistance. decrease in resistance, no response), the number of combinations, equal to 3En, where n is the number of sensors, increases very rapidly with n. So. with only 8 different sensors the number of combinations is 6.561 and goes up to 531,441 with 12 sensors and to 43.046.721 with 16. These are theoretical numbers and do not correspond in practice to the number of molecules, that can be actually discriminated.

Once established that conducting polymers can make sensors able to perform odour discrimination, we can adopt different approaches for choosing and designing the most suitable sensors:

(a) in analogy with the primary odour receptor theory of olfaction, we could make sensors tuned towards each one of the primary odorants; this approach requires the previous identification of the complete olfactory code. which is not an easy task, and also introduces very complicated problems of synthetic chemistry for making polymers with the desired characteristics;

(b) instead of defining the single structures, we can extract the molecular parameters. that are relevant for the codification of odours, then design sensors. for analysing the single parameters in the odorant molecule. These parameters could be:
- all the different functional groups
- double bonds and triple bonds
- aromatic rings
- aliphatic rings
- molecular size
- presence of more than one functional group and their relative distances
- shape of the molecule (elongated, flat, round)

It is impossible to enumerate all the parameters, that are important for defining an odour. because not all of them are yet known. An approach of this type. therefore, will give a machine, that can be made more and more similar to the human nose, by adding sensors and information, as we learn more about the mechanism of odour perception and codification;

(c) following a third approach, we could synthesize a great number of different sensors. test them with a very wide selection of odorants, then choose those that can discriminate among the highest number of odorants , with the smallest number of sensors: a system of this type would not be an artificial olfactory system. but could become a very sophisticated analytical tool to be used whenever reproducible measurements of odour is required.

References

1. Ahlgren,G. and Krische,B.: "Composites of conducting polymers", J. Chem. Soc. Chem. Commun.. 946 (1984)

2. Amoore,J.E.: "A plan to identify most of the primary odors", in "Olfaction and Taste III". Pfaffmann C. ed., Rockefeller Univ. Press, N.Y. (1969)

3. Amoore,J.E.: "Four primary odor modalities in man: experimental evidence and possible significance". in "Olfaction and Taste V", Denton and Cochlan eds.. Acad. Press. N.Y.. London (1975)

4. Amoore,J.E.. Forrester,L.J. and Pelosi,P.: "Specific anosmia to isobutyraldehyde: the malty primary odor", Chem. Sens. and Flav.$\underline{2}$ 17 (1976)

5. Amoore,J.E.. Pelosi,P. and Forrester,L.J.: "Specific anosmia to 5α-androst-16-en-3-one and ω-pentadecalactone: the urinous and musky primary odors". Chem. Sens. and Flav.$\underline{2}$ 401 (1977)

6. Amoore,J.E.: "Specific anosmia and the concept of primary odors", Chem. Sens. and Flav.$\underline{2}$. 267 (1977)

7. Beets. M.G.J.: "Structure activity relationships in human chemoreception" Appl. Sci. Publ. Ltd, London (1978)

8. Bignetti,E.. Cavaggioni,A., Pelosi,P.. Persaud,K.C.. Sorbi,R. and Tirindelli,R.: "Purification and characterisation of an odorant binding protein from cow nasal tissue". Eur. J. Biochem. $\underline{149}$, 227 (1985)

9. Bredas,J.L.. Themans.B.. Fripiat,J.G., Andre,J.M. and Chance,R.R.: "Highly conducting polyparaphenylene, plypyrrole and polythiophene chains: an ab initio study of thje geometry and electronic-structure modifications upon doping". Physical Review B $\underline{29}$. 6761 (1984)

10. Burgmayer,P. and Murray,R.W.: "A ion gate membrane: electrochemicalm control of ion permeability through a membrane with an embedded electrode", J. Am. Chem. Soc. $\underline{104}$. 6139 (1982)

11. Cagan,R.H. and Kare.M.R.,eds. "Biochemistry of Taste and Olfaction" Acad. Press. N.Y. (1981)

12. Carlin.C.M.. Kepley,L.J. and Bard,A.J.: "Polymer films on electrodes. XVI. In situ ellipsometric measurements of polybipyrazine, polyaniline and polyvinylferrocene films". J. Electrochem. Soc. $\underline{132}$. 353 (1985)

13. Chance,R.R.. Bredas,J.L. and Silbey,R.: "Bipolaron transport in doped conjugated polymers". Phys. Rev. B $\underline{29}$. 4491 (1984)

14. Chien,J.C.W.: "Polyacetylene Physics, Chemistry and Science", Acad. Press. N.Y. (1984)

15. Chien,J.C.W.: "Mechanism of electronic conduction in organic metals", Polym. Prepr.. 262 (1985)

16. Cross.M.G. and Lines,R.: "Compositions comprising an electrically conductive polymer", U.K. Pat. 83/20646.080183 (1984)

17. De Paoli,M.A., Waltman,R.J., Diaz,A.F. and Bargon,J.: "Conducting composites from plyvinylchloride and polypyrrole", J. Chem. Soc. Chem. Commun. 1015 (1984)

18. Diaz,A.F.. Kanazawa,K.K. and Gardini,G.P.: "Electrochemical polymerization of pyrrole", J. Chem. Soc. Chem. Commun. 635 (1979)

19. Feldberg,S.W.: Reinterpretation of polypyrrole electrochemistry. Consideration of capacitive currents in redox switching of conducting polymers". J. Am. Chem. Soc. 106, 4671 (1984)

20. Fesenko.E.E.. Novoselov,V.I. and Krapivinskaya,L.D.: "Molecular mechanisms of olfactory reception. IV. Some biochemical characteristics of the camphor receptor from rat olfactory epithelium", Biochim. Biophys. Acta 587, 424 (1979)

21. Ford.W.K.. Duke.C.B. and Salaneck,W.R.: "Electronic structure of polypyrrole and oligomers of pyrrole". J. Chem. Phys. 77, 5030 (1982)

22. Frank,A.J. and Honda.K.: "Visible light induced water cleavage and stabilization of n-type CdS to photocorrosion with surface attached polypyrrole catalyst coating". J. Phys. Chem. 86, 1933 (1982)

23. Genoud,F.. Guglielmi,M.. Nechtschein,M., Genies,E. and Salmon,M.: "ESR study of electrochemical doping in the conducting polymer polypyrrole". Phys. Rev. Lett. 55. 118 (1985)

24. Hikita,M.. Niwa,O.. Sugita,A. and Tamamura,T.: "Patterning of conductive polypyrrole in polymer film", Jpn. J. Appl. Phys. Part 2. 24. L79 (1985)

25. Hill,H.W. and Brady.D.G.: "Polymers containing sulfur", Encyclopedia of Chemical Technology, 18. 793 (1982)

26. Hotta.S.. Hosaka,T. and Shimotsuma,W.: "Electrochemically prepared polythienilene films". Synth. Metals 6, 317 (1983)

27. Kaneto.K.. Yoshino.K. and Inuishi,Y.: "Electrical properties of conducting polymer. ply-thiophene. prepared by electrochemical polymerization". Jpn. J. Appl. Phys. 21. L567 (1982)

28. Kitani,A.. Izumi,J.. Yano,Y.. Hiromoto,Y. and Sasaki.K.: "Basic behaviours and properties of electrodeposited polyaniline", Bull. Chem. Soc. Japan 57. 2254 (1984)

29. Inganas.O.. Erlandsson.R.. Nylander.C. and Lundstrom.I.: "Proton modification of conducting polypyrrole". J. Phys. Chem. Solids 45. 427 (1984)

30. Jakobs.R.C.M.. Janssen,L.J.J. and Barendrecht,E.: "Behaviour of polypyrrole and poly N-methylpyrrole electrodes in acetonitrile". Recl. J. R. Neth. Chem. Soc. 103. 275 (1984)

31. Lecavelier.H.. Devreux.F., Nechtschein,M. and Bidan,G.: "NMR studies in polypyrrole". Mol. Cryst. Liq. Cryst. 118. 183 (1985)

32. Lin.J.W.P. and Dudek,L.P.: "Synthesis and properties of poly(2.5-thienilene)". J. Polym. Sci. Polym. Chem. Ed. 18. 2869 (1980)

33. Lindsey,S.E. and Street,G.B.: "Conductive composites from polyvinylalcohol and polypyrrole". Synth. Metals 10. 67 (1984)

34. Mengoli.G.. Munari.M.T.. Bianco,P. and Musiani,M.M.: "Anodic synthesis of polyaniline coatings onto iron sheets". J. Appl. Polym. Sci. 26. 4247 (1981)

35. Nayak.K. and Mark,J.E.: "Electronic band structures for a p-phenilene benzoxazole polymer". Polym. Prepr. 272 (1985)

36. Niwa.O. and Tamamura.T.: "Electrochemical polymerization of pyrrole on polymer coated electrodes". J. Chem. Soc. Chem. Commun., 817 (1984)

37. Niwa.O.. Hikita.M. and Tamamura.T.: "Polymer-polypyrrole alloy films as semitransparent organic conductors".Appl. Phys. Lett. 46. 444 (1985)

38. Pfluger.P. and Street.G.B.: "Chemical, electronic and structural properties of conducting heterócyclic polymers: a view by XPS". J. Chem. Phys. 80. 544 (1984)

39. Pfluger,P.. Gubler,U.M. and Street,G.B.: "The valence electronic structure of conducting pyrrole polymers: evidence against a linear metallic chaun picture". Solid State Commun. 49. 911 (1984)

40. Pelosi.P. and Viti.R.: "Specific anosmia to l-carvone: the minty primary odour". Chem. Sens. Flav. 3. 331 (1978)

41. Pelosi.P. and Pisanelli.A.M.: "Specific anosmia to 1.8-cineole: the camphor primary odour". Chem. Sens. Flav. 6. 87 (1981)

42. Pelosi.P.. Pisanelli.A.M.. Baldaccini,N.E. and Gagliardo.A.: "Binding of 2-isobutyl-3-methoxypyrazine to cow olfactory mucosa". Chem. Sens. Flav. 6. 77 (1981)

43. Pelosi.P.. Baldaccini.N.E. and Pisanelli,A.M.: "Identification of a specific olfactory receptor for 2-isobutyl-3-methoxypyrazine". Biochem. J. 201. 245 (1982)

44. Pelosi.P.. Pasqualetto,P.L. and Lorenzi,R.: "Synthesis and olfactory properties of some thiazoles with bell pepper like odor". J. Agric. Food Chem. 31. 482 (1983)

45. Persaud.K.C. and Dodd.G.H.: "Biochemical mechanisms in vertebrate primary olfactory neurons", in "Biochemistry of Taste and Olfaction" (Cagan and Kare eds.). 333. Acad. Press, N.Y. (1981)

46. Persaud.K.C. and Dodd.G.H.: "Analysis of discrimination mechanisms in the mammalian olfactory system using a model nose". Nature $\underline{299}$, 352 (1982)

47. Scott.J.C.. Pfluger,P., Krounbi.M.T. and Street.G.B.: "Electron spin resonance studies of pyrrole polymers: evidence for bipolarons", Phys. Rev. B $\underline{28}$. 2140 (1983)

48. Shacklette.L.W.. Elsenbaumer,R.L.. Chance,R.R., Sowa.J.M.. Ivory,D.M. Miller.G.G. and Baughman.R.H.: "Electrochemical doping of poly(p-phenilene) with application to organic batteries". J. Chem. Soc. Chem. Commun. 361 (1982)

49. Stacy.C.J.: "Molecular weight distribution of poly-(phenilene sulfide) by high temperature gel permeation chromatography". Polym. Prepr. 180 (1985)

50. Street.G.B.. Clarke.T.C., Krounbi.M., Kanazawa.K., Lee,V.Y., Pfluger.P.. Scott.J.C. and Weiser,G.: "Preparation and characterization of neutral and oxidized polypyrrole films", Mol. Cryst. Liq. Cryst. $\underline{83}$, 253 (1982)

51. Street.G.B.. Clarke,T.C.. Geiss.R.H.. Lee,V.Y., Nazzal,A.. Pfluger,P. and Scott.J.C.: Proc. Int. Conf. on Low-Dimensional Conduyctors and Superconductors. Les Arcs. 1982. J. Phys. Coll. C3 (1983)

52. Tanaka.S.. Sato.M. and Kaeriyama.K.: "Electrochemical preparation and characterization of poly(2,5-thiophenediyl)". Makromol. Chem. $\underline{185}$, 1295 (1984)

53. Topazzini.A.. Pelosi.P., Pasqualetto,P.L. and Baldaccini,N.E.: "Specificity of a pyrazine binding protein from cow olfactory mucosa", Chem. Senses $\underline{10}$. 45 (1985)

54. Tourillon.G. and Garnier,F.: "New electrochemically generated organic conducting polymers". J. Electroanal. Chem. $\underline{135}$, 173 (1982)

55. Wernet.W.. Monkenbusch,M. and Wegner,G.: "A new series of conducting polymers with layered structure: polypyrrole n-alkylsulfates and n-alkylsulfonates". Makromol. Chem. Rapid Commun. $\underline{5}$. 157 (1984)

56. Yamamoto.T.. Sanechika.K. and Yamamoto.A.: "Preparation of termostable and electric conducting poly(2,5-thienilene)", J. Polym. Sci. Polym. Lett. Ed. $\underline{18}$. 9 (1980)

57. Yu.L.T.. Buvet.R.E.. Vallot,R., Lauwick,B., Lanore.J.C. and Detriche.P.: "Procede de detection et d' analise de gaz notamment pour la prevention des incendies", Eur. Pat. 0.022.028.A1 (1981)

DEVELOPMENT AND APPLICATION OF HUMIDITY SENSORS

P.P.L. Regtien
Delft University of Technology
Department of Electrical Engineering
Delft, The Netherlands

Abstract.

After a short discussion on the application-fields of humidity sensors and the related requirements on specifications, a review of measurement principles is given together with some general characteristics of these methods. Two examples are worked out: absorption sensors and dew-point sensors. A review of the state-of-the-art is given as well as present trends in research on such sensors.

I. INTRODUCTION.

Hygrometry is the art of measuring the water content in substances. The water contained in fluids or solids is usually called moisture. The term humidity is reserved for water vapour in gases, such as air (the atmosphere), natural gas or other industrial gases.

The basic humidity quantity is the number of moles or molecules of water vapour per unit of volume. For practical reasons, other parameters are in use, of which table 1 shows some examples. Unfortunately, there is no common

TABLE 1. Humidity parameters

water-vapour concentration	kg/m^3 wet gas
partial vapour pressure	Pa
humidity ratio	kg/kg dry gas
specific humidity	kg/kg wet gas
mole fraction	ppm
dew-(frost)point temperature	degree C
relative humidity	%
saturation ratio	%
wet-bulb temperature	degree C
precipitable water	m

agreement about all these names and definitions, making the specification of measurement systems rather unclear. Another side-effect of having a lot of different parameters is the need for conversion from one parameter to another. For most conversions it is necessary to know other gas parameters as well, such as temperature and total pressure. As long as clean air is considered, these conversions give no large conversion errors. In other cases, conversion formulas are not very accurate or even nonexistent, because of the lack of experimental data. For instance, in pressurized natural gas the conversion from partial pressure to dew point (both commonly used paremeters) gives very unreliable results.

The importance of humidity measurements follows from the fact that all matter is hygroscopic to a greater or lesser extent. A change in atmospheric humidity does change the moisture content of most materials. The absorbed water in-

NATO ASI Series, Vol. F43
Sensors and Sensory Systems
for Advanced Robots
Edited by P. Dario
© Springer-Verlag Berlin Heidelberg 1988

fluences the characteristic properties of the materials. Some examples: Mechanical properties, such as the strength of adhesives, the stiffness of plastics, the quality of building materials and packing materials; thermal properties of insulation materials; the electrical characteristics of electric and electronic systems (from high-power systems to microcircuits). For instance, moisture penetrates rather quickly through the plastic encapsulation of integrated circuits, resulting in unwanted leakage currents or even breakdown caused by internal corrosion.

Table 2 shows the requirements for humidity sensors, in relation to the application area. For climate rooms one is further referred to Fig. 1, where the shaded areas indicate the ranges for different applications.

TABLE 2. Application areas.

application	temperature	humidity	inaccuracy
meteorology	-60 ... +50 $^{\circ}$C	-70 ... +30 $^{\circ}$C d.p.	0.3 $^{\circ}$C d.p.
climate rooms	-35 ... +30 $^{\circ}$C	10 ... 100 % r.h.	2 ... 10 % r.h.
hot rooms	up to 300 $^{\circ}$C	0 ... 50 % r.h.	2 ... 10 % r.h.
test- and calibration chambers	widest range	widest range	1 % r.h.

Besides accuracy and range, a lot of other requirements are set, depending upon the particular application, such as proof against chemical substances (cigarette smoke, chlorine in swimming pools, nitrates in cow sheds etc.), long-term stability, unattended operation for longer times, electrical output and, it goes without saying, a low price.

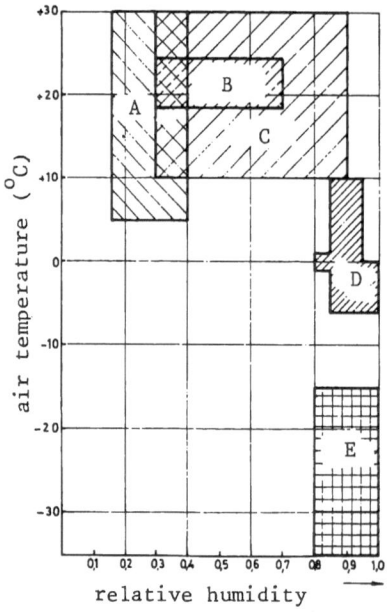

A dry rooms

B residences (comfort)

C climate rooms

D cooling chambers

E freezing chambers

Fig. 1. Temperature and humidity ranges of climate rooms.

Whereas for most other measurement sensors there are sufficient calibration
means available, there is no simple, generally applicable humidity standard.
Some humidity sensors commercially available are delivered with small tins
containing a saturated salt solution, which fit onto the sensor probe. In
this way it is possible to calibrate such sensors at some points on the rela-
tive-humidity scale. Calibration is mostly performed by a humidity generator,
in which dry air is mixed with an adjustable accurately known flow of water
vapour, or by a gravimetric measurement method, which is discontinuous and
time consuming.

II. PRINCIPLES OF HUMIDITY MEASUREMENTS.

To achieve a good understanding of the different measurement methods, a classification will be presented on the basis of the fundamental measurement prin-
cipels. Four classes are defined, covering all types of humidity sensors.
Each will be characterized in terms of accuracy, range and stability.

1. Water-vapour removal methods.

A rather straightforward method for the determination of humidity in a gas
is based on the thorough separation of water molecules from a sample of humid
gas or a gas flow. The mass or mass flux of both humid gas and removed water
or dry gas is measured (Fig. 2).

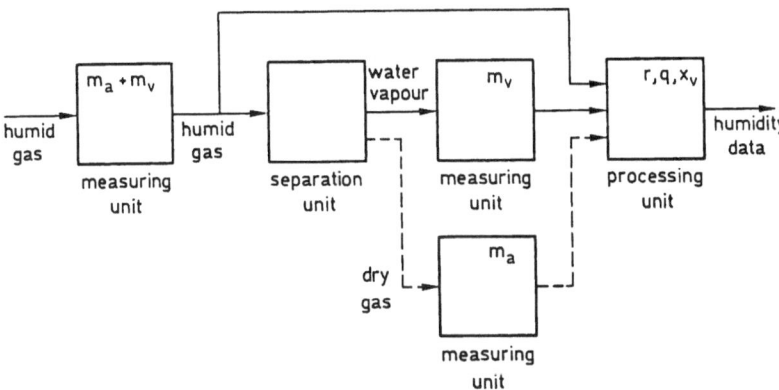

Fig. 2. Functional diagram of the water-vapour removal method.

Separation can be accomplished either by absorption by a desiccant, condensation, cryogenic means, diffusion through a semipermeable plate or membrane,
or by chemical means. The amount of removed water can be determined by weighing or by measuring the decrease in mass flux, volume diminution or pressure
reduction. Other possible means are electrolysis, chemical reactions and the
measurement of the latent heat content.
The dewatering of the gas must be exhaustive, and no water may escape (for
instance absorbed by pipe-walls) before its mass is determined. Both these
difficulties increase with decreasing humidity, limiting the accuracy at low
vapour concentrations and increasing the response time. Of all the methods
in this category only the electrolysis hygrometer is useful outside the laboratory. However, the need for two flow measurements and insufficient dewater-

ing make this instrument rather inaccurate. The gravimetrical method (weighing samples of humid gas and the removed water) and the Karl-Fischer method (chemical separation and measurement based on titration) are considered to be standard measurements, also for industrial gases.

2. Saturation methods.

The uncertainty about the degree of dryness in the previous method contrasts with the easily recognizable maximum vapour density or saturation density. Saturation can be achieved by temperature reduction, compression or adiabatic expansion. These operations should be performed until dew or frost just sets in. A dew detector provides certainty about saturation. Humid gas can also be saturated by adding water, or bringing part of the test gas in close contact with liquid water. The presence of liquid water assures saturation. The rate of condensation or evaporation is a measure for the humidity. Better measurands are the wet-bulb temperature and the dew-point temperature, which are easy to measure. The general block diagram is depicted in Fig. 3. The dew-point principle will be discussed in greater detail in Section III.

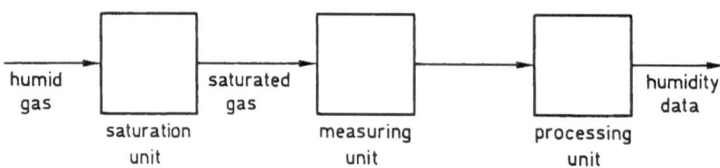

Fig. 3. Functional block diagram of the saturation method.

3. Parameter methods.

The characteristic properties of a gas depend upon its composition, and thus on the water-vapour content. The measurement of one of the gas properties gives information about the humidity. Most gas parameters can be measured continuously without affecting the gas composition or the state (Fig. 4). Properties appropriate for humidity measurements are molecular and acoustical absorptivity, electrical polarizability and molecular mobility.

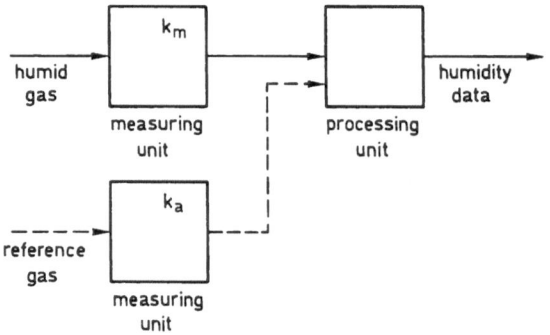

Fig. 4. Functional diagram of the parameter method.

The molecular absorptivity of water vapour depends strongly on wavelength, resulting in absorption lines in the ultraviolet, microwave and infrared regions of the electromagnetic spectrum. The absorption measurement is influenced by temperature variations and by scattering of light due to aerosols and liquid drops.

Humidity measurements based on molecular polarization include capacitance measurements (yielding the dielectric constant of the gas) as well as microwave refractivity measurements (yielding the index of refraction). These quantities of humid gas differ only about 10^{-4} from those of dry gas. Furthermore, the refractivity varies strongly with temperature and (total) gas pressure. These dependences may be eliminated by preconditioning of the humid gas, or by separate measurement of temperature and pressure for correction of the measurement data.

Refractivity hygrometers cover a wide range of humidity and are rather sensitive. However, they are complicated and expensive devices and are therefore primarily suitable for laboratory use.

Acoustic absorption by humid air is a non-monotonic function of vapour pressure. It also depends on frequency and on temperature. This makes humidity measurements based on acoustic absorption rather impractical.

Molecular mobility gives rise to a number of macroscopic phenomena, such as viscosity, thermal conductivity, acoustic velocity and diffusivity. To obtain accurate humidity data from any of these properties, that property should be measured relative to that of a reference gas at the same temperature (Fig. 4). In spite of the low conposition sensitivity and the large temperature dependence, several attempts have been made to construct humidity measurement systems based on this method. However, none of these has resulted in a commercial instrument.

4. Absorption methods.

Absorption methods are based on the relation between characteristic properties of hygroscopic materials and the amount of absorbed water from the gas to which these materials are exposed (Fig. 5).

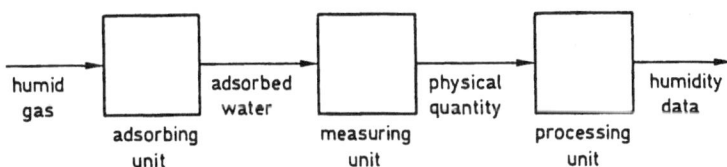

Fig. 5. Functional diagram of the absorption method.

Among the numerous parameters that are sensitive to moisture absorption, the most applicable to humidity measurements are mass, dimension and electrical characteristics. The absorption process itself can also evoke some measurable quantities, such as heat of sorption and concentration gradients due to an unequal diffusivity or solubility of air and water molecules in a particular material.

The relative simplicity of the absorption method compared to the methods described in the previous sections explains the widespread use of humidity sensors based on water-vapour absorption. Moreover, those materials that change their electric properties with absorbed water are very suited to micro-instrumentation. Nevertheless, absorption sensors suffer from a lot of drawbacks, due to some unfavourable properties of the absorption process, such as the temperature dependence and the susceptibility to other environmental conditions. This makes for hygroscopic sensors having a relatively poor long-term stability and a large temperature coefficient. Furthermore, absorption sensors suffer from hysteresis and have a large response time, due to the slow absorption process. To partially overcome this latter drawback, sensors are shaped like a disc, slice, strip or a thin film.

Many hygroscopic solids change their electrical resistance or dielectric properties as a function of humidity in a way suitable for hygrometric applications. Popular materials are porous Al_2O_3 and certain polymers. Among the electrolytes, LiCl is widely used, though other salt solutions are convenient as well. Developments in this field are discussed in Section IV.

III. DEW-POINT SENSORS.

The dew-point temperature of a humid gas is the temperature at which the water vapour is just saturated. The saturation pressure of water vapour is a well-known function of the temperature (Fig. 6). Saturation is confirmed by the

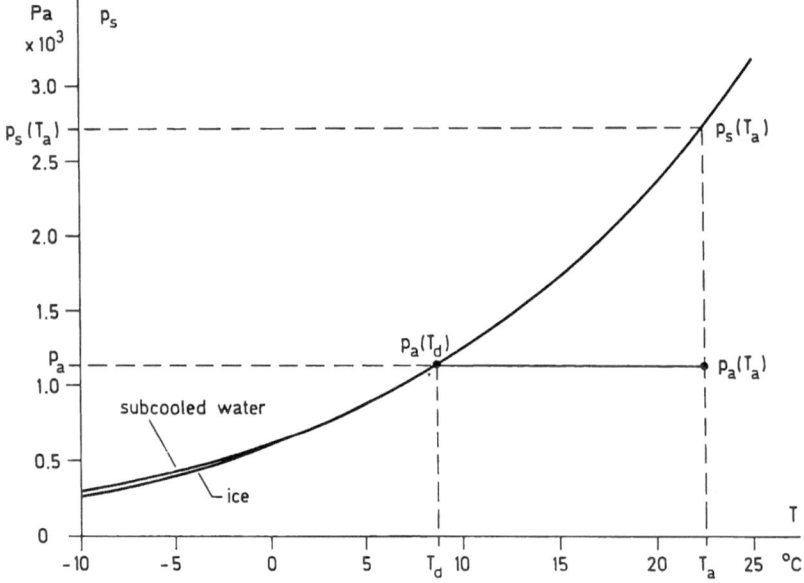

Fig. 6. Saturation pressure of water vapour versus temperature.

presence of a constant amount of condensed water (dew) in thermodynamic equilibrium with the gas under test. A dew-point measurement comprises cooling (preferably thermoelectric), detection of dew and measurement of the cooled-gas temperature (by conventional temperature-sensing devices).

Reproducible dew detection is achieved by a temperature-control loop (Fig. 7). The surface of a flat-shaped piece of smooth material, whose temperature can be controlled by a thermoelectric cooling device, is exposed to the test gas. When the surface temperature falls below the gas dew-point temperature, dew appears on the surface. The output of the dew-detecting circuit, responding to the amount of dew, is compared to a reference value. The difference is fed back to the cooling device. When properly designed, the feedback system keeps the amount of dew at a fixed value. In this situation, evaporation and condensation are in equilibrium, and hence the surface temperature equals the dew-point temperature.

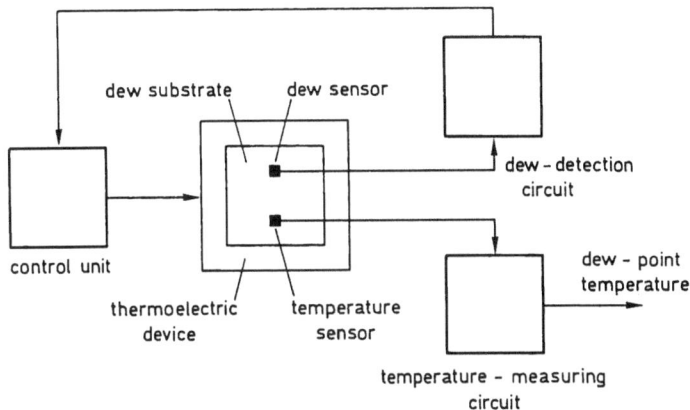

Fig. 7. Schematic diagram of a continuously operating dew-point system.

A change in dew-point temperature of the test gas will act as the input signal of the control system. The surface temperature will change so as to keep the amount of dew at the same, constant value. In this way, the measured surface temperature continuously keeps up with changes in the dew point.
Because of the integrating action within the control loop (transfer from mass flow to mass) the steady-state error is zero. Changing the reference value will only affect the amount of liquid water on the surface; in the steady state, the sensor temperature equals (again) the dew-point temperature.
Dew detection is accomplished either by caloric, visual, optoelectrical, electrical, radiometric or gravimetric means. Caloric detection utilizes the change in heat capacity when dew appears; this method is rather insensitive.
Radiometric detection needs expensive and complex measuring equipment.
Visual detection is influenced by subjectivity and does not allow automatic dew control. Some vibrating-crystal instruments are based on gravimetric dew detection: instead of having a moisture-absorbing layer on the crystal (see Section II), the temperature of a bare vibrating crystal is governed by a fixed deviation from its resonance frequency, corresponding to a constant mass offset by dew.
Optoelectric dew detectors use LED's and photodiodes in conjunction with a mirrored surface, utilizing changes in reflectivity, transparency or scattering. (Fig. 8).
Possibilities for purely electrical dew detection are surface-resistance measurements and capacitance measurements. Figure 9 illustrates the principle of a capacitive dew detector. The capacitance consists of two flat interdigitated electrodes on an electrically insulating substrate, for instance an oxidized silicon wafer. The fixed part of the capacitance is a function of the dimensions and the dielectric constants of the substrate materials. The variable part depends upon the amount of liquid water: when completely dry, the top layer is air (gas), with a dielectric constant of about 1; when dew is present, part of the air has been replaced by liquid water, having a dielectric constant of about 80. Hence, the capacitance corresponds to a

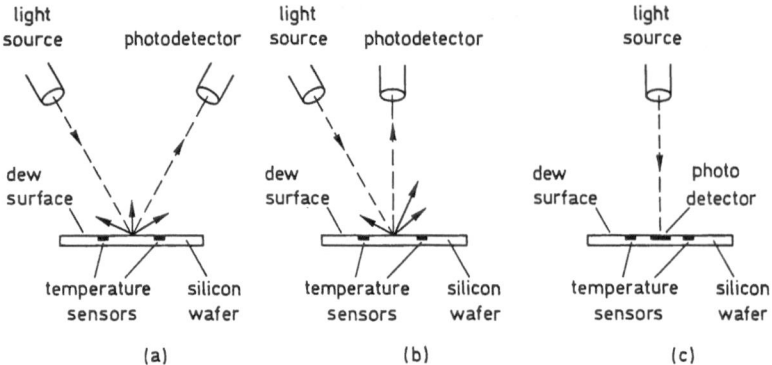

Fig. 8. Optical dew-sensing with integrated sensors, based on (a) reflected, (b) scattered and (c) passed light.

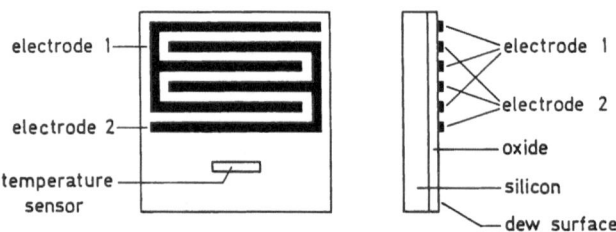

Fig. 9. Capacitive solid-state dew-point sensor.

certain volume of dew on the sensor surface. A proper steady-state capacitance is a compromise between system response time (increasing with increasing water volume) and effect of contamination (decreasing with increasing water volume, see Fig. 10).

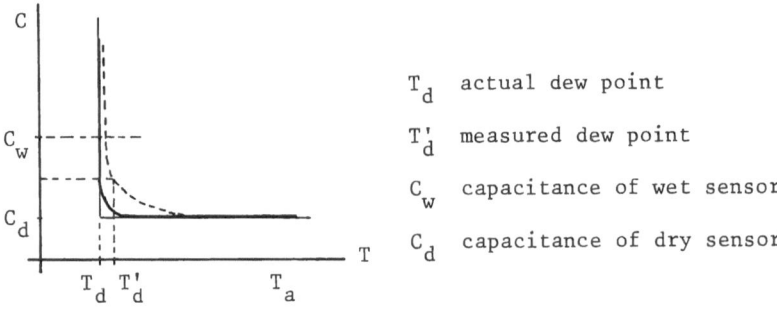

T_d actual dew point

T_d' measured dew point

C_w capacitance of wet sensor

C_d capacitance of dry sensor

Fig. 10. Sensor capacitance versus temperature.

Table 3 shows some specifications of two types of industrial dew-point measurement systems, one with optical dew sensing, and the other with a capacitive
dew sensor. Both systems use Pt100 temperature sensing elements. The specified
inaccuracies apply for clean sensor surfaces, and are attributed to errors in
the electronic processing of the temperature-measurement signals.

TABLE 3. Automatic dew-point measurement systems.

	optical detection	capacitive detection
dew-point range	$-80 \ldots +60$ $^\circ$C	$-15 \ldots +170$ $^\circ$C
temperature difference	75 K (max. 100 K)	35 K
cooling rate	2 K/s	2 K/s
pressure	20 Bar (220 opt.)	10 Bar
inaccuracy	± 0.4 K + Pt100 tol.	± 0.5 K (incl. Pt100)

Trends in dew-point hygrometry.

The essentially high accuracy of dew-point measurement systems is a challange-
ing starting point for further developments, towards tackling some practical
limitations of the method. The main problem is contamination; another is operation below zero degrees dew point.
  Contamination of the cooled surface causes dew formation at a temperature
above the real dew point (see also Fig. 10, where an increase in the capacitance occurs at a temperature above dew point). This results in an indication
which is too high, irrespective of the applied dew-detection method. Therefore, the surface should be periodically cleaned, either by hand or automatically. Some optical systems have a selfcleaning facility, based on heating of
the sensor up to a high temperature, resulting in evaporation of (volatile)
substances. Cleaning is started after a measurement of the reflectance with
cooling switched off. If the scatter by only contaminants has reached a certain level, the cleaning procedure proceeds. During each cleaning cycle, the
measurement of the humidity is interrupted.
With the capacitive dew sensor, the degree of contamination can be measured
during operation. Instead of a capacitance measurement, the real and imaginary parts of the sensor impedance are measured. The imaginary part is related
to the capacitance, and is used for controlling the dew. The real part is related to the conductivity of the water on the sensor surface, and hence to
the concentration of (soluble) contaminants. Figure 11 shows some measurement
results that prove the validity of this idea. The measurements are carried
out on a dew-point sensor, which had been operating for three months in a
pharmaceutical plant. Curves (a) and (b) are the polar plots of the contam-
inated sensor, and curves (c) and (d) of the cleaned sensor. The figure clearly shows that contaminants do not influence the imaginary part of the sensor
impedance in the dry state, and only slightly in the wet state. The real part
of the impedance is strongly influenced by contaminants.
With respect to measurements below zero degrees C, the following aspects are
noteworthy. An optical dew detector can detect liquid water as well as ice
(frost crystals): both scatter the incident light from the light source. However, this system is not able to distinguish between the two phases of water,
resulting in ambiguity in vapour pressure below zero degrees C. It is possible
to have subcooled water on the sensor surface at a temperature as low as -25
degrees C.
The capacitive sensor can, in principle, distinguish between dew and frost,
the static dielectric constant of ice being about 260, that is over a factor
of 3 that of liquid water. A transition from subcooled water into ice causes

a capacitance change which can easily be detected.

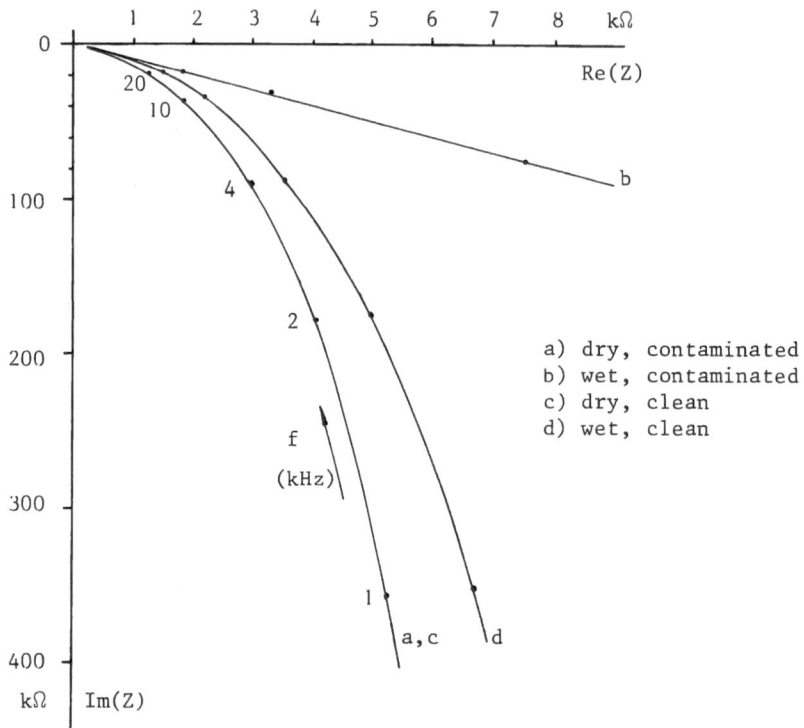

Fig. 11. Effect of contamination on the sensor impedance.

IV. ABSORPTION SENSORS.

The basic principle and the general characteristics of absorption sensors
have already been outlined in Section II. Practical absorption sensors are
based on either surface absorption (sometimes referred to as adsorption) or
volume absorption.
Surface absorption originates from the Van der Waals forces and the electrostatic forces between vapour molecules and the molecules at the surface of
the absorber. The amount of adsorbed water is minute. Therefore, the only
parameter with a sufficient sensitivity to moisture variations is the electric surface conductivity, changing rapidly with humidity.
The effective absorbing area is enlarged by a porous or fibrous structure.
This increases the sensitivity, enabling the measurement of characteristics
other than surface conductivity, and diminishing the sensitivity to contaminants. Methods to further improve the characteristics usually concern the
chemical treatments during and after manufacture.
Porous sensors and volume-absorption sensors are relatively slow. Their response time cannot be given by a single parameter because of non-linearity.
Figure 12 shows two response curves of a single sensor, but with different
time scales. Practical response times to within the specified accuracy is in
the order of minutes, but may also be much more. At very low humidity, stabilization may be reached after hours or even days. Not only the sensor itself

is responsible for such long times, but also supply tubes, which exhibit surface absorption on the inside as well.
Figure 13a shows an example of a porous Al_2O_3 sensor. The flat substrate has

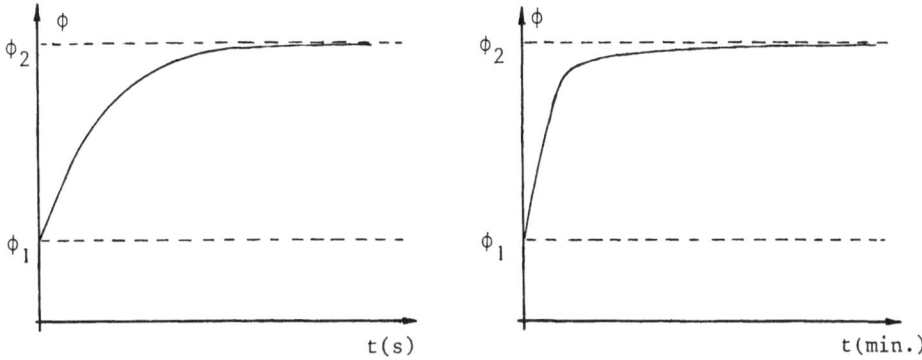

Fig. 12. Time responses of an absorption sensor.

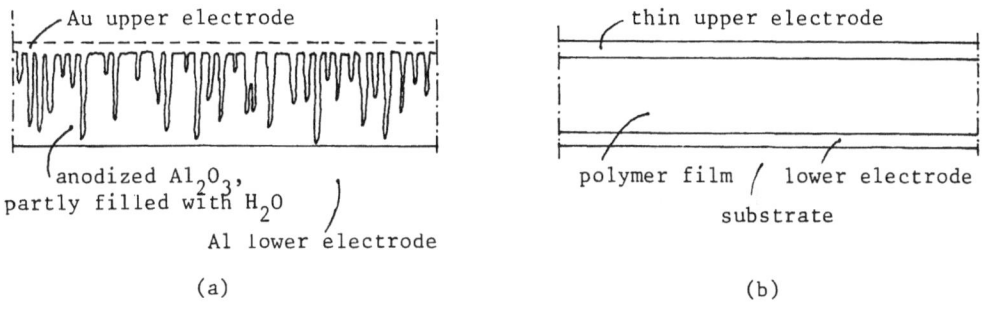

Fig. 13. Two examples of absorption sensors: (a) Al_2O_3 (b) polymer film

been made from aluminium. Its surface is anodized such that the oxide layer
has a porous structure. A thin top layer of gold, permeable to water molecules,
completes the flat capacitor structure of the sensor. Water molecules are adsorbed on the walls of the pores, filling the pores gradually by liquid water
as the humidity increases. Liquid water is formed mainly by capillary condensation. That is also the reason why such sensors respond mainly to dew-
point temperature. Some instruments use the capacitance of the sensor, while
others employ the resistance or a combination of both, as a measure for the
dew point.
The absorption sensor of Fig. 13b is based on volume absorption; it is also
constructed as a capacitor, but now with a hygroscopic dielectric film, for
instance some polymer. These types of sensors respond mainly to the relative

humidity. Table 4 gives some specifications., illustrating the state-of-the-art and the differences between the two sensor types.

TABLE 4. Specification example of absorption sensors.

	Al_2O_3- type	polymer film	
dew-point range	-80 ... +20 °C	humidity range	0 ... 100 % r.h.
temperature range	-20 ... +70 °C	sensitivity	0.1 pF / % r.h.
pressure	up to 350 Bar	temp. coefficient	0.05 % r.h. / °C
inaccuracy	1 °C (at 25 °C)	non-linearity	1 % r.h.
		hysteresis	2 % r.h.

A high accuracy of absorption sensors is only achieved by individual calibration, because the fabrication process is not very reproducible. In modern instruments, sensors are delivered with a matching PROM, which corrects for sensitivity, offset and non-linearity.

Trends in absorption-sensor development.

Despite the tremendous efforts to improve the overall characteristics of absorption sensors, the results have been marginal. Large improvements may only be expected by the application of other techniques or another technology. Some examples are given below.
A general measure for improving the transfer of a system is feedback. This principle may also be applied to humidity sensors, and is illustrated in Fig. 14. Some sensor output quantity, which can be the measurand itself, or another quantity, is fed back to control a particular sensor input quantity,

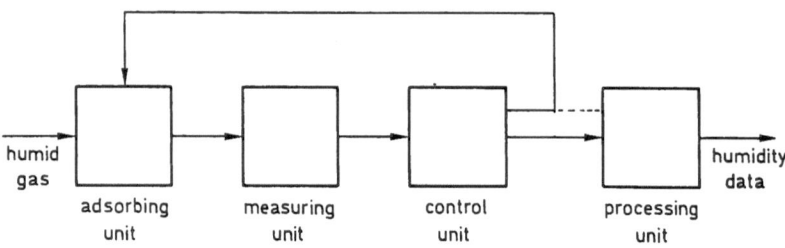

Fig. 14. Functional diagram of a feedback system

usually the temperature. This principle is applied in the LiCl dew-point sensor, in which the conductivity of a LiCl salt solution is kept constant by controlling the temperature of the salt. In this particular case, the temperature is a measure for the humidity (expressed in dew-point temperature). Another application of the feedback principle is a simple temperature control, to eliminate the temperature dependence of the sensor, or to avoid unwanted condensation at high humidity. In general, feedback shrinks the range of some disturbing quantity, thus eliminating its influence.

Improvement of the sensor characteristics may also be obtained by another
technology, in particularly the IC-technology. Figure 15 shows two examples
of silicon absorption sensors, one based on surface adsorption (measuring the
surface conductivity) and the other based on porous Al_2O_3, acting as the gate
isolation in a MOS-structure.

For the time being, these attempts have only resulted in a reduction of sensor dimensions. However, the possibilities for on-chip compensation of temperature effects, correction for non-lineatity etc., and consequently a lower
price, have been opened up.

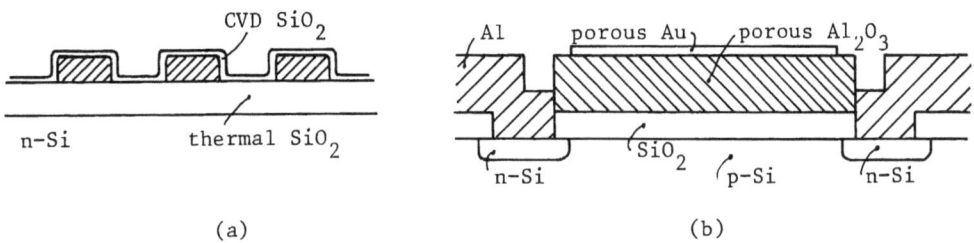

(a)   (b)

Fig. 15. Two examples of silicon absorption sensors.

V. CONCLUSIONS.

It is obvious that the microprocessor will play an important role in future
developments of humidity sensors. There are already quite a lot of instruments equipped with a microprocessor, mainly intended for conversion purposes.
It must be realized that a microprocessor or any other signal-processing circuit cannot add information lost during the transduction process. Improvement
of the overall characteristics of the sensor is only possible if there is
more knowledge about the transduction process taking place within the sensor.
Such information concerns for instance non-linearity, temperature dependence,
hysteresis, dynamic properties etc.

Up to now, sensor manufacturers (and not only of humidity sensors) tended to
improve sensor behaviour by ingenious constructions or the application of
contrived materials. This kind of "internal feedback" is being replaced more
and more by feedforward compensation: **knowing** the **non-linearity** of a **transfer**
function, it can be digitally linearized; knowing the temperature dependence,
measurement results can be corrected by adding a temperature transducer; hysteresis may be corrected by storing former data; the response time can be
(virtually) speeded up by prediction algorithms, and so on. The only requirement is exact knowledge about the sensor behaviour with respect to these
characteristics, thus reproducibility of the transfer properties. This requirement cannot always be fulfilled. In particular, environmental transducers such as humidity sensors, are susceptible to contamination, and have unpredictable and unmeasurable changes in their transfer functions.

Future research efforts will largely move from the mechanical and chemical
field to the electrical field, relieving the requirements on the sensor itself, and with an outlook to partly or even full integration of sensor and
signal processing.

P. Bergveld
Twente University of Technology
Box 217
7500 AE Enschede, The Netherlands

Introduction

In order to measure the chemical constituents of a liquid, such as
various ionic concentrations, dissolved gases, biochemical species
etc. by electronic means, such as various types of recording equipment, a large variety of electrodes has been developed. Focussing
attention on these electrodes, there will always be an interface, or
a system of interfaces, in which the "ionic world" communicates with
the "electronic world".

The simplest example is a direct contact between an ionic conductor
and an electronic conductor. Analysing such a system will directly
show the essential problems of chemical sensing in liquids, namely
how the requirements of stability and reversibility can be met.

In order to investigate this problem in more detail, the significance of characteristic chemical definitions on the one hand and the
corresponding solid-state ones on the other have to be discussed.
The relation between electrochemical potentials and Fermi energy
levels has to be clear, before possible thermodynamical equilibrium
between ionic conductors and electronic conductors can be understood.

After these theoretical considerations, the various ion-sensitive
materials can be classified and typical applications can be discussed. In this way the possibilities of glass membranes, metal-
oxide systems, coated wires and insulators can be lined up, from
which it can be deduced that the use of ion-sensitive insulators is
a much more generally applicable concept, than the other materials.

The application of ion-sensitive insulating layers compels the use
of a field-effect transistor as the basic element of a quite new
class of chemical sensors, the so-called chemically sensitive electronic devices, CSED's.

In the case of chemical sensors of the potentiometric type, the reference electrode is just as important as the sensor itself, especially with respect to the stability. However it is remarkable that
the development of solid-state reference electrodes has not kept
pace with the development of CSED's, although some new concepts are
being investigated at the present time.

Development and applications of CSED's are making progress nowadays,
especially in the biomedical field, although the original problems
of measuring in a liquid are intensified in this case due to the

NATO ASI Series, Vol. F43
Sensors and Sensory Systems
for Advanced Robots
Edited by P. Dario
© Springer-Verlag Berlin Heidelberg 1988

additional requirement of biocompatibility. However at present this is the furtherst developed application. One of the future developments will be the integration of a CSED and its corresponding actuator, in order to obtain measuring systems which can be calibrated in situ. This is especially necessary in the case of implantable sensors for physiological control of artificial organs. The modern sensor technology, which is a combination of IC-technology, micro machining and thin-film technology is in any case a strong approach towards the construction of complete integrated sensor-actuator systems which can perform chemical micro-analysis in vivo.

Definition of energy levels and electric potentials in solid- state physics and physical chemistry.

If we consider the thermal equilibrium condition of a solid state junction, for instance a p-n junction, the zero net electron and hole currents require that the Fermi level E_F must be constant throughout the sample. This results in a diffusion potential, or built-in potential $\Delta\phi$, which follows simply from:

$$E_F = E_{F_1} + q\phi_1 = E_{F_2} + q\phi_2 \tag{1}$$

where E_{F_1} and ϕ_1 are respectively the Fermi energy and the electrical potential of phase 1 while E_{F_2} and ϕ_2 are the same for phase 2, resulting in:

$$\Delta\phi = \phi_1 - \phi_2 = \frac{1}{q} (E_{F_2} - E_{F_1}) \tag{2}$$

So the contact potential between two electronic conductors is completely determined by their Fermi levels, which can be approximated in the case of semiconductors with Boltzmann's distribution as:

$$E_F = E + kT \ln F(E) \tag{3}$$

where E is the kinetic and potential energy of the charge carriers, k the Boltzmann constant, T the absolute temperature and F(E) the Fermi-Dirac distribution function.

The contact potential is thus the result of a redistribution of charge carriers over the junction, maintaining the state of equilibrium, and can therefore not be measured. The system can not deliver energy. Only if energy is applied to the system, a measurable contact potential will be available, for example a thermo voltage due to heating (ΔT) or a chemically sensitive voltage due to a specific chemical reaction (ΔE).

Contact potentials are stable and reversible because the concentrations of charge carriers, electrons and/or holes, are well defined and are exclusively the only movable species across the junction.

Studying physical chemistry or electrochemistry it is remarkable that principally the same laws are applied, as those in solid-state physics, but with completely different symbols and easily misunder-

stood interpretations. The reason is that the two scientific fields
have developed independently of each other. While the physicist
counts per charge carrier with a charge q (equations 1 and 2), the
chemist counts per gram molecule, containing N ions each having a
charge q, resulting in the Faraday constant F=Nq. Also the factor kT
(equation 3) is multiplied by N, resulting in the gas constant R=Nk.
Therefore the physical factor kT/q will be described by a chemist as
RT/F, both being 26 mV at room temperature (300 K).

Furthermore the definition of Fermi energy E_F is for a chemist the
same as the electrochemical potential, a slightly misleading notation for electrically oriented people, because it refers to the
energy level of an electron or an ion i in a solution or solid:

$$\bar{\mu}_i = \mu_i + zF\phi \tag{4}$$

where μ_i is the chemical potential, z is the valence of the particular ion and ϕ is the electrical potential.

If two solutions 1 and 2 with different concentrations of a particular ion i are separated by a membrane which is permeable for that
ion, a voltage can be measured over the membrane, from which the
value can be calculated according to equation (1):

$$\bar{\mu}_i = \mu_{i_1} + zF\phi_1 = \mu_{i_2} + zF\phi_2 \tag{5}$$

Therefore, according to equation (2) for the solid-state case, a
similar expression for the liquid-liquid junction yields:

$$\Delta\phi = \phi_2 - \phi_1 = \frac{1}{zF}(\mu_{i_1} - \mu_{i_2}) \tag{6}$$

The chemical potential μ_i can be written as

$$\mu_i = \mu_o + RT \ln a_i \tag{7}$$

where μ_o is the standard chemical potential and a_i is defined as
the ion activity, related to the concentration c_i by
$a_i = f_i c_i$. The factor f_i is the so-called activity coefficient
which depends on the overall concentration of the liquid and equals
1 when the solution is infinitely dilute. For not very concentrated
solutions $f_i = 1$ is often used and a_i can be interpreted as the
concentration c_i.

Because in this case $\mu_{o_1} = \mu_{o_2}$ it follows from equations (6)
and (7) that

$$\Delta\phi = \frac{RT}{zF} \ln \frac{a_{i_1}}{a_{i_2}} \tag{8}$$

$\Delta\phi$ is a measurable voltage because the diffusing ions can be delivered by chemical reactions at the measuring electrodes. This voltage
will be constant as long as a_{i_1} and a_{i_2} are constant.

The analogy between solid-state physics and physical chemistry regarding the interfacial equilibria is summarized in Figure 1.

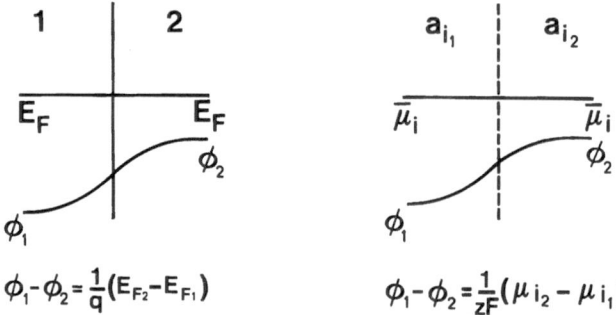

$$\phi_1 - \phi_2 = \frac{1}{q}(E_{F_2} - E_{F_1}) \qquad\qquad \phi_1 - \phi_2 = \frac{1}{zF}(\mu_{i_2} - \mu_{i_1})$$

Figure 1. Comparison between an interfacial equilibrium at a semiconductor-semiconductor and a liquid-liquid interface.

The difference between the two types of interface is not only the difference in symbols, the availability of chemical energy in the liquid-liquid type, as well as the possibility that then also the diffusing charge carriers can have a charge larger than unity ($|z| > 1$), but that there are also as many different ions as there are chemical elements and molecules which can be involved in the process. In practice a liquid-liquid junction potential is seldom determined by only one type of ion, but instead it is usual that all ions which can contribute do so. For all participating ions equation (5) will be valid, for example

$$\bar{\mu}_{H_1^+} = \bar{\mu}_{H_2^+}, \ \bar{\mu}_{Na_1^+} = \bar{\mu}_{Na_2^+}, \ \bar{\mu}_{Cl_1^-} = \bar{\mu}_{Cl_2^-}, \ etc.$$

This leads to more complicated equilibrium equations.

Starting with the definitions given above, we have now to determine the expressions which describe the contact potential between a solid and a liquid, or in other words how Fermi levels and electrochemical potentials will combine.

The metal-electrolyte interface.

If an electrode of a certain metal M is immersed in an aqeous solution with contains M^{z+} ions, the conditions as given in equation (6) will also be valid. Because the standard chemical potentials are however not equal in this case ($\mu_{o_1} \neq \mu_{o_2}$), equation (8) will contain now an additional term $1/zF(\mu_{o_2} - \mu_{o_1}) = E_o$, resulting in an electrode potential:

$$\Delta\phi = E_o + \frac{RT}{zF} \ln \frac{a_{i_1}}{a_{i_2}} \tag{9}$$

It will be clear that the electrode potential will be constant as in both phases the concentration of the potential determining ions is constant. Because in this particular case $a_{i_1}=1$ by definition (the metal phase), $\Delta\phi=E_o$ for the condition that also $a_{i_2}=1$ (metal ions). E_o is called the standard electrode potential, from which the value can be found in the IUPAC table for half-reactions, defined with respect to the hydrogen electrode. For example:

$$Ag^+ + e \rightleftharpoons Ag(s) \qquad E_o = +0.799 \tag{10}$$

With this redox reaction the electrochemical potential $\bar{\mu}_{Ag^+}$ of Ag^+ ions in the solution is related to the Fermi level of the electrons in the solid $Ag(s)$. Hence in the case of a silver electrode in a diluted silvernitrate solution the electrode potential is:

$$\Delta\phi = 0.799 + 0.059 \log c_{Ag^+} \tag{11}$$

which means that the electrode acts as a Ag^+ ion sensitive electrode, giving 59 mV per decade $[Ag^+]$. Unfortunately a metal electrode will seldom only be used in pure solutions of its own ions. Many other ions will exist and also dissolved gases as for example O_2. This means that other redox reactions may also occur at the electrode, interfering with the original reaction such as given by equation (10). For example the reduction of dissolved oxygen according to:

$$O_2 + 4 e \rightleftharpoons 2 O^{2-} \tag{12}$$

may change the value of $\Delta\phi$ as given in equation (11) tremendously. Therefore metal electrodes are, in general, not applicable for ion sensing in liquids. A solution to this problem is the introduction of an intermediate layer between the metal and the liquid, of which the Ag-AgCl electrode is the most well-known example.

Ag-AgCl and related electrodes.

An Ag-AgCl electrode consists of a silver disk or wire, on which a hardly solvable layer of silverchloride is electrolytically deposited in a sodium chloride solution. Two interfaces can now be distinguished, which both can be in perfect equilibrium, because one and only one charge carrier will be responsible for determining the interfacial potential. At the Ag/AgCl interface it is the Ag^+ ion and at the AgCl/liquid interface the Cl^- ion, which ensures that the electrode is only stable in solutions which contain a fixed concentration of Cl^- ions. The overall potential is:

$$\Delta\phi = 0.2224 - 0.059 \log a_{Cl^-} \tag{13}$$

which shows that the Ag/AgCl electrode is a perfect Cl⁻ ion sensitive electrode and it is also applied as such. The introduction of
the intermediate layer results in this case in a well-defined
well-defined coupling mechanism between an electronic and an ionic
conductor and is generally applicable for all silver-salt membranes.

A similar approach is employed by Fjeldli [1] who constructed a
fluoride electrode, in which the fluoride sensitive membrane of
LaF_3 is connected to the Ag electrode substrate by an intermediate
layer of ion-conducting AgF. The Ag substrate may also be a silver
conducting epoxy on any other metal substrate. The sandwich construction in this case again enables the rapid adjustment of a state
of thermodynamic equilibrium throughout the whole system. Unfortunately such a simple sandwich construction for a stable contact between electronic and ionic conductors is not available for all membranes, unless the classical approach of membranes with an inner
liquid and a Ag/AgCl electrode can be seen as such.

Ion-sensitive membranes

All ion-sensitive membranes have in common that they combine an intrinsic or extrinsic ionic conducting property with specific ion
sensitivity at the interfaces. The most logical application is
therefore to mount such a membrane between two liquids, from one of
which the ion concentration has to be measured and the other with a
constant ion concentration. Both liquids can be contacted by means
of Ag/AgCl electrodes to provide a connection with an electronic
measuring device.

The classical glass membrane electrode is the most well-known example of such a liquid-contacted membrane. The operational mechanism
is very obvious: the glass hydrates to a certain extent and the
electrochemical potentials will balance according to equation (5)

$$\bar{\mu}_{H^+_{sol}} = \bar{\mu}_{H^+_{glass}} \tag{14}$$

resulting in very stable interfacial potentials, linearly related to
the pH of the solution. As the inner solution has a constant pH, the
inner interfacial potential delivers a constant contribution to the
output voltage of the electrode, as is the case with the contributions of the Ag/AgCl contact electrodes. Hence, a glass electrode at
300 K delivers an output voltage of:

$$V_{out} = V_{const} + 59 \text{ mV/pH unit} \tag{15}$$

This approach is generally applicable for all ion-conducting membranes, in whatever form or construction is chosen, such as sintered
pellets, polymers in which electroactive compounds are embedded,
etc. This type of electrode is usually called a solid-state electrode [2] and is illustrated in Figure 2.

It will be obvious that many attempts have been made to obviate the
relatively awkward internal solution and replace it by a solid-state
contact. Le Blanc et al. as well as Cobbe et al. used a pH-
sensitive polymer membrane, dip coated over a pregelled Ag/AgCl
electrode [3,4]. In fact they still make use of an internal "wet

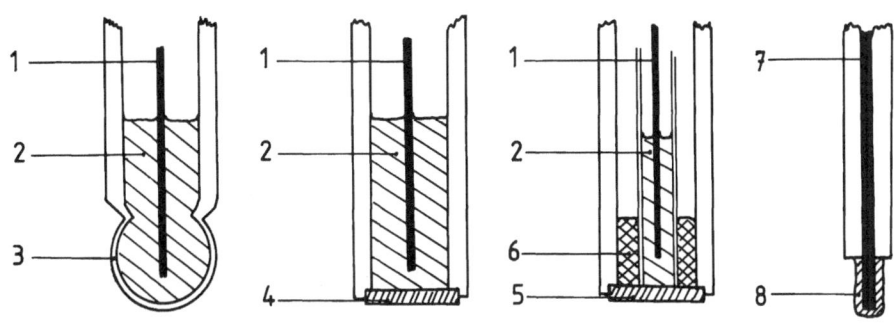

a   b   c   d

Figure 2. Various performances of ion-sensitive electrodes.
a: glass membrane electrode, b: solid-state membrane
electrode, c: liquid membrane electrode, d: coated
wire electrode.
1 = Ag/AgCl electrode, 2 = internal electrolyte, 3 =
glass membrane, 4 = solid-state membrane, 5 = porous
substrate containing liquid membrane, 6 = supply of
liquid for liquid membrane, 7 = metal wire, 8 = membrane coating.

contact" although with miniscule dimensions. A completely dry internal contact was realized by Afromowitz, who deposited pH glass by
means of the thick-film screening technique directly on top of a
layer of silver, resulting in a very rugged sensor construction but
unfortunately with rather poor characteristics [5]. The reason is
that no thermodynamic equilibrium is present at the inner contact.
The materials chosen for the sandwich construction are more or less
dictated by the applied technology, rather than by the theoretical
considerations regarding interfacial phenomena between ionic and
electronic conductors.

A better solid-state contact with a pH glass-membrane is recently
reported by Fjeldly et al. [6]. The inner contact of the glass bulb
of a commercial glass electrode was established by the reaction of
silver fluoride while heating in a gas flame. Auger depth profiling
analysis indicates that in this way a suitable, gradual junction is
established, resulting in a stable electrical contact.

The same arguments can be used with respect to the development of
the so-called coated wires [7]. They are very easy to produce by
simply dipping the tip of a metal electrode in a membrane polymer,
which, besides the plasticizer, also contains a specific ion exchanger. If the coated metal electrode is directly (on-chip) connected to the gate of a MOSFET, the sensor is called an extended-

gate FET [8]. A stable potential will exist at the membrane solution interface, based on the equilibration of the electrochemical potentials of the ion which is specific for the ion exchanger. At the inner contact the same problem as mentioned above will occur. Relatively good measuring results are achieved with respect to the sensitivity and selectivity, factors which are determined by the membrane properties. The stability however, is rather poor because no special measures are taken with respect to the membrane/metal interface. The various successes or drawbacks mentioned in the literature should be interpreted in such a way that the reported stable electrodes will most probably have an intermediate layer between the membrane and the metal, although not intentionally so designed. An oxide layer can perform this function, in which case a charged particle, which up to now has not been identified, could determine the equilibrium of the corresponding electrochemical potentials. That oxidized metals behave as very stable electrodes has already been proved, resulting in the application of metal-metal oxide electrodes.

Metal-conducting metal oxide electrodes

A special class of ion-sensitive electrodes is formed by the metal-metal oxide electrodes such as PtO_2, IrO_2, Sb_2O_3, etc. [9].
The oxides appear to behave as electronic conductors, which means that the metal-metal oxide interface is determined by the equilibration of the Fermi level of the electrons. The interface is in fact a normal ohmic contact and the oxides can also be deposited on other substrate metals than their own metal ions [10]. The oxide-solution interface is pH sensitive due to the hydroxyl groups, as is known from the glass membrane electrodes. However, a difference with the latter oxide is that the electronic conduction of the metal oxides involves a strong redox·sensitivity, resulting in interference with oxidizing and reducing agents, shifting the electrode potentials by $\pm$ 100 mV, which is a serious drawback for the application of this type of electrode. The internal mechanism, where the two interfacial phenomena, one from the "ionic world" and one from the "electronic world" are coupled is still a subject of research. Various models are under investigation at present of which the so-called oxygen intercalation reaction is probably the best. This mechanism assumes a change in the oxide state as function of pH, according to:

$$MO_x + 2\delta H^+ + 2\delta e^- \rightleftharpoons MO_{x-\delta} + \delta H_2O \qquad (16)$$

from which reaction the electrode potential can be calculated by equilibration of the corresponding electrochemical potentials in the solid and the liquid phase as described in previous sections. The state of the oxide may also result in a specific colour of the electrode, which has been observed and investigated for instance in the case of IrO_x and applied in the design of a flat display [11].

The application of ion-sensitive insulators

It is known from colloid and organic chemistry that various insulating oxides and polymers in contact with a liquid are ion sensitive, but the question arises whether this property can be employed
for the construction of a chemical sensor. The ion sensitivity of
various oxides such as SiO_2, Ta_2O_5, Al_2O_3 etc. was originally determined by applying potentiometric titration of a suspension of the metal oxide, with either acid or base, for solutions of
specific ionic strength. Since the ions added must either go to
change the pH of the solution or be adsorbed on the inherent large
metal oxide surface, it is possible to determine the surface charge
density as a function of the pH for several concentrations of supporting electrolyte. A characteristic feature is the pH at which no
net charge is adsorbed or the concentration of adsorbed OH^- and
H^+ ions is equal [12]. Further addition of metal oxide to the
solution does not change the pH in this case. Colloid chemists have
traditionally called this point the pH_{pzc}, the pH at the point of
zero charge. At other values of the pH, the net surface charge σ_0
can be measured by titration experiments, from which the corresponding surface potential can be calculated according to:

$$\Psi_0 = \sigma_0/C_{DL} \tag{17}$$

where C_{DL} is the double-layer capacitance as function of electrolyte ionic strength. This potential can also be calculated by the
approach as mentioned in the previous section because in equilibrium
it yields:

$$\bar{\mu}_{ad} = \bar{\mu}_{sol} \tag{18}$$

where $\bar{\mu}_{ad}$ is the electrochemical potential of the charged
surface groups, either OH_2^+ or O^- and $\bar{\mu}_{sol}$ of the H^+
ions in the solution. Another theoretical model which explains the
existence of a surface potential at the insulating oxides as function of pH is the site dissociation model [13], which describes the
equilibrium between the surface OH groups and the H^+ ions located
directly near the surface which are related to the concentration of
H^+ ions in the solution by Boltzmann's equation. It is essential
to note that the verifying measurements in colloid chemistry are in
fact time-consuming charge measurements, from which surface potentials can be calculated. It will be clear that this method of surface potential determination is in no way applicable for a direct
measurement of the pH.

The same problem holds for the measurement of the surface potential
of specific ionophores, such as crown ethers, of which a large
variety can be synthesized, each type having a specific affinity for
one type of ion [14]. The uptake of ions can be measured as a dilution of the surrounding liquid but the corresponding surface potential can only be calculated.

The basic problem of a direct measurement of the surface potential
of an insulating layer or membrane is the fact that the bulk of the
insulator forms a disconnection in the measuring circuit, as shown

in Figure 3.

The input potential of such a measuring circuit drops across the insulator and not over the amplifier input. At the same time this potential drop over the insulator sets up an electric field inside

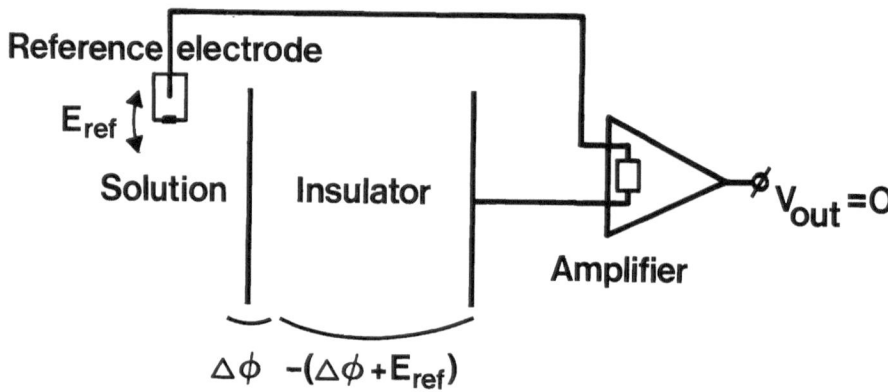

Figure 3. Schematic representattion of the measurement of a potential Δφ at an insulator surface by means of an amplifier.

the insulator. This means that, if the insulator can be interfaced with a semiconductor, a surface charge in the semiconductor will be induced according to Gauss's law. Fortunately much experience has been gained during recent decenia with insulators superimposed on semiconductors, especially with respect to SiO_2-Si structures, which are also provided with leads to measure the surface charge as function of the oxide field. This is known as the class of MOS field effect transistors. Used as an electronic device the internal insulator field is modulated by an applied voltage to the metal gate M, but in the light of the previous discussion this can also be a surface potential caused by contact between the oxide and an aqeous solution. These devices are called ISFETs or CHEMFETs [15], which provide the user with a direct method of measuring the surface potential at an insulator/electrolyte interface.

The Ion Sensitive Field-Effect Transistor.

Although the basic operation of an ISFET lies in the interfacial phenomena at the insulator-electrolyte junction, the solid-state part of the device is such an integral part of the sensor, that comprehension of this part is essential for a complete description. The operation of an ISFET should therefore be compared with that of the pure electronic analog, the MOSFET. Figure 4 illustrates the similarities as well as the differences between this well-known MOSFET and the ISFET. The metal gate of the MOSFET in figure 4 is replaced by the metal of a reference electrode, while the liquid in

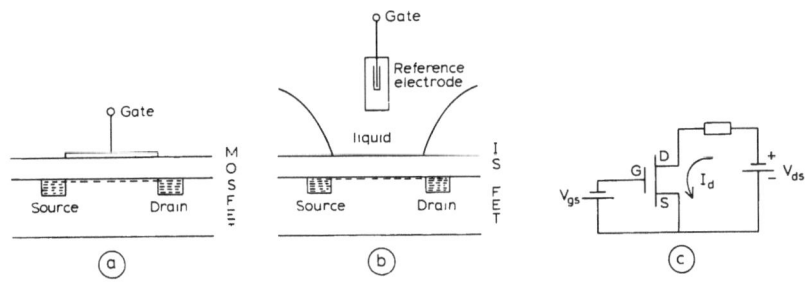

Figure 4. Schematic diagram of a MOSFET and an ISFET and corresponding electrical diagram.

which this electrode is present makes contact with the original gate
insulator (Fig. 4b). Both devices have the same electrical equivalent circuit, which is symbolized in Fig. 4c in which V_{gs} is the
applied gate-source voltage and V_{ds} the applied drain-source voltage. The following equation is valid for the non-saturated region
of both devices (below pinch-off):

$$I_d = \beta(V_{gs} - V_T - 1/2\ V_{ds})V_{ds} \tag{19}$$

in which β is a parameter, determined by the mobility of the electrons in the inversion layer, the gate insulator capacitance per
unit area C_{ox}, and the width-to-length ratio of the channel W/L.

$$\beta = \mu\ C_{ox}\ W/L \tag{20}$$

Besides differences in the performance of the two devices, there is
a difference in the value of the threshold voltage V_T which is
constant in the case of a MOSFET, but is a function of the pH of the
liquid for the ISFET, according to equations 21 and 22:

$$V_T = V_{FB} - Q_B/C_{ox} + 2\ \phi_F \tag{21}$$

with:

$$V_{FB} = E_{REF} - \psi_o + \chi^{sol} - \frac{1}{q}\ \phi_{si} - (Q_{it} + Q_f)/C_{ox} \tag{22}$$

Here V_{FB} is the so-called flat-band voltage, Q_B is the depletion
charge in the silicon, ϕ_F the Fermi potential, E_{Ref} the reference electrode potential relative to vacuum, χ^{sol} the surface dipole potential of the solution, ϕ_{Si} the silicon work function,
Q_{it} the surface-state density at the silicon surface and Q_f the
fixed oxide charge. The surface potential at the oxide-electrolyte
interface as mentioned in the previous section, is denoted as ψ_o,
being the parameter which makes the flat-band voltage a function of
the pH, resulting in the ion sensitivity of the device. Therefore
this parameter will be investigated in more detail.

The interaction between an inorganic insulator and an adjacent electrolyte is based on the assumption that the surface contains a dis-

crete number of surface sites which may dissociate. Therefore an expression for ψ_0 will be derived on the basis of the site-dissociation model, the model commonly used and already cited in the previous section.

Insulators which are widely used for the ISFET process technology
are SiO_2, Al_2O_3 and Ta_2O_5. The surface of these oxides
contain hydroxyl groups which act as discrete sites for chemical reactions of the surface, when it is brought into contact with an
electrolyte solution. It is usually considered that only one type of
site is present, having an amphoteric character. This means that
each surface site can be neutral, act as a proton donor (acidic reactions) or as a proton acceptor (basic reaction). This surface
property is schematically represented in Figure 5.

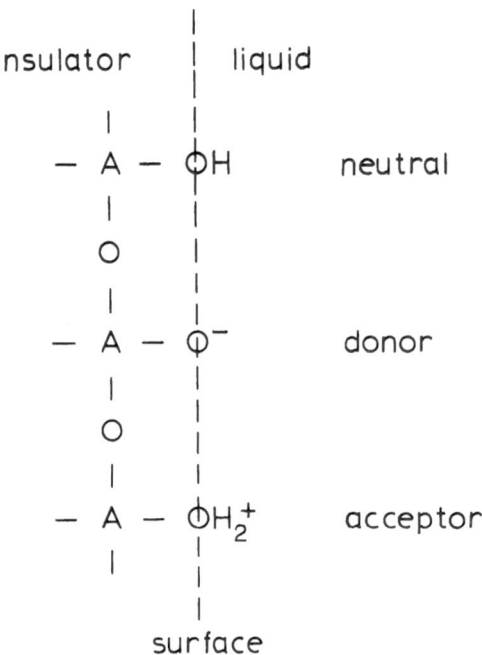

Figure 5. Schematic representation of the site-dissociation
          model.

The corresponding acidic and basic reactions are characterized by
their equilibrium constant K_a and K_b.

The resulting surface potential ψ_0 can be calculated from the total number of surface sites N_s which will be partly charged to a
surface charge σ_0, depending on the pH as determined by the equilibrium constants K_a and K_b.

$$\psi_o = 2.3 \; \frac{kT}{q} \; \frac{\beta}{\beta+1} \; (pH_{pzc} - pH) \tag{23}$$

where:

$$pH_{pzc} = -\log \; (\frac{K_a}{K_b})^{(1/2)} \tag{24}$$

being the pH for which $\psi_o=0$ and $\sigma_o=0$,
and

$$\beta = \frac{2q^2 N_s (K_a K_b)^{1/2}}{kT \; C_{DL}} \tag{25}$$

being a surface reactivity parameter in which $C_{DL}=\sigma_o/\psi_o$, being the double layer capacitance.

Substitution of equation (23) into equation (22), (21) and (19) results in a fixed relation between the drain current of an ISFET and the pH of the measuring solution [16].

In order to obtain a stable ISFET operation, the ISFET is always applied in a feedback circuit [17]. This will always result in an operation mode where the drain-source voltage has a constant preset value and the feedback control also ensures a constant drain current. This will result in a gate-source voltage that is adapted to the value of ψ_o. As the gate voltage is actually the voltage of the reference electrode which is usually the ground connection of the feedback amplifier, it means that the source voltage with respect to ground exactly follows the pH dependent surface potential ψ_o as given in equation (23).

It will be clear that, based on the original pH-sensitive ISFET's, many other FET-based ion-sensitive sensors can be and, in fact, have been constructed [18]. The concept is generally applicable for all kinds of insulating ion-sensitive materials, such as polymers with added ionophoric substances, for example crown ethers, which are attached to the original inorganic gate material. In fact the problem of a stable contact for the inner side of the membrane with the "electronic world", as mentioned in all previous sections, is now solved. The galvanic contact is replaced by a capacitive one, which is stable due to the stable SiO_2-Si interface, a positive contribution of micro-electronic research.

The reference electrode.

Up to now it was assumed that the potentiometric ion sensor, whether is was of the glass membrane, the metal-metaloxide, the coated wire or the ISFET type, are applied in an aqeous solution which also contains a stable reference electrode. Considering the stability of the ion sensor, it was assumed that the stability of the reference electrode was much better. This is also the case with the use of the conventional saturated calomel electrode, essentially a

Hg/Hg_2Cl_2 electrode in a saturated KCl solution, contacting the measuring solution with a porous plug. In fact this liquid-filled system is not suitable for the modern all solid-state construction of ion sensors. It is remarkable that the adaptation of reference electrodes has not kept up with modern sensor concepts. Only a few new approaches are being investigated at the present time.

What has been realized up to now is a miniaturization of the classical reference electrode concept. Catheters containing an ISFET chip which contacts blood via the side wall may have a reference electrode in the tip, where the usual sintered glass pellet is replaced by a poly-HEMA plug and the salt bridge contains a gelled Ringer's solution [19]. This construction is shown in Figure 6A. It appears to be a very good reference electrode for in vivo use, with an average stability of $\pm$ 0.92 mV for at least a period of 6-8 hours, independent of the pH.

An approach that uses thin-film technology, which is also compatible with the IC technology in use for ISFETs, is Prohaska's micro-electrode design [20], as shown in Figure 6B.

Although developed for the measurement of low-frequency EEG signals in the brain, the electrode design is essentially a very small and flat reference electrode, comparable with the catheter type shown in Figure 6A. However, a salt bridge is still necessary to make the internal Ag/AgCl electrode insensitive to variations in the Cl^--concentration of the measuring solution. No data are available for the lifetime of this electrode, but it will be obvious that this design, as well as all the other designs mentioned above, is not suitable for definite implantable use.

A similar small reference electrode, although not produced using IC-technology, is described by Janata [21], who mounted a liquid junction system on an ISFET with a closed pH-buffered compartment. This construction is shown in figure 6C. The reference ISFET can be applied, together with a normal ISFET, in a differential amplifier circuit.

A reference electrode construction as integral part of the ISFET chip and as such making use explicitly of IC technology and micro-machining was developed by Smith and Scott [22]. They etched a cavity in the silicon chip, in which they evaporated a Ag/AgCl electrode and covered it with a layer of porous silicon, with pores varying from 1.0 nm to 1.0 μm. The cavity can then be filled with KCl. This example of IC compatible processing of a reference electrode as shown in Figure 6D still retains the disadvantages of a liquid-filled internal cavity, which will certainly present problems for long-term measurements.

The best solution is the use of a liquid free, all solid-state reference electrode, most probably consisting of an ion-insensitive FET

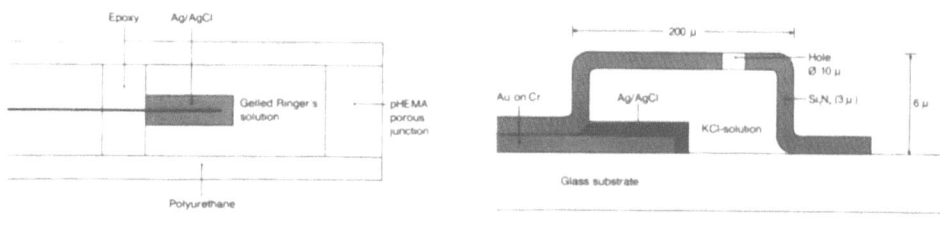

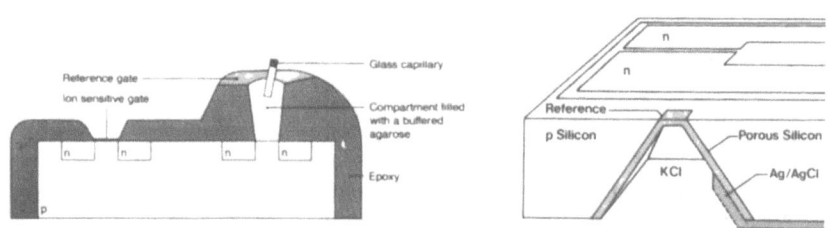

Figure 6. A. Margules type of reference electrode [19].
         B. Prohaska type of reference electrode [20].
         C. Janata   type of reference electrode [21].
         D. Smith   type of reference electrode [22].

sensor, a so-called REFET [23]. The REFET is originally an ISFET, but covered with a fully inert membrane. The development of REFETs is still under investigation, but if it is possible to produce a chemically stable REFET with identical electrical properties as an ISFET, while the two can be integrated in one chip, a simple evaporated noble metal can act as a pseudo-reference. The system is schematically given in Figure 7.

Although the interface metal/solution will never be in thermal equilibrium as argued in the first sections of this paper, the instable electrode potential is not measured by the differential circuit of the ISFET and the REFET. The same system will however fully measure the pH-sensitive surface potential of the ISFET as shown in Figure 7. Therefore most probably the all solid-state combined electrode in the future will consist of FET-based sensor pairs, which again underlines the impact of the ISFET approach.

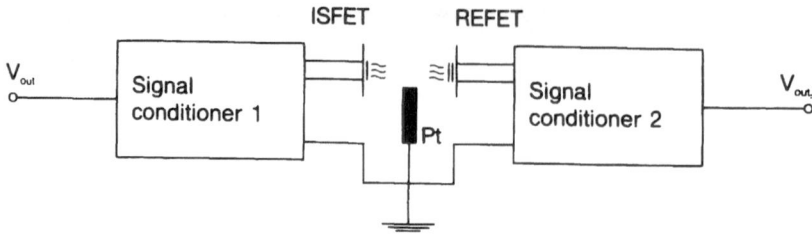

$$V_{out_1} - V_{out_2} = f\,(pH)$$

Figure 7. Schematic representation of ISFET/REFET measuring set-up with common platinum (pseudo) reference electrode.

Future developments.

A serious problem of all sensors described in this paper up to now is the necessity of calibration and recalibration after a certain period of time. The time lag between the initial calibration and the second one depends on experience with the particular sensor in use, especially concerning the occurrence of drift and loss of sensitivity. Also in this case the modern sensor performance results in new possibilities, namely the integration of an actuator near to the corresponding sensor.

An approach, at present under investigation at our university, is the integration of a micro-coulometric cell with the ISFET chip

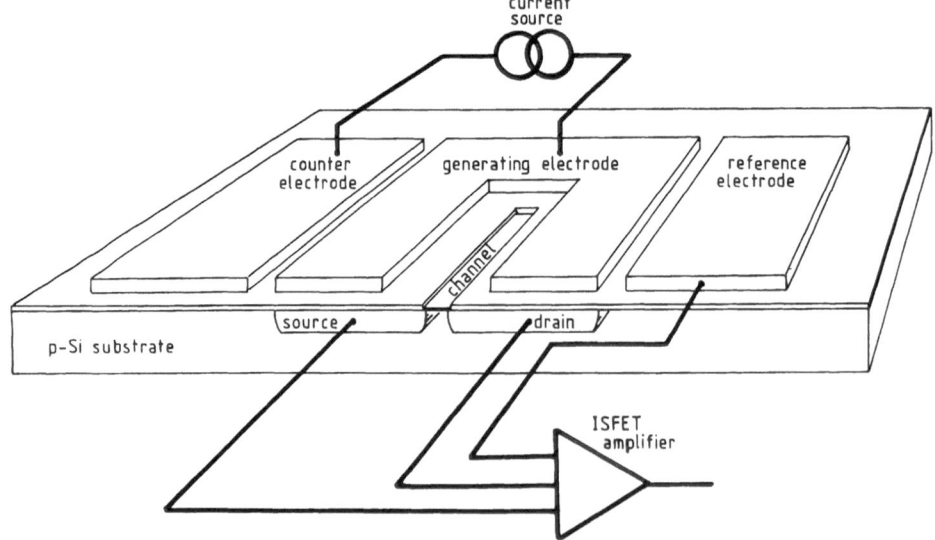

Figure 8. Basic elements of an integrated chemical sensor-actua-tor system for titration.

[24]. This cell contains a working electrode around the ISFET gate and a counter electrode at a certain distance from the gate as schematically drawn in Figure 8.

It is now possible to induce very locally a ΔpH by applying a current pulse through the actuator and the counter electrode as a test signal for the ISFET. Generation of a complete titration curve enables the actual pH to be calculated from this localized chemical experiment. Such systems are possibly the solution for sensor application where in-situ calibration is an absolute necessity, such as in the case of implanted sensors in the human body.

REFERENCES

[1] Fjeldli, T.A., Nagy, K.: Fluoride electrodes with reversible solid state contacts. J. Electrochem. Soc. 127 (pp. 1299-1303) 1980.

[2] Buck, R.P., Thompson, J.C., Melroy, O.R.: A compilation of ion-selective membrane electrode literature. Chapter 4 of Ion-Selective electrodes in analytical chemistry. Vol. 2, H. Freiser (Ed.) Plenum Press N.Y. 1980.

[3] Le Blanc, O.H., Brown, J.F., Klebe, J.F. et al.: Polymer membrane sensors for continuous intravascular monitoring of blood pH. Journal of Applied Physiology, Vol. 40, No. 5, (pp.644-647) 1976.

[4] Cobbe, S.M., Poole-Wilson, P.A.: Catheter tip pH electrodes for continuous intravascular recording. Journal of Med. Eng. & Techn. Vol. 4, Nr. 3, (pp. 122-124) 1980.

[5] Afromowitz, M.A. and Yee, S.S.: Fabrication of pH sensitive implantable electrode by thick film hybrid technology. Journal of Bioengineering, Vol. 1, (pp. 55-60) 1976.

[6] Fjeldi, T.A., Nagy, K.: Glass electrodes with solid-state membrane contacts and their application in differential potentiometric sensors. Sensors and Actuators, 8 (pp. 261-269) 1985.

[7] Freiser, H.: Coated wire ion-selective electrodes. Chapter 2 of Ion-selective electrodes in analytical chemistry, Vol. 2, H. Freiser (Ed.) Plenum Press N.Y. 1980.

[8] Van der Spiegel, J., Lauks, I., Chan, P. et al.: The extended gate chemically sensitive field effect transistor as multi-species microprobe. Sensors and Actuators, Vol. 4, (pp. 291-298) 1983.

[9] Fog, A., Buck, R.P.: Electronic semiconducting oxides as pH sensors. Sensors and Actuators, Vol. 5, (pp. 137-146) 1984.

[10] Katsube, T., Lauks, I. and Zemel, J.N.: pH-sensitive sputtered iridium oxide films. Sensors and Actuators, 2 (pp. 399-410) 1982.

[11] Gottesfeld, S. and McIntyre, J.D.E.: Electrochromism in anodic iridium oxide films. J. Electrochem. Soc. 126 (pp. 742-750) 1979.

[12] Butler, M.A. and Ginley, D.S.: Prediction of flatband potentials at semiconductor-electrolyte interfaces from atomic electro negatives. J. Electrochem. Soc., Vol. 125, no. 2 (pp. 228-232) 1978.

[13] Yates, D.E., Levine, S. and Healy, T.W.: Site-binding model of the electrical double layer at the oxide/water interface. J. Chem. Soc. Faraday Trans., Vol. 70 (pp. 1807-1818) 1974.

[14] Vögtle, F., Weber, E.: Neutrale organische komplex liganden und ihre alkalikomplexe I - Kronenäther, cryptanden, podanden. Kontakte, Vol. 1, (pp. 11-28) 1977.

[15] Sibbald, A.: Chemical-sensitive field-effect transistors. IEE Proceedings, Vol. 130, Pt. 1, No. 5, (pp. 233-244) 1983.

[16] Bousse, L., de Rooij, N.F., Bergveld, P., Operation of chemically sensitive field effect sensors as a function of the insulator-electrolyte interface. IEEE Trans. on Electron Dev. Vol. ED-30, nr. 10 (pp. 1263-1270) 1983.

[17] Bergveld, P., The operation of an ISFET as an electronic device. Sensors and Actuators, 1 (pp. 17-29) 1981.

[18] Bergveld, P.: The impact of MOSFET based sensors. Sensors and Actuators, 8 (pp. 109-127) 1985.

[19] Margules, G.S., Hunter, C.M., MacGregor, D.C.: Hydrogel based in vivo reference electrode catheter. Med. & Biol. Eng. & Comp. 21, (pp. 1-8) 1983.

[20] Hochmair, E.S. and Prohaska, O.: Implantable sensors for protheses. Chapter 4 Implantable sensors for closed-loop prosthetic systems. Ed. W.H. Ko. Futura Publ. Comp. Mountkisco, New York 1985.

[21] Janata, J., Huber, R.J.: Ion-sensitive field effect transistors. Ion-Selective Electrodes Rev., Vol. 1, (pp 31-79) 1979.

[22] Smith, R.L. and Scott, D.C.: A solid state miniature reference electrode. Proc. of the Symposium on Biosensors, IEEE: 84CH2068-5 (pp. 61-62) 1984.

[23] Tahara, S., Yoshii, M., Oka, S.: Electrochemical reference electrode for the ion-selective field effect transistor. Chemistry Letters, (pp. 307-310) 1982.

[24] v.d. Schoot, B.H. and Bergveld, P.: An ISFET-based microlitre titrator: integration of a chemical sensor-actuator system. Sensors and Actuators, 8 (pp. 11-22) 1985.

Section 5

Integration of Sensory Systems

Integration of robot sensory systems

B. V. Jayawant, University of Sussex, Brighton, England

1. Introduction:

The views expressed in this paper have been formed as a
result of attempts to fabricate array tactile sensors.
Whilst an array of 20 x 15 elements of 1 mm pitch has been
successfully fabricated, it is now clear that shape
recognition, which is all that this sensor is capable of, is
only a limited facet of tactile sensing. It is, therefore,
valuable to examine the nature of tactile sensing in human
beings.

2. Nature of tactile sensing:

Tactile recognition in human beings is a dynamic process
dependent on hands equipped with dense cutaneous and
kinesthetic sensors and controlled by a central unit which
reconstructs tactile images and then formulates strategies for
manipulation. A gripper with force, torque, and displacement
sensors suitably disposed over the gripper jaws would be the
minimum requirement.

The principal touch receptors in human skin are Pacinian
Corpuscles, Merkel Cells and Meissner Cells. They are
differently located throughout the depth of the skin and
confer to it spatial resolution and capability to respond both
to slow and fast stimuli.

Although the total number of sensitive cells in a finger
tip is large (> 60000) due to extensive branching near the
skin surface, an array of 600 or so points covering an area of
5 sq. cm. (at 1 mm pitch) is effective. The target
specification is, therefore, an array of 20 x 20 elements of 1
mm pitch surface resolution and a force resolution of 10^{-2} N
to 10 N.

NATO ASI Series, Vol. F43
Sensors and Sensory Systems
for Advanced Robots
Edited by P. Dario
© Springer-Verlag Berlin Heidelberg 1988

Taction, vision and proximity are probably some of the most desirable features of a robot. If in taction one were to include shape recognition, torque, force slip and temperature, this presents a fairly challenging task, not only of fabrication and disposition of the various sensors but that of signal processing in a coordinated manner. The objectives of any research in this area must, therefore, be to try and separate the various measurements, carry out the signal conditioning as close to the measurement as possible, and then organise the signal processing to give a coordinated output.

3.   Development of shape recognition sensors:

It will be obvious from the previous two sections that the author has considered shape recognition as an essential but only one part of the process of "tactile sensing". However, his experiences with methods described elsewhere with conductive rubbers etc. were far from encouraging and he had to embark on a programme of developing his own array sensors. Mainly due to previous experience with capacitance sensors which can be constructed to have immunity from extraneous signals, temperature drift, humidity, dirt and vibrations, but more importantly lack of hysteresis, it was decided to explore configurations involving capacitance sensors. Although a coaxial capacitor has linear displacement - capacitance characteristics, the difficulties of providing connection to all the moving pins of an array of some 300 elements was considered as impractical. A configuration (Fig. 1) which required two fixed concentric conducting cylinders with a dielectric pin moving in the space between the cylinders was found to be fairly successful (MDCT) and a 4 x 4 array was constructed. Experience with this array led to one possible application (Fig. 2) in which the use of a 32 x 4 elements of 1 cm pitch arranged in a vernier fashion could be used on a conveyor belt to determine the position of parts of assembly to within 2 mm. This information is fed to the assembly robot further down the line giving a saving on assembly time.

The operation of each element of the sensor, i.e. the measurement of the capacitance, is dependent upon an amplifier and it is unrealistic to provide, say, 300 to 400 amplifiers. A novel method of scanning each node of the array with, say, only 20 amplifiers was, therefore, evolved (Fig. 3). In order to read the force applied to a node of the array and thus its displacement, it is necessary to measure the capacitance of the MDCT at that location. To reduce the number of circuits required, the approach adopted uses one low capacitance FET switch per MDCT with multiplexed excitation signals to allow well isolated transducer row selection. Column current is sensed by current amplifiers which feed an analogue multiplexer. This provides a selected column signal to a sample hold circuit. The ramp excitation is switched to the same row as the currently selected 'enable' FETs. A call address is first set up. This opens an excitation path to one row of the array and switches on the enable FETs in that row. The column address field selects the current amplifer output to be fed to the S/H circuit where its level is stored prior to digitisation. After a delay during which transients are allowed to decay, the excitation (ramp) generator is triggered. This provides a voltage signal with constant dv/dt to the selected row. The induced current is transduced to a voltage signal and sampled after a further settling delay. This is subsequently acquired and logged by the micro.

Although the method of construction proposed is capable of miniaturisation down to a 2 mm pitch, it was felt that less than this, i.e. the target specification of 1 mm pitch, the complexities of construction might make the costs rather high. Considering that applications such as edge or corner detection were being proposed and required the 1 mm resolution, another 4 x 4 array using a slightly different method of construction was fabricated (Fig. 4). This construction consists of two pcbs with through plated holes and dielectric coated pins moving in these holes. The movement of the pins changes the capacitive coupling between the two metal cylinders formed by the through plated holes. The return force for the pins is provided by compressed air on top of the array. The compressed air gives the added advantage of lubrication for the pins. An array of 20 x 15 pins spaced at 1 mm pitch has been constructed and appears to work satisfactorily.

4. Signal processing electronics:

With both forms of construction it will be necessary to mount the signal conditioning and signal processing electronics on the back of the sensors. Here, of course, one comes back full circle to examining the philosophy of sensory requirements of robots. It would be a very expensive exercise to design integrated electronics for just shape recognition. As indicated in Section 2, force and torque are also important constituents of a "tactile sensor", alongside shape recognition. It is the author's belief that some practical experience is now necessary with implementation of the shape recognition sensors. It will then be necessary to attempt to classify operations and their sensory requirements. After a period of study and research it is hoped to arrive at a reasonable classification and thus characterise the signal processing and signal conditioning. The LSI/VLSI integration of electronics can then proceed, at least on an experimental basis.

5. Conclusions:

It is shown in this paper that shape recognition, although an important component of tactile sensing, must be supplemented by other measurements. These will be determined more by the application in hand and the separate measurements will need to be coordinated as close to the robot gripper performing assembly tasks to produce the requisite response. The advantages of complementing vision, for example, in assembly of various measurements, including shape, could be many but much further work is necessary to establish these advantages. The first step in producing reliable shape recognition sensors has now been implemented.

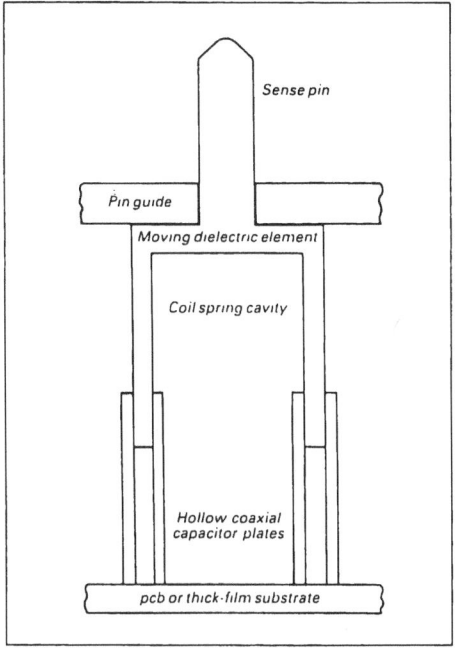

Figure 1 Suspension mechanism

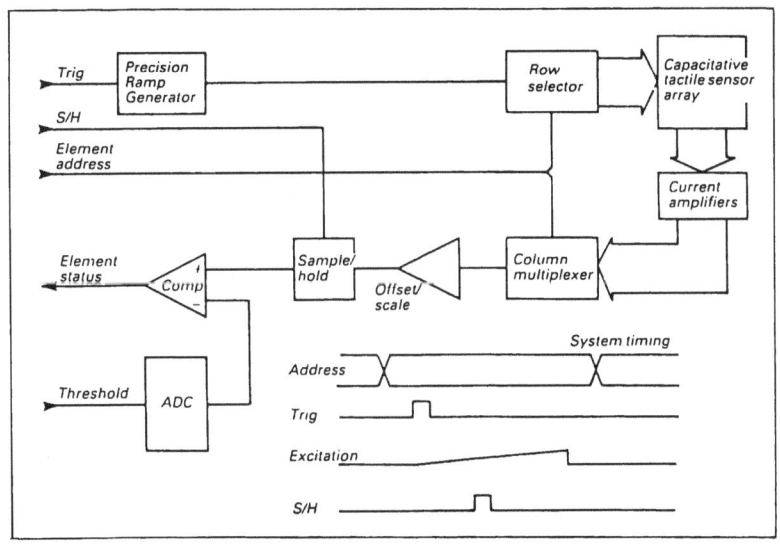

Figure 2 MDCT array scanner/data
acquisition unit block
diagram

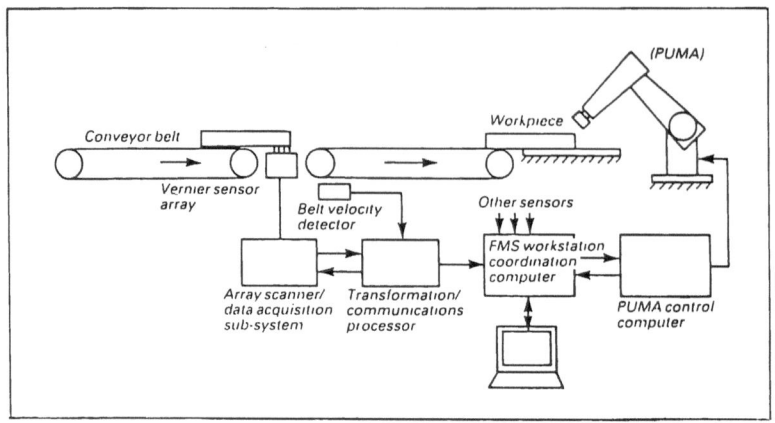

Figure 3 FMS workstation system with 'Vernier'
array conveyor sensor

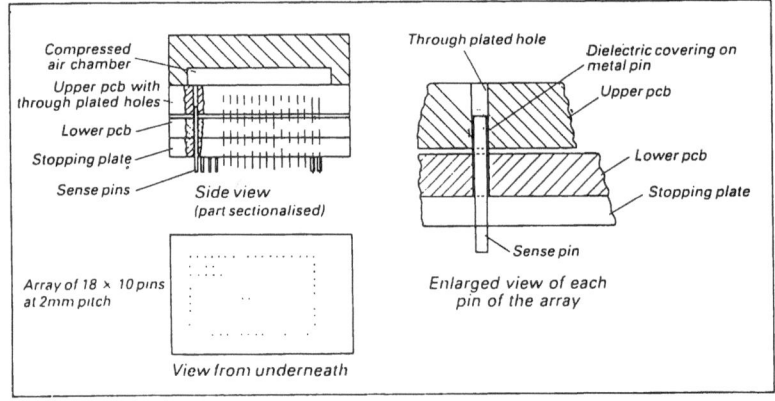

Figure 4 Mechanical construction of 18 x 10
orthogonal matrix array tactile
sensor using capacitance method

MULTISENSORY FEEDBACK INCLUDING COOPERATIVE ROBOTS

G. HIRZINGER, J. DIETRICH

DFVLR Institute for Flight Systems Dynamics
D-8031 Wessling, West Germany

ABSTRACT

For on-line sensory feedback in robotics several major prerequisites are necessary:

a)   the design of small, cheap gripper sensors with decentralized signal processing

b)   the availability of systematic procedures for generating adequate feedback structures

c)   tight man-machine interfaces

This paper outlines the recent sensory developments in the DFVLR robotics lab, as are different force-torque sensors, laser range finders, inductive sensors and sensor balls for robot and 3D-computer-grafic control. With the example of our preferred multisensory arrangement implying vision, range sensing, force-torque and speech, the fine-motion planning and path generation techniques as developed in our lab are discussed. The special case of a two-arm cooperative robot using two force-torque-sensors is treated in more detail. An outlook to space robotics in terms of an experiment proposal is given.

INTRODUCTION

On-line sensory feedback gains increasing importance in future intelligent robots. Actually quite a number of "low-dimensional" sensors for integration into grippers and tools have been developed in the last years that allow quasi-continuous feedback at rates of 20 msec and less. Thus uncertainties and tolerances in the robot's mechanics or its relation to the environment are balanced in real-time similar to the behaviour of the light human arm.

NATO ASI Series, Vol. F43
Sensors and Sensory Systems
for Advanced Robots
Edited by P. Dario
© Springer-Verlag Berlin Heidelberg 1988

On the other hand there are TV-object-recognition and scene analysis systems available today that in general need more than 100 msec for part recognition in a scene with several parts and thus do not allow dynamic feedback. However we feel that this is not necessary, if the output of the object recognition system is just used as a rough information source for global path planning, while in the proximity of e.g. an object to be grasped a handcamera or optical fibers in the gripper present a scene that is slightly displaced or rotated compared to the nominal scene the robot "saw" in a learning phase. There is good evidence that these fine corrections, often necessary due to the limited precision of robots and "global" TV-systems are computable in real-time and amenable to the same fine-motion algoritms as used for "low-dimensional" sensors. However we can not discuss these vision techniques here which are of course different from any inefficient correlation methods.

Obviously the development of "smart" sensors is only one step towards more intelligent robots. The other one is the development of efficient techniques for programming and automatic execution of complicated sensor-based tasks.

Our sensor-based fine-motion-planning concept called "sensor programming" has been outlined in different papers (e.g. /6, 7, 12/). It aims not only at a unified sensory feedback structure but also at an efficient method for programming sensor-controlled tasks in a very short time so that the robot may repeat them immediately in a changing environment. Fig. 1 shows a basic configuration of our programming system using fixed and hand-mounted cameras, different gripper sensors, sensor ball (see next section) and voice input-output as well as (optionally) 3D-computer graphics. The main features of our approach - briefly recalled - are:
"Rudimentary" teach commands Δp_T are derived either on-line

from a human teacher operating a sensor ball (in simple tasks) or from a path-generator connecting preprogrammed points (e.g. by leading the robot to these relevant points via the sensor-ball as discussed later). They are interpreted in a dual way as "force or positional commands" in case of sensory contact. They are projected into the "position controlled subspace" to yield the position feedforward control component and into the orthogonal sensor controlled subspace to either be neglected in case of nominal sensory patterns or to be counterbalanced with the sensor data to provide compliant motion with arbitrary stiffness (Fig. 2). Sensor data, too, are interpreted in a dual way as (pseudo-)force or positional information. A main goal of our work was to automate the generation of "C-frame"-subspaces (introduced in /4, 5/) in each point of a trajectory using information on the type of task and sensor data.

This paper focusses on two major topics

a) a survey of "low-dimensional" sensors as developed in the DFVLR-lab,

b) a discussion of how our fine-motion-planning concept applies to cooperative two-arm robots, teleoperators and the
learning of complicated multisensory tasks.

THE ROBOT AS A SAMPLED DATA CONTROL SYSTEM

The notation "continuous feedback of sensory signals" has to
be interpreted in a way that sensory information in terms of
position or force is sampled with a preferably fixed sampling
rate (typically between 10 and 60 msec) and processed in the
robot control system, e.g. on the level of the "interpolation"
cycle. It is thus convenient to treat the robot as a sampled
data control system independent of the fact that in such a
system several processors have to collaborate in order to keep
processing times reasonably low. Fig. 3 shows a possible structure of sensory integration into a modern robot control
system. The different blocks may be realized by separate processors.

As in each control system it is important to know the plant's
dynamics. A robot is a dynamical system that cannot respond to
a command in zero time but in the simplest case reacts with a
simple time constant. In fig. 4 and 5 the elementary structure
of force feedback is shown. Following fig. 4 the robot is
commanded to a position vector $\underline{p}$ behind a "wall" or surface
$\underline{p}_w$; thereby it is slightly bended, as of course it cannot
penetrate the wall, and via its stiffness $\underline{S}$ its exerts a
force-vector $\underline{f}$ *).
If one constructs the control loop as shown in fig. 5 and
feeds back the error $\Delta\underline{f}$ with respect to a nominal vector $\underline{f}_{nom}$
via simple proportional gains yielding incremental motion
commands $\delta\underline{p}_{com}$, then we may be sure that the robot will
permanently "knock" onto the wall, as in general we need some
"lead compensation" to stabilize a system with non-negligible
time-delays. From our experience an additional differential
term may induce remarkable improvements.

The transformation of a force control problem into a position
control problem is distinguished by its feasibility using
commercial robots. A high gain $\underline{S}$ when working with a very
stiff robot can be compensated by correspondingly low control
gains, however the limits are in the restricted accuracy of
the position measuring devices in the arm joints. If the
smallest executable motion increment is $\delta\underline{p}_{min}$, then via the
robot's stiffness $\underline{S}$ a minimal force variation $\delta\underline{f} = \underline{S}\,\Delta\underline{p}_{min}$ is
achievable. Remedy is attained only by proceeding to more
precise angular measurement devices or the direct transfer of

*) Note: In this paper "force" stands for force and torque,
        "position" stands for position and orientation.

sensed wrist force-errors into motor torques. This direct
motor torque control by the robot user is possible only in
laboratories until now and - of course - does not solve the
problem, that the robot may destroy its environment when sud-
denly hitting it due to its mechanical inertia.

Furthermore it is important to realize that the simple equa-
tion $\underline{f} = J^T \underline{j}$, (where J is the Jacobion and $\underline{j}$ the torque vector
generated in the joints, while $\underline{f}$ again is the force-torque
vector at the end-effector) is only a static relationship. If
the robot is not perfectly stiff, small arm motions have to
occur for exerting external forces and then we have to solve
the complete set of dynamical equations. Finally we feel that
direct force feedback to the joint torques is very problematic
with robots in the gears and harmonic drives of which up to 40
% of notes current torques get lost due to friction.

Essentially less problematic in general is the feedback problem with non-tactile sensors, which by nature measure position
differences.

A somewhat more detailed representation of a sensor feedback
loop is given in fig. 6. The relative position $\underline{p}_{rel}$ between
robot $\underline{p}$ and environment $\underline{p}_{env}$ is expressed via some force
vector $^H\underline{f}_{sens}$ or pseudo-force-vector in case of non-tactile
sensing. The "external" control block yields commanded posi-
tion increments $\delta \underline{p}_{com}$, or as possible in the special case of a
force sensor, motor torque changes $\Delta \underline{m}$ (dashed line in fig. 6).

The cartesian increments are transformed ito joint angle increments $\Delta \underline{q}$ and fed into the internal joint control system. In
fig. 6 the structure of a decoupled feedback control law is
indicated, which for computation of the inverse model needs
derived values as joint accelerations etc.. As it makes sense
to let the internal joint control system run with a higher
sampling frequency than is used in the outer loops, we have
indicated for simplicity continuous variables q, $\dot{q}$, $\ddot{q}$. From
the author's opinion it should be the goal of the internal
joint control to provide the robot in all positions with a
cartesian decoupled, fast and sufficiently known (not necces-
sarily position-independent) linear dynamics. The design of an
external controller then becomes a standard problem of digital
control and thus is practicable for users that are no control
specialists, too.

The question of the sampling interval relevant in an external
sensory loop normally depends strongly on the time needed for
computing the coordinate transformation (typically 10 - 30
msec for present microprocessors). This comes clear when
realizing that the sensor values occur in the robot gripper
and so their interpretation in inertial space is based on a
knowledge of the hand-orientation as given by the hand-"frame"
or orientation matrix (fig. 6). Furthermore the sensor-derived

corrective commands have to be transformed into joint increments; i.e. the operations to be performed within an "interpolation cycle" for transforming a cartesian difference into a
joint angle difference, are needed for sensory corrections,
too. Thus it seems reasonable to run the sensor control program in parallel with the coordinate transformation synchronous
with the interpolation cycle as indicated in fig. 7. It does
not matter here whether we use an analytic or differential
procedure for calculating the joint increments Δq out of the
cartesian increments δp. Fig. 7 is to show that from sampling
the sensor signal to issuing the corresponding joint increments two interpolation cycles may elapse. It is important to
note that this dead time has not been reduced by the parallel
processing scheme of fig. 7, only the sampling frequency has
been increased.

In how far the cartesian interpolation cycle is sufficient for
sensory feedback cannot be answered generally. It would be
feasible - if necessary - to run the computation of e.g. the
Jacobian and its inverse in the background. Then the computation of hand-frame and the control algorithm itself might run
more fequently; however the coordinate transformation would be
updated less frequently and thus not alway refer to the newest
joint positions. Taking into account that sensory feedback
normally means modest motion speed and that the principle of
sensory feedback implies the instantaneous correction of
errors this seems a feasible strategy. Together with faster
micorprocessors it should be possible in the near future to
get down from the presently realized sampling and interpolation cycles (28 msec with PUMA and VAL2 language, 30 msec in
an early DFVLR robot control system) to 10 msec and less, e.g.
synchronous with the socalled "fine interpolation" cycle on
the joint level. Whether this is really necessary depends on
the special control task. If one keeps to the idea of a robot
as just an inertia to be moved around with motors of restricted torque, then representative time constants of 80 msec and
more are prevailing and sampling rates of 20 msec are fast
enough. In the early years of digital control people have
often neglected that increasing the sampling rate up to more
than four or five times the dominant time constant does not
really improve system performance.

However if the robot has mechanical contact with its environment and the dynamics of the motor torques lying in the msec-
range are essential e.g. in deburring applications, then
higher sampling frequencies seem inevitable.

Fig. 7 also indicates that the dead time T_t in the feedback
system "sensor controlled robot" not only depends on the computation times but also on the duration of sensor preprocessing (e.g. computation of forces/torques out of voltages), and
of course on the data transmission time. To the author serial
links with baud rates of 9600 seem inadequate, as just a
single 6D-force-torque-sensor needs 25 -30 msec transmission
time for providing three forces and three torques in real

format. In contrast the preprocessing times of low-dimensional sensors are normally below 5 msec and thus contribute to the overall dead time mainly in case of very fast robots.

While in the past we favoured parallel sensory links, our proposal using e.g. the new sensor generation as developped in our lab (see next section) is a high-speed serial bus system connecting different types of sensors with only 2 wires, thus reducing also the bundle of cables normally coming with sensors in the robot gripper. Fig. 8 indicates that even the joint controllers mounted directly at the joints might be connected to such a serial bus. We estimate a 6D-force-torque-measurement including computation and transmission in the range of 2-3 msec, the same for an array of about 10 range finders in and at the robot gripper.

THE DFVLR-SENSOR DEVELOPMENTS

Although on a first glance items like "the price of a sensor" do not seem to have anything to do with scientific work, it is - after a decade of exciting and yet sometimes frustrating work in the field of sensor development - our dedicated opinion that it is a real challenge for the research community to invent cheap, robust and yet precise sensors without "black boxes" (i.e. all electronics including computations and bus interface inside) to help sensors "conquer" the robot scene in a much faster way than in the past. This idea has strongly characterized our recent developments that are presented in the sequel.

a) Force-torque sensors for six components

The DFVLR-force-torque-sensors are based on strain-gauge-measurements. The original mechanical design is shown in fig. 9. It has been commercially available for a couple of years now. Nevertheless there were recent developments in our lab that have aimed at the following goals

1) to integrate all the electronics (including the single-chip microprocessor that computes forces/torques out of voltages) in the sensor.

2) to find other mechanical configurations that

- are of small size
- have an integrated overload protection
- are fully symmetric
- are easy to manufacture
- allow for simple and precise mounting of strain-gauges (prewired on carrier-foils)
- provide full-bridges
- show up high linearity and good decoupling

We found an arrangement resembling a double-maltese-cross (see fig. 10 and /10/) that seems to have all of these features.

The elastic deformations are measured by 20 strain-gauges, paired in half-bridges. Each strain component causes equivalent deformations in at least 4 spokes, so that there are always two half-bridges to make up a full-bridge. Thus we arrive at a double signal-to-noise-ratio. We evaluate 7 voltages (10 half-bridges resulting in 7 full-bridges). In our first experiments linearity and de-coupling were extremely high, so that maximal load in one component had no measurable influence on the others.

For fabricating the sensor it is very helpful that the gauges are in only four planes. So one may use carrier foils with five prewired gauges each and a center hole for precise mounting. This makes the sensor more precise and cheaper - a strong argument for a wider use of sensors. Furthermore the sensor's overload protection consists of only a few bolts adding nearly no additional weight.

b)   The DFVLR-sensorballs

For a very "natural" 6-degree-of-freedom-control of robots (and 3D-computergraphic objects) with only one human hand two basic types of a sensor-ball were develo-ped. Fig. 11 shows the first version with the original force-torque-sensor. From the forces translational com-mands and from the torques rotational commands are deri-ved for control of any object that can move in up to 6 degrees of freedom.

A new version of the sensor-ball developed in our lab works fully optical (fig. 12). It transforms forces/torqu-es into translational and rotational displacements Δx, Δy, Δz and $\Delta\phi z$, $\Delta\phi y$, $\Delta\phi z$ which are measured. Its design was based on several requirements:

- simple mechanics
- no calibration
- simple electronics
- no lenses
- good decoupling between the 6 components
- resolution ≈ 1 %
- low cost

The basic measuring arrangement (fig. 12) is composed of a LED, a slit and perpendicular to it a linear PSD which is mobile against the remaining system. Six of such systems (rotated by 60 degrees each) are mounted in a plane, whereby the slits alternatingly are vertical and parallel to the plane. The ring with PSD's is fixed inside the ball and connected via springs with the LED-slit-basis. The springs bring the ball back to neutral position when no forces/torques are exerted.

There exists a particularly simple and unique transformation from PSD-signals U_1 ... U_6 to the unknown displacements.

In our experiments this arrangement showed up excellent performance and the manufacturing costs are very low.

c) Laser-range-finders

New range-measuring sensors were developed /14/, to meet the following requirements.

- large measuring range (0-500mm)
- very small size for integration into robot gripper or tools
- high precision particularly at low distance
- good spatial resolution of the measuring spot
- independence of slant angle and surface
- low cost

The principle used is laser-triangulation (fig. 13). With a laser diode and focussing elements we generate a fine laser beam that is reflected from a surface and is received via focussing lenses on a position sensitive photodetector (PSD). The PSD is mounted in a certain angle with respect to the optical axis to provide a well-focussed image of the spot. We use a pulsed laser beam to enable the suppression of disturbing light from the surroundings in the signal processing unit.
In the sensor head only the laserdiode with control electronics for temperature-independent operation is integrated, together with the PSD including the optical components and two transimpedance-amplifiers that transform detector currents into two voltages U_{ML1} and U_{ML2}. The computed voltage

$$U_L = (U_{ML1} - U_{ML2}) / (U_{ML1} + U_{ML2})$$

is proportional to the reflected light spot's position. The quotient of this equation is formed very accurately by a new analog-circuitry (denominator range 1 - 1000 compared to a range of 1 to 50 with hither to existing analog-dividers).

A main problem was to design a control system that adapts the light transmitter's intensity depending on the reflected light intensity. The goal was to ensure accurate operation for extremely different surfaces and a wide range of slant angles. A nonlinear control system was designed that allows to vary the emitted power in a range of 1 to 10000 within 10 μsec. This enables the sensor to measure surfaces with strongly and quickly changing reflection characteristics in a reliable, precise and fast way. For short distances (0 - 25 mm) a four-sensor-array was integrated into small robot fingers. So with eight beams in a two-finger gripper we got a nice sensor system for precise grasping of objects.

d)   Inductive 3-D sensor

For determining the position of holes, screws or bolts in metallic plates an inductively measuring sensor system was developed /13/. It allows to measure the three cartesian displacements X, Y (planar to the surface) and Z (vertical to the surface) between sensor center and hole or screw center. The sensorhead consists of five coils arranged as drawn in fig. 14. The center coil serves as transmitter while the four receiver coils are arranged in a concentric and symmetric way around it. The measuring principle is based on the asymmetric deformations of the magnetic field, caused by conducting surfaces in the proximity of the sensor and picked up by the receiver coils. As an interesting matter of fact, by extensive experiments and a least squares approximation we found the equations

$$Z = A \cdot \ln^{(B/Uz)}$$
$$X = D \cdot U_x \cdot e^{(Z/C)}$$
$$Y = D \cdot U_y \cdot e^{(Z/C)}$$

to fit very accurately. Ux, Uy, Uz are three voltages resulting from the preprocessing scheme fig. 15 and A, B, D, and C are constants. The sensor was successfully applied for inserting rubber plugs into car bodies.

PROGRAMMING OF MULTISENSORY TASKS

In this paragraph the fast programming of multisensory tasks e.g. for precise grasping and assembly of parts is pointed out.

The learning-phase is based on leading the robot through the task from point to point via the sensor ball. When using a fixed camera the operator typically is asked (using voice output) to present the objects or work pieces in all possible kinds of support and show the robot the corresponding points where to move e.g. for gripping. In the execution phase the path planning system collocates the appropriate sequence of points due to the actual support situations, modifies them by the registrated (approximate) positions and orientations thus finding the rudimentary path, which is not sufficient e.g. for precise grasping.
When teaching the different points, one or several of the different "local" gripper sensors may be activated and the robot registrates and stores the sensory patterns together with "a priori" informations (fig. 16):

a) whether these nominal sensory patterns have to be refound
    in an approach process or whether they have to be kept
    constant along the subsequent path segment (in this
    latter case we prefer to keep the projected positon-feed-
    forward component constant in its magnitude).

b) the type and dimension of sensory information as well as
    the type of sensorcontrolled subspace (e.g. dimension,
    orientation in hand coordinates - if it is fixed - or the
    rule how to construct it automatically when it varies
    e.g. in case of crank turning, see /5, 7/).

Parallel to storing the relevant nominal sensory patterns the
robot must know the mapping that relates pattern errors (later
in the automatic mode) to estimated positional/orientational
errors (fig. 16). In simple cases - e.g. in case of a one-
dimensional range sensor - the mapping may be given by the
calibration characteristics; or it is part of extensive preprocessing e.g. with a hand-camera. In other cases - e.g. when an
array of range sensors in the robot finger as constructed in
our lab generates a pattern of 6 or 8 range data - a training
generating an associative memory may be the appropriate technique (see /8, 9/).To each of the stored robot points and
sensor systems active at that point such a mapping is assigned.

Note that the absolute measurement capabilities are not important here, as in the automatic repetition phase the robot
recalls the originally measured value above the corresponding
surface and the errors are servoed to zero.

Let us give an example for the automatic path generation after
programming. In the proximity of a preprogrammed point (possibly modified by a global vision or scene analysis system) we
assume a hand camera and an array of range finders to become
active (fig. 17). The 3D-subspace $\underline{C}_{f1}$ controllable via the
range finders might be z, ϕ_x, ϕ_y while the subspace $\underline{C}_{f2}$
controllable by the hand camera in case of small differences
might be up to 6-dimensional using advanced real-time image
processing techniques. Then the sensor-controlled subspace $\underline{C}_f$
is $\underline{C}_f = \underline{C}_{f1} \cup \underline{C}_{f2}$, while the estimated 6-dimensional positional error $\Delta\underline{p}$ may compute as

$$\Delta\tilde{\underline{p}} = E\ (\Delta\tilde{\underline{p}}_1,\ \Delta\tilde{\underline{p}}_2)$$

where $\Delta\tilde{\underline{p}}_1$ and $\Delta\tilde{\underline{p}}_2$ are the error estimates given by the two
sensor systems, while E is an overall estimator that takes
into account the different precision or reliabilities of
single estimates affecting the same dimension.

In this example the sensor-controlled subspaces are fixed in
hand coordinates while in the turning crank case the subspace

varies with the actual sensordata (i.e. the force-controlled subspace, a 2D-plane, may be generated by a filtering algorithm from subsequent forces).

COOPERATIVE TWO-ARM ROBOTS

Let us now investigate a two-arm-robot, and have a closer look at its special fine-motion planning problems.

Fig. 18 shows the configuration of experiments performed in our lab. The operator is using a sensor-ball that controls either robot 1 (pressing some key 1) or robot 2 (pressing some key 2) or both robots simultaneously when they have both grasped an object (pressing both keys); we call this the "cooperative mode", where the sensorball commands automatically refer to the center point $\underline{c}$ between the two robots. Let us denote by $\underline{T}$ the transformation from a vector in the robot 2 coordinate frame to a vector in the robot 1 system. Then initially we move the robots to positions where we not only read their positions $\underline{x}_{10}$ and $\underline{x}_{20}$ in their own frames but also the connection vector $\underline{r}_0$ from robot 1 to robot 2 in frame 1. Then for all situations occurring later the actual connection vector $\underline{r}$ is in frame 1

$$\underline{r} = \underline{T} (\underline{x}_2 - \underline{x}_{20}) + \underline{r}_0 - (\underline{x}_1 - \underline{x}_{10}) \tag{1}$$

Let the operator's rudimentary commands $\Delta\underline{p}_T$ be composed of translational components $\Delta\underline{x}$ and rotational components $\Delta\underline{\phi}$.

The rotational "rudimentary" commands $\Delta\underline{\phi}$ are immediately transferred to the supervisory control system, while the translational commands $\Delta\underline{x}$ have to be augmented by translations caused by rotating the robots around their center point $\underline{c}$.

So the overall rudimentary translational commands to the robots in their own frame are

$$\Delta\underline{x}_1 = \underline{r}/2 \times \Delta\underline{\phi} + \Delta\underline{x}$$

$$\Delta\underline{x}_2 = \underline{T}^{-1} (-\underline{r}/2 \times \Delta\underline{\phi} + \Delta\underline{x}) \tag{2}$$

The special problem with two-arm robots is that we need additional force control loops even if the grasped object has no contact with the environment. This is due to mechanical inaccuracies that cause internal forces and torques.

We investigated both alternatives, only one robot having a force control system, or both robots having force sensors, the supervisory processor closing two control loops. This two-sensor-case is the more interesting one and we focus on it in the sequel.

How are the different forces and torques distinguishable by the controller? Fig. 18 shows the gravity force vector $\underline{g}$ of the grasped object causing some torque $\underline{t}_g$ in each of the grippers (we assume the same compliance in both grippers). Let us assume that the grasped object has no contact with the environment, then the force-torque balance is

$$\underline{f}_{s1} = \underline{g}_1 + \underline{f} \tag{3}$$

$$\underline{f}_{s2} = \underline{g}_2 - \underline{f} \tag{4}$$

$$\underline{t}_{s1} = \underline{t}_g + \underline{t}_f + \underline{t} \tag{5}$$

$$\underline{t}_{s2} = \underline{t}_g + \underline{t}_f - \underline{t} \tag{6}$$

where $\underline{f}_{s1}$, $\underline{f}_{s2}$, $\underline{t}_{s1}$, $\underline{t}_{s2}$ are the sensed forces and torques in robot 1 and 2 (at the gripping points), $\underline{g}_1$ and $\underline{g}_2$ denote the forces due to the gravitational load, $\underline{f}$ is the internal force generating an internal torque $\underline{t}_f = {}^-1/2 \, (\underline{f} \times \underline{r})$ at each gripper, and $\underline{t}$ is an additional internal torque not based on reaction forces.

With both robots equipped with a force sensor it is indeed possible to immediately discriminate between the different influences.
From adding the sensed forces equ. (3 + 4) we get the gravity vector

$$\underline{g} = \underline{f}_{s1} + \underline{f}_{s2} \tag{7}$$

But we do not yet know the shortest connection $\ell \cdot \underline{z}$ between $\underline{r}$ and $\underline{g}$ where $\underline{z}$ is a unit vector perpendicular to $\underline{r}$ and $\underline{g}$, generated via the crossproduct $\underline{r} \times \underline{g}$ (its base point on $\underline{r}$ has the unknown distance r_ℓ from gripper 1). We split up $\underline{g}$ into a vector parallel to the connection line $\underline{r}$ and a vector $\underline{g}_v$ vertical to it, and we characterize the associated coordinate system by the indices r, v, r x v.
Then we proceed as follows:
By subtracting equ. 6 from equ. 5 we get the "pure" internal torque $\underline{t}$. Furthermore as there cannot be any force-generated torque $\underline{t}_{f,r}$ around $\underline{r}$ we may add equ. 5 and 6 to get for the r-component:

$$(\underline{t}_{s1} + \underline{t}_{s2})_r \;=\; 2 \cdot \underline{t}_{g,r} \;=\; \underline{g}_v \times (\ell \cdot \underline{z}) \tag{8}$$

So we have $\underline{t}_{g,r}$ and length ℓ.
Now with

$$2\,\underline{t}_{g,v} \;=\; \underline{g}_r \times (\ell \cdot \underline{z}) - \underline{g}_{1,rxv} \times \underline{r} \tag{9}$$

$$\underline{t}_{s1,v} \;=\; \underline{t}_{g,v} - \underline{f}_{rxv} \times \underline{r} + \underline{t}_v \tag{10}$$

$$\underline{f}_{s1,rxv} \;=\; \underline{g}_{1,rxv} + \underline{f}_{rxv} \tag{11}$$

We easily compute

$$\underline{t}_{g,v} = \underline{g}_r \times (\ell \cdot \underline{z}) - (\underline{f}_{s1,rxv}) \times \underline{r} - \underline{t}_{s1,v} + \underline{t}_v \tag{12}$$

As $\underline{t}_{g,rxv} = 0$ we have all "gravitational" torques. Thus we may compute the internal force-generated torque $\underline{t}_f$ from equ. 5 or 6. The next steps then are to compute $\underline{f}_{rxv}$ from $\underline{t}_{f,v}$ (via $\underline{r}$), $\underline{f}_v$ from $\underline{t}_{f,rxv}$ and $\underline{f}_r$ (by assuming $\underline{g}_{1r} = \underline{g}_r \cdot 0,5$ and applying equ. 1).

Finally we get $\underline{g}_{1,v}$ and $\underline{g}_{2,v}$ with known $\underline{f}_v$. This gives us r_ℓ, the last unknown parameter by

$$\underline{g}_{2,v} \cdot /r/ \;=\; -\,\underline{g}_v \cdot r_\ell \tag{13}$$

Thus we have all internal forces $\underline{f}$ and torques $\underline{t}$ and the supervisory controller can immediately (e.g. with position control based on a compliance model) react to reduce them to zero.

Another observation is quite interesting. From a first measurement we may immediately start the controller but we do not yet know the center of gravity in hand frame of e.g. robot 1. Yet we know via the parameters ℓ and r_ℓ its line of attack in hand frame and if we now move the object into a different orientation, we get a second line, that must intersect the first one just in the center of gravity. Of course iterative estimation procedures are immediately applicable for more precision. So moving the object first in free space allows the robot to determine all gravitational loads for any orientation a priori and by subtracting them from the measurements we may admit

external forces and yet find out the internal forces/torques
following the above algorithm. The external ones (simply replacing the gravitational loads in the above equations) determine the sensor-controlled subspace and have to be compared
with the rudimentary teach command projections. Thus we are
treating the two-arm robot just like a single-arm system, i.e.
sending both robots identical control commands in addition to
the internal force/torque corrections.

CONCLUSION AND OUTLOOK FOR SPACE ROBOTICS

The paper presents a review of the robot sensors and the
programming of sensorcontrolled tasks as developed in the
DFVLR-lab.

Apart from a number of laboratory experiments (e.g. assembly
operations including TV-object recognition force and range
sensing) there are now industrial applications with robots
that use the above sensors and methods in factories for example in deburring castings, inserting rubber plugs into car
bodies etc.

There is one feature of our sensory control structure that
seems worth to be reemphasized. If a human operator issues
"rudimentary" commands via the sensorball and the gripper
sensors are active in the refining feedback loop, then the
human operator is a kind of "supervisor" who performs advanced
teleoperation. The point here is that these supervisory commands (provided with additional information on the type of
task fixing the C-frames) may be slow and might be transmitted
via delay-lines as used in ground-spacecraft communication,
while the gripper sensory loops are fast, closed directly at
the robot and provide it with subtask autonomy ("macro-com-
mands"). In fact we have proposed a robotic experiment of this
type to fly with D2, the next "German" spacelab mission,
though at the moment it is not clear when this might be. Fig.
19 shows the proposed configuration with stereoscopic cameras
(fixed and hand-integrated), force and proximity-sensing,
voice input output and 3D-computergraphics for displaying sen-
sorinformation and (for the teleoperator on earth) precompen-
sating transmission dead times in the display of the robot's
kinematics.

REFERENCES

/1/ Hirzinger, G.: Robot-Teaching via Force-Torque-Sensors. Sixth European Meeting on Cybernetics and Systems Research, EMCSR'82, Vienna, April 13-16, 1982.

/2/ Salisbury, J.: Active Stiffness Control of a Manipulator. 19th IEEE Conference on Decision and Control, Albuquerque, Dec. 1980.

/3/ Whitney, D.E.: Force Feedback Control of Manipulator Fine Motions. Journal of Dynamic Systems Measurement and Control, June 1977, 91-97.

/4/ Craig, J.J. and Raibert, M.H.: Hybrid Position/Force Control of Manipulators. Transaction of the ASME, Journal of Dynamic Systems, Measurements and Control, Vol. 102, (6/1982) 126-132.

/5/ Mason, M.T.: Compliance and Force Control for Computer Controlled Manipulators, IEEE Trans. on Systems, Man and Cybernetics, Vol. SMC-11,. No. 6, (1981) 418-432.

/6/ Hirzinger, G.: Robot Learning and Teach-In Based on Sensory Feedback. Proc. 3rd ISRR, Oct. 85, Paris.

/7/ Hirzinger, G.: Sensory Feedback in the External Loop. Proc. IFAC Symposium on Robot Control - SYROCO'85. Nov. 85, Barcelona.

/8/ Albus, J.S.: A New Approach to Manipulator Control. The Cerebellar Model Articulation Controller (MAC), Journal of Dynamic Systems Measurement and Control, Vol. 97 (1975), No. 3, pp. 228-233.

/9/ Arbter, K., Heindl, J., Hirzinger, G., Landzettel, K., Lange, F.: New Techniques for Teach-In, Acceleration and Learning in Sensor-Controlled Robots. IFAC'84 9th World Congress, Budapest, Hungary, July 1984.

/10/ Bejczy, A.K.: Smart sensors for smart hands. AIAA paper No. 78-1714, AIAA/NASA Conf. on "Smart" Remote Sensors, Hampton, Virginia. 14-16 Nov. 1978.

/11/ Schmieder, L.: Mettin, F., Vilgertshofer, G.: Kraft-Drehmoment-Fuehler. Patent EO 1L1/22.

/12/ Hirzinger, G. and Landzettel, K.: Sensory Feedback Structures for Robots with supervised Learning. IEEE Int. Conf. on Robotics and Automation, St. Louis/Missouri, March 1985.

/13/ Dietrich, J. and Gutjahr, J.: Inductive Sensor ... Patent P 34 20 330.3-52, ert. 5.12.85.

/14/ Dietrich, J.: Optisch-elektronischer Entfernungsmesser. Patent P 35 02 634.0, pending 26.1.85.

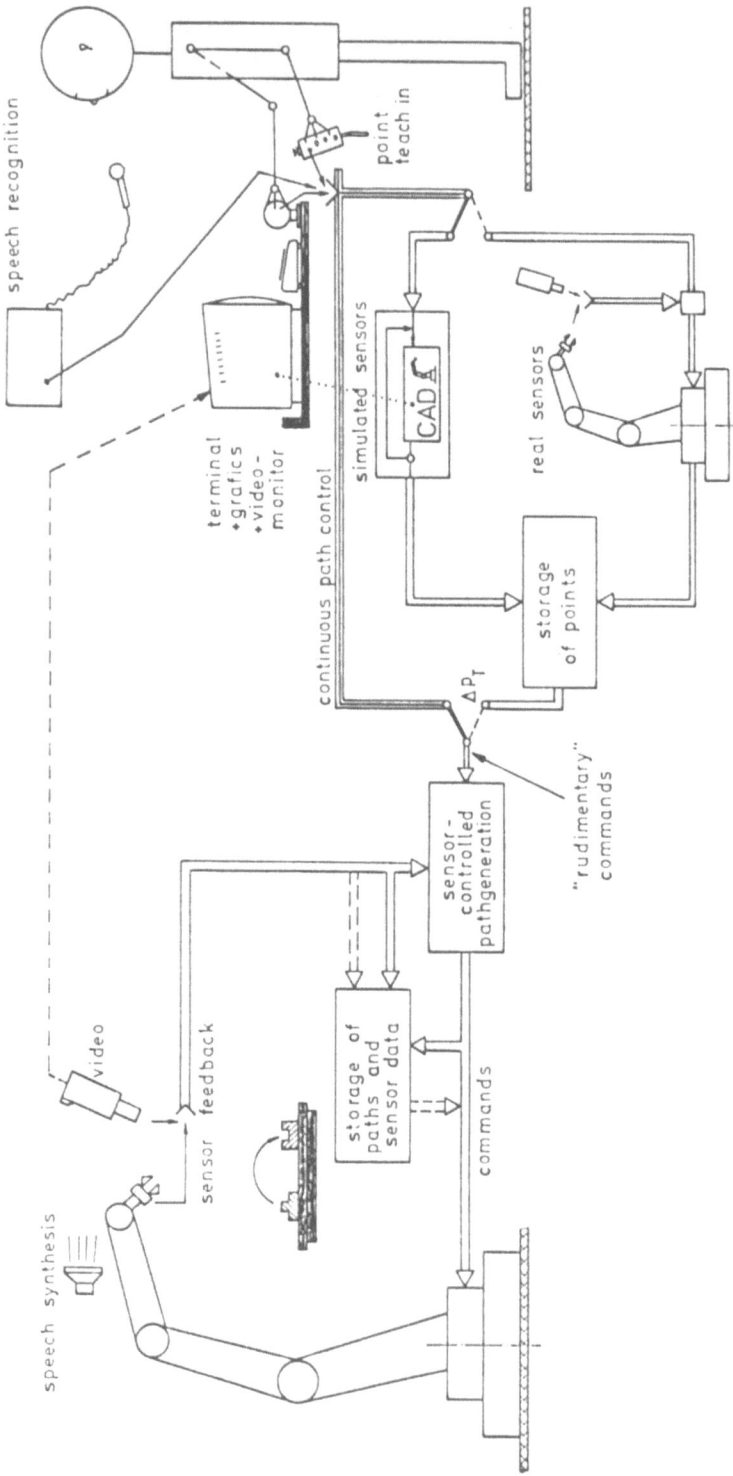

Fig. 1 Overall loop structures for the "sensor-programming concept"

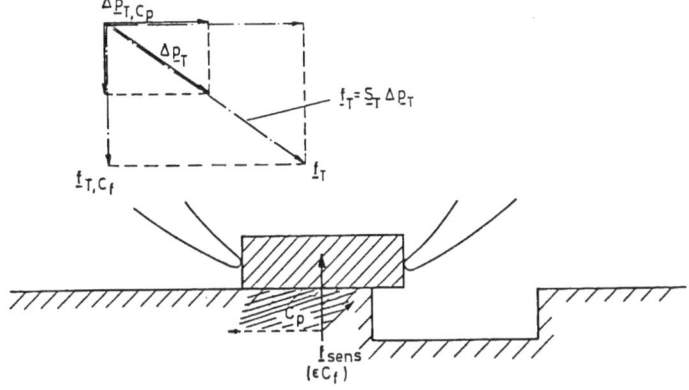

<u>Fig. 2</u> Projection of a teacher's command $\Delta \underline{p}_T$ and its "dual" force $\underline{f}_T$ (related by an artificial "teacher's stiffness" $\underline{S_T}$) into the sensor-controlled subspace C_f and the position controlled subspace C_p

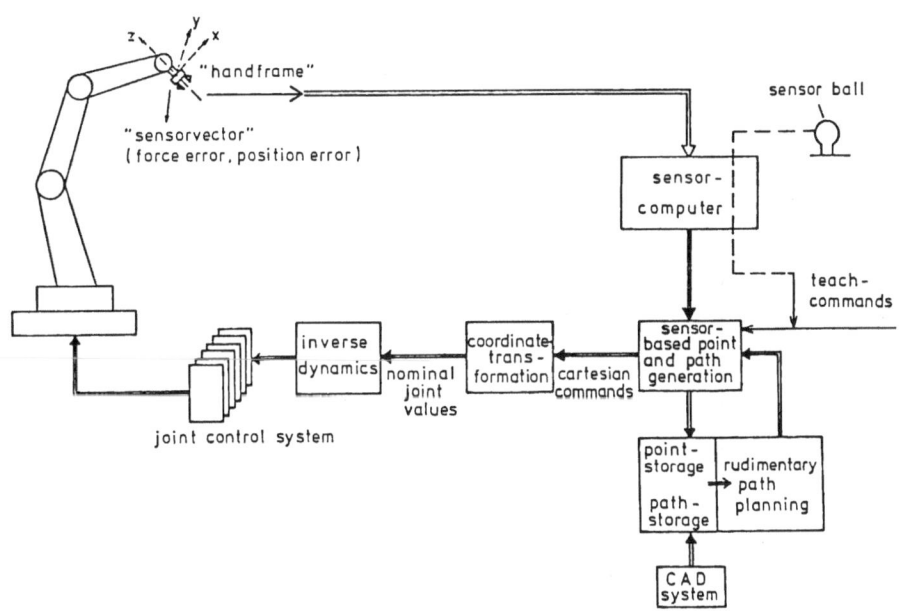

<u>Fig. 3</u> Blockstructure of a sensor-based robot control system

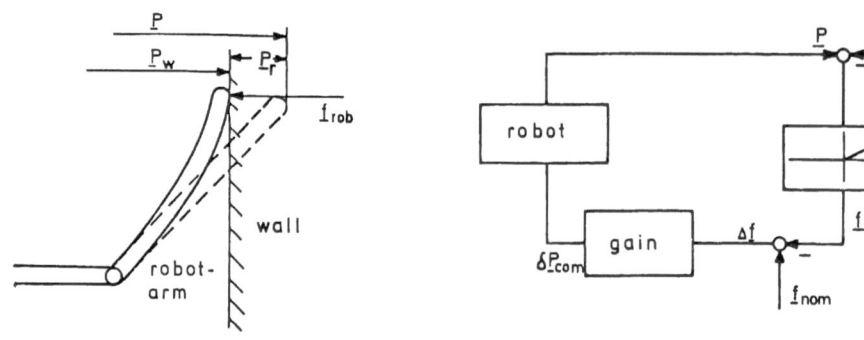

Fig. 4 Mechanical defor-
mation of a robot
arm when hitting
an obstacle

Fig. 5 The simplest force control loop for fig. 4

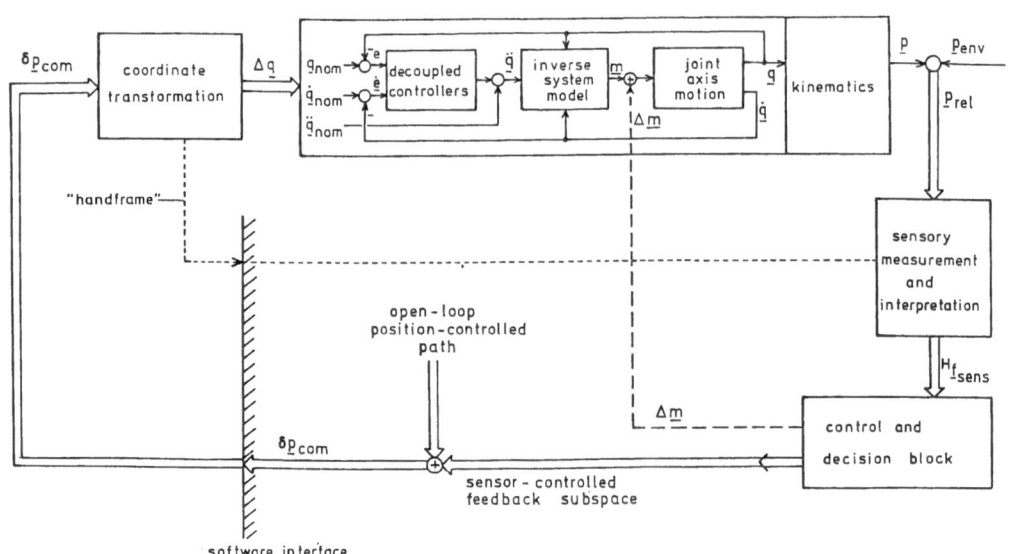

Fig. 6 Separation of external and internal control loops

441

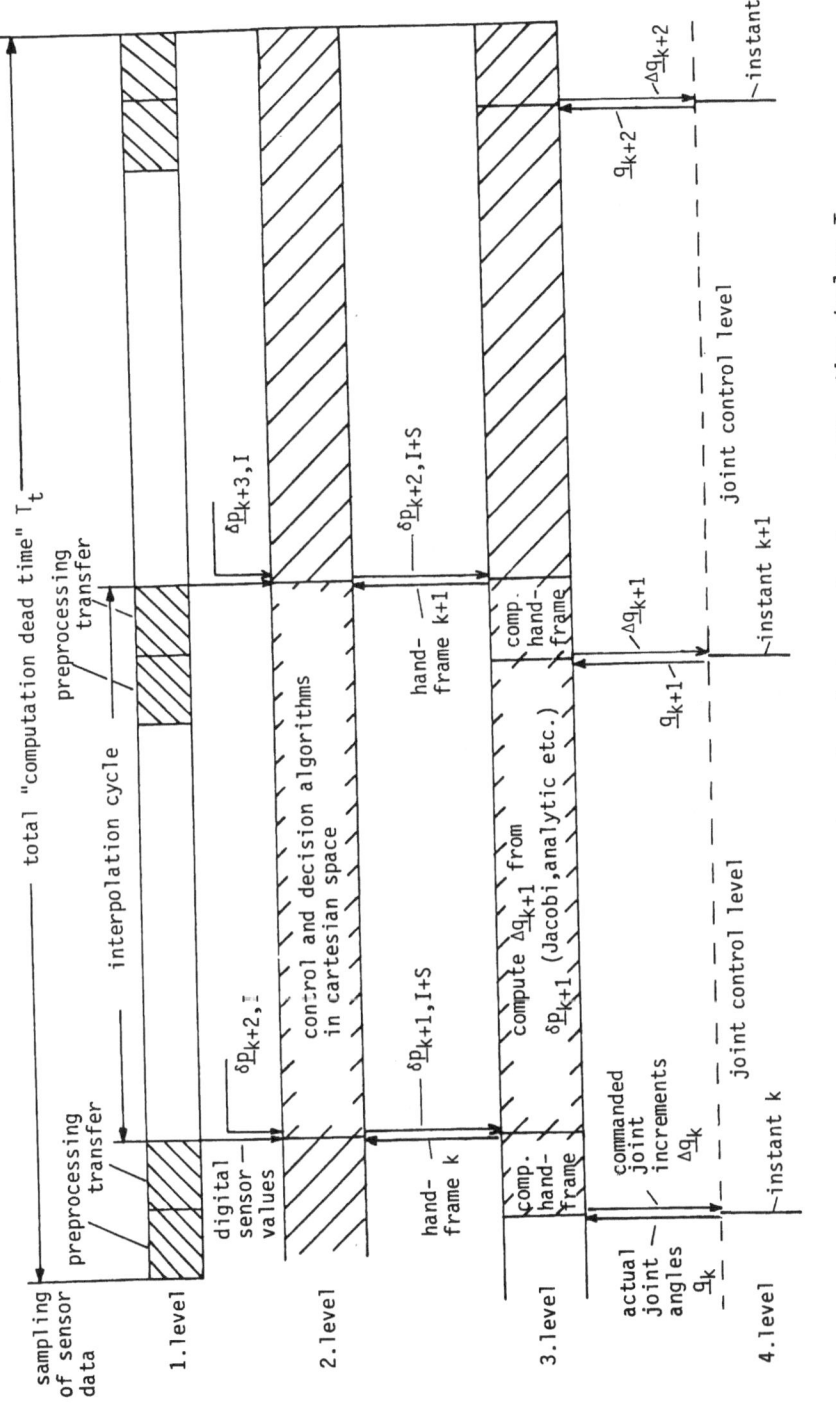

Fig. 7 Level hierarchy in a sensor-based robot control system, the index I denotes "from path interpolator", S "from sensor"

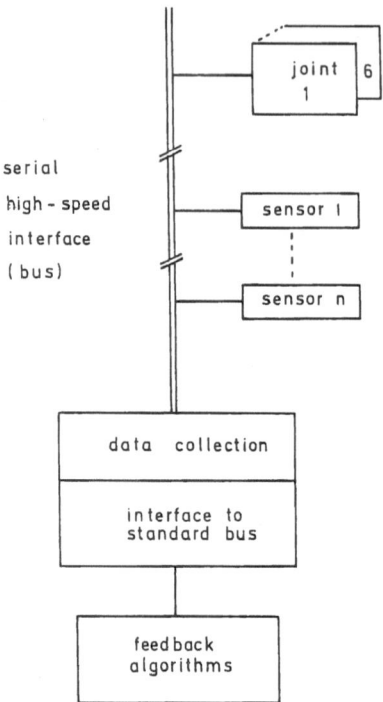

<u>Fig. 8</u> Connecting sensors and dezentralized joint
controllers via serial bus systems

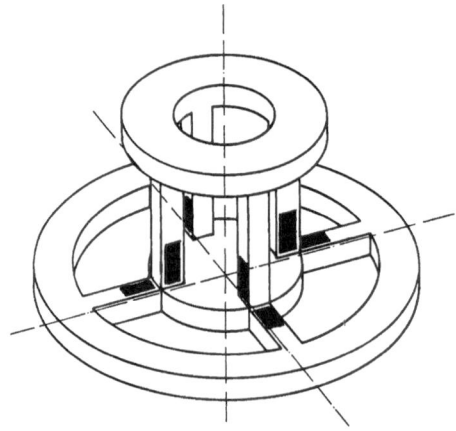

<u>Fig. 9</u> The original DFVLR-force-torque-sensor

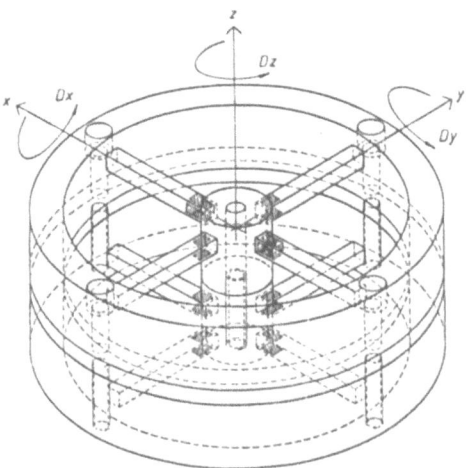

Fig. 10 A new force-torque-sensor arrangement

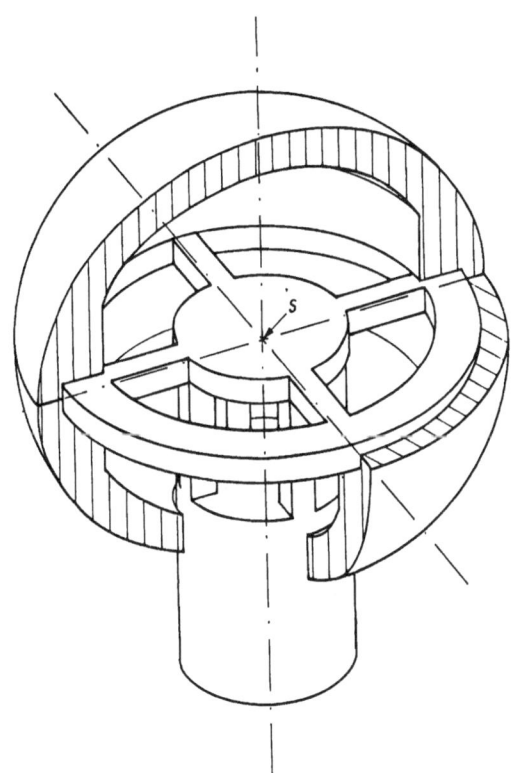

Fig. 11 The original force-torque sensor-ball (6D-joystick)

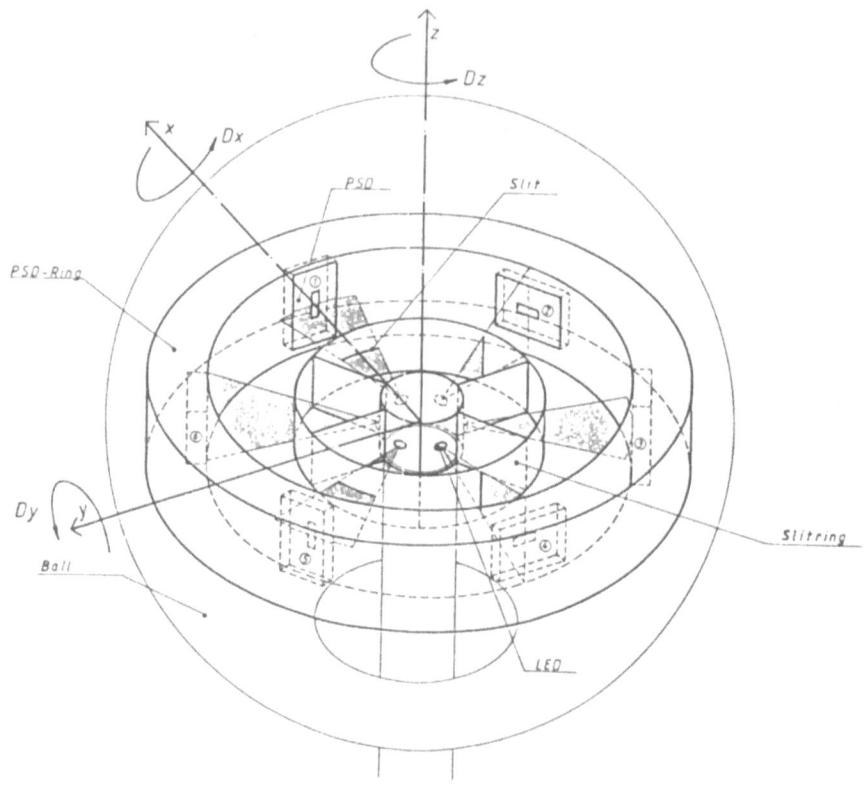

Fig. 12   Optical sensor ball

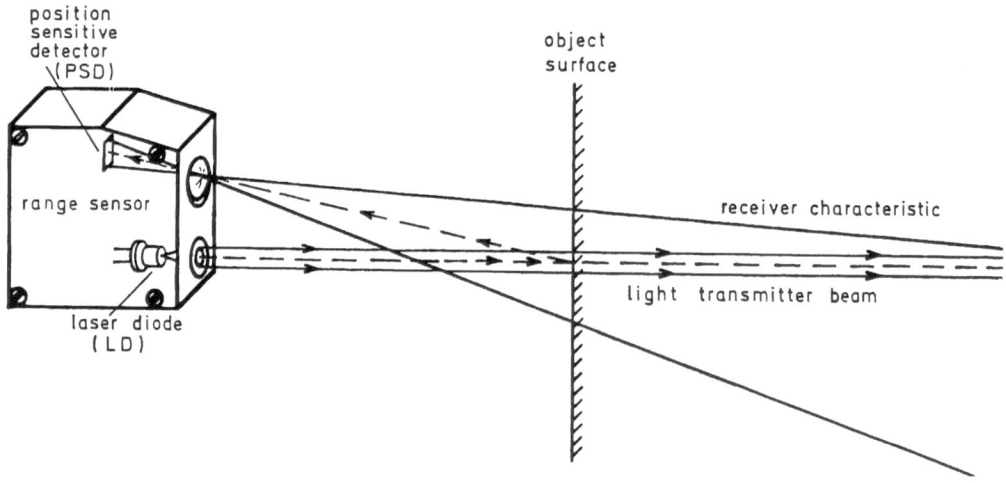

Fig. 13   Laser range finder based on triangulation

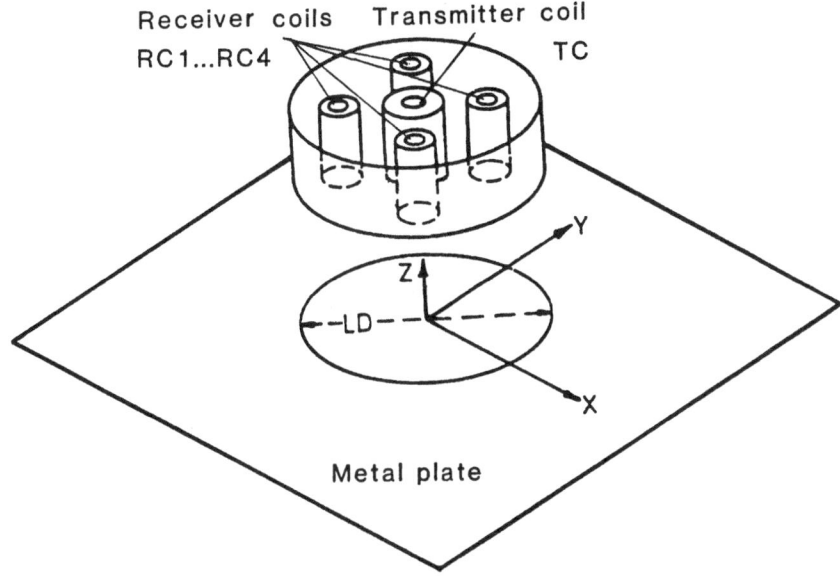

Fig. 14   Inductive 3D-sensor

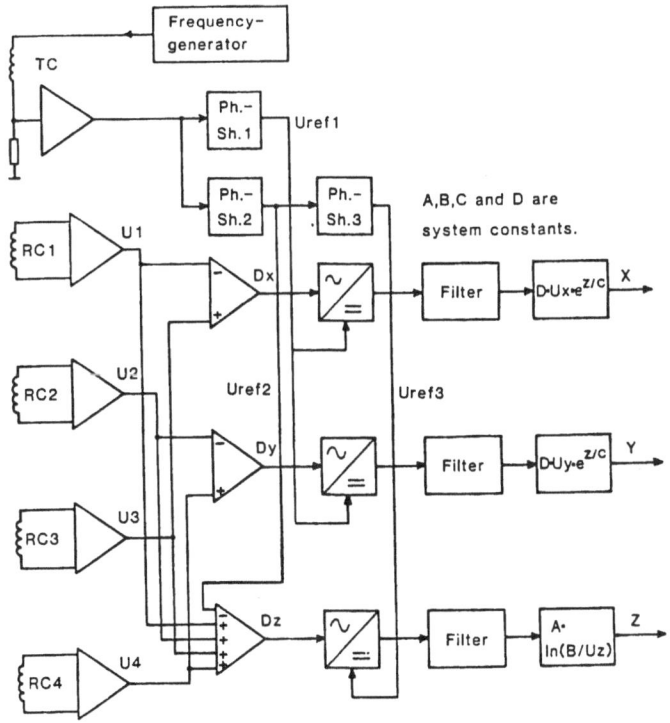

Fig. 15   The inductive sensor's preprocessing scheme

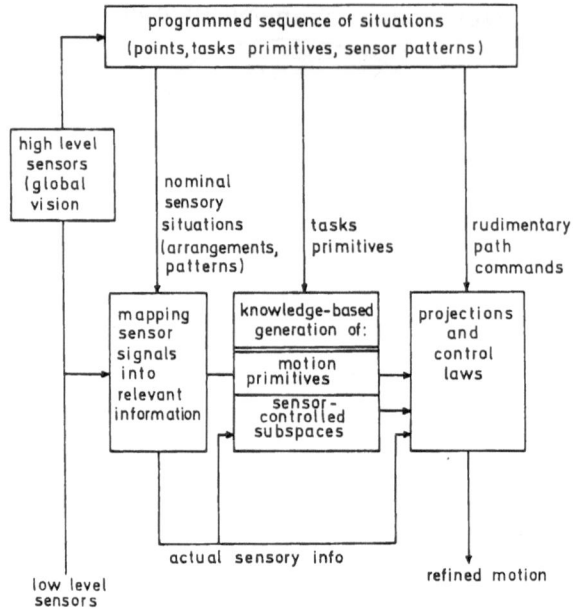

<u>Fig. 16</u> The block structure of the pattern matching concept
has some similarity with one level in J. ALBUS's
sensor processing hierarchy /8/. The dashed arrows
refer to the learning phase.

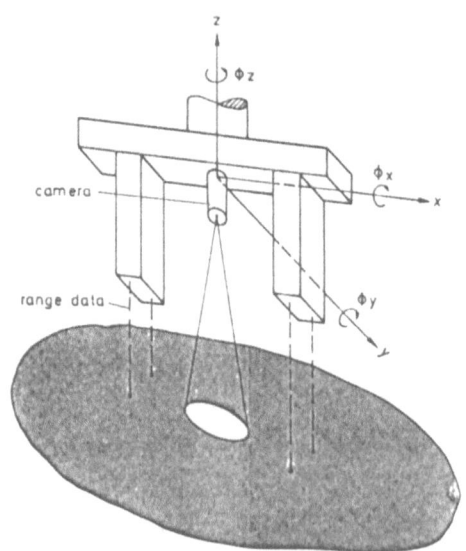

<u>Fig. 17</u> Combining an array of 4 range finders and a hand-
mounted camera for improved estimates of the deviation
from a "nominal" situation.

447

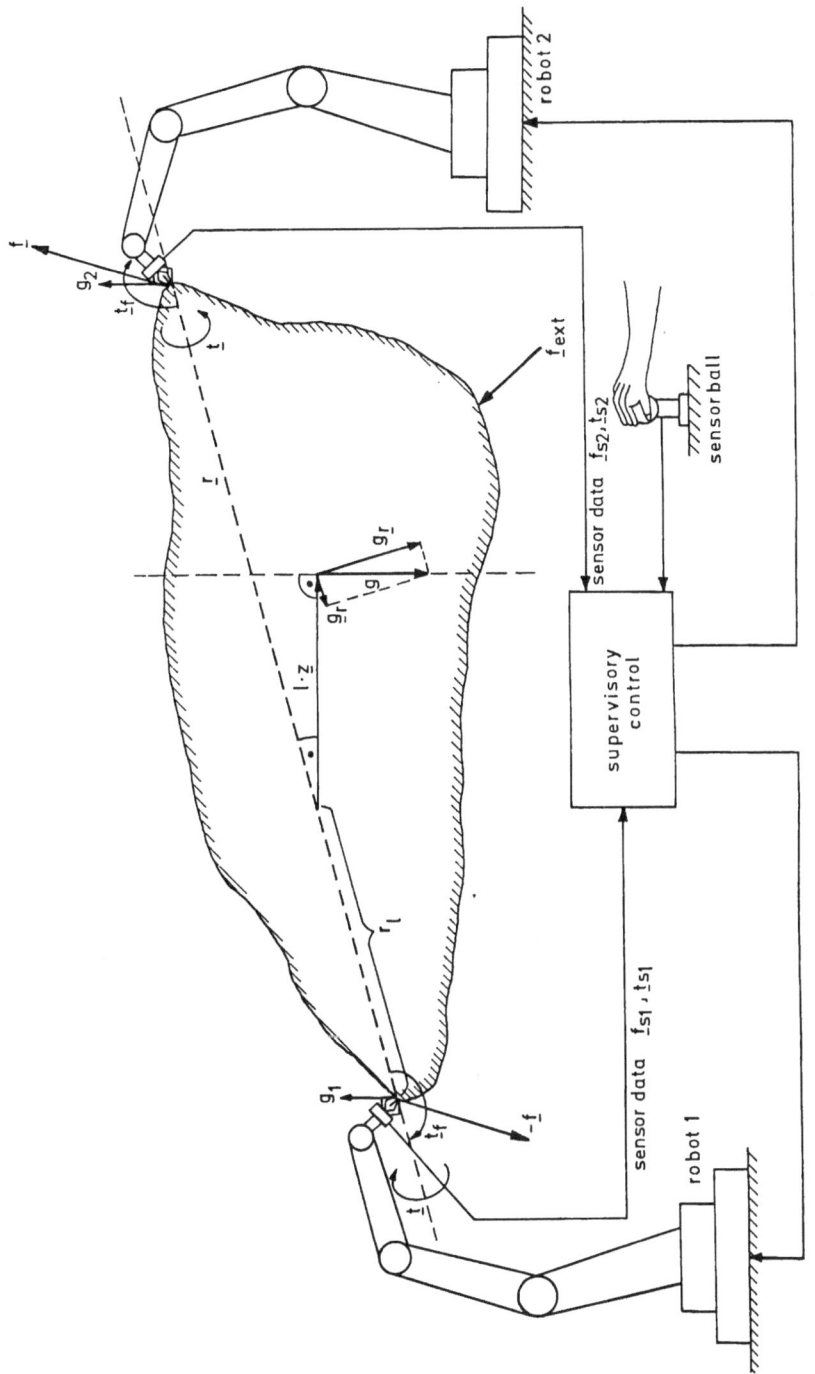

Fig. 18 Two-arm-robot with a force-torque sensor in each gripper is moving an object according to the sensor-ball commands

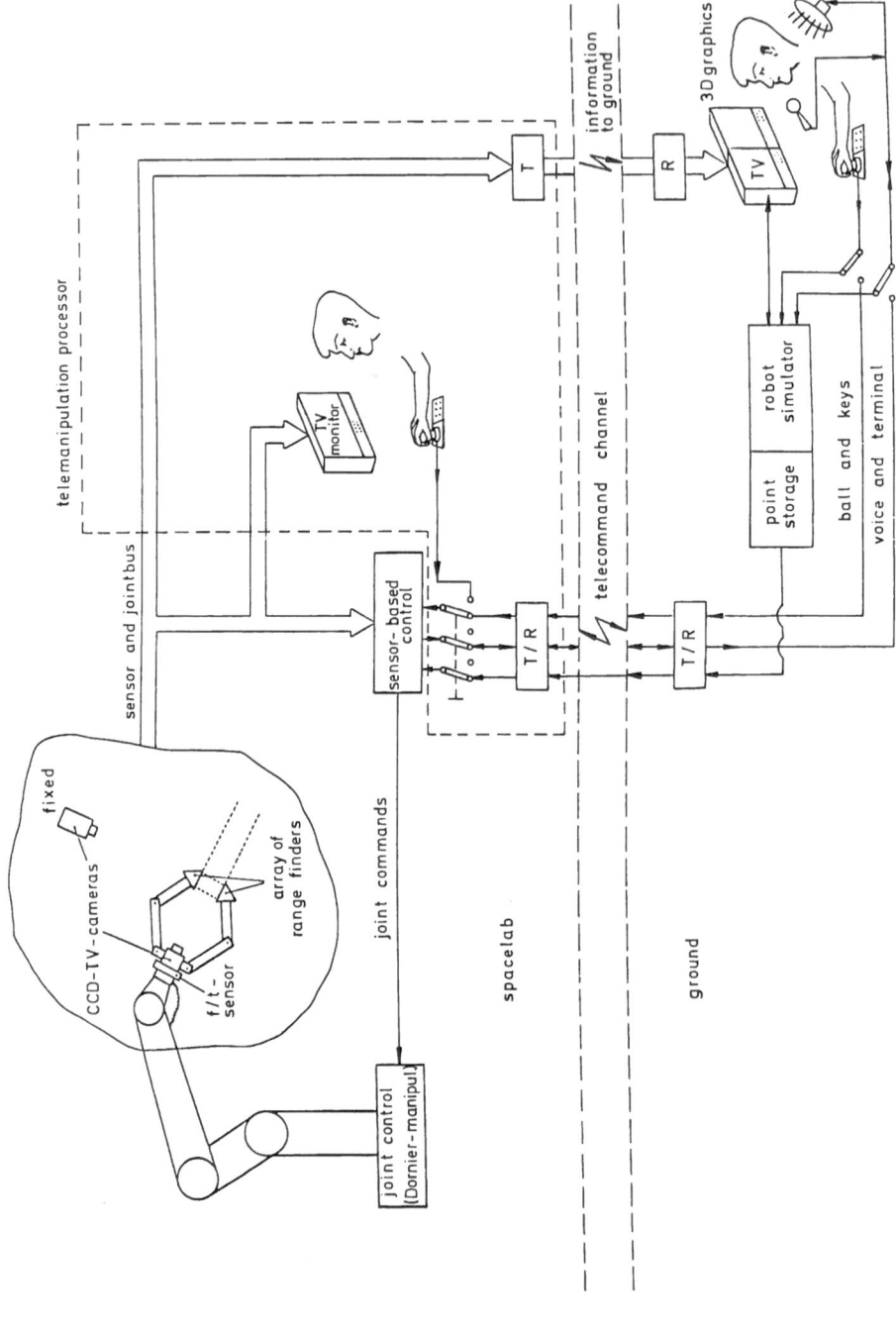

Fig. 19 The DFVLR proposal for a robotics experiment in spacelab mission "D2"

ACTIVE VISION

INTEGRATION OF FIXED AND MOBILE CAMERAS

P. Morasso, G. Sandini, M. Tistarelli

University of Genoa
Department of Communication, Computer and Systems Science
Via Pia 11A — 116145 Genoa

The integration of vision and movement is investigated. Active control of egomotion is a powerful source of constraining perceptual processes. Stereometric strategies related to movement are compared: triangulation strategies (binocular and kinetic) and volumetric strategy (based on occluding contours). Volumetric representations are considered as candidate structures for accumulating a flow of stereometric data. Software tools and experimental results are presented.

1. Introduction

Purposive behavior — the behavior commonly displayed by primates in the real world and the goal for the next generation robotic systems — is intrinsically a *data flow problem*. Actions and perceptions are immersed in a spatio—temporal continuum that requires appropriate computational structures.

The size of the input flow is impressive. We can estimate it in a crude way by looking, for example, at the cerebral cortex: about 1.5 square feet of tissue with a density of 10^5 per square millimiter. The primary visual cortex alone covers about 15 square centimeters which means 150 million neurons, i.e. a flow of several *Gigabits per second*. A similar figure can be estimated if we consider man—made high—resolution imaging devices.

Data reduction can be performed immediately at the sensor level in high—speed servo—mechanisms. An example is given by the insect—catching behavior of frogs, that destroy the greatest part of the available information — like common industrial machine vision systems. Mechanisms of this type are essentially *synchronous*. They are able to *react* but are unable to *understand*. A reaction—type behavior is ego—centric, it cannot discriminate between situations that, although similar from the purely sensory point of view, are significantly different from the environmental point of view and require, accordingly different behaviors.

Eco—centric behavior cannot afford to destroy information during data reduction. On the contrary, it requires to *accumulate* it, in order to build up knowledge about the environment. Knowledge acquisition and knowledge interrogation are uncoupled, they can proceed asynchronously: in this way, action is freed from "ego—centric slavery". The cost is speed: humans do not come even close to frogs in insect—catching.

In order to perform data reduction we need some sort of *accumulator*, i.e. an internal representation where we *accumulate* invariant features of the input flow. Two major sources of invariance can be singled out: one is *Eco— centric*, the other is *Ego—centric*.

NATO ASI Series, Vol. F43
Sensors and Sensory Systems
for Advanced Robots
Edited by P. Dario
© Springer-Verlag Berlin Heidelberg 1988

The former one relates to the *rigidity* of objects. Objects have a *shape* that does not change[1] when we move with respect to them or when they move with respect to us.

Shape is both a topological and symbolic concept. A shape representation can be expressed, at any level of detail, both in an *analogic* way — using some kind of homomorphism with the outside world — and in terms of explicit descriptions. However, there is a significant difference: It is difficult to perform an eco—centric integration of a flow of ego—centric descriptions, whereas an analog representation of space is a natural way to accumulate the input flow.

The second major source of invariance of the input flow is *active movement*: movements of the eyes (with respect to the head), of the head (with respect to the body), of the body (with respect to space). Active movements, i.e. movements purposively planned and measurable either in terms of efference or afference, induce powerful structures in the input flow that constrain the perceptual processes and produce in a straightforward way *shape—in—space*. A direct example of this statement is the shape_from_occluding_contour paradigm: When we move around objects, the sequence of occluding contours "envelopes" the object inside a volume that is computable from the performed motion.

The pursuit of the topic outlined above — *active vision* — requires an integration of different disciplines that have been developed essentially along independent paths: shape—from—binocular stereo [Marr and Poggio 1979], shape—from—motion [Ullman 1979, Hildreth 1983], shape—from—contour [Martin and Aggarwal 1983], modeling of solid objects [Requicha 1980, Ballard and Brown 1982], etc.. In the following sections we attempt to outline the *fil rouge* that may join them.

But what is the meaning of *active*? In our formulation, active vision is a shape formation process which is based on both iconic data and movement data. It is a rather general way to approach natural and artificial vision. This is to be contrasted with the current notion of active vision as the set of techniques that use structured light patterns[2] to get depth information. However, it is possible to look at the active control of illumination as a special kind of (virtual) movement[2] and then the current notion of active vision would be only a subtopic of the general topic of integration of vision and movement.

2. Stereometric Strategies related to Movement

A basic form of stereometric strategy is *triangulation*. The same object point is observed from two different vantage points and its position in space is evaluated by intersecting the two rays which are generated by the corresponding picture elements. Stereometric triangulation, applied systematically to a carefully selected set of picture elements, induces a depth map, i.e. a bundle of line segments emanating from either picture plane and passing through the corresponding center of projection (fig. 1).

>From the *computational point of view*, stereometric triangulation requires the knowledge of different items:

---

[1] Except for catastrophic events, such as explosions, even in the case of non— rigid objects their shape can only change smoothly over time.

[2] Common structured light patterns are rays and planes of light (usually generated by means of lasers) that produce spots or lines with controllable positioning parameters. This gives triangulation paradigms similar to those characteristic of binocular stereo or kinetic stereo.

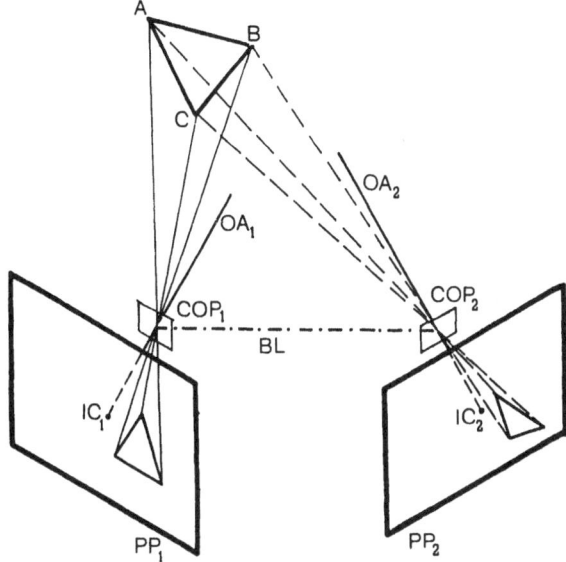

Figure 1. The use of Triangulation in Stereometry. BL is the Base Line; PP is a Picture Plane; COP is a Center of Projection; IC is the Center of an Image; A,B,C are points of an object. A typical triangle used for solving depth is (COP_1 COP_2 A).

► *Relative positioning* of the two cameras. This implies 6 parameters: 3 for translation and 3 for rotation of two Cartesian frames rigidly connected to the two cameras[3]. The overall geometric transformation can be conveniently represented by means of a 4x4 homogeneous matrix.

► *Optics.* This implies several optical parameters: One is the focal length (the distance of the center of projection from the picture plane); Two other parameters identify, in the picture plane, the center of the picture; Another parameter is a gain factor, that relates picture coordinates with world coordinates[4]. Totally this makes 4 or 5 optical parameters.

► *Iconic correspondence.* This implies the identification of pairs of picture elements that correspond to the same object point. Although this problem is quite complex, it does not require a search in two dimensions but only in one: The geometry of stereometric triangulation, in fact, constrains candidate correspondent picture points to stay on conjugate *epipolar lines*[5] (fig. 2) [Horn 1986].

---

[3] The origin of a frame is usually taken in the center of projection. In this case, the translation component is the so called *base line* vector.

[4] In practical cases, e.g. when using vidicon cameras, two gain parameters (one for each axis) are needed to take into account the anisotropy of the imaging device.

[5] Given two cameras, identified by picture planes and centers of projection, conjugate epipolar lines are the family of lines which are obtained by intersecting the two picture planes with the so called epipolar planes. An epipolar plane is a plane that contains the line connecting the two projection centers (the so called base line) and the family of epipolar planes is obtained by rotation around the base line. Note that, by construction, the family of epipolar lines on either picture plane meets at the point in which the base line intersects the picture plane (this point is called epipole). This can be understood easily if we consider a ray on one of the cameras (it generates a single picture element Q1) and we move a point P on it from the center of projection to infinity: an infinity of picture elements Q2 are generated on the other picture plane. The initial point Q2 is, by construction, the epipole. The final point Q2 (generated by the point P at infinity) is, by definition, the *vanishing point* of the ray (the point of the second picture plane obtained by projecting through the second center of projection a line parallel to the ray). The same operation can be repeated for a ray from the second camera, coplanar with the first one, and the result is the couple of conjugate epipolar lines.

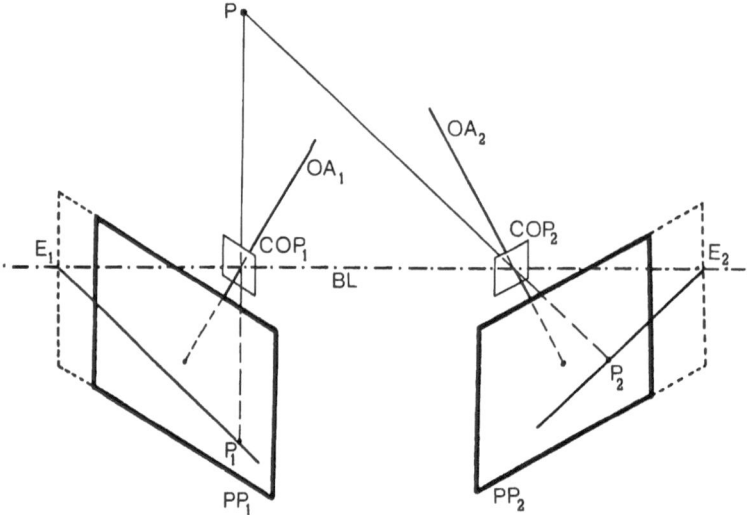

Figure 2. Conjugate epipolar lines in stereometric triangulation. COP is a Center of Projection; PP is a Picture Plane; E is an epipole; P is an object point.

>From the *operational point of view*, stereometric triangulation can be implemented in two ways:

◆ *Binocular*. In this case, the couple of corresponding icons is provided by two cameras with a base line of (approximately) constant length and the relative movement of the two cameras is constrained in such a way that the two optical axes and the base line are coplanar. In binocular stereometry the relative positioning parameters are reduced to two rotations only — rotations around two parallel axes perpendicular to the base line. Moreover, the correspondence pattern between the two images, expressed by the family of epipolar lines, has a prevalent "horizontal" structure[6] (figure 3).

◆ *Kinetic*. In this case, the couple of corresponding icons is provided by one camera that moves in space — the egomotion. The base line is generated by movement. In qualitative terms, there is a complete analogy with the binocular case. The difference is that the movement may generate an arbitrary relative positioning of the two frames of the triangulation. Therefore arbitrary patterns of correspondence may occur that require complex computations. Special patterns of movement, however, may generate special correspondence patterns that reduce the complexity (see figure 4): If the camera movement is in the horizontal plane, the effect is similar to that of the binocular case[7]; If the movement is in the direction of the optical axis, then the epipoles are in the center of each image and the patterns of correspondence have a polar structure; If the camera movement is in the vertical plane, the epipoles stay on the vertical axes and the patterns of correspondence are vertical—like, and so on.
Egomotion generates an apparent motion of iconic features on the picture plane that is characterized by a focus of compression or a focus of expansion: This focus is the epipole.

---

[6] "Horizontal" and "vertical", here and in the following, is related to the images, the x—axis corresponding to the stereometric base line.

[7] For lateral motion without rotation of the optical axis, the epipolar lines are horizontal trend. In all the other cases (a motion component in the direction of the optical axis and/or a rotation of the same) the epipolar lines will converge to the two epipoles, on some point of the x—axis of each image.

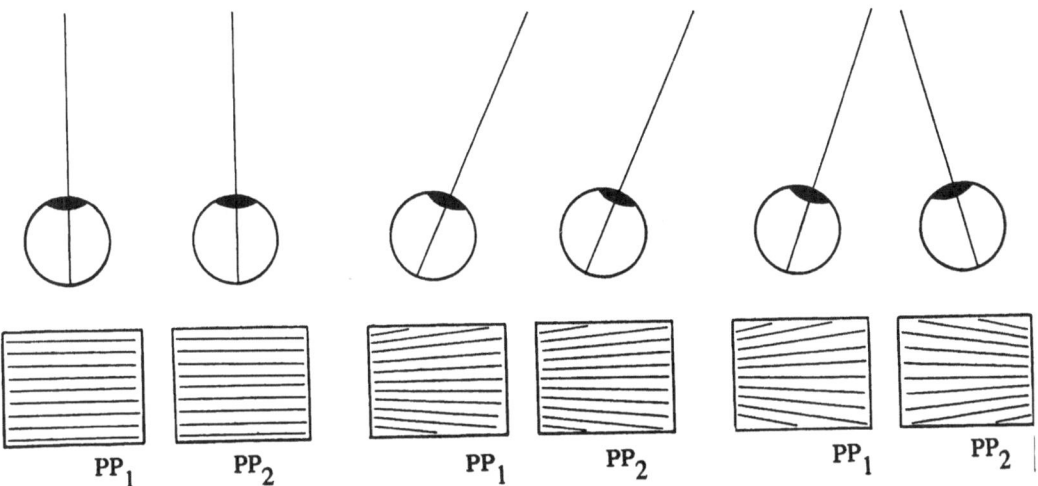

Figure 3. Patterns of epipolar lines for binocular stereo. Left: parallel optical axes orthogonal to the base line. Center: parallel optical axes both rotated on a side. Left convergent optical axes.

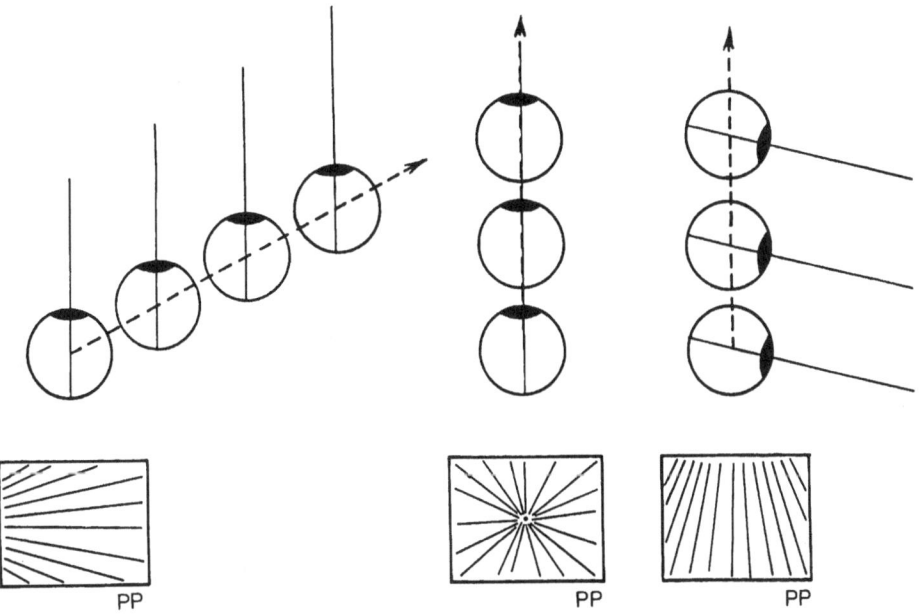

Figure 4. Patterns of epipolar lines for kinetic stereo. Letf: horizontal motion. Center: axial motion. Right: vertical motion.

>From the *functional point of view*, a basic problem to be faced, by triangulation as well as by other methods, is *Spatial Integration*. The depth−maps generated by stereometric triangulation induce a sort of *bas−relief*, a solid structure in space that has a know "forward" face and may have infinite "thickness". In order to integrate the depth maps obtained from sequences of stereometric operations, it is necessary to use solid modeling primitives (this is a topic considered later on) and to have knowledge of *Absolute Positioning Parameters* of each stereometric operation. This requires 6 more parameters (3 translations and 3 rotations) that relate one of the cameras, taken as a reference, to an environment Cartesian frame.

The shape – formation role of triangulation is based on an explicit computation of depth, the depth of visible points of the face of an object, up to the occluding contours[8]. When approaching this contour, however, the straight triangulation may become inaccurate unless the occluding contours correspond to sharp edges of the object surface. For smooth occluding surfaces, it is better to use a weaker method than triangulation, a method that uses rays without computing distance explicitly. For this purpose, a good candidate is a method that uses a volumetric representation of occluding contours and performs explicit *solid intersections* among them without passing through a depth map[9].

Let us now compare the pros and cons of the different stereometric strategies outlined above:

- Binocular stereo is good for detecting the depth of edges which are close to vertical, e.g. the legs of a chair. This is due to the fact that the epipolar lines are about horizontal and edges can be detected best if they are orthogonal to the direction of search.

- Kinetic stereo on the horizontal plane gives information of the same nature as binocular stereo because the epipolar lines follow a similar pattern. Therefore this method is a surrogate of binocular stereo for a monocular observer and is simply redundant for a binocular observer.

- Kinetic stereo on the vertical plane gives a performance which is complementary of that given by binocular stereo, i.e. it detects best horizontal edges like the seat of a chair because the pattern of epipolar lines is about vertical.

- As regards kinematic stereo in the axial direction, the same complementarity that characterizes kinematic stereo on the vertical plane holds only in the central part of the visual field and not in the two lateral parts, according to the radial structure of the epipolar lines.

- The occluding contour method has two roles. One is to improve the estimate of the object shape in the neighborhood of the occluding contour and the other is to provide the 3D support for the integration of sequences of bas – relief evaluation by means of the other methods.

The knowledge of the weak and strong points of every method may suggest a way of integrating them, for example by means of a weighted average where the weights measure the distance, for each method, from its best or worst operating condition.

Another point concerns the limits of the analogy between binocular stereo and kinetic stereo. In the former case, couples of corresponding images are the source data, whereas in the latter case the source data is a time varying image. The movement of the observer in a 3D world generates an optical flow field on his retina [Pradzny 1980] and it is this optical flow the starting point of any computational formulation of kinetic stereo. In kinetic stereo, the problem consists of processing the optical flow *while controlling the egomotion*, in such a way that patterns of correspondence can be established for images at different instants of time. We should point out that this kind of problem is different from the problem usually faced by people working with optical flow, i.e. the problem of deducing egomotion from time – varying imagery. Of course this problem, that concerns passive (unknown) movements, is much harder than the problem formulated in this paper, that concerns active movements, because knowledge of egomotion is an extremely powerful source of perceptual structure.

---

[8] The occluding contour of an object is a contour that separates a figure from the ground. Across an occluding contour there is usually a step in depth and/or other other features which makes them detectable.

[9] A volumetric representation of an occluding contour of an object is a *conic half space*, i.e. the infinite portion of space constrained inside the bundle of emi – lines originating from the observer location and

A final point concerns sampling in kinetic stereo. Two sampling rates are needed: One (related to the computation of the optical flow) must be fast enough to allow the local operators to sense the passing edges; The other (related to triangulation) must be slow enough to provide a sufficiently large base line (comparable with the binocular base line). If the two rates are fixed, then it is possible to formulate an optimization problem, i.e. to look for the speed of egomotion which best matches the timing of the perceptual processes − a psychophysical topic that did not receive enough attention.

3. Experimental Activity

This section overviews the experimental activity that was carried out in the framework outlined above. First of all, we present a software tool that was developed for dealing in a general way with the multiple representations generated by early processing of visual data. Then, we discuss connections with volumetric representations. Finally, we present some specific algorithms related to binocular and kinetic stereometry.

3.1. The Virtual Image System

The Virtual Image System (VIS) is being developed as a tool for investigating low− level, or "early", visual processes [Sandini and Vernon 1986] in an interactive way. This tool allows a user to build an "image system", i.e. a hierarchical structure of representations where the computational transformations are implied by the "transfer" from one representation to another. This hierarchical structure is generically referred to as a Virtual Image Structure.

The global Virtual Image Structure managed by VIS may contain multiple component Image Structures: each one is identified by such parameters as the time instant (for an image sequence), by the left/right specification of stereo pair, etc.. Image Structures are organized into one or more color channels according to different color coding schemes (RGB, HIS, HSV etc). Each color channel may be organized into different spatial frequency channels using a pyramidal structure[10], with image resolution varying from 1024x1024 to 64x64. The Virtual Image Structure is simply a hierarchy of these pyramids. Moreover, each pyramid can store three levels of representations: iconic, regional, and boundary.

Each Virtual Image Structure hierarchy may comprise many iconic representations: framestores, intensity images, convolution images, etc. Lower in the hierarchy, the iconic representation of a channel may be transformed into a tree−structured regional representation which makes explicit the topology of "regions" of uniform convolution sign. >From the regional representations it is possible to build, further down the hierarchy, several contour representations which make explicit, for the boundary of a region, such features as the zero−crossing slope and direction, boundary motion, stereo disparity, etc.

A Virtual Image Structure can be configured so that it comprises as many pyramids as necessary and representations may be added "with" selective parameters and selective

---

tangent to the object.

[10] A pyramidal structure is a set of iconic representations of the same image with variable resolution. The "pyramid" is a metaphor of such arrangement − a sort of stack of icons.

windows, which are specified dynamically, at run time. This feature, together with the possibility of deleting representations, provides the basic framework for a cognitively tunable early perceptual processing − a computational formulation of "attention".

The Virtual Image System (VIS) uses menu−based user/system dialogue and a facility is provided which allows a user to teach the system a sequence of menu selections which may be saved on file and recalled later for replay. Several such files may be generated and a directory of these "command" files is available upon request. All menu selections are supported by on−line help texts. Further, the system is designed in such a way as to enable a processing and analysis session to be suspended and resumed at a later date, saving in a file the current virtual image structure.

A data−flow paradigm was adopted to allow the user to manipulate images merely by specifying source and destination representation, the types of these representations defining an implicit transformation. For example, if the source image is an intensity image type and the destination image is a convolution image type (i.e. one that has been convolved with a Laplacian of Gaussian mask) then the appropriate transformation is effected implicitly by the system as part of the transfer, prompting the user for auxilliary information, as appropriate. In addition, the system allows extensive windowing in both source and destination images so that, in addition to being transformed, the destination image may be a scaled and translated version of the source window.

A user can build a virtual image structures from scratch, beginning with an empty structure and subsequently adding, arborising, and/or deleting representations as desired and as the situation demands.

Formally, VISs can be defined, using a BNF−like notation, as follows:

<Virtual Image Structure>::= {<Image Structure>}*
<Image Structure>::= {<Color Channel>}*
<Color Channel>::= {<Spatial Resolution Channel>}*
<Spatial Resolution Channel>::= <iconic rep>|<iconic rep><regional tree>

It should be noted that a sequence of BNF syntactic constructs, e.g. <...><...><...>, does not necessarily imply temporal or spatial ordering (as it would in the linguistic formulation) but, more importantly, that the constructs are mutually associated, i.e. logically related. A Virtual Image Structure may contain several Image Structures, each of them being identified by a specification of time parameters, spatial parameters (coordinates of the camera, identification of right/left camera etc.), and optical parameters. An Image Structure is possibly split into color channels and each one may be divided into several spatial frequency channel. The spatial frequency channels are organized in a pyramidal structure (i.e. using a multiple resolution scheme).

Each pyramid, in turn, contains different types of iconic representations (organized as 2D arrays), whose topological arrangement can be made explicit by means of a tree−like structure of regional representations:

<iconic rep>::= <intensity image>|
                <intensity img><convolution img>|
                <intensity img><convolution img><region crossing img>
<regional tree>::= tree of <regional representations>
<regional representation>::= <regional descriptor>|
                <regional descriptor><contour representation>

where the <contour representation> is a rich specification of the shape of region boundaries (stored by means of boundary chain codes − BCC) and of the distribution along these boundaries of several image features. The result is a set of "profiles": slope profile, orientation profile, velocity profile, disparity profile, etc.. The slope profile stores

TV framing rate of 25 frames per second) and oculomotor sampling (nonuniform and not exceeding the rate of 3—4 saccadic eye movements per second). Between a saccadic eye movement and the next one, the brain receives a flow of smoothly variable early visual representations, coupled with a flow of information about ego—motion. A working hypothesis that we formulate is that these two flows are used to set up and refine a stereometric representation of object shape.

In its simplest meaning, a 3D icon may be conceived as a 3D ternary array, one value indicating free voxels, one value occupied voxels and one value non—classified voxels. We already studied in a preliminary way some aspects of this problem developing algorithms that use the occluding contours for several views and the depth maps (from stereo matching) for the same views. Figure 5 gives an example of volumetric representation from occluding contours [Massone et al 1985] and figure 6 gives an example of volumetric representations from binocular depth maps [Morasso and Sandini 1985].

In perspective, however, it is possible to conceive VSSs (Visual Solid Structures) which parallel the VISs discussed above: structures in which multiple iconic representations are linked with relational representations. The flow of VISs over time would operate on the VSS for the duration of an "ocular fixation" because VISs are changing faster that VSSs (a VSS "accumulates" a coherent stream of VISs within an ocular fixation time). We may also add that possibly "Cognitive Accumulators" may be conceived for integrating over space solid structures acquired for each fixation interval.

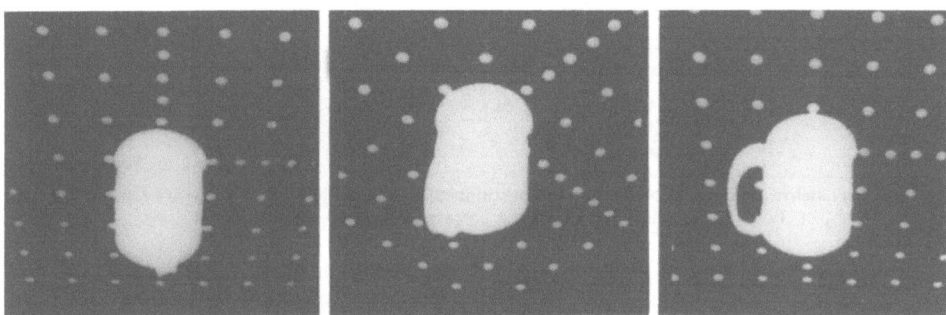

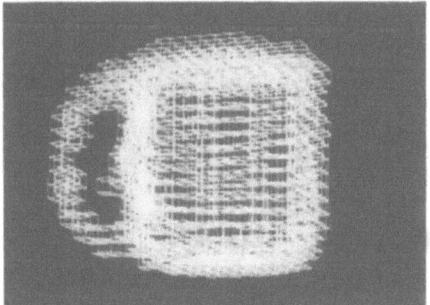

Figure 5. Volumetric representation from multiple occluding contours. Top: Multiple images of an object used for the extraction of the occluding contours. Bottom: Volumetric reconstruction of the object obtained by solid intersection.

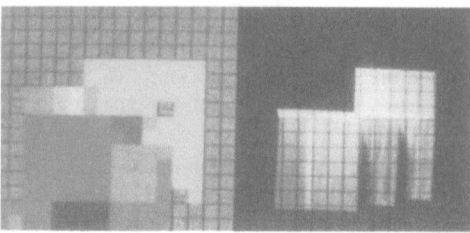 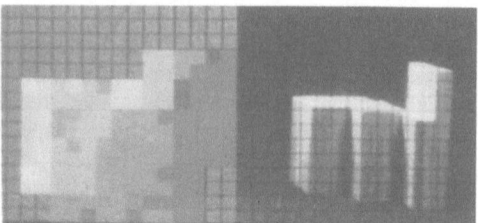

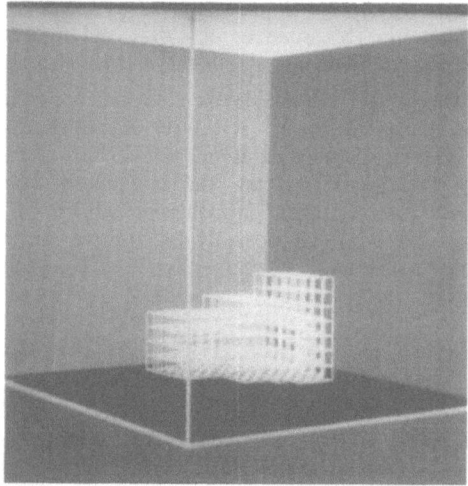

Figure 6. Volumetric representation from multiple binocular stereo pairs. Top: Disparity maps for two views of an object. Bottom: Solid intersection of the corresponding bas–reliefs.

The coherence between the two types of structures (or, better, between the current VSS and the last VIS) may be obtained by means of "relational icons", i.e. icons of pointers similar, in concept to the region–crossing images defined above. In particular, it is possible to conceive two types of such relational icons:

2D icons of pointers to the 3D icon voxels, or
3D icons of pointers to the 2D icon pixels.

Which one of such computational alternatives is better to choose is too early to say, however we may already outline the interface between data driven processes (that produce VISs and VSSs) and cognitive perceptual processes that must interpret and use such representations according to a goal.

What is important is that the interface must be bidirectional: the cognitive processes not only "read" the perceptual data but must also "actively guide" the data gathering process. As regards the former aspect, a VSS and the current VIS may be considered as a data base which may be interrogated in order to obtain qualitative or quantitative answers (P419 has investigated some parts of this aspect, as reported in the next section). Possibly more important (and certainly in a more primitive stage of conceptualization) is the dual aspect of interaction the other way around. Two main topics may be outlined: One is tuning and guiding the data gathering process in order to limit the complexity explosion by "focusing the attention" on a limited part of the world (this can be obtained using for example the Selection Mechanisms and Windowing Mechanisms already included in VIS);

the other is using conceptual hypotheses in order to "modify" the visual data base, either as regards 2D or 3D representations. In this way it is possible, for example, to complete contours or merge regions (in the 2D representations) or to "carve" volumes in the 3D representation on the basis of expectation or gestaltian concepts of continuity.

3.2. Optical Flow Computation in Kinetic Stereo

In both binocular and kinetic stereo, a basic computational task, preliminary to triangulation, is correspondence. While triangulation is essentially the same in both cases, correspondence has significant differences that suggest different processing strategies.

The lines of correspondence (the epipolar lines) are known for both binocular and kinetic stereo on the basis of geometry and motion parameters. But this is not enough: The basic obstacle is the possibility of false matches[11] and this can be overcome by means of some kind of prediction that allows to estimate, at least roughly, where the matching point is likely to be, in order to restrict the search in that neighborhood.

In the binocular case, the transformation from one image of the triangulation pair to the other is discontinuous, whereas in the kinetic case it is a continuum. Therefore, in the latter case we can use some sort of *tracking*, whereas in the former we cannot.

The solution for the binocular case is to exploit the different spatial resolution channels[12] that are known to operate in early visual processing, according to a classical *divide−and−conquer strategy*: For the low resolution channel, even a straightforward correlation is unlikely to give false matches, although it can only give rough estimates of disparity; for the intermediate resolution channel the matching is started using the previous match, and so on for the high resolution channel. The disparity map of figure 5, for example, was computed by means of a variation of this basic strategy [Sandini et al 1986].

In the kinetic case, time varying images are available and local operators can be conceived that detect variations of illumination intensity patterns and from these variations attempt to estimated their motion. It is quite evident that operators of this type will operate best in areas of high contrast that correspond to edges (the zero−crossing contours of the spatial resolution channels) and that they will only be able to detect the component of the motion orthogonal to the contour[13].

Summing up, in kinetic stereo two kinds of information are available:

► The orthogonal component of the motion of edge−elements, provided by local iconic operators.

► The parameters of egomotion. From this, the location (on the picture plane) of the focus of expansion (or compression) can be estimated, as shown in a previous

---

[11] False matches would generate false triangulations and then false estimates of depth, a type of error that sometimes happens in visual illusions.

[12] Selective spatial resolution channels are obtained by filtering an image with band pass spatial filters of an appropriate frequency response. Laplacian of Gaussian masks are commonly used for the filters and the three channels commonly used carry information about low, middle, and high resolution, respectively.

[13] A motion parallel to a contour element does not obviously change the intensity pattern viewed by a local operator. This is the so called *Aperture Problem*. It can be shown that the orthogonal component of the motion is directly proportional to the time derivative of the local image intensity and inversely proportional to the magnitude of the local gradient. The orthogonal component becomes undefined for a vanishing value of the gradient.

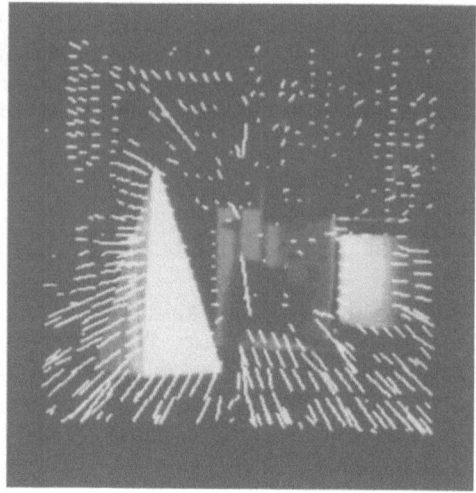

Figure 7. Kinetic stereo for an axial egomotion (in the direction of approach). Left: The first two images of an image sequence (top), with the corresponding output of the intermediate resolution channel (bottom). Right: Tracked edge−elements over the sequence (of 9 images).

section. What is important is that the direction of the motion of any edge−element is constrained to the line that joins such element to the focus of expansion.

What remains to be done is to solve a simple problem of vector algebra: To find the magnitude of a vector when we know its direction and its component on a known direction. In such a way, we get an estimate of the disparity due to motion (that can be further refined performing a local search) and then we can "track" the transformation of an image due to egomotion until we get a base line large enough to perform a significant triangulation. Figure 7 shows an example of kinetic stereo for an axial egomotion and figure 8 for a horizontal movement that keeps a fixation point.

Acknowledgement. The research reported in this paper is partially covered by an Esprit project (P419) of the EEC and by two projects on Bioengineering of the Ministry of Education.

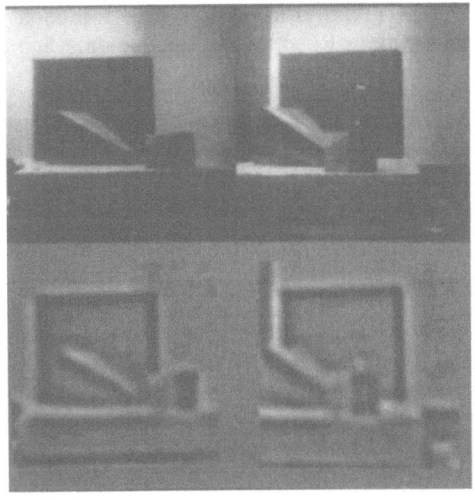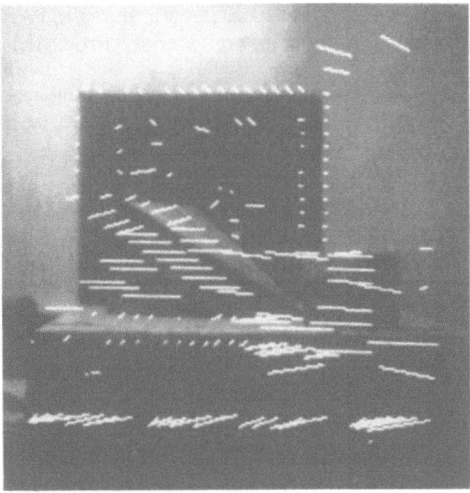

Figure 8. Kinetic stereo for an horizontal egomotion (toward the right of the picture. Left: The first image of an image sequence with the corresponding output of the intermediate resolution channel. Right: Tracked edge—elements over the sequence (of 9 images).

References

BALLARD DH, and BROWN CM (1982) Computer Vision. Prentice Hall:Englewood Cliffs,N.J.

HILDRETH EC (1984) The Measurement of Visual Motion. Cambridge:MIT Press.

HORN BKP (1986) Robot Vision. Cambridge:MIT Press.

MARR D, and POGGIO T (1979) A Computational Theory of Human Stereo Vision. Proc. Royal Society London B 204:301−328.

MASSONE L, MORASSO P, and ZACCARIA R (1985) Shape from Occluding Contours. SPIE Proceed. 521:114−120.

MARTIN WN, and AGGARWAL JK (1983) Volumetric Description of Objects from Multiple Views. IEEE Trans. Pattern Analysis Machine Intelligence PAMI−5:150−158.

MORASSO P, and SANDINI G (1985) 3D Reconstruction from Multiple Stereo Views. III Intern. Conf. Image Analysis and Processing (Rapallo, October 1−2).

PRAZDNY K (1980) Egomotion and Relative Depth Map from Optical Flow. Biol. Cybern. 36:87−102.

REQUICHA AG (1980) Representation of Rigid Solids: Theory, Methods and Systems. ACM Computing Surveys 12:437−475.

SANDINI G, TAGLIASCO V, and TISTARELLI M (1986) Analysis of object motion and camera motion in real scenes. Proc. IEEE Intl. Conf. Robotics & Automation. San Francisco.

SANDINI G and VERNON D (1986) Software Tools for Integration of Perceptual Data.Esprit Technical Week, Brussels.

ULLMAN S (1979) The Interpretation of Visual Motion. Cambridge:MIT Press.

SENSORS FOR MOBILE ROBOTS

G. Drunk

Fraunhofer-Institute for
Manufacturing Engineering
and Automation (IPA)

Stuttgart, West-Germany

Mobile robots represent a new generation of automation. Technologies developed for industrial robots in the field of Production Automation are still being generalized for application outside the factories. Mobility is a characteristic feature for this kind of robots assigned for a lot of different applications extending from space to underwater purpose, from construction to agriculture and from mining to fire fighting.

A key element to obtain this ambitious goal is sensor-guidance. More intelligent controllers which need sensor input for their autonomous route planning and navigation will be used.

This paper gives a short perspective of sensor tasks for mobile robots, of basic possibilities for sensor principles and contains some remarks how to use them. Finally, some examples for complete sensor combinations to guide mobile robots are presented.

NATO ASI Series, Vol. F43
Sensors and Sensory Systems
for Advanced Robots
Edited by P. Dario
© Springer-Verlag Berlin Heidelberg 1988

1.   Introduction

1.1.   What is a mobile robot?

Very different systems are named mobile robots. This term is
common for remote teleoperated vehicles, walking machines,
automated guided vehicles and research platforms for artifi-
cal intelligence. A detaited overview on mobile research platforms is given in /1/ and /2/. International developments
in the field of teleoperated vehicles are listed in /3/, and
/4/ discusses the control of walking machines.

When talking about sensors for mobile robots the intelligent
autonomous operating devices, fullfilling the following conditions, are the most interesting:

     o free manoeuverability
     o autonomy
     o manipulator arm equipment

Free manoeuverability not only means to be mobile but also
not being bound to tracks or wire guide paths. An autonomous
controlmode implies an own intelligence for path planning
based on sensor signals which is not given with teleoperated
vehicles. Finally a mobile robot needs at least one robot
arm to justify the name. As the sensor problems for mobile
autonomous platforms and vehicles are the same, this group
shall be regarded as well. Figure 1 gives an overview on the
components of a mobile robot.

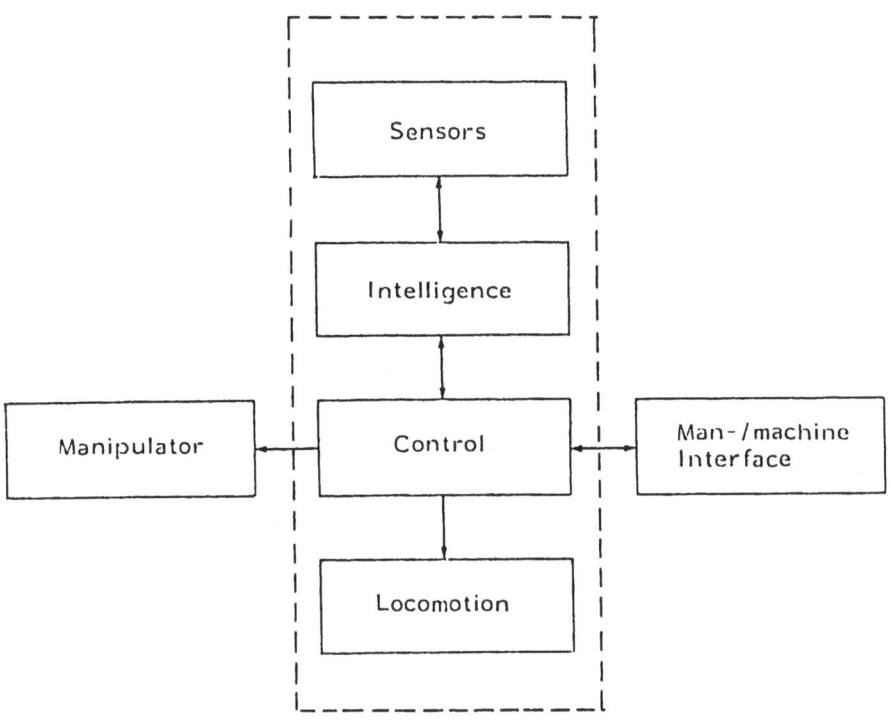

Figure 1: Components of a mobile autonomous robot

So the exact title would be "Sensors for mobile autonomous robots and mobile autonomous platforms"

1.2.   Sensor tasks in mobile robots

The autonomy of mobile robots and platforms cannot be reached with one single sensor, but only with a complex system of complementing sensors for different tasks, ranges and working areas.

In general there is a need for a solution for the following
sensor tasks:

- NAVIGATION:

  The actual absolute position has to be found without taking in account any obstacles..

- COLLISION DETECTION:

  The robot has to be prevented from damages by detection
  or avoidance of collisions with obstacles by stopping in
  time.

- ENVIRONMENT PERCEPTION:

  Information about the environment is needed for building
  a world model and for matching the environmental scene
  with the world model. This is necessary for task and path
  planning in a real world with known and unknown obstacles.

- MANIPULATOR ARM GUIDANCE:

  The movement of the robot arm has to be corrected according to uncertainties and inaccuracies of the gripping
  process. This problem is similar to the sensor guidance
  of industrial robots.

- APPLICATION RELATED SENSORS:

  This group of sensors depends on the robot mission and
  is not important for the movement control but for the

application process. Examples that might be stated are the sensing of a welding current or of an environment radiation.

The following chapters refer to the sensors for navigation, collision detection and enviroment perception. Sensors for manipulator arm guidance and application processes are not special problems of mobile robots but are the same as in fields of industrial robots.

2.   Sensors for navigation

The coordinates of the absolute robot position can be obtained by external or internal navigation. External position finding is handled by beacons or marks in the environment.

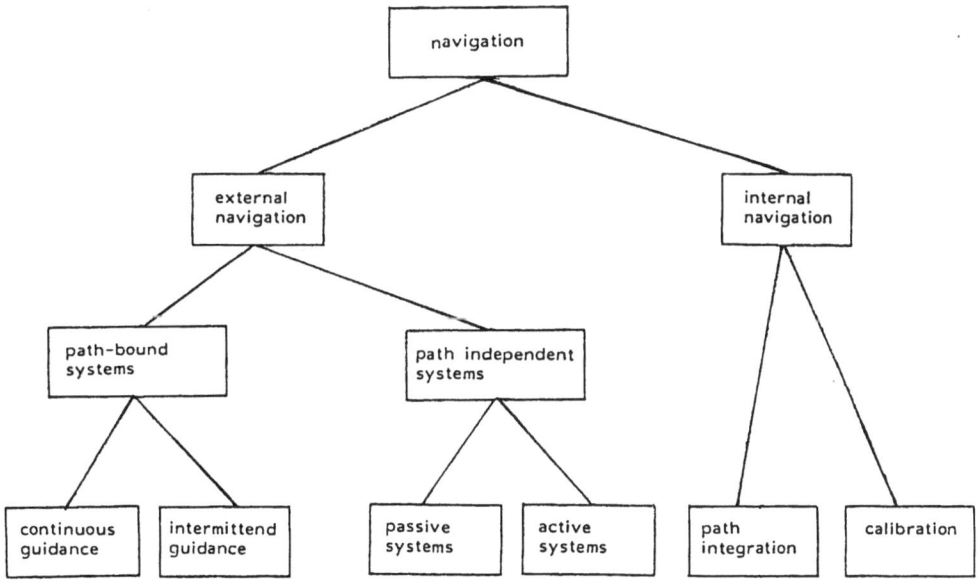

Figure 2: Classification of navigation sensors for mobile autonomous robots

Internal navigation applies only sensors mounted inside of the robot without any relation to marks or features in the environment. A disadvantage is the unavoidable drift of internal position sensing. Figure 2 shows a possible classification of navigation sensors.

2.1.   External navigation

External navigation systems can be either path-bound or path independent. Path-bound navigation can be of continous or intermittend type. Examples for continuous path-bound navigation systems are floor-embedded electromagnetic wire guidepaths, tracking of painted lines or laser beam guidance. These techniques are known from automated guided vehicles for flexible manufacturing systems.

Intermittend guidance systems mark the path in constant distances with point features like floor-embedded magnetic marks. Between these marks the robot is moving freely until the next mark gives a new fixpoint. An interesting intermittend guidance solution is a system developed by Komatsu in Japan. The reflected light of small glas bead marks mounted on the floor in robust plastic cases is detected by a camera on the moving robot travelling from mark to mark (figure 3). A survey on path-bound navigation systems for AGV application in Japan is given in /5/.

Of greater importance for mobile autonomous robots are path- independent navigation systems which can be classified into active and passive principles. Active robot navigation systems perform direction and range finding by electromagnetic optical or acoustical waves transmitted by external beacons.

Figure 3: "Spot Mark" navigation system of Komatsu, Japan for
          intermittend guidance with a robot based camera and
          glas bead marks on the floor

Certain disadvantages with this kind of navigation are the
additional costs of these active marks or beacons and the
inflexibility in the working area. Possible solutions for
active path independent navigation systems are:

- floor embedded active grid of wires or induction coils
  providing a magnetic field of position dependent intensity

- direction finding of radio beacons

- direction finding of ultrasonic beacons by phase comparison between pairs of acoustical receivers. A navigation
  system of the MEL in Tsukaba, Japan. uses three indepen-
  tently rotating pairs of receivers on the top of a mobile
  robot /6/. Each of the receiver pairs performs a direc-

tion finding to one of the three ultrasonic beacons transmitting on different frequencies (figure 4).

- laser scanner beacon with angle data transmission detected by 3 optical receivers on the vehicle /5/

- locating of light beacons by a scanning optical receiver. The University of Cincinnati uses a camera with a fisheye lens to detect the position of simple external light beacons /7/.

- vehicle recognition with a fixed vision system mounted on the ceiling.

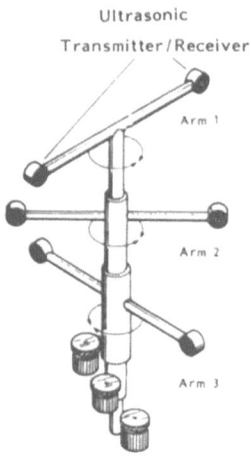

Figure 4: External Navigation by direction finding of three ultrasonic beacons

In passive navigation systems the vehicle based sensors recognize external passive marks for position finding. The position sensing of passive marks in the environment is cheaper than the active methods and easier to change, but a certain inflexibility in the working area remains. The most important approaches are listed below:

- rotating optical telemeter which locates the direction and distance of an optical reflector

- rotating laserscanner which locates the direction of 3 external vertical reflector stripes

- two dimensional direction finding of diamond shaped optical marks by vision. The distance can be calculated by the size of the mark

- An edge distribution matching navigation system has been developed by MEL for the mobile robot MELDEIC /8/. In the panorama picture of a rotating camera characteristic edges and corners of the room are matched with a stored edge pattern to compute the robot position

2.2.   Internal navigation

Internal navigation systems consist of the internal position sensor itself and an additional calibration device. The basic principle of internal position sensing is the integration of directions distances, velocities and accelerations. Inaccuracies of the sensor and other influences like wheelslip when using odometry are summing up too. The drift has to be compensated from time to time at calibration stations whose position is exactly known. At this station the position of calibration marks with respect to the mobile robot is measured by additional calibration sensors.

Most common for internal position sensing is the use of odometry and mechanical gyroscopes. In Japan Honda and Toshiba
have developed special gas rate gyros for mobile robots. Other
forward-looking developements direct towards very precise multidimensional acceleromenters and fiber optical gyroscopes.

In general, calibration can be realized with any of the known
external navigation principles or by using the environment
sensors for comparison with an a priori world model.

Other solutions use a multi dimensional arrangement of optical, ultrasonic, inductive or even tactile range finders.
In this case neither external marks nor world models are needed.

In /9/ an internal navigation system using a gyroscope for
direction finding and an optical encoder for odometry is presented in an application of Mitsubishi for a wireless operating
AGV. Calibration takes place at the loading/unloading stations
by a system of two optical proximity switches for the lateral
displacement and an optical range finder for the distance
correction using three reflectors on each station.

3.   Sensors for collision detection

Actually it's not possible to exclude any occurance of collisions by environmental sensing. In the case of a collision a
mobile robot has to stop immediately to prevent further damage.
For that reason, collisions have to be detected. The mobile
robot Shakey of SRI /10) used a system of whisker bump detectors for that purpose. Modern solutions are tactile ribbons
around the vehicle or area contact switches on the robot s surface.

Even better than collision detection is collision avoidance
by contactless proximity switches. Systems of ultrasonic or
photoelectric range finders around the contour of the robot
stop the vehicle when an obstacle is detected within the security range.

4.   Sensors for environment perception

Environment perception is the most complex and complicated
sensor task in the field of mobile robots. The decisions of
task planners, path planners and motion control are based on
their informations. Sophisticated mobile robots use world models for a priori knowledge of the environment. The world model
has to be completed and corrected by additional information
and the environment of the robot has to be matched with the
world model. Normally this is a three dimensional task. The
following main groups of sensor types are used for environment
perception:

- Vision systems

- Optical range finding systems

- Ultrasonic range finding systems

These groups of sensors are discussed in detail in the following chapters. Other sensor principles like the tactile stick-
pointer sensor in /11/ are of minor importance.

4.1.   Vision sensors

There are three general vision approaches which are usefull
for mobile robots:

- scene analysis on grey scale vision

- stereo - Vision
   * 2 or more cameras
   * slider camera
   * rotating camera
   * motion stereo

- active imaging (vision with structured illumination)
   * grid of lines
   * grid of light points
   * parallel lines

Scene analysis methods extract objects from the two dimensional picture of a single camera by searching for characteristic
areas or lines and matches them with known objects. The mobile
robots Shakey /12/ and Hilare /13/ use such sensors.

Stereo vision takes several pictures of the same scene from
different positions. Exposed points, lines or areas are selected. In a second step, features deriving from pictures of
different cameras have to be matched. By means of triangulation the three dimensional position of these exposed features can be determined.

Stereo pictures can be received by two or more cameras viewing
the same scene, by a rotating or sliding camera or by motion
stereo while the robot is moving. Examples for use of stereo
vision are the CMU-Rover /14/, the JPL-Rover /15/ and the mobile robot of the University of Tokyo /16/.

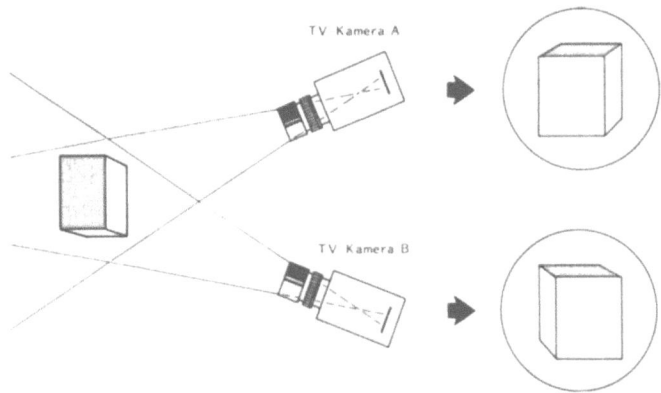

Figure 5: Basic principle of stereo vision

The computing amount for stereo vision as well as for scene analysis is very high and therefore only recommended for very sophisticated and expensive systems. The third approach is active scene illumination by structured light as a line grid, a point grid or under a known projection angle to the direction of a camera. Again by using triangulation the 3D position of each lightpoint can be computed.

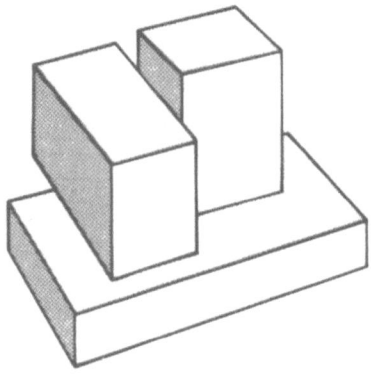 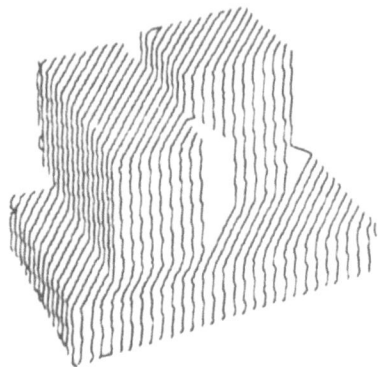

Figure 6.7: Principle of active imaging using parallel line
          projection

4.2.   Optical range finders and scanners

Laser based optical range finders and scanner produces a very
precise three dimensional image of the environment. The laser
beam can be focussed down to spot of 1 mm in diameter and the
range finding principles allow a clear interpretation. Unfortunately, the time required for complete scans is very high,
so the number of measuring points has to be reduced.

Three levels of complexity can be distinguished:

- optical range finder
  * triangulation
  * time of flight
  * interferometer
  * focusing

- 2D optical scanner
  * vision with light-section-projection
  * one-dimensional scanned distance measuring

- 3D optical scanner
  * vision with laser spot pointer
  * two-dimensional scanned distance measuring

Fixed range finding may be used for exact contour following of a complicated course of walls and obstacles.

The triangulation principle is only suiteable for a limited measuring field whereas the time of flight principle up to now is very inaccurate, so, it can only be used for greater distances. Another possibility is the measurement of the diameter of a projected laser spot by a CCD-line-camera.

Two-dimensional optical scanning can be realized in two basic ways. The first is to compute distances from a projected light-section in the field of a camera by triangulation. The second possibility is to deflect the transmitted laser beam and the reflected light. Most common for deflection are rotating or free-swinging mirrors.

Two-dimensional scanners can be profitably used in environments with prismatic obstacles and walls. A number of mobile robots uses three-dimensional scanners with a camera for recognition of a laser spot deflected by two step motor driven mirrors.

The JPL-Rover /17/ and the Hilare /13/ of LAAS use 3D laser scanners. The device of Shakey /10/ uses an optical deflection for the vertical direction. For horizontal scanning the complete device is rotated.

A two dimensional deflected range finding sensor has been developed by the Fraunhofer-Institute IPM in West-Germany, but not so far used with a mobile robot.

5.   Ultrasonic Sensors

To use ultrasonic sensors for environment perception is much cheaper than using vision or optical range finders for that purpose. Acoustical range finders work on the time of flight principle. An ultrasonic pulse is transmitted by an acoustical transducer and reflected back to the ultrasonic receiver by the nearest obstacle. The time of flight gives the distance to the obstacle.

It is very important to keep in mind that ultrasonic distance measuring produces another image of the environment as the image of optical range finders would look like:

- The ultrasonic beam cannot be focused as good as a laser beam. In spite of the extension of the beam the distance to the next obstacle is known but not the exact lateral position of the obstacle.

- The lateral resolution of ultrasonic scanners is very poor.

- Acoustical waves projected on a wall under an angle are deflected like light by a mirror. In this case no echo returns to the receiver. Edges and corner reflect in all directions.

Understanding this behavior of ultrasonic waves and systematically using them, the following types of acoustical environment perception are possible:

- single acoustical range finder

- geometrical arrangement of multiple distance measuring elements

- array of acoustical transducers

- acoustical scanning with rotating elements

- acoustical phased array for electronical scanning

5.1.   Acoustical range finder

Depending on the transmitting frequency, distances from about 300 mm up to 1000 mm can be measured with ultrasonic transducers from 40 - 200 kHz. For closer ranges, sensors with separate transmitting and receiving elements have to be used.

Applications for use with a mobile robot could be the range finding to a known obstacle or to a wall on the side.

The ultrasonic range finding system of the Micromouse I from the University of Kaiserslautern (figure 8) measures the distance to a system of rectangular walls of a labyrinth.

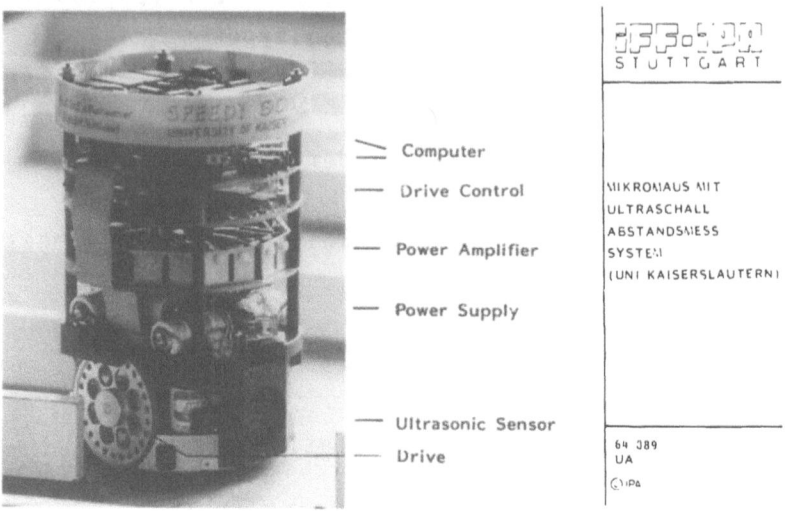

Computer

Drive Control

Power Amplifier

Power Supply

Ultrasonic Sensor

Drive

STUTTGART

MIKROMAUS MIT
ULTRASCHALL
ABSTANDSMESS
SYSTEM
(UNI KAISERSLAUTERN)

64 389
UA
(C) IPA

Figure 8: Small mobile robot micromouse I "Speedy Gonzales"
of the University of Kaiserslautern, Germany, using
ultrasonic range finders

5.2.   Geometrical arrangement of multiple elements

In this case, several independent ultrasonic range finders are
mounted on the mobile robot in a fixed geometrical relationship, so that certain special features can be detected. For
example, it is possible to measure the distance and the orientation of a wall simply with an arrangement of two parallel
range finders. With two elements at the side and one in front
of the vehicle, the X-, Y- and C-position of the vehicle relative to a rectangular corner can be determined. It is necessary for this type of sensor, that the single elements have to
be multiplexed by time or by frequency.

A fairly complex arrangement of multiple ultrasonic range
finders is used by the robot Hilare /18/ of the LAAS. With 14
elements not only collision avoidance is secured but also
closed loop navigation around obstacles (figure 9).

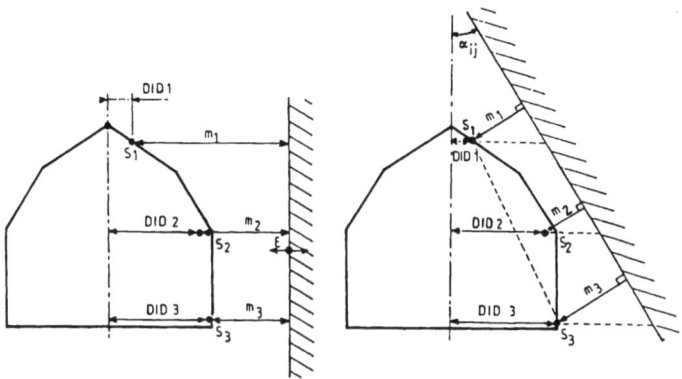

Figure 9: Use of a multiple arrangement of ultrasonic range
finders for contour following (Hilare)

5.3.   Array of acoustical transducers

By a line or matrix of receiving or transmitting ultrasonic
elements a complete surface contour can be obtained and then
processed.

A well known example is the main perception system of the
guide dog robot MELDOG /19/ (Figure 10) of the Japanese "Mechanical Laboratory" of Miti. One transmitter and three recei-

vers on each side form a linear array of seven transducer elements. With the pattern of the received echos distance, direction and orientation of walls and poles can be determined with
only one acoustical transmitter burst.

Figure 10: The "guide-dog" robot MELDOG uses a linear array
           of one ultrasonic transmitter and 6 receivers

5.4.   Rotating acoustical scanner

By rotating one or more ultrasonic range-finders on the top
of a mobile robot a complete map of the environment can be
built. Several devices such as Jason of Berkeley /20/ Yamabiko of Tsukuba /21/ (figure 11). VESA /22/ of LATEA in France

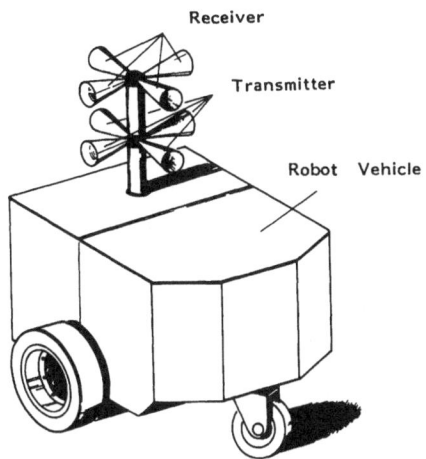

Figure 11: Environment perception with rotating
          ultrasonic range finders (principle of the percep-
          tion of Yamabiko)

and Microbe of Munich University /23/ use this approach. It
is a very simple way of getting much information about the
surrounding world. Disadvantageous is the long time necessary
for a complete scan.

5.5. Electronic scanning with an acoustical phased array

Without any mechanical movement an acoustical beam can be directed in a chosen direction by phased array technique. By
phase shifted driving of an array of acoustical transmitters
a deflected and focused ultrasonic beam can be produced as
a result of the superposition of the acoustical waves from the
different transducers. The deflection angle depends on the
pattern of phase-shiftings.

Figure 12: Ultrasonic phased array scanner head with 16 transmitting elements

Such an ultrasonic phased array scanner with 16 elements has
been developed at the IPA /24,25,26/. The phase angle and the
number of pulses for the transmitted ultrasonic burst can be
programmed independently for each element under computer control. The figures 12 and 13,14 show the scanner-head and the
beam characteristic for different angles.

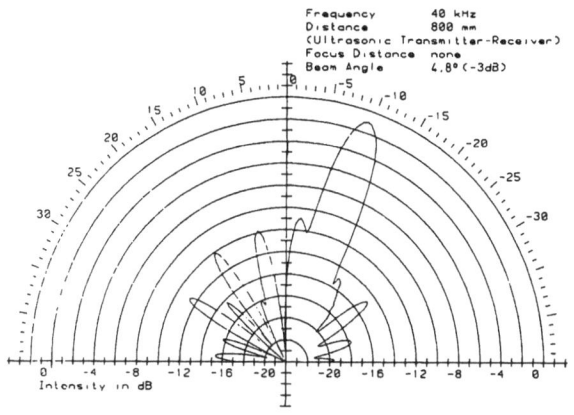

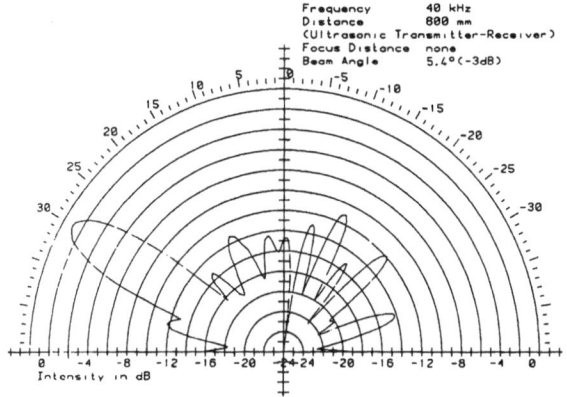

Figures 13,14: Directivity characteristic of the phased array
transmitter head

A still unsolved problem remains the interpretation of the
received signals. Compared with optical scanners, such a device
is still low cost and the integrating characteristic of the
measurement results are usefull for the application with mobile
robots to give a rough perspective of the environment. This is
in most cases absolutely sufficient for path planning.

6.   Combined sensor systems for mobile robots

A historical example for an autonomous mobile robot using a complete system of complementing sensors is Shakey /10/ which has been developed at SRI from 1968-1973. For navigation odometers on the two driven wheels are used. Several tactile cat whisker sensors perform a bump detection. The environment perception is done by grey scale vision in combination with scene analysis and by a laser scanner. Both systems are mounted on a rotating platform on the top of the robot.

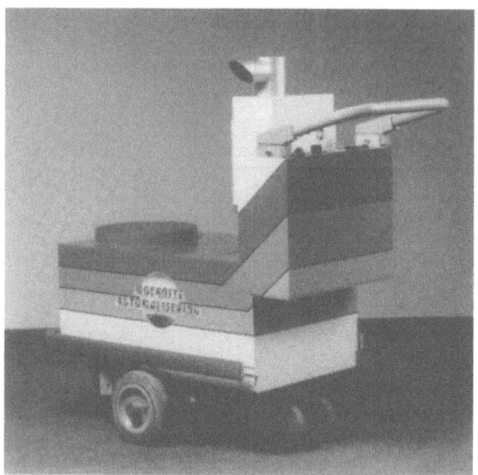

Figure 15: Sensor guided mobile robot of Moekotte Automatise-
ring (NL) with optical and ultrasonic sensors

The following examples derive from industrial developments of the recent years and have been presented to the public as product oriented prototypes:

The Dutch company Moekotte has developed the mobile robot
"Scrobbie" (figure 15) with a rotating laser scanner on top
for navigation. The vehicle's position is computed by the
angles of reflections from optical reflector stripes. In addition to this external navigation, by integration of motion
direction and odometry an internal navigation is performed.
Collision avoidance is done by five ultrasonic proximity switches in the front.

Figure 16: Mobile cleaning robot from Toshiba

A Japanese development from Toshiba (figure 16) uses a gas
rate gyroscope for navigation a tactile area switch in the
front for collision detection and laser based optical range
finders of the side as well as an ultrasonic range finder in
the front for environment perception.

Hitachi developed the automatic vacuum cleaner robot "Robby"
(figure 17). Position finding is done by two complementing navigation systems: a vibration gyroscope and odometers for the
two rear wheels. A tactile ribbon around the vehicle is for

Figure 17: Household vacuum cleaner robot of Hitachi

collision detection while the front contains an ultrasonic range finder and two photoelectric proximity switches on the side for collision avoidance. For environment perception a rotating ultrasonic scanner which makes use of a parabolic mirror is situated on top.

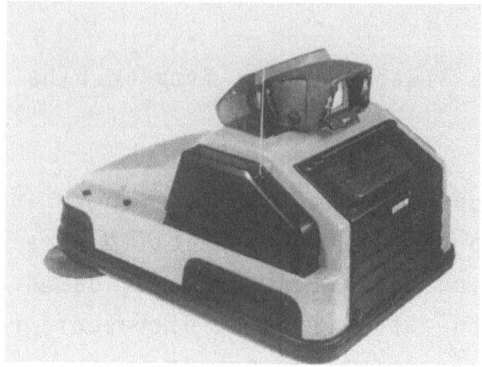

Figure 18: Mobile Robot of Automax, Japan

A similar device has been built by Automax. Again, the navigation system consists of the combination of a gas rate gyro with
odometry. A tactile ribbon and several ultrasonic range finding sensors are for collision detection and environment perception.

All of these mobile sensor guided systems are limited in their
application possibilities. Perception capabilities and intelligence and still to be improved, but the existing developments
mark the way towards future.

7.   References

/1/   Giralt, G. "Mobile Robots". NATO ASI Series, Vol. F11
      Robotics and Artificial Intelligence edited by M. Brady
      et al, Springer-Verlag, Berlin, Heidelberg, pp. 365-393
      (1984).

/2/   Place, H., Julliere, M. and Marce, L. "Qu'en est-il des
      robots mobiles?". Le Nouvel Automatisme, pp. 31-39 (Ja-
      nuary-February 1983).

/3/   Fischetti "Robots do the dirty work". IEEE Spectrum, pp.
      65-72 (April 1985).

/4/   Raibert, M.H., Brown, H.B. and Murthy, S.S. "Machines
      that walk". NATO ASI Series, Vol. F11 Robotics and Arti-
      ficial Intelligence edited by M. Brady et al, Springer-
      Verlag, Berlin Heidelberg, pp. 345-364 (1984).

/5/   Tsumura, . "Recent Development of Automated Guided Ve-
      hicles in Japan". Robotersysteme, ol.2, No.2, pp.91-97
      (1986).

/6/ Arai, T. and Nakano, E. "Development of Measuring Equip-
ment for Location and Direction (MELODI) Using Ultrasonic
Waves". Transactions of the ASME, Vol.105, pp. 152-156
(September 1983).

/7/ Ehtashami, M., Sung, J.O. and Hall, L. "Omnidirectional
Position Location for Mobile Robots". Intelligent Robots
and Computer Vision, SPIE Vol. 521 (1984).

/8/ Nakamura, T. "Edge Distribution Understanding For Locating A Mobile Robot". Proceedings of the 11. ISIR.
Tokyo, 1981, pp. 195-202.

/9/ Fujiwara, K., Kawashima, Y., Kato, H. and Watanabe,
M. "Development Of Guideless Robot Vehicle". Proceedings of the 11. ISIR, Tokyo, 1981, pp. 203-211.

/10/ Rosen, C.A. "An Experimental Mobile Automaton". 18.
Conforence on Remote Systems Technology, Washington,
D.C., November 1970.

/11/ Julliere, M., Leflecher, J., Marce,L. Perrichot, H.
and Place, H. "Detection simple d'obstacles: bras tactile pour robot mobile", Le Nouvel Automatisme, April
1981.

/12/ Duda, R.O. and Hart P.E. "Experiments In Scene Analysis". Proceedings Of The 1. National Symposium On Industrial Robots, Illinois, April 2-3, 1970.

/13/ Briot, M, Talou, J.C. and Bauzil, G. "The Multi-Sensors
Which Help A Mobile Robot Find Its Place". Sensor Review, January 1981.

/14/ Moravec, H.P. "The Stanford Cart And The CMU Rover". Proceedings of the IEEE, Vol. 71, No.7, July 1983.

/15/ Lewis , R.A. and Bejczy, A.K. "Planning Considerations For A Roving Robot With Arm". Proceedings of the 3. IJCAI, 1973, pp. 308-316.

/16/ Fujii, S. and Yoshimoto, K. "Computer Control Of A Loco- motive Robot With Visual Feedback". Proceedings of the 11. ISIR, Tokyo, 1981, pp. 219-226.

/17/ Lewis, R.A. and Johnston, A.R. "A Scanning Laser Range- finder For A Robotic Vehicle". 5. IJCAI, Cambridge, Mass., USA, 1977.

/18/ Bauzil, G., Briot, M. and Ribes, P. "A Navigation Sub- System Using Ultrasonic Sensors For The Mobile Robot Hilare". Proceedings of the 1. RoViSeC, 1981, pp. 47-58.

/19/ Tachie, S. and Komoriya, K. "Guide Dog Robot". Procee- dings of the 2. International Symposium of Robotics Research, Kyoto, Japan, August 20-23, 1984, pp. 325- 332.

/20/ Coles, L.S., Robb, A.M., Sinclair, P.L., Smith, M.H. and Sobek, R.R. "Decision Analysis For An Experimental Robot With Unreliable Sensors." Proceedings of the 4. IJCAI, 1975, pp. 749-757.

/21/ Kanayama, Y., Iijima, S., Ochiai, H., Watarai, H. and Ohkawa, K. "A self-contained robot Yamabiko". Procee- dings of the 3. USA-Japan Computer Conference, 1978, pp. 246-250.

/22/ Monchaud, S., Merguen, H. and Lemaire, B. "A self apdap- ting low cost sonair for use on mobile robots". Sensor Review, October 1981, pp. 180-183.

/23/ Freyberger, F., Kampmann, P, Karl, G. and Schmidt, G. "Microbe - ein autonomes mobiles Robotersystem". VDI-Z, Vol. 127, No.7, pp. 231-236 (April 1985).

/24/ Ahrens, U. and Langen, A. "Ultraschallscanner, ein neues System zur Roboterführung". Proceedings, Steuerung und Regelung von Robotern, Langen, West-Germany, May 12-13, 1986,

/25/ Ahrens, U. "Möglichkeiten und Probleme der Anwendung von Luft- Ultraschallsensoren in der Montage- und Handhabungstechnik". Robotersysteme, Vol.1, No.1, pp. 19-28 (1985).

/26/ Ahrens, U. "Möglichkeiten und Grenzen des Einsatzes von Luft- Ultraschallsensoren in der Montage- und Handhabungstechnik". Robotersysteme, Vol.1, No.4, pp. 203-210 (1985).

MULTIMEDIA SENSORY SYSTEMS FOR THE
SOLUTION OF THE ERROR RECOVERY PROBLEM

V. Caglioti
R. Simino
M. Somalvico

Milan Polytechnic Artificial Intelligence Project
MP-AI Project
Dipartimento di Elettronica
Politecnico di Milano
Milan, Italy

Abstract

The advanced role of sensory systems is very important in
intelligent assembly robots.

Two special-purpose techniques are developed for
sensory perception and planning, by exploiting the
particularity of the problem they are meant to deal with:

to provide a robot with both the capability in
understanding the causes of errors in its activity and the
capability in correcting the error situation and in
recovering its fully operational functionality.

A robot, while is executing a working cycle, may be
stopped within an assembly process by an error caused by a
defective object, by a wrongly positioned object, by the
absence of an object, or by an unpredicted collision
between objects.

Therefore, when such unpredicted (and undesired) events
happen, the correct execution of the cycle may be
compromised or even stopped. Sensors can then be utilized
to extract knowledge about the actual (error) situation.
This knowledge can then be employed in order to plan the
correction of the occurred error.

In the sequel, we shall call this set of activities as
the activity of error recovery.

The purpose of this paper is to show how the
integration of multimedia sensory data may become
functional for the solution of the automatic error recovery
problem.

Before a planning a strategy intended to error
recovery, the elements to be considered within the planning
problem have first to be identified. These elements are:

- the starting state (in this case the error state);

- the target state, which has to be determined according
  to the task, that is assigned to the robot.

NATO ASI Series, Vol. F43
Sensors and Sensory Systems
for Advanced Robots
Edited by P. Dario
© Springer-Verlag Berlin Heidelberg 1988

Therefore a comprehensive correcting activity can be arranged along three phases:

1. the deduction of the task , starting from the assigned program and from possible further information;

2. the detection of the error situation, through the use of sensors;

3. planning a correcting strategy, whose execution allows the correct accomplishment of the robot task, in spite of the error that has occurred.

Our approach is based on the the adoption of a three-modules architecture.

1. A first (off-line) module, starting from the program assigned to the robot and using some knowledge about physical and geometrical features of the objects, makes a knowledge based **task deduction,** via a simulation of the program execution ("dynamic" action's analysis), in order to provide a wide representation of the desired evolving world state (in terms of positions, contacts, and constraints among the objects).

2. A second (on-line) module, after an error has occurred, plans a **dynamic** sequence of sensory detections, in order to acquire a sufficient amount of knowledge to construct a useful description of the actual (error) situation. The attribute "dynamic" refers here to the fact that the **next** sensory detection of the sequence depends on the **current** world model, that has been constructed exploiting the result of the current sensory detection.

3. A third (on-line) module plans a correcting strategy, using as starting state the description of the error state (provided by the second module), and using as target state one of the desired evolving states (provided by the task deduction module).

This approach allows to reduce the weaknesses indicated above, but has the drawback of a supplementary processing charge, especially during the task deduction activity.

However, the fact that the task deduction is executed off-line reduces the importance of this shortcoming.

1. INTRODUCTION

In many cases, artificial intelligence techniques can be employed advantageously, in order to improve the independence of robot systems from humans.

In additon, robotics can be viewed as the study of systems which are able to solve complex problems, by means of a cooperation of "primitives". Primitives are referred to as to elementary problems that are automatically solvable by the system. Among the primitives of a robot there are both the processing ones (e.g., data transfer between memory registers)and the "interacting" ones.

The "interacting" primitives can be divided schematically into:

- sensing primitives (inputs for the robot);

- actuating primitives (outputs for the robot).

A typical problem for a robot consists in the attainment of a set of desired modifications of the surrounding world. A plan is constructed and assigned to the robot in order to accomplish this task.

The world knowledge, involved in the robot plan, can contain inaccuracies or errors. Therefore a feedback mechanism, providing information to the robot about the actual world evolution, is needed in order to accomplish correctly the assigned task.

A feedback mechanism can be based on the use of sensors. Analysis of sensory data can be performed within three different scopes:

a. Off-line, during plan formation in order to increase the detail of the employed world model.

b. On-line, during the execution of the assigned program, whenever a sensory detection is prescribed to identify the actual world situation among the various possible situations provided in the program.

c. On-line, after an unpredicted event has happened during the execution of the assigned program (and a situation has been reached, that was not provided in the program), in order to acquire knowledge about the actual (error) situation.

A robust task-level language interpretation system should provide all these three feedback mechanisms.

The purpose of the system, that is being presented here, is to employ the third feedback mechanism, after an error has been detected, in order to plan the attainment of a desired situation, starting from the actual (error) situation.

Therefore error correction involves a planning activity; moreover two more activities prelude to it, namely:

1. the acquisition of knowledge about the error situation (error state perception);

2. the construction of a useful task description, in order to allow the selection of a target state, belonging to one of the desired evolutions (task deduction).

The robot task can be expressed both in terms of actions to be performed by the robot (explicit programming language) or in terms of the desired modifications of the world state (implicit programming language).

For the accomplishment of the task deduction, it would be convenient to have available a task-level description of the robot plan. In fact this description would then be expressed in terms of the desired evolving world state, providing directly the states that are needed as target for the planning of a correcting strategy.

At present, no implicit (task-level) language is implemented.

Nevertheless, a certain amount of research have been carried on about the error recovery problem.

Some works have been produced in which it is assumed that plan descriptions are available to the system, including high-level statements, like, for instance, "Screw S1 with Block2 and Block3" [Gini,Gini,Somalvico 75; Gini,Gini,Somalvico 80]

In some approaches, the task deduction activity is expedited in one of the following ways:

- the robot action's repertoire has been limited [Friedman 77];

- a **static** action's analysis is performed (i.e., the result of an action is assumed to be independent from the situation to which the action is applied) [Gini,Gini,Somalvico 75; Gini,Gini,Somalvico 80]

A three-module architecture for a correcting system has been proposed in [Gini,Gini,Somalvico 75; Gini,Gini,Somalvico 80]
a task deduction module, an error situation detection module, and a correcting strategy planning module are the proposed components of the correcting system.

A monitoring system has been proposed [Gini,Smith 86], in which some monitoring actions are interposed among the actions of the originary assigned program. Therefore an "augmented program" is executed.

Among the monitoring actions there are:

- sensing actions, that verify the correct proceeding of the execution;

- actions, that transfer the control to the emergency recovery module;

- actions, that create statements (facts) informing about the actual proceeding of the plan execution.

In many of the proposed paradigms [Gini,Gini,Somalvico 75; Gini,Gini,Somalvico 80, Friedman 77] the robot program, written in an explicit programming language and therefore expressed in terms of actions, constitutes the main part of the information, that is employed for the task deduction activity.

The knowledge about the task is impoverished during the preliminarly activity of translating a task-level (implicit) plan description into an explicit program, or during programming.

When a static analysis of the action is performed, during the task deduction activity, a weak representation of the evolving world state is constructed, in which some important physical aspects are ignored, like contacts and constraints among objects.

In spite of this weakness, several errors, like dropped objects or unpredicted collisions, are often correctable by the proposed systems.
These systems show good correcting performances in the following conditions:

- when if the actions are **reversible**; this means that the "ante-action" situation is reconstructable starting from the "post-action" situation, by means of the execution of a (sequence of) back-action (independent from the "ante-action" situation, which can be possibly unknown);

- when actions cannot have a variety of possible results, each one of these bringing to a different branch of the program.

When the above mentioned conditions are verified, errors are correctable: i) by retrieving some of the last executed actions, until one of the past situations is reconstructed, in which the error cause was not present yet; and ii) by redoing the retrieved actions, after a new intervention of the error cause has been warded off (e.g., by substituting a defective object).

An example of a non-reversible action consists in a motion action, in which an object is **pushed** along the table (while not being held by the gripper): in this case, the "ante_action" situation cannot be identified by using only both the "post_action" situation and the action performed; in particular the position that the pushed object assumed in the "ante_action" state is not unambiguously identifiable.

An example of an action, having a **variable** result, can consist in taking an object from a box, that contains objects of many types: in this case, if during an error correction a substitution of an object results in a change of it, then the task description available to the system does not allow an intelligent adaption of the correcting strategy.

When the above mentioned conditions are not verified, a more efficient state representation is needed.

For this aim, a formalism for representing physical world situations has been explored, that could be expressive for the general class of problems assigned to the robot: assembly tasks. A situation representation formalism, including positions of objects, contacts and constraints among objects has been chosen, because of its simplicity in describing desired (goal or intermediate) states for assembly tasks [Lieberman,Wesley 77; Lozano-Perez,Winston 77; Mazer 83].

It has been shown [Ambler,Poppelstone,Bellos 78; Poppelstone, Ambler 81; Poppelstone,Ambler,Bellos 80; Brooks 81; Binford 82; Brooks 83] that different **object** representation formalisms are the most convenient ones for different activities like sensing and planning. It has been suggested [Poppelstone,Ambler,Bellos 80; Poppelstone,Ambler 81], that an automatic exigency-directed selection of the object representation formalism should provide the best choice.

The work, being presented here, has started with an identification of a **task** representation formalism, in which variable result actions have been taken into account by means of a multi-situation **state** representation.

A three module architecture is employed in this approach:

1.   The task deduction module operates off-line, utilizes the explicit program, and constructs the set of the desired evolving world states, according to the chosen state representation formalism. In order to accomplish this goal a knowledge based simulation of the program (dynamic action's analysis) is performed. Moreover a second information source is needed, including a description of the initial state of the execution and a description of physical (weights, friction coefficients) and geometrical (shapes and dimensions of object's subparts) features of the objects.

2.   The error situation detection module operates on-line, after an error has been detected. It plans the use of sensors in order to construct a physical model of the error situation. No scene analysis is performed "ex novo" within this module: the model of the expected (desired) state, coming out from task deduction, plus some information about the proceeding of the execution (including the error event, e.g. collisions during trajectory, collisions at the arrival point, dropping parts during trajectory a.s.o.)are used to direct the sensory detections .

3.   The (on-line) module plans a correcting strategy starting from the error situation, whose description has been constructed by the preceding module. Before

starting again with tackling "ex novo" with the generic planning problem, whose elements are the starting (error) state and the target state (belonging to one of the desired evolutions), some partial solutions are evaluated. These partial solutions, (namely solutions of grasping problems, solutions of collision-free path finding problems, solutions of cronological actions ordering problems), are extracted from the explicit version of the assigned program, during the task deduction activity. This can sometimes expedite the solution of the planning problem.

With this approach some of the previous mentioned shortcomings are overcome, because the more expressive situation representation, which has been employed, allows the system to cope with problems resulting both from non-reversible actions and from variable result actions.

It has also been shown that the availability of both implicit plan description and explicit plan description is convenient in the following cases:

- the first one for task deduction;
- the second one for expediting the planning of a correcting strategy,

The chosen formalism for representing physical situations is expressive but difficult to manage. Therefore an additional processing charge weights on the three phases of error correction. Although this additional processing charge is not an important problem for the task deduction activity, since it takes place off-line, the error situation detection is slowed down, because a sequence of several sensing-understanding steps are required, in order to construct a description of the error situation, expressed in terms of defects in objects, positions of objects, contacts and constraints among objects, according to the chosen situation representation formalism.

Two more architectures are proposed:

1. An architecture for a task-level robot programming language translator, that constitutes matter of an ongoing ESPRIT research, is proposed, in which all of the three previous mentioned sensory feedback levels are provided, and in which the use of a planning module is synergically shared between i) implicit to explicit translation and ii) error recovery planning.

2. An architecture, in which an arbitering module manages information coming out from many parallel operating multimedia sensory systems (that are disposed in a cell-architecture) in order to expedite the translation of sensory data into "high-level" symbolic statements.

The paper is organized as follows: Section 2. illustrates the notion of Abstract Robot Machine. In Section 3. an architecture is proposed for a task-level language translator, which planning capabilities, sensory feedback levels and error correction capabilities are provided with. Section 4. deals with multimedia sensory systems, a hierarchic architecture for a single multimedia sensory block is proposed. In Section 5. a (two media) LINCE sensory system is illustrated, in which an ultrasonic proximity sensor is integrated with a CCD camera. In section 6. the problem of automatic error correction is analyzed. The paper goes on (Section 7.) with a description of the logical architecture of an automatic correcting system and with an illustration of the component modules. Finally some conclusions are outlined.

2. ROBOT AS AN ALGORITHMIC MACHINE

The continuous improvement in the functional flexibility in robots, due to the application of artificial intelligence techniques to robotics, makes the manipulator-based robot model obsolete.

2.1. Robot as an Algorithmic Machine

In order to take into account the employ of robot programming languages, robot can better be represented as an algorithmic machine. The components of an Abstratct Robot Machine are the followings:

a. A control unit.

b. A memory unit.

c. A processing **and interaction** unit.

d. An actuatory unit.

e. A sensory unit.

2.2. Relationships with a Computer

Therefore an abstract robot machine is an enlargement of a computer. In particular, an interacting unit presides over the interaction between the robot machine and the external world, through both the sensory unit and the actuatory unit.

Within this representation, the robot machine is the executor of an algorithm, which can be viewed as a cooperation of "primitives". Among these primitives there are both processing ones (typical of the computer) and interacting ones (that are typical of the robot). Furthermore, the interacting primitives can be divided into:

  - actuatory primitives (e.g., "move right_arm to position **P1** ");

- sensory primitives (e.g., "measure the inclination angle of the plan under the Structured Light Vision Machine").

A task can be assigned to a robot, through a program which describes the algorithm that contains the solution of the complex problem. There is a considerable amount of already operational robot programming languages, in which the actions are described, to be executed by the robot.

A further way to specify the task to the robot consist in transmitting to it the complex problem to be solved, instead of providing the elementary problems (i.e., the primitives) that are immediately solvable by the robot itself.

2.3. Knowledge and Meta-algorithms

In this case some knowledge about the problem domain must be available to the robot. The robot can then employ this knowledge, in order to plan the needed actions for solving the assigned problem. The operations to do, in order to construct the executable algorithm starting from the complex problem, can be described within a **meta-algorithm,** since an algorithm is produced during this activity. The activity, that is performed by processing **knowledge** is called **inferential** activity.

2.4. Types of Algorithmic Machines

The physical executor of the algorithm can differ, in general, from the physical executor of the meta-algorithm. For instance the **man** can construct the algorithm that a computer is going to execute; the man can construct an algorithm to be executed by a (blue-collar) robot; a (white-collar) robot can costruct an algorithm to be assigned to a man (as, for instance, in decision-support-inference systems).

Sensory systems are important for both the blue-collar and the white-collar robot.

2.5. Notion of Error Recovery

The robot can also construct algorithms to be executed by itself, as in the following case: when a robot, while is executing a working cycle, is stopped by an error caused by a defective object, or by a wrongly positioned object, or by the absence of an object, or by an unpredicted collision between ojects, the following problem arises:

to rescue from the actual error situation, and to recover the fully robot's operational functionality. This problem is called error recovery problem.

The robot can then use various sensors both to detect the error cause and to extract knowledge about the actual (error) situation. Successively the robot can use both the permanent knowledge and the on-line extracted knowledge to plan the correction of the error situation.

3. ERROR RECOVERY AND ROBOT PROGRAMMING

The error recovery activity can be set against the background of a more comprehensive activity: the translation of a task-level programming language. Although the translation involves a planning activity, the employ of sensory feedback is also needed to cope with problems deriving from the fact that the world knowledge, involved in the planning activity, is not exhaustive and therefore can contain inaccuracies and errors.

3.1. Components in a Programming System for Assembly Robot

The components of a task-level language translator are the followings:

a. An implicit-language interpreter. Its main input consists in the implicit-language program (expressed in terms of the complex problem to be solved). Its output consists in the explicit-language program (expressed in terms of the elementary primitives). This module is constituted by a set of interacting knowledge bases, and an inferential mechanism. Some feedback paths provide further inputs to this module:

 - the extracted explicit solution program that is used (by the skeleton extractor) to foresee the world evolution that can derive from the execution of the explicit program; this knowledge can then be used both to modify eventually the current explicit solution or to provide knowledge which will be used while planning a strategy intended to correct an error that may occur during the program execution.

 - information coming out from sensory detections, that allow the module to acquire knowledge about the actual world situation.

b. A compiler, that translates the explicit program into a virtual machine program.

c. An interpreter, that translates the virtual machine program into a virtual machine language.

d. The robot actuatory unit, that physically executes the virtual machine instructions.

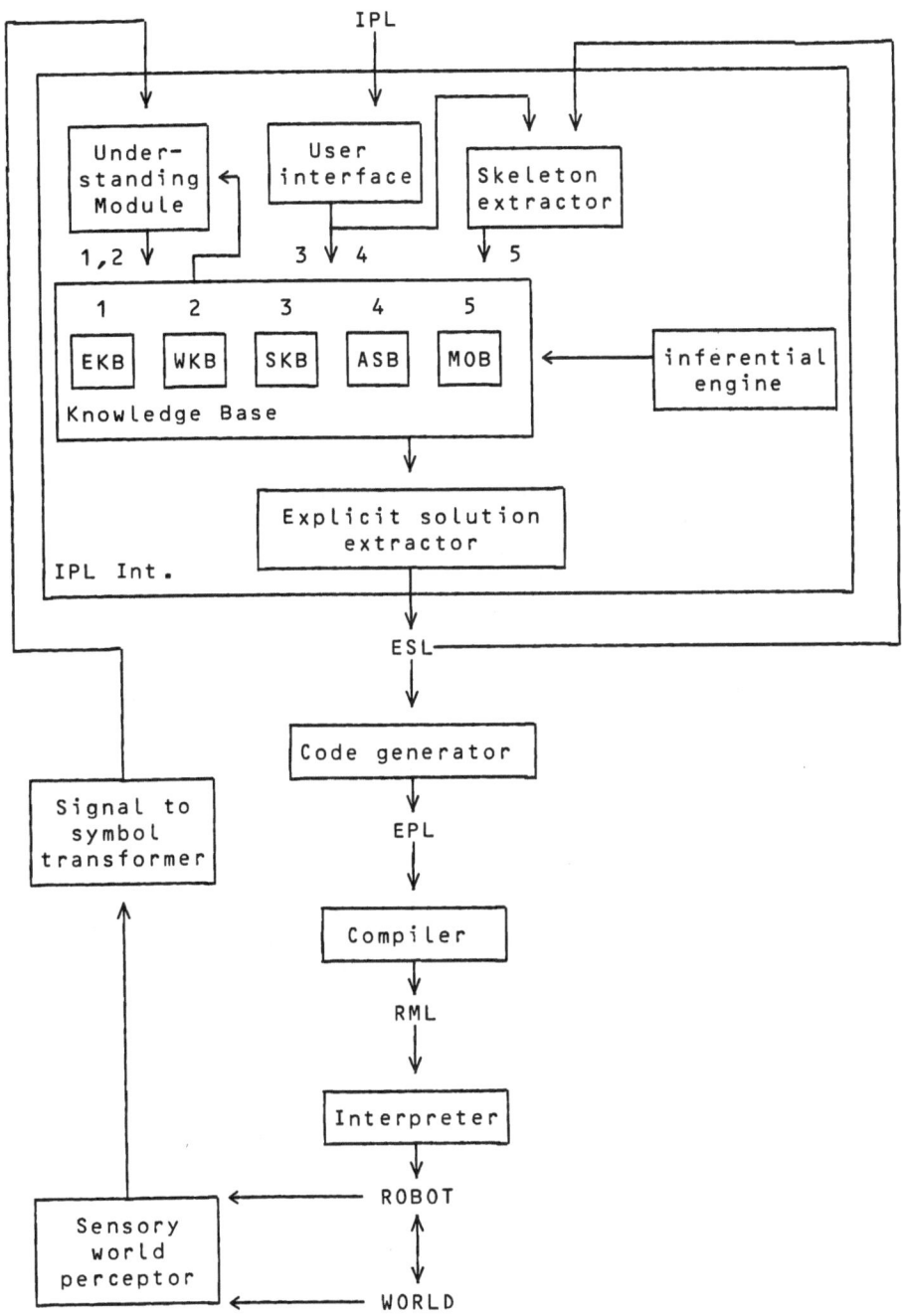

Fig. 1. Components in a Programming System for Assembly Robot

e. The robot sensory unit, that trasduces physical quantities into electrical signals.

f. The signal to symbol transformer, that performs the following activities:

- cleanses the signal from noises;

- eventually converts analog signals to digital ones;

- performes some preordained processings, in order to express information in such a way, that it can be easily translated into a meaningful form for the employ by knowledge bases.

    An example of this preordained processing can consist in the extraction of geometrical characteristics of objects, by programs that process the object image, constructed by a camera.

    This information is far from allow employing by a planning knowledge base, since for instance the position of the object and both the contacts and the constraints with other objects are not immediately extractable without an **understanding** inferential activity.

    Moreover, the variability of the ways, this symbolic information is susequently processed before a relative item of knowledge can be created, makes unpractical the arrangement of a further processing of this symbolic information within the signal to symbol transformer.

    Therefore an especialized inferential module is provided, that translates symbols coming out from signal to symbol transformer, into knowledge items. This activity is performed by using a partial world model (since it is easier to complete or to update than to construct "ex novo"). Furthermore, since an exhaustive perception is often not needed, the particular scope of the robot activity directs both the sensory perceptions and the sensory interpretations.

3.2. Role of Sensory Systems within the Interpreting System

Acquisition and interpretation of sensory data can be performed within three different scopes:

1. off-line, during plan formation, in order to increase the detail of the employed world knowledge.

2. on-line, during the ordinary execution of the assigned program, whenever a sensory detection is prescribed to identify the actual world situation among the various possible situations provided in the program.

3. on-line, after an error has occurred, in order to acquire knowledge about the actual (error) world situation, that does not belong to the situations provided in the program.

3.3.   Types of Knowledge Bases in the Implicit Language Interpreter

The knowledge base of the implicit language interpreter can be divided into many interacting knowledge bases:

1. The world knowledge base.

2. The skeleton knowledge base (from explicit program to the evolution).

3. The understanding knowledge base (from sensory symbols).

4. The sensory detection planning knowledge base.

5. The error detection knowledge base, that employes skeleton knowledge to direct the sensory detections to be planned, and to transfer the control to the next base.

6. The error correction knowledge base, that employes skeleton knowledge and error situation knowledge to plan a strategy intended to correct the occurred error.

4. ERROR RECOVERY AND MULTI SENSORY PERCEPTION

4.1.   Problems of Sensorial Perception

Many transformation-steps are to be executed, before the information that comes out from a transducer can be utilized within a knowledge based inferential activity. This information is expressed in terms of an array of time-varying (electrical) quantities, as it comes out from a transducer. This signal has to be sampled, and converted to digital. If the target of the sensory perception is the construction of a **static** world model, then the average value of each time-varying signal has to be extracted. A cleansing of noise has also to be performed, either before the analog-to-digital conversion or after it.

Since after these activities has been performed information is expressed by symbols, it can now be processed by a computer.

A sequence of processing steps of the sensorial data remains to be done, before they can be expressed in terms of statements that are utilizable as items of knowledge within one of the previous mentioned knowledge bases (e.g., within the world knowledge base for planning activity).

Some of these steps (the "low level" ones) are
performed within dedicated blocks, by executing preordained
programs. Other steps (the "high level" ones) are performed
within an especialized understanding inferential module.

The inference activity, performed within the
understanding module, depends both on the available partial
model of the current world situation and on the scope of
the system activity. Therefore, when a certain data
processing is needed, either the understanding module
orders a dedicated block to perform it (if there is one
dedicated to it) or the understanding module performs it
itself .

It is convenient to provide dedicated blocks, which
perform those processings that are needed to transform data
into a meaningful form for being used by the understanding
module.

4.2.   The Fusion of sensory Data

Often a fusion, of data coming out from different
sensors, is convenient for providing sensory information
with a suitable physical meaning. For instance a block,
that processes data coming out from a CCD camera and
integrates them with data coming out from an ultrasonic
proximity sensor, provides a useful geometrical features of
objects (like dimensions, moments a.s.o.), expressed in
metrical (instead of angular) dimensions.

If recognizing objects basing on their elasticity is
not beyond the scope of robot activity, a block integrating
information about the robot gripper span (coming out from
an encoder) with force sensors data could be useful, as a
further example.

4.3.   The Production of a World Model as an Inferential
     Activity

During the construction of a world model, within the
scope of error recovery, the understanding module can
exploit the current world situation model and can update it
basing on the interpretation of the results of sensory
detections. During interpretation the current world model
is compared with the results of both the present sensory
detections and the past sensory detections (that were
performed in the actual world situation).

Whenever a contradiction is discovered, between the
current model and the result of sensory detections, a set
of possible situations is then constructed (that is
compatible with the observations). An especialized
inferential module is then called into operation, that
plans further sensory detections in order to update the
current set of possible situations (possibly reducing the
cardinality oh this set).

When a selection becomes necessary (e.g., because no more useful sensory detections are executable) the world situation can be adopted, that is the nearest one to the expected situation.

4.4. The Arbitering between construction and utilization of a World Model as an inferential Activity

An arbitering module decides whether further inference is necessary to complete the current interpretation, or further processing of the sensory data can be suitable, or a new sensory detection is advisable, or an actuating action can be executed.

In the case of error recovery, the arbitering activity is performed by the so called "directing" module. This module exploits the knowledge about the task to be accomplished, and exits from the detection-interpretation cycle as soon as no more detail is needed (about the actual world model) for constructing a correcting strategy.

Within more general scopes, if the total time needed both to perceive the world situation and to plan a consequent action to be executed can be limited, then the system can cope in real time with a world, that can be modified, even by an external agent, with a speed related to the above mentioned time.

In this case, an expected world model is hardly exploitable, and therefore the interpretation activity becomes harder.

5. UNDERSTANDING SENSORY INFORMATIONS

The concept of sensor is generally associated to a dipole where electric characteristic at its clamps changes in function of a physical agent belonging to the ambient where it is situated.

This way to see sensors was exact until the semiconductor technology hasn't allowed the development of new VLSI components. They have had the following consequences for these system:

1. The necessity to interface them with a computer for a correct management of the automation process;

2. The development of sensors where the information is constituted by aggregate of the signals coming from a high number of physical transducers (CCD cameras).

Now many types of sensors are been developed and we can classify them on the basis of the type of information they supply:

1. Binary sensors;

2. Sensors for a middle informative complexity;

3. Sensors for a high informative complexity.

The sensors of the first type don't require particular explanations because they are components with two logic states, like microswitches, photocells, etc.

The sensor of the second type supply a signal that continuously changes within a determinate range of values, and that requires a time sampling and an analog-digital conversion.

The sensors of the third type are generally constituted by a group of sensors of the first or second class. For example in CCD cameras where we have a matrix of photosensitive elements. In this case, besides the eventual process of signal conversion in binary code, it is necessary to store the single information to allow an eventual processing of the whole data collected.

At this point we have to deal with the problem of the best utilization of the information. Before doing it, we want to supply a new classification of the sensors on the basis of their use [Cassinis 79; Cassinis 81]:

1. Sensors for the correct operation of the machine;

2. Sensors for the security of the robot and of the surrounding ambient;

3. Sensors for the correct operation of the executive programs.

The devices pointing out velocity, position of the moving components and the ones for the actuation of the machine, belong to the first class.

The sensors of the second class are usually binary, nevertheless they are in some case substituted by sensors for a middle informative level to foresee (and not to verify) eventual emergency situations (for instance, the ultrasonic distance gauge).

The vision systems, the artificial skins and others form part of the third class. In this class there are the sensors more used in artificial intelligence.

In figure 2 we can see the employ of the sensors in robotics on the basis of type of information that they supply.

Let us refer to both the classification we have used. The researches have been directed in particular to the development of the management of third type sensors. In fact they are the most qualified for automatic resolution of the robot problems.

However often we have to integrate the information coming from many kind of sensor (also belonging to second and first class) to obtain an information that an expert

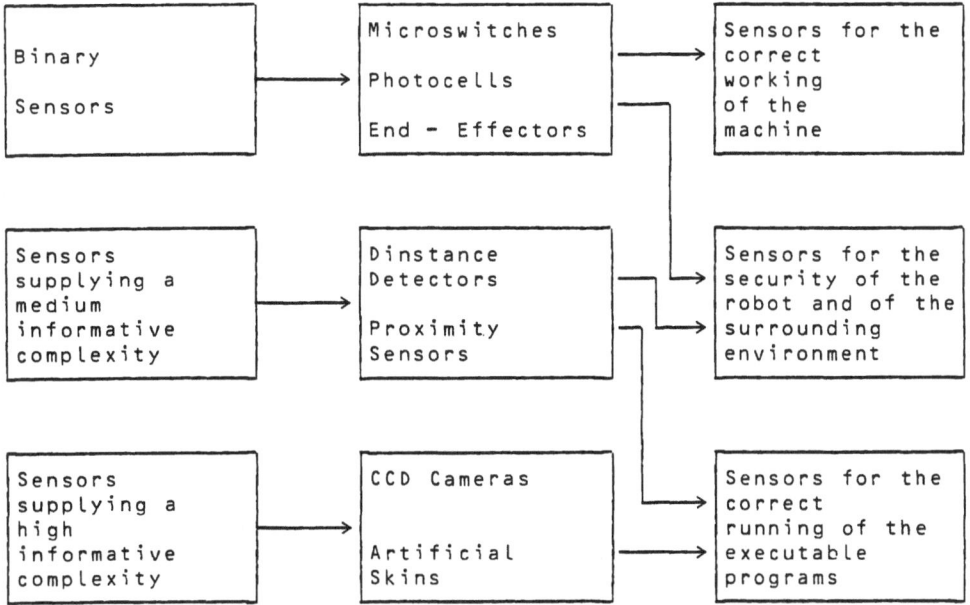

Fig. 2.

system can utilize. For instance to have a three - dimensional vision system we have to integrate the CCD camera image with sensors that have to detect the lens-object distance and the configuration of the object's surface.

The resolution of these problems has led us to develop a new architecture (see figure 3) to manage a multimedia sensory system. This architecture, who is shared in three hierarchic levels, allows a right subdivision of the task and, as consequence, a real-time response of the machine that it manages. The levels are determined by the kind of information in them elaborated.

5.1. Third level

At this level the sensor information, also if binary-coded, can require a lot of elaborations before to be utilized by the computer that manages the robot.

To hasten the operative capacity of machine, therefore it is necessary join to the sensor one or more microcomputers that develop the device function suppling to the upper levels some data that doesn't require further dedicated elaborations. Such a work of preelaboration can appear complex and it creates hierarchic microcomputer structures inserting in the main architecture and constituting a sensor.

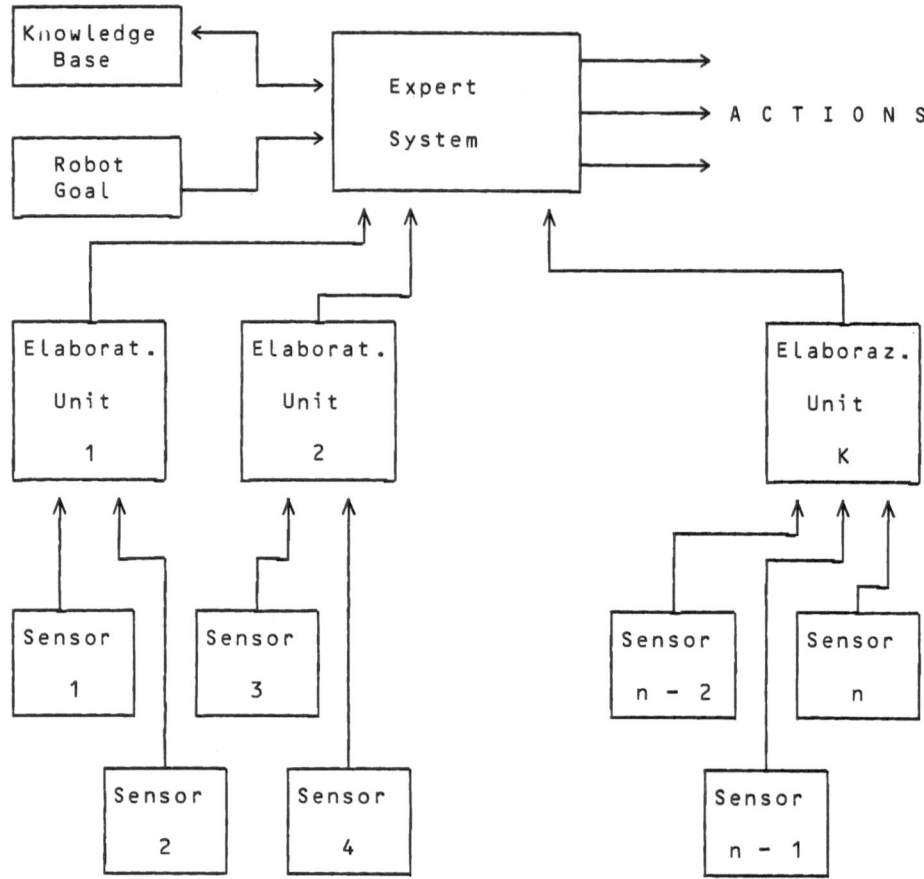

Fig. 3. Hierarchic architecture on three levels

5.2.   Second Level

At this level the diversification is no more so evident, because data are generally expressed in binary code according to the used communication protocol. The diversification is to be searched on the basis of the evaluation of the following parameters:

a.   Object of the sensor;

b.   Meaning of data supplied by the sensor;

c.   Sensor technology.

In our system the sensors are grouped in set on the basis of the aim which they are destined to.

The information supplied by the ultrasonic device has naturally a different meaning from that supplied by the camera. In fact it is destined to allow the calculus (in metric measure system) of the real dimension of the object to manipulate, which would otherwise be expressed in pixel.

The information obtained by the camera is extremely complex since it is composed of a large number of binary digits (typically one for each pixel) which must be processed as a single datum. Inside the aggregate considered in the example, therefore there must be the integration of all the data provided by its sensors.

The necessity to execute such tasks in real time, requires on this level some multiprocessor structures working in parallel to process the image (binary-coding, characteristics extraction, calculus of the geometric dimensions expressed in pixel end after expressed in real dimension thanks to the information provided by the ultrasonic sensor). For this reason besides their own memory they must dispose of a shared memory to store the common data.

5.3.   First Level

At this level the machine must perform a comparison between the information provided by its sensors and its experience to be able to execute correctly the planned tasks. This operation is performed by one or more expert systems which must match the received signals and knowledge base to supply a decision (workings movement,data request from other sensors). Besides it is opportune the machine is able to get up to date its knowledge base at any time on the basis of the information provided by the sensor.

So the architecture of figure 3 doesn't result relatively simple as it appears, but we must be remember that each its block can contain some complex structures.

It is very important to analyze the architecture of these modules to understand as a simple sensor has evolved in function of the new exigences of the expert systems. For this reason we will tell the project an the realization of the first sensor of the kind above mentioned.

6. ULTRASONIC SENSOR

The ultrasonic signals, whose frequency is included in a 20 kHz and 100 kHz range, are used for distance detection, thanks to their capability to be reflected by obstacles along the propagation. In the case the signal travels through two different media divided by a infinite

flat surface compared with the wave-length of the signal.
The signal percentage reflected is function of the <u>Acoustic</u>
<u>Impedance</u> R, [Carlin 49] a parameter that is characteristic
of the propagation medium.

So we have the following relations:

$$R_1 = {}_1c_1$$
$$R_2 = {}_2c_2$$

= medium density
c = velocity propagation in the medium

The amplitude of the reflected wave is exposed as:

$$A_r = \frac{R_1 - R_2}{R_1 + R_2} A_i$$

Therefore there is total reflection when the wave
travels from a gas to a solid. In this case the direction
of the reflected beam results to be symmetric in comparison
with the normal of the incident plain. Instead, when the
surface presents irregularity superior to 1/20 of the wave
length, the reflection happens along infinite directions.
Moreover it is possible to establish trough experimental
trials, dependently by the shown parameters, that some
obstacles with dimensions like the wave length of the
utilized signal, have some frequency for which the
reflection is high and others for which it is very small.

Furthermore it is possible, with experimental trials
depending on the previous parameters, that the objects
having dimensions next to the wavelength of the used signal
have some frequencies for which the reflection is high and
others for which it is practically non-existent.

6.1.   The Polaroid ultrasonic ranging system

Initially we used the kit that Polaroid planned for its
cameras. This kit consists of two circuit boards, the first
is a receiver-transmitter device; this accomplishes the
task of generating, upon order, a burst of 1 ms, that is
composed of a series of wave bursts having frequencies
between 50 kHz and 60 kHz.

The transducer is used both for transmitting and for
receiving, so after transmitting the signal, it can detect
a possible reflected wave.

The second circuit measures the elapsed time between
the beginning of the transmission and the moment the first
echo arrives. Then this time is used to compute and to
display the distance from the object.

This sensor has been used in several trials that
emphasized some faults which made it useless for the
purposes which it was intended for:

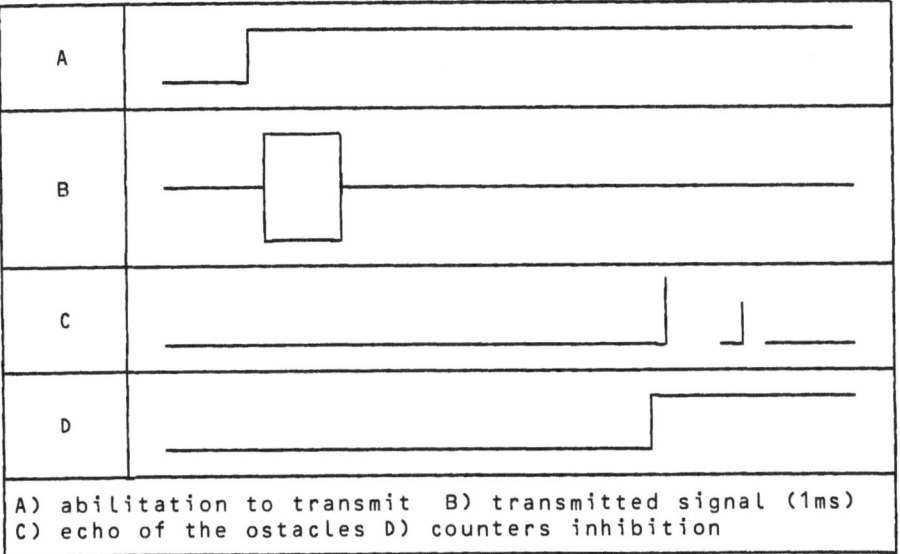

A) abilitation to transmit   B) transmitted signal (1ms)
C) echo of the ostacles D) counters inhibition

Fig. 4. Polaroid receiver-transmitter signals

1. Measure resolution of 3 cm. So it causes unacceptable mistakes in the computation of the objects dimensions to manipulate;

2. Since the sensor has to be able to detect any object, independently of shape, material, kind of surface, it uses different exploration frequencies; this implies a rather long transmission time (1 ms.). We have to add to it the time of inhibition. In such a way it is possible to damp the transducer oscillations.

   In order to save space on the robot gripper, we use a single ultrasonic capsule acting both as a transmitter and as a receiver; this involves that it can't receive any signal in the mentioned interval.

   Since we know the propagation velocity of an ultrasonic wave, we calculated that the minimum measurable distance is 30 cm.;

3. The bad directional characteristic of this device. In fact if there are two objects inside the emission solid angle, the transducer detects the distance of the nearest of them, so we can not know which object it belongs to.

   We thought to obviate the first fault replacing the second circuit board with another one having a much higher clock frequency of the counters. In such a way we thought to obtain an accuracy of about 1 mm. However it would be corresponded to the real distance only for a single kind of objects.

In fact, the counters are enabled at the same time when the transmission begins; but this time doesn't correspond to the beginning of the emission of all the used frequencies.

Let's examine this case with an example. Let's assume to have two objects different for material, shape, surface and call them 1 and 2. Let's call f_1 and f_2 the frequencies, emitted by the receiver-transmitter device, that can detect these objects. Let's put the object 1 at a previously fixed distance m from the transducer. If t_0 is the trigger time of transmission, f_1 will be transmitted at:

$$t_0 + x$$

where:

$$0 \leq x \ll 1ms$$

If t_d is the time that f_1 emploies to reach the object 1 and go back to the transducer, we'll have the counters will be stopped after a time:

$$t_1 = x + t_d.$$

Let's put now the object 2 at the same distance m, then f_2 will be transmitted at:

$$t_0 + y,$$

where:

$$0 \leq y \ll 1ms \; e \; x \neq y.$$

We can assume the same propagation velocity for f_1 and f_2, therefore the counters will be stopped after a time:

$$t_2 = y + t_d.$$

We can deduce that t_1 isn't equal to t_2; therefore the accuracy is made worse. In figure 5 there are the plot of the signals versus time for the previous example.

Since it is impossible to modify the Polaroid receiver-transmitter device, which is made with custom devices, we decided to design and to realize a new one with devices easy to find; it had to have the following requirements:

- emission of a single frequency (so the emission time is reduced)

- chance of choosing the frequency to use dependently on the object to inspect.

6.2. Computing the distance between lens and object

For distances from the object to inspect between 10 and 150 cm., with the new receiver-transmitter device we can obtain a maximum accuracy of about 1 mm, depending on the rise-time of the first received echo. The resolution may be equal to this minimum value or not, according to the clock

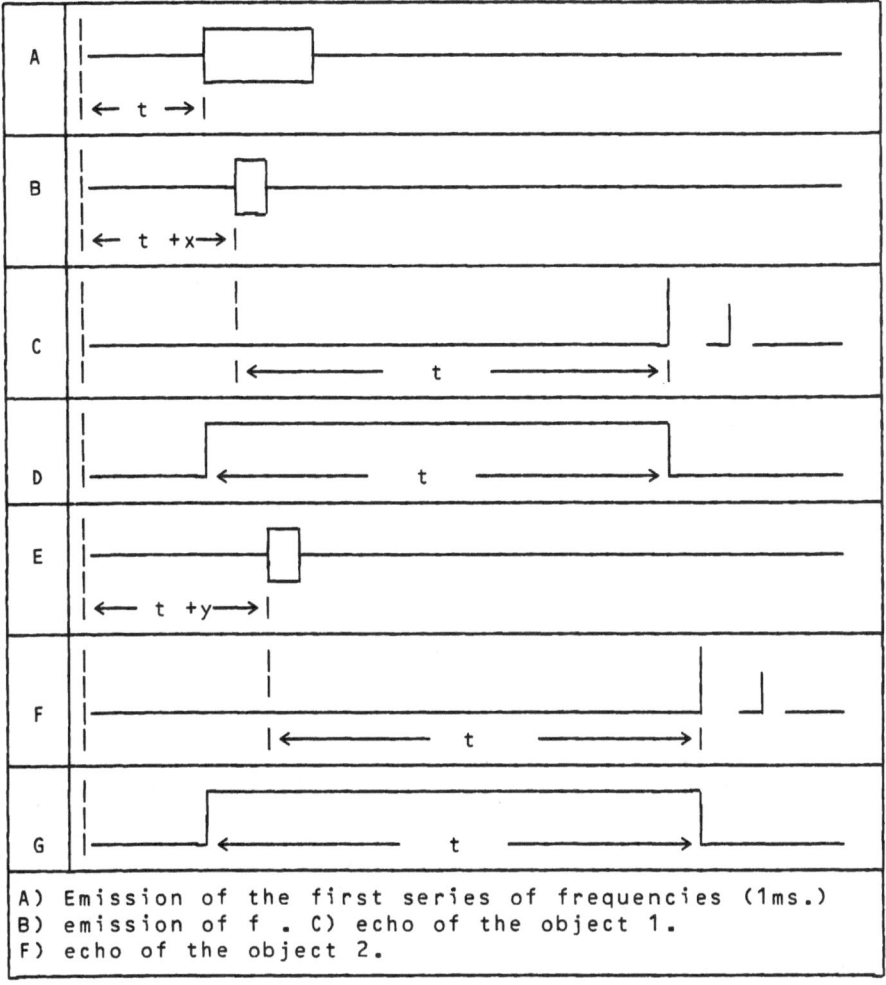

A) Emission of the first series of frequencies (1ms.)
B) emission of f . C) echo of the object 1.
F) echo of the object 2.

Fig. 5.

frequency we have chosen for the counters. The frequency we chose allows us a resolution of 2 mm.

In order to make the measure even more accurate, we used the Intel 8748 microcomputer (MICRO1) [MCS-48 Microcomputer User Manual 78], having the task to manage a series of measures, saving them in its memory and performing on them a series of statistical operations; these give a value sufficiently independent on the noise that may happen during the acquisition time.

The use of a single frequency has furthermore improved the capability for directing the ultrasonic waves. After a series of trials we have been able to decide that the emission solid angle has a value lower than 10 degrees.

6.3. Intelligent selection of exploration frequency

Let's define now the criteria for the choose of the exploration frequency, which is based on its associated echo.

First of all, the system must be able to detect objects whose echo may not taken. On this subject there are two possible cases:

1. The object to manipulate is present, but the echo is not detected;

2. The object to manipulate is present and the echo is detected.

The first case, the simplest to deal with, occurs when the used frequency corresponds neither to that of the object nor to that of the surface which the object is on. We have therefore to test the working area with samples of different frequencies and to choose the best signal considering these following factors:

a. The shortest interval of elapsed time between the transmission and the reception of the first echo;

b. The maximum echo amplitude.

The factor a is certainly more important than the factor b, since the echo may be reflected either by an object or, more probably, by the surface which the object is on. In fact, if the object is a solid of practically infinite size placed perpendicularly to the emission axis, the percentage of the reflected wave, for every frequencies, would be 100 %. From this fact, we can argue what we have said before, because we can consider the plane with infinite dimension compared with the ultrasonic wavelength.

The b factor, partially included in case a, would induce us to think to a correct system operation, for the reason just described, we are sure of the right working only if the object is of the same kind of that one manipulated in the previous operation.

Therefore, when the system is turned on, we must scan all the exploration frequencies selecting the better one to employ. This operation must be remade every time we examine a new object.

Another 8748 microcomputer (MICRO 2) drives a programmable frequency generator (made using programmable counters) [Borghini, Fasola, Simino 85] and acquires the amplitude of echo by means of an a/d converter. Like MICRO 1 also MICRO 2 examines a set of several acquired values in order to calculate the measure of the echo.

6.4.   Sensor architecture and its operating mode.

MICRO 1 drives a receiver-transmitter device when measurement is requested. It must give a data usable both for focalization and for computation of the absolute object dimension.

Another 8748 microcomputer (MICRO 3), placed at same hierarchical level of the others, controls the stepping motor which drives the fiberscope focusing system.

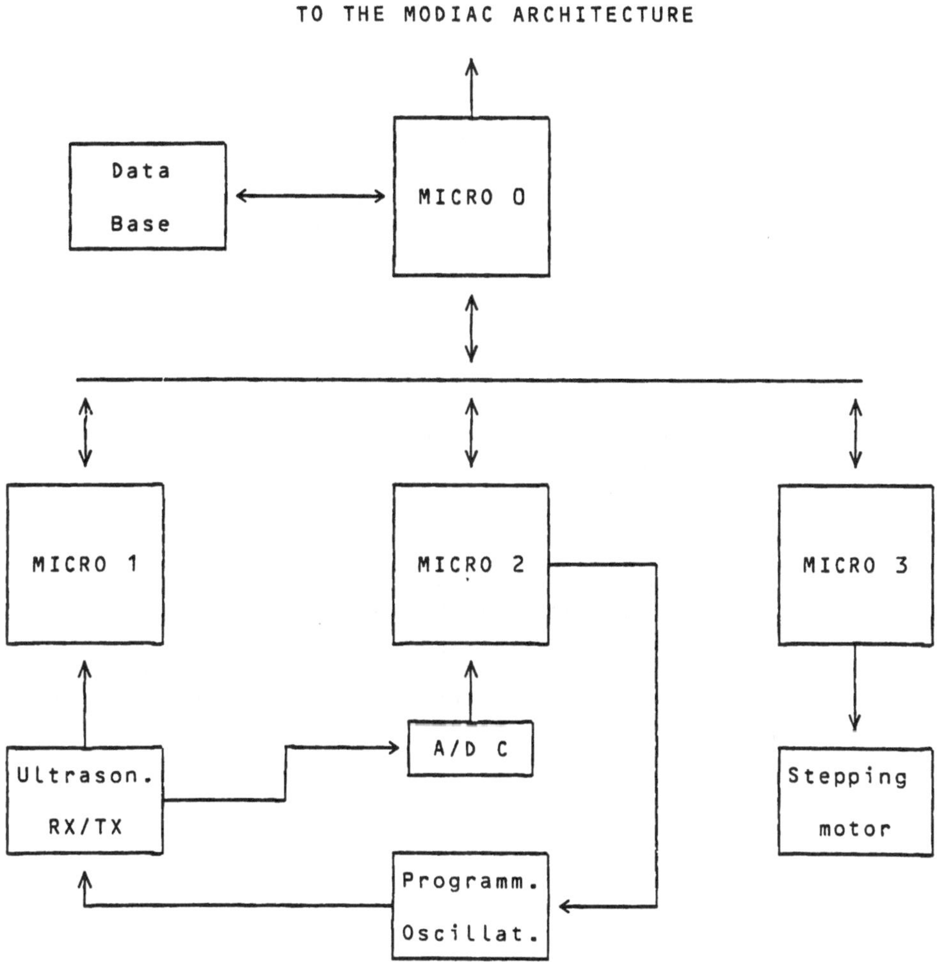

TO THE MODIAC ARCHITECTURE

Fig. 6. Ultrasonic sensor architecture

The Z80 microprocessor (MICRO 0) is placed in a hierarchical level above the previous ones and it handles the communication bus of the 8748 processors; more exactly:

- It handles the communication between the sensor and the upper level to which it is subjected;
- It selects the better ultrasonic frequency for a certain object on the basis of the transmitted parameters from MICRO 3 like mentioned in par 4.3;
- It manages the data base in which are saved the usual operating frequencies of the robot.

We can see the scheme of the whole ultrasonic sensor structure in figure 6.

7. THE PROBLEM OF AUTOMATIC ERROR CORRECTION IN ROBOTS

An approach to the problem of Automatic Error Correction in Robots is illustrated. A peculiarity of this approach is that the Correcting System is designed in order to deal with programs (assigned to the robot) which can sometimes meet, during its execution, variables based on world conditions, which are hardly predictable during the off-line programming activity. Therefore the computation of the program cannot be foreseen at the time of the compilation, since several control structures depend on predicates which have, within their scopes, some of the above mentioned unpredictable physical variables.

An introduction to the Problem of Automatic Error Correction now follows, then the global Architecture of the designed System is illustrated, successively the single modules of the System are examined as well.

During execution of an assigned program by a robot, some undesired and unpredicted events may happen (e.g., due to wrongly positioned or defective objects). These events may jeopardize the final result of the execution or may even stop it.

The goal of avoiding human intervention for recovering from emergency and of enabling the robot itself in automatically rescueing out from emergency has motivated the present work.

In any robot a program is provided in order to perform an interaction with the physical world, which consists in a sequence of desired modifications both of the physical state and of the relationships of the objects contained in the robot's working area.
When the programming activity is made by a human, the desired modifications can reside in an abstract idea as well as in some natural language statements (eventually integrated/substituted with mechanical drawings).

In the case of a selfprogramming robot the desired modification is described in terms of Implicit (Task-Level) Language instructions.

The execution of the assigned program produces an evolution of the state and of the relationships of the objects in the robot's working area. This evolution can differ from the desired one, because of the following facts:

- The world's representation involved in the generation of the assigned program contains error and/or inaccuracies, due to the limited amount of world-knowledge employed;

- Robot's actuators haven't an unlimited precision.

Sensory feedback can be used in order to opposite these deviations of the world evolution.

The following fact could give the feeling of an unusefulness of an Automatic Correcting System. In principle, the same world-knowledge employed in order to correct errors, by the correcting system, could be used by the **programmer** (or by the explicit-program-extracting module), in order to set up a more complex algorithm able of avoiding or of recovering from errors.

In order to avoid errors, some pre-action controls have to be introduced into the program. In order to recover from errors, some post-action controls have to be introduced into the program.

The justification of the project of an Automatic Error Correcting System can be argued as follows:

1. Given a desired modification and an associated executable robot program, some undesired events may happen during the execution of this program. There is a subset, of these undesired events, for which the following facts are true:

   a. the occurrence frequency is high enough to make them relevant;

   b. the occurrence frequency is low enough to make attractive the idea of saving programming-time, by leaving out of the assigned program the following items:

      - the pre-action controls (that allow to avoid these undesired events) that give almost always negative result;

      - the post-action controls (that allow to recover from these undesired events);

      - the code describing the avoiding actions, that follow pre-action controls giving positive result;

      - the code describing the correcting actions, that follow post-action controls giving positive result;

2. The Automatic Correcting System is designed in order to operate with a **variable** desired modification (to which a variable assigned program is associated). So the programmer's time-saving - due to the entrusting (to the correcting system) of the responsibility of the management of the undesired events - can be multiplied using the Correcting System with many assigned programs,

The assigned program constitutes one of the off-line inputs of the System.

8. THE AUTOMATIC CORRECTING SYSTEM: LOGICAL ARCHITECTURE

We can sketch the performance of the system as follows:

1. An off-line module processes some information (a part of which may be constituted by the assigned program), in order to determine the set of the desired evolutions of the working area;

2. An on-line module, once an error during the program execution has been detected, executes the following actions:

   a. Interrupts the execution of the program;

   b. Suggests the necessary sensorial detections in order to achieve information about the current physical situation;

   c. Tranfers the control to the next module.

3. Another in-line module selects a target-state belonging to one of the desired evolutions and plans the way to reach this target-state starting from the error-state.

If an Automatic Correcting System is available, the programmer is able to avoid the definition of those instructions that usually he neglects because he relies on the human supervision of the robot [Gini,Gini,Somalvico 75].

The involved knowledge in the program, that has been written in the above-mentioned way, differs from the whole knowledge available to the programmer. Practically the employed knowledge is equivalent to the knowledge that is available to the programmer, plus some "good working" hypotheses.

For instance:
the knowledge

(if(peg P is inserted in holes H1,H2,H3 of blocks B1,B2,B3)
    then(in the reached situation constraints
        are created among P, B1, B2 and B3))

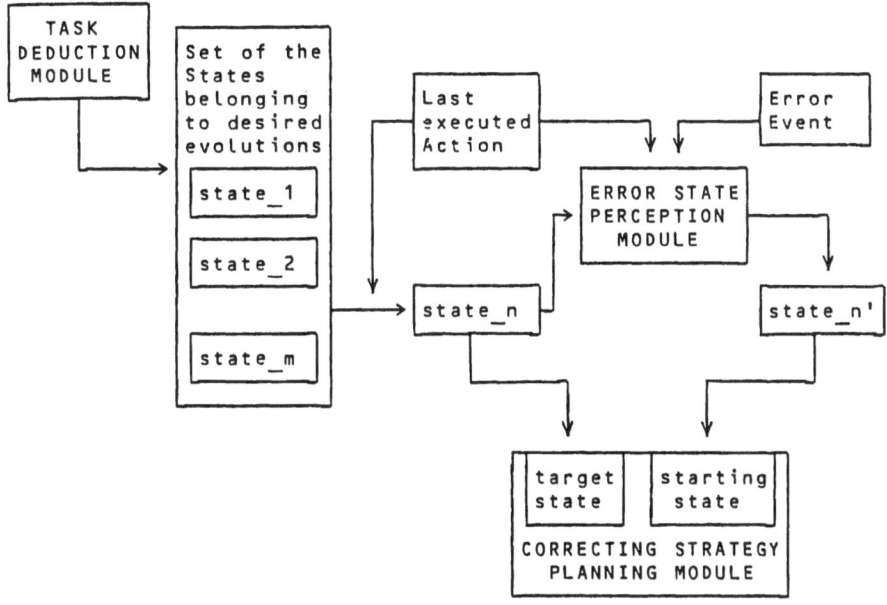

Fig. 7. Logical Architecture of the Correcting System

is equivalent to the knowledge

```
(if(peg P is inserted in holes H1,H2,H3 of blocks B1,B2,B3)
   then (in the reached situation
        (constraints are created among P, B1, B2 and B3)
          or
        (P partially inserted and P is defective)
          or
        (P partially inserted and B1 is defective)
          or
        (P partially inserted and B2 is defective)
          or
        (P partially inderted and B3 is defective)
          or
        (P partially inserted and P,H1,H2,H3
        are wrongly aligned))
 )
```

plus the hypotheses

```
  ((P  is not defective and B1, B2, B3  are not defective)
   and
   (P, B1, B2, B3 are allined))
```

Let 's suppose that a programmer's target is to create
a constraint among the peg P, and the blocks B_1, B_2 and B_3.
Let 's suppose that the programmer relies on one of the

following facts:

- human supervision of the robot is available;

- automatic correcting system is available;

- hypotheses, he is going to assume, are valid.

In this case the programmer has just to introduce, in his assembly program, the instruction "move P in position P_1 (where P_1 is the position where the holes are supposed to be).

8.1. The Module, that Constructs the Set of the Desired Evolutions

In order to plan a correcting strategy two elements are functional:

1. The starting state (in this case the **error** state);

2. The target state (belonging to one of the desired evolutions).

Information, about the current (error) state, can be extracted through the use of sensors. This activity is performed following some "criteria" described in the next Section.

Before the selection of the target state, the set of the desired evolutions must be constructed [Gini,Gini,Somalvico 80] . In order to do this, a **simulation** of the program is performed, starting from the initial state of the execution.

In order to accomplish the simulation further knowledge is needed describing the characteristics of all the objects that will be manipulated and the situation of the initial state of the execution.

If physical knowledge is employed during the simulation, then the set of the **possible** evolutions is constructed. If the physical knowledge is integrated with "good working" hypotheses (in the performing of the simulation) the set of the **desired** evolutions is constructed.

If a "task-level" language were employed to assign the program to the robot, the determination of the desired evolution would be an easy task. This is because high-level statements are expressed in terms of desired modifications of the world state, instead of in terms of actions to be performed in order to attain these desired modifications.

At present no high-level language is available. Therefore, in order to ascertain "what may happen", physical world knowledge must be applied to the knowledge about physical and geometrical characteristics of the objects to be manipulated.

In order to extract, out of the **possible** evolutions, the **desired** evolutions, the world knowledge is integrated with some "default" hypotheses. Examples of default hypotheses are:

- Movements do not stop before the target position is reached.

- Objects are not defective.

- When objects, belonging to particular groups, come into contact with each other, particular constraint are created (e.g. among the peg and the blocks in an above mentioned example constraints are created that impede relative translations along x-, x+, y-, y+, z-).

-

In order to disable some of the default hypotheses, two options are available for the programmer :

1. to use an "ON <condition> EMERGENCY" instruction, in order to signal that even a right terminated action can produce an undesired effect;

2. to use an "ON EMERGENCY CONTINUE" instruction, in order to signal that even a bad terminated action can produce a tolerable effect (e.g., using robot arm to look for an object by means of a collision with it).

The ultimate goal of the module is not the exhaustive prediction of the evolution of the physical state of the world. The precision required is needed in order to be able :

- to distinguish among error-situation and desired situation;

- to reach a desired situation, starting from an (error) situation.

The error situation is very similar to the situation to be reached, if compared with a random situation. In fact the objects in the working area are almost the same for the two situations, and almost all objects are in the same position.

Difficult physical problems like:

- to determine the displacement of an object $Block_3$, due to the collision with object $Block_2$ being moved by the manipulator;

- to determine, for an object $Block_2$ **pushed** by another object $Block_1$ (that is moved by the manipulator), the tangential component (to the contact surface between $Block_1$ and $Block_2$) of the displacement of $Block_2$;

are tackled, relying on the fact that no robot programmer pretends to move simultaneously more than an object in this way. Qualitative physical reasoning is then reproduced by this module.

For instance, it is assumed that:

- no collision is desired before a movement is finished.

- motion, along a direction parallel to the contact surface between $Block_1$ and $Block_2$, is transmitted from $Block_1$ to $Block_2$ if $Block_2$ does not have other contacts (see fig.6).

- if $Block_2$ has one contact with $Block_1$ (moved by the manipulator) and another contact with $Block_3$ (motionless), $Block_2$ follows the motion of the block that supports the main part of its weight.

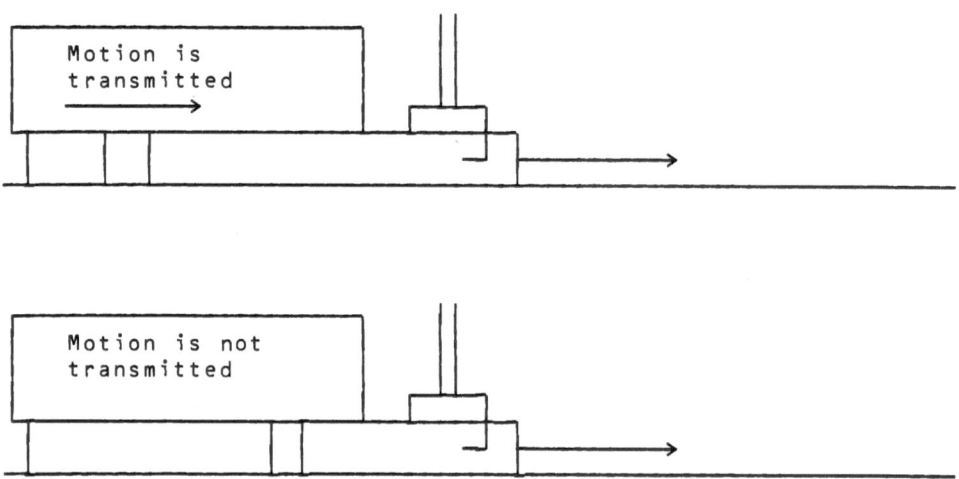

Fig. 8. Motion transmission by friction

The assigned program can contain **branches** (like "if <condition> then ...") and **loops** (like "while <condition> do ... "). The conditions, included in these control structures, may be dependent on some variables describing the physical state of the world. In this case the **flow** of the program execution will be dependent on the physical situation, that is evolving during run-time.

Therefore the execution flow may be off-line unpredictable.

The assigned program can be represented as a net, in which the links correspond to actions, and the nodes correspond to states.

During the program execution, whenever an action is finished , the reached node can contain a set of **variable** situations. The variability of a node-situation can depend on:

- the variability of the initial state situation;
- the variability of the characteristics of the objects, that are coming into the working area;
- the variability of the effects of some actions.

Therefore different desired evolutions can result from the execution of the program. In turn, these desired evolutions lead, in general, to different final situations.

Cosider a particular evolution, during the ordinary execution of the assigned program. When an error occurs, this evolution is interrupted, and, after an inferential activity, the execution of a correcting strategy starts.
If, during the strategy execution, some conditions change that prevent the actual evolution to meet the interrupted one, a problem arises:

how can the system decide which, out of the remaining desired evolutions, are the **valid** ones to be pursued?

The concept of **task** has to be specified for the particular class of problems (assembly problems) we are dealing with. A definition of task is proposed.

The part of the world, including the robot and the objects that can be manipulated is considered as a **system**, the so called "working area". Among the "communication channels" between the working area and the exterior there are:

a. Those behaving as inputs, e.g. lines carrying signals from the exterior, boxes and tapes carrying objects into the working area.

b. Those behaving as outputs, e.g. lines carrying signals from the working area to the exterior, tapes or boxes carrying objects assembled by the robots.

Actions are not considered as inputs, because they are **internal** to the working area (i.e. applied, by the robot, to the rest of the working area).

It is assumed that, in order to define a task, it is necessary and sufficient to specify the desirable behaviours (of the evolving working area), as viewed from the external point of view. More precisely, the **input** (plus the initial state) - **output** (plus the final state of the working area) relationships, are the needed specifications for defining a task.

For instance, neglecting the system outputs, a task, for which the needed relationships are specified, is accomplished correctly if :

- given the particular configuration, that is assumed by the initial state,

- given the particular configuration, that is assumed by the input variables during the execution,

the configuration, that is assumed by the final state of the execution, corresponds to the initial state configuration and to the input configuration, according to the specified relationships.

If the assigned program is available, then the set of the corresponding desired evolution must be considered in order to identify the needed task specifications. For every allowed (initial state configuration - input configuration) pair, the subset, of the desired evolutions, assuming these initial plus input configuration is then constructed. Finally the set of final state configurations, that are assumed by the evolutions belonging to the previous constructed subset, is bound with the single (initial configuration - input configuration) considered.

If, during the execution of the correcting strategy, the input configuration changes (e.g. due to a substitution of a variable object) a new final (or intermediate) state configuration is selected as strategy target state.

The criterium, that directs this selection, is the **compatibility**, of the new target configuration, with the new input configuration.

8.2.   The "Directing" Module

This module operates in-line, and performs the following activities:

1. Monitors the program execution, and stops it if an error is detected;

2. Integrates, with eventual information about execution (acquired during monitoring) information coming from the error event (e.g. the position where a movement has been interrupted),
    and plans sensing operations, in order to extract more information about the error situation;

3. calls the third module into operation, the output of this module being a correcting strategy to be execute;

4. monitors the execution of the planned correcting strategy, and performs the activities 2. 3. 4. again if the strategy was not executed correctly;

5. returns the control to the execution of the assigned program, starting from a point, that depends on the state that was reached by the execution of the strategy.

A description follows of the way, in which the system is designed to perform the first (plan monitoring) and the second (information extracting about the error situation) of the above mentioned activities.

The program execution monitoring

Before the way can be designed, in which the system should perform the monitoring of the program execution, the following fact must be considered:

Errors do not always cause an interruption of the program execution.

For instance, an insertion of a pin into a too large hole, is an undesired but "transparent" event.

In principle some of these "transparent" errors are detectable, through the use of sensors, because the task representation, constructed by the off-line module, is expressed not only in terms of the desired positions of the objects but also in terms of both the desired contacts and the desired constraints among the objects.

In the previous example the error is detectable, by verifying that the relative motion between pin and hole is possible, though orthogonal to the pin-hole axis.

Therefore, an "augmented program" may be constructed and executed [Gini,Smith 86], in which some "post-action" sensory detections are added to the originally assigned program. These "post_action" tests could have one of the following targets:

a. to verify the whole state;

b. to verify only that part of the evolving state, that had to be modified by the action (verifying the "effect" of the action).

In a first implementation of the system, the "augmented program" has been left equal to the originally assigned program (i.e., no action is executed, that does not belong to the assigned program, during monitoring). This choice takes into account the fact, that in industrial applications the actual time needed to execute the cycle has an economic relevance.

Therefore, in a first implementation, the program execution is interrupted only when a movement action is stopped before its regular termination ("emergency error") or when a condition, contained in an "ON <condition> EMERGENCY" instruction, proves true.

The extraction of information about the error situation

The following fact is taken into account:

It is unpractical to attempt to verify all the positions, the contacts and the constraint among the objects in the working area. In fact some information sources are examined before sensory detections are executed:

- ordinary execution information (about well terminated actions and about the results of eventual programmed sensing actions);

- exception excution information (about past bad terminated actions);

- the description of the expected world situation;

- information about the error event (i.e., the present bad terminated action).

The examination of the above listed information sources restricts the number of the nedeed detections. Furthermore, both the ultimate goal (the correction of the occurred error) and the intermediate goal (the detection of "mismatches" between the actual and the desired (expected) situation) causes an additional restriction of the number of the nedeed detections.

An incremental information acquisition is planned within this module. This incremental information acquisition is based on the reiteration of a cycle consisting of three steps:

1. The currently available information about the actual world situation is analysed, and a comparison is made between the result of this analysis and the expected world model. The possible "mismatches" between the expected world model and the possible models of the actual world situation, are identified. Among all the currently possible mismatches, the ones that have not been verified yet are considered and employed in order to construct the set of the (currently) possible actual world situations.

2. In order to verify some of the previously considered mismatches, some sensory detections are selected basing both on the relevance of each of the considered mismatches for the correction of the occurred error and on the "cost" of the needed sensory detection corresponding to each of the considered mismatches.

3. The selected sensory detections are executed and the results are classified among the following classes:

    a. results proving the presence of an expected "mismatch";

    b. results proving the absence of an expected "mismatch";

    c. results proving that no one of the two previous cases can be excluded;

    d. unexpected results (for which a more complicated interpretation is needed, that leads often to the detection of a new, unexpected mismatch).

The control is then returned to the first step, where the actual world model is updated.

The sensory detection interpretation-proposal-execution cycle starts by analysing the above listed information sources, looking for contradictory groups of information items. One of these groups consists certainly in the (expected-model, error-event) one.

Each non-consistent group of information items gives rise to a multiplication of the possible actual situations. For instance, the non-consistent group

(peg-diameter =4)
(hole-diameter=4)
(hole-and-peg aligned)
(stop-movement-during-insertion-peg-hole)

gives rise to different world situations. Among these there are, for instance:

1. (peg-diameter>4)
or
2. (hole-diameter<4)
or
3. (wrongly-aligned peg-hole)
or
4. (wrongly-grasped peg)
or
5. (defective peg-shape)
or
6. (defective hole-shape)
and many others.
Some of these statements (1.,2.,3.) contradict some items of knowledge, that explicitly belong to the description of the expected situation;
Other statements (4.,5.,6.) contradict knowledge items, that are involved implicitly in each of the expected situations.

Subgoal of the second phase (proposal) is the planning of sensory detections, whose result allows to prune some of the possible world situations deriving from non-consistent groups of information items.

During this phase some facts are considered:

- a sensory detection , that involves a disassembly of already assembled objects, has a high cost;

- if an alignment has been successfully performed, and a sensory detection involves the mis-alignment of the previous aligned objects, then this sensory detection has a high cost;

- a strategy, consisting in a substitution of a not yet assembled object, has a low cost;

- a strategy, consisting in a substitution of an already assembled object, has a high cost;

- a sensory detection, consisting in verifying a shape, has a high cost;

- a sensory detection, consisting in verifying a diameter dimension with an accuracy of 2mm., has a low cost;

- if a low cost strategy can lead to the same information as a sensory detection, then an execution attempt of the strategy is preferred;

Before a sensory detection is proposed, a trade-off is made between information power (reduction of the world situation multiplicity), detection cost, and synergy with strategy attempts (low cost strategies, whose execution lead to significant information, are examined).

In the current example, the substitution of the peg is proposed as first. If the new trial gives no results, the defective-peg hypothesis is discarded.

When a "mismatch" has been detected (for instance the holed object is wrongly positioned), then a contradiction between error event and current world model can be eliminated (i.e., either a cause of the error has been identified) or simply the multiplicity of the world situations can be reduced.

If a cause of the error has not yet been found (for instance why the holed object had been wrongly rotated around the hole axis), then new non-consistent groups may be created because of the detection of this mismatch: in fact, during the part of the execution, that has taken place after the mismatch creation, errors could have been propagated to other objects, while never having been detected. Therefore an "error propagation analysis" is performed within the first (interpretation) phase, leading to an additional multiplicity of the possible actual world situations.

The error propagation analysis is based on the fact, that a position error may propagate from an object A to other objects if:

- A is moved near other objects,

- other objects are moved near A,

- an object, to which a position error was propagated, is moved near other objects, or vice versa.

The new possible mismatches, caused by the propagation of errors among objects, can be created in some of the past states of the program execution. The actual presence of these mismatches can then be verified, only if they can be observed from the present state. Therefore, the effects over the actual state , of the possible propagated errors (that have occurred in past situations), have to be calculated before a sensory detection can be prescribed in order to verify the presence of these "propagated" mismatches.

If a strategy can be attempted, whose execution could involve either the movement of uncertainly positioned objects or their substitution, then those sensory detections, whose goal is to ascertain the position of these objects, is useless.

In the current example, if the holed object has been found to be both wrongly positioned and defective, then it has to be substituted, and therefore eventual error propagations to objects placed on it are innocuous, in that these objects have to be displaced before the substitution can take place.

The sensory detection interpretation-proposal-execution cycle is stopped, as soon as the analysis of the currently available information about the actual world situation results in the detection of a "mismatch", that can constitute the cause of the error to be corrected.

When this happens, an implicit correcting strategy, corresponding to the (error-event, cause) pair, is chosen, planned and executed.

An example of an (error-event, cause) pair versus implicit-strategy correspondence consists in the (stopped_movement_near_arrival_point, defective_object) pair versus "substitute_defective_object" correspondence.

During all the three phases of the system activity, namely simulating perceiving and planning, a model of the physical objects is employed, in which data are structured in an object representation tree.

The node of the tree represents a subpart of the object, and includes some attributes of the represented subpart (like **name** of the subpart [e.g., "cyl_1"], **type** [e.g., cylinder], **dimensions** [e.g., height and diameter], **consistence** [e.g., hole or wood], and the names **component subparts**). The links, between a (super)part and one out of her components, is labelled with the frame of the component, expressed in the superpart frame.

8.3.   The Module That Plans The Correcting Strategy

This module operates in line, after an error has been detected, and after information about the actual situation (eventually detecting other, transparent errors) has been extracted.

Goal of this module is the attainment of one of the final state configurations, that are compatible with the particular initial configuration and input configuration assumed by the actual evolution.

In order to avoid the duplication of some of the reasoning activity, that has been already performed before the assigned program was written, the following remarks are made:

it is sufficient to attain a situation, belonging to an intermediate state of the program execution instead of belonging to a final state, which is compatible with the particular input and initial configuration assumed. The "nearer" (to he actual error situation) the target state is, the lower is the amount of the needed inference activity.

Calling **A** the action, during which the error is been detected,
S the (never reached) correct arrival state of the action

A, S is chosen as the first (provisional) target state of the correcting strategy.

If, during the execution of the correcting strategy, input configuration changes, it is possible that also the target state must be changed: the first state, that is intermediate between **S** and a final state and that is compatible with the new input configuration, is chosen as the new target state.

At the end of the correcting strategy execution the execution of the assigned program starts again. The new starting point depends on the target situation, that has been reached by the correcting strategy.

In order to take into account the variability of the target situation, all the correcting strategy branches, eachone of those leading to a different target state, are incrementally constructed.

In order to plan the correcting strategy, the module operates along two phases:

1.

- The first target state is **S**;

- the II module has detected some mismatches (due to both the "main" error and the "transparent" errors);

- some elementary implicit strategies are collected, in order to eliminate the mismatches;

- a provisional global strategy, resulting from this collection, is planned;

This global strategy will constitute a branch of the definitive correcting strategy. The current strategy, that will be referred to in the next phase, is set initially to this global strategy.

2.

- The module identifies possible input configuration changes (typically due to substitutions of variable objects), that may occur during the execution of the current strategy;

- for each possible change, a target state (compatible with the changed input configuration) is chosen, and a new branch is constructed, that leads to this target state;

- for each new branch, the just above described "planning-branching" cycle is reiterated (i.e., the possible input configuration changes, that can occur during the execution of the considered branch, are first examined, then new corrisponding target situations are selected, and finally new branches are constructed, that lead to the new selected target states).

In this manner, the definitive correcting strategy is constructed.

In order to plan the correcting strategy some physical knowledge is needed. However, in many cases, a further powerful knowledge source is available.

Let us consider an example, in which a defective object, belonging to an assembled group, has to be substituted. Access, for the robot arm, to this object must first be made. Then the object must be substituted. Finally the previous (or another) assembled group must be recomposed.

In order to accomplish this final part of the strategy some problems must be solved:

- the identification of the grasp-points;

- the determination of free paths;

- the construction of a time order for the assembly operations.

Part of these problems have already been solved by the programmer, and the solutions are described in the instructions, that prescribe the assembly operations!

In order to allow , to the planning module, the exploitation of these solutions, one brief analysis of each action is performed by the off-line module.
The **effect** of the action, described in terms of the part of the arrival state that differs from the starting state, is identified
(e.g. for the action "move block1 to position pos1 over block2" the identified effect is
"((position block1 pos1)(contacts block1 block2))").

In general, during correction, an action is applied to a situation, that is different from that, for which the action has been programmed. Therefore some further information is extracted by the analysis of the actions.
The **preconditions** , for the action effect to be the same as the programmed effect, are identified
(eg. for the previous mentioned action effect the identified preconditions are
"((block2 in pos1)(position pos1 free from other objects))"
).

This information allows the planning module to ascertain if a needed action is:

1. useful for the current state;

2. applicable to the state, that is evolving during the correcting strategy execution.

9. CONCLUSIONS

With the present approach some of the common shortcomings can be overcome, because the more expressive situation representation employed allows the system to cope with problems resulting both from non-reversible actions and from the presence of actions having a variable result.

It has also been suggested, that the availability of both implicit plan description and explicit plan description should be convenient, within a perspective architecture of a **task-level** language **interpreter**.

The chosen formalism for representing physical situations is expressive but difficult to manage. Therefore an additional processing charge weights on the three phases of error correction (namely task deduction, error situation perception, and correcting strategy planning). Although this additional processing charge is not an important problem for the task deduction activity, in that it takes place off-line, the error situation detection is slowed down, because a sequence of several sensing-inference steps are required, in order to construct a description of the error situation, according to the chosen situation representation formalism.

It is still a very difficult task to analyze a scene, taken from a completely non-structured environment, without a scope other than perceiving.

On the contrary it may be fruitful to use sensors, in order to detect differences between the real scene and a prestored expected model. The presence of the expected model allows inference, in order to "prune" the main part of the possible interpretations.

The presence of a well defined scope (in this case error correction) and a stored knowledge about the expected "history" of the environment, allow the existence of a history directed "attention focusing" mechanism.

At present the LINCE Sensory System has been implemented, the off-line module of the Correcting System is being implemented and the (Yaps-written) knowledge base of the two on-line modules of the Correcting System is being constructed.

REFERENCES

G. Gini, M. Gini, M. Somalvico, "Emergency Recovery in Intelligent Robots", Proceedings of the 5th International Symposium on Industrial Robotics (1975)

L.I. Lieberman, M.A. Wesley, "AUTOPASS, An Automatic Programming System for Computer Controlled Mechanical Assembly", IBM Journal of Research and Developement (1977)

T. Lozano-Perez and P.H Winston, "LAMA, A Language for Automatic Mechanical Assembly", Proceedings of the 5th

International Joint Conference on Atrificial Intelligence (1977)

A.P. Ambler, R.J. Poppelstone, J.M. Bellos, "RAPT, A Language for Describing Assembly", Industrial Robot, Vol.5 (September 1978)

R.J. Poppelstone, A.P. Ambler, "A Language for Specifying Robot Manipulation", Hull Summer School Robot Technology, Hull University (1981)

E. Mazer, "Geometric Programming for Assembly Robots", Proceedings of the International Meeting of Advanced Software in Robotics, Liege (1983)

G. Gini, M. Gini, M. Somalvico, "Program Abstractin and Error Correction in Intelligent Robots", Proceedings of the 10th International Symposium on Industrial Robotics, (1980)

R.A. Brooks, "Symbolic Reasoning Among 3-D Models and 2-D Images", Artificial Intelligence, vol. 17 (1981)

T.O. Binford, "Survey of Model-Based Image Analysis Systems", International Journal of Robotics Research, vol.1 n.1 (1982)

R.A. Brooks, "Model-Based Three-Dimensional Interpretation of Two-Dimensional Images", IEEE Transactions on Pattern Analysis and Machine Intelligence, n. 5 (1983)

R. Cassinis, "Sensing System in Supersigma Robot", Proceeding IX ISIR, Washington (1979)

R. Cassinis, "La Sensorialita' nei Robot: Problemi e Prospettive", in Automazione e Strumentazione, ed. Associazione Nazionale Italiana per l' Automazione, vol. 29 n. , Milano (aprile 1981)

Benson Carlin, Ultrasonics, ed. McGraw-Hill Book Company, Inc. New York (1949)

MCS-48 Family of Single Chip Microcomputers User's Manual, ed. Intel Corporation, Santa Clara, CA (July 1978)

G. Borghini, P. Fasola, R. Simino, "Progetto AUREO, Cambio di Frequenza di Esplorazione Ultrasonica Intelligente per il Sistema di Visione LINCE", Sistemi Integrati per l' Automazione, Convegno Nazionale Anipla, Genova (Dicembre 1985)

R.E. Smith and M. Gini, "Robot Tracking and Control Issues in an Intelligent Error Recovery System", Proceedings of the 1986 IEEE International Conference on Robotics and Automation, S.Francisco (1986)

Pattern Recognition in Multidimensional Perception:
Robots and Humans

Gerardo Beni and Susan Hackwood

Center for Robotic Systems in Microelectronics
University of California, Santa Barbara, Ca 93106

Abstract

We discuss several aspects of the theory of *multidimensional human and robot perception* which is important for developing robotic sensory capabilities beyond the ones currently in use, i.e. beyond black & white vision and force/tactile sensing. Understanding Multidimensional Perception (MDP) is necessary for developing sensory capabilities for color vision, for general cases of force/torque/tactile sensing, and for thermal/chemical sensing. In this paper we describe *a new method* of pattern recognition in robotic MDP:*Class Connectivity Analysis* (CCA). We also describe *a new effect* in human MDP: the *Color Void* effect. Finally, we propose a *new design criterion* for MDP in robots. The CCA method allows fast inspection of complex images, in particular textured images and random patterns. The Color Void effect clarifies the mechanism of human pattern recognition in MDP and suggests a bionic design criterion for MDP. This criterion could allow the description and recognition of patterns in sensory fields with arbitrary time-dependent signals.

1. Introduction

The design of more advanced robots requires significant progress in sensor technology. Besides developing new types of sensors, progress in sensor technology for robots requires developing new methods of using sensors efficiently. The efficient use of sensors depends on developing a satisfactory theory of the mechanism of perception and on developing design criteria for applying the theory in practical robotic applications. The theory of the mechanism of perception in humans has been developing for more than a century. In robots (or computers) the theory has been developing for at least two decades. In spite of this, much remains to be understood for developing design criteria to apply the theory in practical robotic applications.

The subject of this paper is one aspect of the theory of the mechanism of perception. This aspect, *multidimensional human and robot perception*, is particularly important for developing robotic sensory capabilities beyond the ones currently in use, i.e. beyond black & white vision and simple force/tactile sensing. Understanding Multidimensional Perception (MDP) becomes necessary for developing sensory capabilities for color vision, for general cases of force/torque/tactile sensing, and for thermal/chemical sensing. Of immediate practical importance is color vision which has recently become commercially affordable for industrial inspection and other applications.

NATO ASI Series. Vol. F43
Sensors and Sensory Systems
for Advanced Robots
Edited by P. Dario
© Springer-Verlag Berlin Heidelberg 1988

In this paper we describe *a new method* of pattern recognition in robotic MDP. We call this method *"Class Connectivity Analysis"* (CCA). We also describe *a new effect* in human MDP. We call this effect *"Color Void "*. Finally, we propose a *new design criterion* for MDP in robots. The CCA method allows fast inspection of complex images, in particular textured images and random patterns. The Color Void effect clarifies the mechanism of human pattern recognition in MDP and suggests a bionic design criterion for MDP. This criterion could allow the description and recognition of patterns in sensory fields with arbitrary time-dependent signals. The outline of this paper is as follows:

Section 2: Definition of concepts and statement of the MDP problem
Section 3: Method of Solution
Section 4: Application to Human Perception
Section 5: Application to Robot Perception

2. The Problem of Multi-Dimensional Perception (MDP)

We define a *sensory field* as a rectangular array of *sites* in real space. The theory developed in this paper applies similarly to higher dimensional arrays and non rectangular arrays. Thus we have no loss of generality in simplifying the description using 2-dimensional rectangular arrays. Each site of the sensory field has a *value*. The value is a number or a set of numbers. In general, the value is a function of time, either discrete or continous. A *pattern* in the sensory field is defined as a set of values each assigned to one site, i.e. a pattern is a mapping from the space of values to the array of sites in real space . The values discussed in this paper are either magnitudes of physical properties (e.g. amplitudes of a signal) or representations of perceptions (e.g. hues). Corresponding to these two types of values we refer to Patterns as *Physical Patterns* or *Perceptual Patterns.*

Another essential concept is that of *object*. An object is defined as a maximally connected set of sites with the *same* value. *Connectivity analysis* of a pattern is the process of labeling objects in a pattern and classifying them topologically, i.e. as *parents, children, siblings and neighbours.* All these concepts are part of the standard conceptual framework of pattern recognition and image analysis [1].

The crucial problem in practical applications is to recognize the "real objects" in a pattern. The objects as mathematically defined above, in general, do not correspond to physical objects, i.e. to "real objects". In fact, generally, the same physical object contains a distribution of values, not just one single value. The distribution of values is generally relatively narrow around an average value. Nevertheless, determining this distribution of values, and distinguishing it from other factors such as noise, is a major practical difficulty. In order to find a method of recognizing real objects we introduce the following definitions.

A *"real object"* is defined as a maximally connected set of sites with the same *class-value*. To define class-value we must first define *class*. A class is defined as a set of maximally connected values, with the same real-space density, in the space of values. (This essential concept will be discussed in great detail later on. For the time being it is sufficient to think of a class as a set of values) .

Finally the class-value is defined as the mean (or another suitable function) of the values of a class. Thus, the first, and most critical, problem in segmenting the pattern into a set of real objects is defining the classes of values. After this crucial first step, the segmentation can be carried out in either of two ways: by *edge detection* or *connectivity analysis.*

The latter two techniques are both well known segmentation methods [2]. Our interest here is not in the technicalities of the two methods but in their limitations in defining classes of values, and thus, ultimately, in finding the real objects in the pattern. The limitations appear clearly in the case of *textured* patterns. In this case, edge detection fails to achieve segmentation essentially because it detects too many edges. Connectivity fails because it detects too many objects. It would not fail, however, (except in pathological cases to be discussed later) if suitable classes of values had been defined. Similarly for edge detection. Thus we see that the problem of defining classes is the critical step in segmentation.

Actually defining classes presents little difficulty in one-dimensional perception. For example, in black & white vision, histograms of light intensity can be used to divide values in classes, usually with very good results. Mathematically, the problem is basically finding the extrema of a one-dimensional function. The difficulty arises when the space of values is multi-dimensional. Then the histogram becomes a distribution in a multidimensional space and finding extrema becomes complex.

In the next section we describe a method of defining classes in multi-dimensional spaces of values and thus, in general, of finding real objects in the pattern. The patterns which require class definition in multi-dimensional spaces appear in MDP to which we now return in order to complete its definition.

What then is MDP? Formally, it is perception via more than one sensor. Thus, in general it is perception of different physical properties, such as light, force etc. Since robots are being engineered with more and more sensors, clearly it becomes important to combine the perceptual information gathered by these multi-sensory systems into one representational unit. The key problem here is to find physical properties which are related so that their combined representation has physical meaning. This is then, primarily, a physics problem, not a perception problem. Furthermore, this combination of sensory data can be very useful but it is not necessary. Indeed, it is always possible to add sensor to sensor and combine the representation of their sensory data, but this multidimensional combination, although useful, is not intrinsic to the perception, i.e. *it is not intrinsic MDP.*

In contrast, intrinsic MDP arises when a single physical property is sensed by more than one sensor, i.e. when the multidimensionality is intrinsic in the perception. The simplest example is color vision. In this case, an electromagnetic wave (single physical property) is sensed by three detectors (for Red, Green and Blue) so that the sensory space is intrinsically 3-dimensional.

In general, if the physical pattern is in the form of an array of time dependent signals, the perceptual pattern can have values of any dimension, depending on the number of sensors. If only e.g. the energy (or the amplitude) of the signal is sensed, we have 1-dimensional perception. If all the Fourier components of the signals are sensed (eg via a spectrum analyzer), we have "infinite"-dimensional perception. Generally, if N frequencies are detected, we have N-dimensional perception. For example, for color vision, we have 3-dimensional perception.

In one-dimensional perception, as we have seen, it is easy to define classes and thus find real objects, but too much information is lost, and so too few real objects are resolved. In contrast "infinite-dimensional" perception contains too much information so that no class definition is possible and thus real objects cannot be found at all. Between these two extremes lies the most interesting and practically relevant case: finite MDP. This is also the case that biological evolution appear to have selected for advanced sensory perception. In finite MDP, real objects can be found efficiently in the given pattern if suitable algorithms are devised. In the next section we describe an efficient method of pattern recognition in MDP and the associated algorithms.

3. Method of Pattern Recognition in MDP

We call this method "*Class Connectivity Analysis*" (CCA). The basic idea of CCA is a way of defining classes in the multidimensional space of values. (Since we are dealing with MDP, by definition, the space of values is multidimensional). As we have seen in the previous section a class is defined as a set of maximally connected values (with the same real-space density) in the space of values. Thus the following duality exist between objects and classes.

Object	Class
Real Space	Value Space
site	value
value	real-space density

Superficially, we are driven into a difficulty since we have introduced the classes in order to avoid the "*same* value" problem in real space and we are now defining the classes via the "*same* real-space density". Fortunately, "same real-space density" can be approximated reasonably well by a thresholding operation. Thus, the real-space density distribution becomes binary and classes can be defined as sets of maximally connected values (with real-space density above threshold) in the space of values.

Thus, finding the classes has been reduced to a binary connectivity analysis problem in multidimensional space. This is not a simple problem since even the definition of connectivity in multidimensional spaces had not been established until recently [3] when efficient algorithms were developed for multidimensional connectivity analysis.

After having found the classes, finding the real objects in the pattern requires labeling the real objects. Several methods have been developed for this labeling algorithm, but until recently, all the methods were based on sequential multiple binary connectivity analysis. This procedure is very time consuming, and clearly becomes prohibitively slow when the number of classes is very large. The multi-valued connectivity analysis developed in [4] solves this problems by obtaining the labeling of multivalued objects in the same time as that required for the labeling of binary objects.

Basically then, by using the recently developed multidimensional and multivalued connectivity analysis algorithms of refs [3,4], it is possible to achieve CCA efficiently in a short processing time. The CCA method is outlined below:

The CCA method

1) Obtain real-space density of values and plot it in value-space.
2) Set threshold for real-space density and obtain binary image in value-space.
3) Do multi-dimensional connectivity analysis on this binary image and thus find 'objects' in value-space, i.e. the classes.
4) For each class, determine class-value by averaging values in the class.
5) Obtain a new pattern from the given pattern by using class-values instead of values.
6) Do multi-valued connectivity analysis on the new pattern to label, i.e. find, the real objects.

The main advantage of the method is that CCA finds real objects undetectable by edge detection, e.g. random color patterns and certain textured patterns. For example, a real object appearing as a textured or random pattern of colors whose values are closely clustered around an average can be trivially detected by CCA. The other obvious advantage is that it can be applied to any dimension of value-space.

On the other hand, there are some real objects which escape CCA detection. For example, patterns formed by saw-tooth type distribution of 1-dimensional values. In this cases edge detection is better. Thus, clearly the best method is a combination of CCA and edge detection.(With the latter extended to multi-dimensions).

Conceptually, the important aspect of CCA is the emphasis on the classes as topological entities in value-space. Although the basic CCA method as outlined above depends only on defining the classes and not on their topological structure, many interesting questions maybe raised by examining this structure. One such aspect will be examined in section 4.

4. Application to Human Perception

In examining the topological structure of classes a new effect has been found. This effect shows that human beings have limitations in color perceptions, and perhaps in other forms of perceptions, related to the topological structure of classes in value-space.

To discuss this effect it necessary to specialize our discussion to a particular case of MDP, namely color vision.

A vast amount of knowledge has been accumulated on the capacity of human beings to discriminate colors. It is well known that normal trichromats discriminate between wavelengths that differ by ~ 20 Å across the visible spectrum (and by less than 10 Å in various spectral locations) [5]. It is also well known that normal humans can detect at least 10 million different colors [6].

In spite of this high degree of sensitivity we have found that the human color perception system fails to perceive certain large color discontinouities which appear as topologically well defined class structures in color space (i.e. the value-space of color perception). Below we describe the experimental demonstration of this effect. First, however, we must describe in some detail the representation of color perception.

4.1 Color Perception Theories

The perception of color is a psychophysical phenomenon dependent on the frequency and intensity of light and the visual sensation perceived by an observer. Color detection is due to the cones in the retina of the eye. Different cones react to different wavelengths of light. It is known that there are three types of cones, each type reacting to a different color stimulus. In spite of many hypotheses, the exact mechanism of color perception is still a mystery [7].

One of the earliest theories of color perception [7] suggested that red, green and blue are detected separately by the three different types of cones. The sensory information is then sent via the optical nerve to the brain for processing. This model is similar to that of a typical robotic vision system. In this system, light impinges on the array of photodetectors of a color camera. In this array, at each site there are separate detectors for the red, green and blue (RGB) components of the light. The sensory information is then sent to a computer for processing.

The RGB theory of color perception was first introduced by Young in 1807 (and reintroduced later by Helmholtz), and places all vision processing in the brain. Other theories place part of the processing elsewhere. For example, the Dominant Modulator theory [7] groups the cones together as pre-processing units. Each group sends luminosity as dominant information and chromaticity as modulated information. Other theories suggest more color processing in the eye itself, e.g. the Opponent Process theory suggest that groups of cones send out pairs of responses (white or black, yellow or blue, green or red). In spite of their large number, none of the theories developed so far have been proved completely satisfactory.

4.2 Representation of Color Perception

Many models have been proposed to represent color perception. The *Commission Internationale de l'Eclairage* (CIE) was the first committee to study color, and to attempt to standardize its representation[8]. The representation is based on three primary attributes of color :*hue, saturation and brightness* [9]. A geometric model of these three attributes is shown in Figure 1.

Referring to the figure, the Hue (Ø) is the azimuthal angle and represents the sensation associated with the dominant wavelength of the visible spectrum. Saturation (r) is the distance from the center axis and
represents the absence of white content in a color. A pure color has a saturation of 100%, e.g. vivid red, whereas a pale color, e.g. pink, has a saturation of e.g.~ 50%. Any color with 0% saturation is a shade of gray. Brightness is the distance on the z-axis and represents the perceived brightness. The maximum brightness is perceived as pure white and the minimum as pure black. The conical shape of the model limits the brightness of vivid colors.

The model based on hue, saturation and brightness is an example of many color perception representations which have been developed during several decades. The essential common feature of all the representations is that each color perception is represented by a three-component value.

In fact, according to the tristimulus theory [5], any color can be represented as combination of three primary colors. For example, if we chose the three primaries as red (R), green (G), and blue (B) , then the following equation holds for any color C.

$$C = rR + gG + bB.$$

where r,g, and b are appropriate coefficients . The CIE committee in 1931 [5] selected the primaries as red, green and blue at wavelengths 700nm, 546.1nm, and 435.8nm respectively.

A very convenient representation of color perception is based on these primaries and is commonly referred to as *RGB color space* (Figure 2). The values of the three primaries are represented on the positive orthogonal axes. Any color can be represented as a point inside this octant of color space. Thus, each value in color space is a 3 dimensional vector whose components are the values projected on the directions of three primary colors R, G, and B.

As representation of color perception, color space is convenient because of its linearity[10]. In other words, if color A matches B, then by multiplying by a constant c, color cA matches cB. Furthermore, if color A matches B, and color C matches D, then color A+C matches B+D. Actually this holds true in any color space (e.g. RGB), regardless of which colors are chosen as primaries. Color spaces can be mapped one into another via linear transformations. Because of this property, color spaces are the most convenient representation of color perception.

Since no color space is optimal for all applications, much work has been done to find which color space is best for a particular application [8]. Besides RGB space, some of the more common color spaces are the U*-V*-W* color coordinate system, the S-∂-W*, the L-a-b, the Karhunen-Loeve and Ohta space [11]. Each color space has properties that are optimized for a particular application. We use the RGB model in this work since it is simple to use and easy to visualize. We have no loss of generality since our discussion is based on topological considerations, which are unaffected by linear transformations. Thus they apply equally to any color space [12].

4.3 Experiments

The hardware, as described in refs. [13,14,15] is based on an AT&T color digitizing board [16] (Image Capture Board) installed in an IBM-PC AT computer. The system configuration is represented schematically in Figure 3. A solid state color camera [17] provides 384x485 pixel resolution. The color digitizing board has a 256x256 pixel array memory, with 15 bits of resolution of color for each pixel. The board provides separate RGB video output. With 15 bits of color (5 each for R, G and B) it is possible to display 32768 different colors. An RGB monitor is used to display the image.

The color camera outputs primary R, G, and B signals. These values are then digitized and quantized. A color is represented with these three values. Using these values as vector components, each color can be plotted as a point in RGB color space.

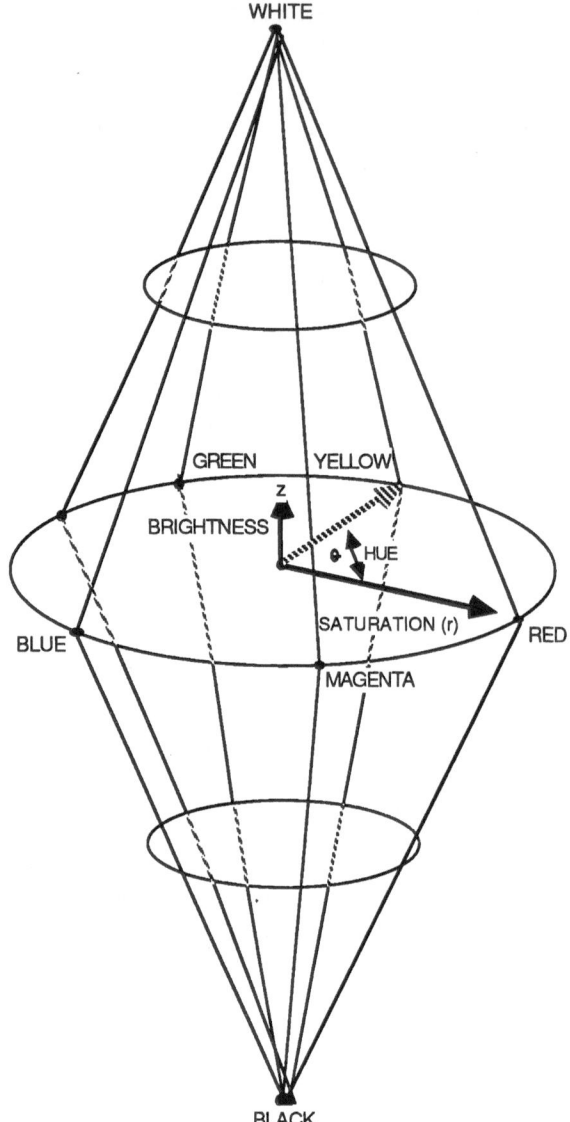

Figure 1. Model for hue, saturation and brightness.

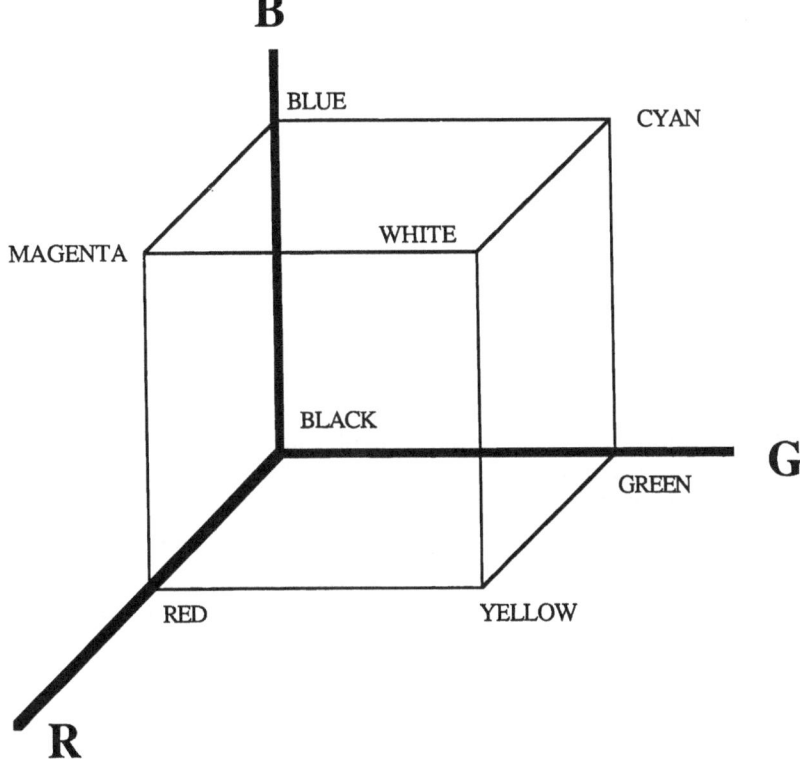

Figure 2. RGB color space.

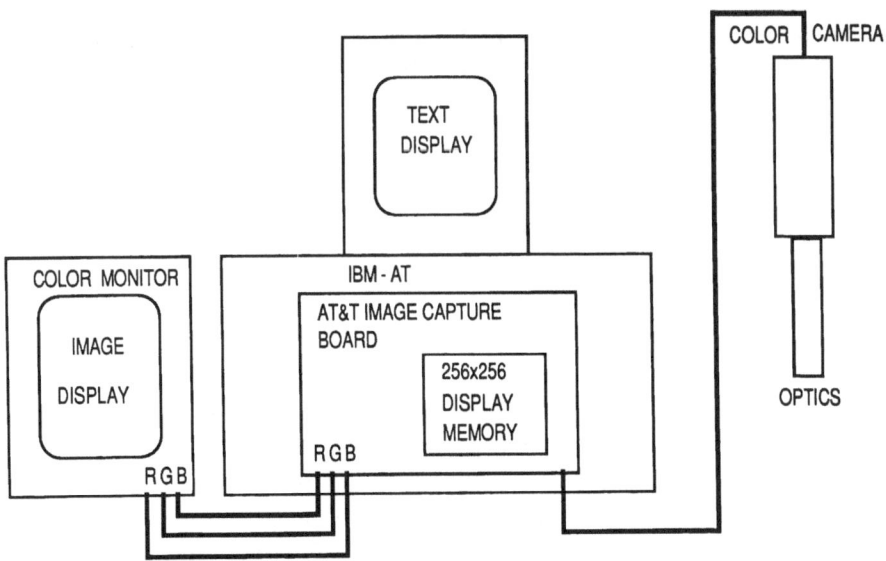

Figure 3. Color vision hardware configuration.

Since in our color vision system, R, G, and B have a 5 bit resolution, the RGB color space is a 32x32x32 cube. The corners of the cube represent black, red, green, blue, cyan, magenta, yellow and white, and have the coordinates (0,0,0), (31,0,0), (0,31,0), (0,0,31), (0,31,31), (31,0,31), (31,31,0), and (31,31,31) respectively.

In the following experiments we focus on topological relations among maximally connected sets of points in color spaces, i.e. among classes. The two topological relations discussed here are shown schematically in Figure 4a. The figure shows three classes H_1, H_2, and H_3. All the points in the classes have the same real-space density value (we assume that thresholding of the real-space density has already been done). By binary connectivity analysis we determine that H_1 and H_3 are topological siblings whereas H_1 and H_2 are topological parent and child, respectively.

A somewhat more precise (and more useful for the analysis of the experiments) description of these topological relations can be given as follows. For simplicity assume that the classes are in the forms of spherical shells. Let H_1 be a spherical shell with a major axis R_H and a minor axis R_h. Similarly, let H_2 be a shell with major axis r_H and a minor axis r_h. The union set ($H_1||H_2$) is disconnected if the two sets have no intersection ($H_1||H_2 = 0$) i.e. no common points. The disconnectedness is *internal (external)* if H_2 is *internal (external)* to H_1, as shown in Figure 4a. Internal and external disconnectedness correspond to topological child and topological sibling respectively. (Note that H_3 has external disconnectedness from both H_1 and H_2). The histogram projections in Figure 4b demonstrate that the concept of disconnected is dependent on the dimension of the class space. The disconnected between H_1 and H_2 is not conserved in the one-dimensional projections. We will see that this is an important concept in MDP.

Experimentally we have created internal and external disconnectednesses in RGB space and tested their perception by human beings and robots. We have found that, generally, external disconnectednesses (i.e. siblings) are perceived as color contrast, whereas internal disconnectednesses (i.e. children) are not, *even in the cases of very large discontinouities.* We have called these topological structures of unperceived color contrast **color voids.**

Experiment 1. 'Flowers in the Meadow'

First, a shell H_1 was created in RGB space around the center (16,16,16) with outside radius $R_H = 11$, and inside radius $R_h = 10$ as shown in fig.4a. There are 1,436 colors within this shell. Second, we selected a single color **C** not belonging to the set H_1. A rectangular image (Figure 5a) was generated composed of 2,560 (40x64) squares (5x4 pixels each). The color of each of these squares was selected from the union of colors of H_1 and $H_2 = C$ (for a total of 1,436 + 1 colors) by a random number generating algorithm. Twenty normal trichromats (i.e. people with normal color vision) were asked to say (from <50cm distance, to avoid averaging colors among neighboring squares) if, after looking at the rectangular image for 60sec, they could perceive any squares standing out from the background. It would be as seeing "flowers in a meadow".

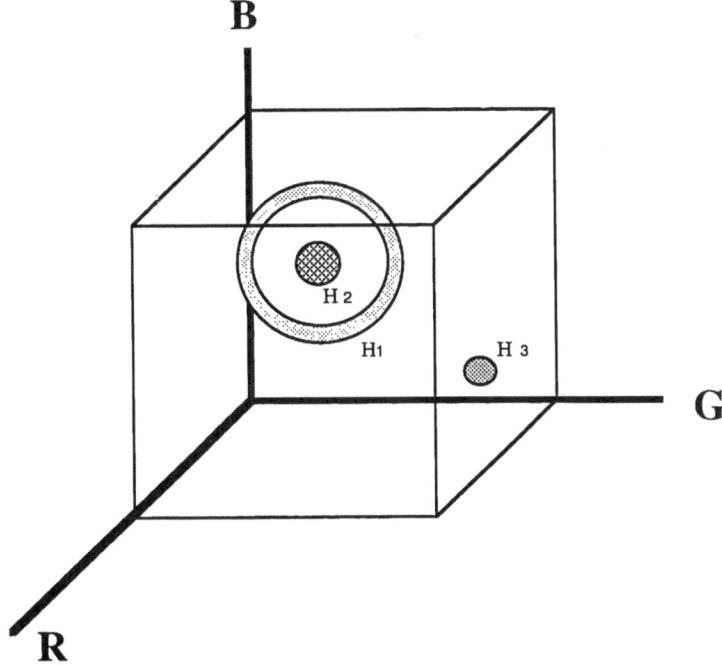

Figure 4a. Classes in color space.

The experiment was carried out with the color of H_2 chosen at various locations in the color space. Figure 5a shows red (31,0,0) as the discontinuous color H_2 (H_2 is outside the shell H_1). In this case it is relatively easy to pick out the red squares from the background. Figure 5b shows the two classes in RGB space.

H_2 was then moved inside the shell of H_1. Figure 6a is similar to Figure 6a, except the color of H_2 is gray (16,16,16). Now it was impossible for the 20 observers to pick out the unique squares. Again, Figure 6b shows the RGB space (with the shell H_1 cut open in order to see H_2).

The experiment suggests that human perception of discontinuities in color space is a function of the distance d (in color space) of C from the center of H_1. We define

$$|\partial| = d - (R_h + R_H) / 2$$

as a measure of color contrast perception, since we find that when $|\partial| < 2$, color contrast is not perceived. Thus, we can take $|\partial| = 2$ as a rough estimate of perceptual resolution, i.e. the expected limit of contrast discrimination for this context [7]. (It ishould be noted that this is only an estimate since $|\partial|$ is also a function of the center of H_1 in color space. If we shift the center of H_1 towards the green and draw H_2 out towards the red, better resolution is achieved as expected from color vision theory).

What is significant is that for $|\partial| > 2$ the results depend on the sign of ∂, i.e. on whether **C** is internal or external to the shell. For $\partial > 2$, **C**-colored squares are singled out by an increasing number of observers as ∂ increases (by 90% of them for $\partial = 4$) as intuitively expected. In contrast, for $\partial < 0$, **C**-colored squares are *never* singled out, even for $\partial = -(R_h+R_H)/2$, i.e. *even when **C** is at the center of the shell.*

The shell acts as a screen for the colors internal to it. This screening effect can be further demonstrated by expanding or shrinking the radii of H_1 and by changing its center. The experiments show that as long as **C** lies internally to H_1, the color of **C** does not stand out from the background.

Experiment 2. 'Snake in the Grass'

This inability of human beings to perceive such a large discontinouity is further demonstrated by a second experiment, in which a 'pattern' is concealed in a background. Although this is not a case of pure color discrimination since a geometric pattern biases the recognition process[18], the non-detection of the pattern would further corroborate the results of the previous experiment.

We have contructed a second shell H_2 with radii $r_h = 4$ and $r_H = 5$, and concentric with H_1 (defined as in the previous experiment). In the15 bit color space, 264 colors fall with shell H_2. The background is again colored randomly out of the 1,436 colors that fall within the shell H_1 ($R_H = 11$, $R_h = 10$). Then, out of the 2650 squares of the image, we select a subset of 100 contiguous squares forming a snake-shaped pattern of twelve straight lines (Figure 7). This 'snake' is then colored randomly with colors selected from the 264 colors of H_2. The observer was then asked " can you perceive any pattern concealed in the background as a 'snake in the grass'?"

No recognition of the pattern occurred for this case, as well as after any shrinking of H_1 or expansion of H_2 (with $r_H < R_h$). Expansion of H_1 results in some individuals detecting the longer segments of the pattern. More and more detections occurred as H_2 became larger. The distance between r_H and R_h was 6 for the first detection of at least part of the pattern. An example case is shown in Figure 8a, with $r_h = 4$, $r_H = 5$, $R_h = 14$, and $R_H = 15$. Figure 8b shows a cutaway view of shell H_1 with H_2 imbedded inside.

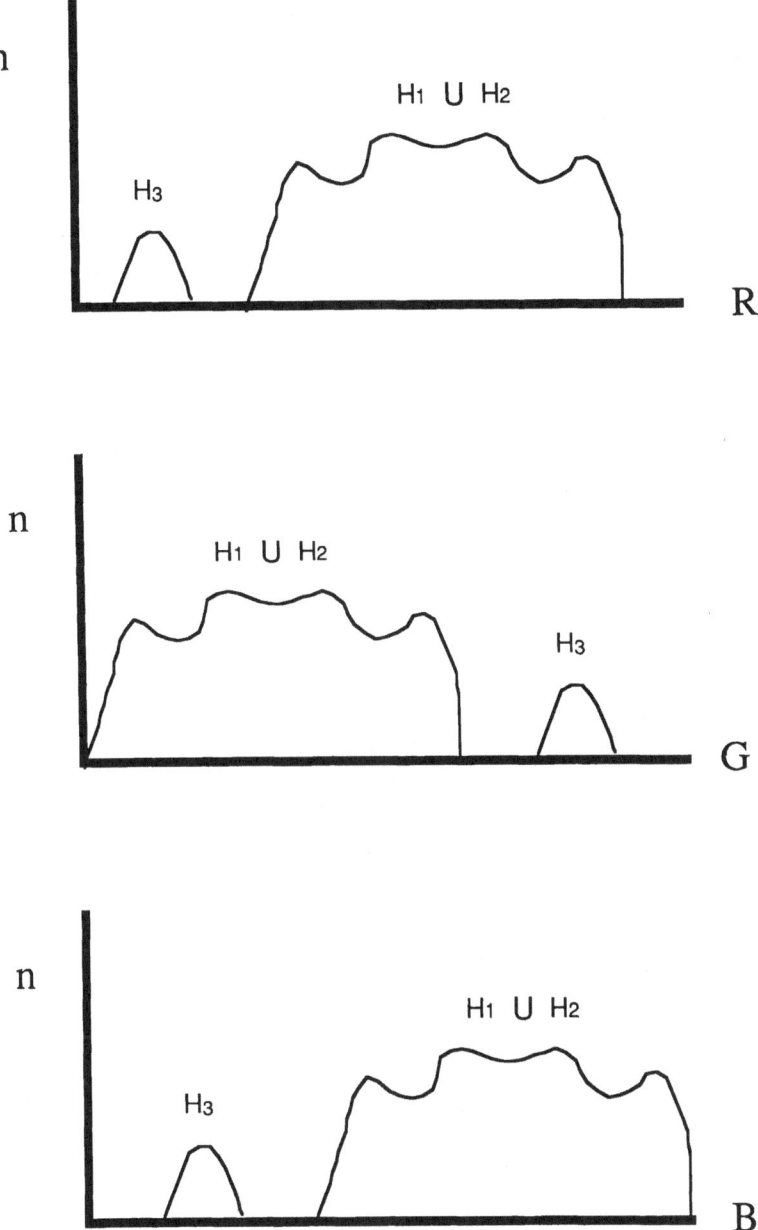

Figure 4b. 1-D projected histograms of color classes of Fig. 4a.
The ordinate axis represents the real-space density **n**.

Figure 5a.   "Red Flowers in the Meadow"

Figure 5b.   Color space representation of "red flowers in the Meadow"

Figure 6a. "Gray flowers in the meadow"

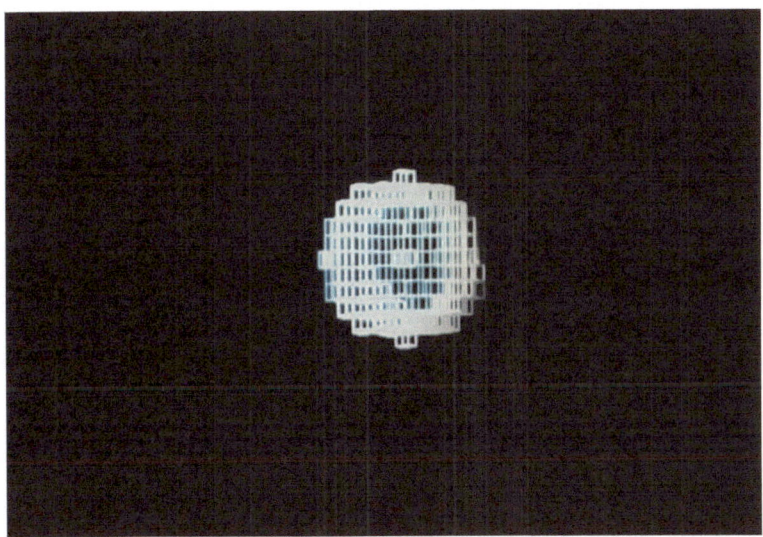

Figure 6b. Color space representation of "Gray flowers in the Meadow"

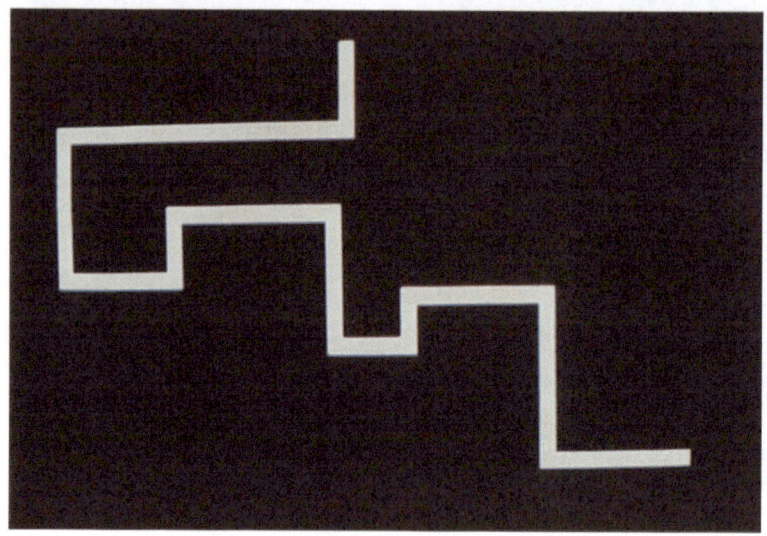

Figure 7. Shape of 'snake'.

Figure 8a. 4::5 'snake' in 14::15 'grass'

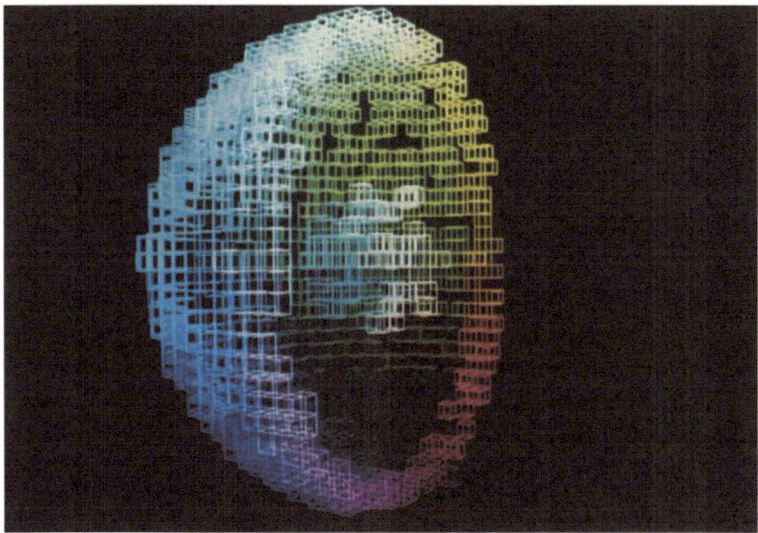

Figure 8b. Cutaway view of color space representation of 'snake in the grass'

4.4 Interpretation of experiments

First, let us contrast human performance with computer vision for the two recognition tasks 1) "flowers in the meadow, and 2) "snake in the grass". To do this it is necessary to note the differences in the detection of disconnectness between histogramming and CCA algorithms.

To perceive the"flowers in the meadow" a crucial step is necessary, i.e. to 'understand'that C is not part of the same set of points as H_1. Similarly, to perceive the "snake in the grass" it is necessary to "understand" that the points in H_2 are distinct from the points in H_1. A 'smart' computer is one that 'understands' this concept, i.e. the concept of disconnectedness. CCA is a computer algorithms that can perform this type of "understanding" for a robot.

Human beings also generally perceive sensory discontinuities as special events and thus single them out [10]. Why then do they fail to single out such large discontinuities in color space?

First let us contrast CCA with histogram analysis. Histogram analysis [2] is an algorithm (much simpler than CCA) which is used routinely in robot vision. Applied to the present task, it consists in plotting the number of colors as a function of their R, G and B components. Discontinuities are singled out as gaps between portions of the histogram.

The first important point is that histogram analysis is 1-dimensional, whereas CCA is multi-dimensional. The second important point is that histogram analysis singles out

3-dimensional discontinuities (i.e. disconnectedness) only if they are external since only external disconnectedness is projected as discontinuities of the components as shown in Figure 4b. Internal disconnectedness does not appear as discontinuities when projected on the 3 component axes.

We suggest that the human failure to perceive the internal disconnectedness (color voids) in color space is due to the inability of the human brain to 'analyze' the topology of color space, as e.g. in CCA. It appears that the human brain, although perceiving the three color components as a sensory unit, can only 'analyze' them separately, i.e. 1-dimensionally as in histogram analysis.

This hypothesis is confirmed by the observation, in the "snake in the grass" experiment, that the internal disconnectedness is perceived if the two shells are sufficiently separated. In this case, if we project the two shells on the three axes and form the R, G and B histograms it is easy to see that the histograms have a 'bump' in the center, corresponding to the overlap of the two shells. The 'bump' becomes prominent only as the outer shell expands sufficiently. When the "bump" is prominent, it is conceivable that it is singled out by the histogram analysis.

One-dimensional sensory analysis seems to be the rule in human perception. Indeed, the difficulty in analyzing (even with conscious reasoning) multi-dimensional spaces (other than the physical space) is well documented. Pattern recognition [19] requires high-level intellectual functions. The unique aspect of the color void effect is that the 3 sensory components (R, G, and B) of any color are perceived as one perception (i.e. one single color is 'seen') in a 3-dimensional space, but appear to be processed as three quantities in three 1-dimensional spaces.

In conclusion, the previous interpretation leads to the hypothesis that humans do not map the sensed light in the 3D color space; therefore they can only do 1-D analysis, as discussed above.

But this interpretation might not necessarily be true. In fact, we might speculate as to why we have 3 color sensors in our cones. Why would we have *three* sensors, and not 2 or 4, if we could not use some form of 3D perception space?

Still, if humans do not perform MDP pattern recognition as the experiment show, what is the reason? There are basically two possiblities: either 1) the brain cannot do it or 2) we have not learned how to do it.

We might speculate that possibility 1) does not appear valid because we know the brain can do 3D connectivity analysis in real space. So it seems plausible the same neurological apparatus could be used by the brain to map other MDP spaces with 3 dimensions, as the color space.

As to possibility 2) we might speculate that we might not have learned it because our environment does not provide color frames of reference as it provides geometrical frames of references. Our environment does not often provide *continous* variations of colors as does for continous variations of lengths. If this speculation were correct, humans could possibly be trained to detect patterns of the type used for the experiments. Obviously, these considerations are highly speculative and rigorous work is necessary to clarify the situation.

5. **Applications to Robot Perception**

In speculating about the interpretation of the previous experiment we have discussed the concept of mapping the physical stimulus into a perceptual space. For biological systems, this mapping is fixed by evolution, but for robots it is an engineering design issue. The issue can be simply restated as to what is the optimum (for pattern recognition) mapping from the physical stimulus to the perceptual space. Thus we must return to the analysis of MDP which we introduced in section 2.

One interpretation of the color void effect suggest that the effect is due to the mapping of the light stimulus into a 1-D perceptual space (rather than 3-D). Because of this, the 3-D topological relation of parent-child is not present, resulting in the color void effect.

In general the topology of the class structure depends on the mapping, performed by the sensors, from physical stimulus to perceptual space. An example of how a topological structure could create parent-child 2-D structures from a monotonic 1-D pattern is given by the following function:

$U = (A/n) \sin (x)$
$V = (B/n) \cos (x)$

$2(n-1)\pi < x < 2n\pi$; n= 1,2,3,....

Concentric ellipses are generated in U-V space as x increases continously over the real axis.

To achieve this hypothetical mapping, a 2-D perceptual space could be engineered by constructing two sensors so that the response to the signal x is given by U and V. CCA performed in the perceptual U-V space would identify classes and their topology.

But why engineering such an artificial perceptual space ? It seems that the classes are not "real", in the sense that they do not correspond to real objects.

The answer is that "real objects" are always a result of our perceptual structure. (Of course "real" is defined as in section 2, not in a metaphysical sense, so that we are not entering into a philosophical discussion about reality and the ability to perceive it. "real" for us means "physical", and perceptual means "representation of the physical" as mapped, i.e. measured, by the sensory apparatus).

When we say that 'real objects are always a result of our perceptual structure', we mean that only if a particular perceptual structure is in place we can segment a pattern of physical signals into real objects. Otherwise the pattern is just a 'meaningless' array of time dependent signals. To extract 'real objects' out of this array of signals is possible only if we engineer the perceptual structure (i.e. sensors and their functions) in a particular way. In which way we should engineer this perceptual structure so that we obtain the most convenient segmentation of the pattern in "real object', is a critical design problem.

Toward this design task, the study of the color void effect, and in general of the human color vision perceptual structure, provides some 'bionic' guidance.

First we assume, as in section 2, that we are dealing with a 2-dimensional array of time dependent signals. The perceptual pattern can have values of any dimension depending on the number of sensors. If only e.g. the energy (or the amplitude) of the signal is sensed, we have 1-dimensional perception. If all the Fourier components of the signals are sensed (eg via a spectrum analyzer), we have "infinite"-dimensional perception. Generally, if N frequencies are detected, we have N-dimensional perception. In one-dimensional perception too much information is lost, and so too few real objects are resolved. In contrast "infinite-dimensional" perception contains too much information so that no class definition is possible and thus real objects cannot be found at all. Between these two extremes lie the most interesting and practically relevant cases. These are also the cases that biological evolution appear to have selected for advanced sensory perception. But, given that *finite* MDP is the method of choice that allows both biological systems and robots (using CCA) to reduce arrays of physical signals to patterns segmented into real objects, the question remains as to what dimensionality of perceptual space is the optimal one.

A rigorous answer can only be given on further modeling the system by including, in particular, sensor response times and computational processing capabilities. This analysis lies outside the scope of this paper. Tentatively, however, we might use biology as guidance to our answer and speculate that 3-dimensionality might be a best compromise between too little information and too much complexity.

Thus, a possible design criterion (which could be applied to arrays of chemical stimuli, thermal radiation, mechanical vibrations etc.) is as follows:

1) Transform the physical input at each array site by Fourier analysis or other suitable frequency analysis.
2) Convolve the spectrum with three sensory functions. The choice of this functions depend on the range of stimuli and, generally, can be optimized. Obtain 3 numbers u, v, w.
3) Map u,v and w in a 3-D perceptual space and perform CCA.
4) the physical input is now interpreted as a pattern segmented in real objects.

An advantage of this method is its robustness. Since the method is based on the topological structure of classes, it is insensitive to relatively large mapping distortions which are likely to be created by variations in the sensory structure. These variations are almost unavoidable (and they certainly are in biological systems) so that a method which is not robust against these variations would be scarcely useful for pattern recognition.

Apart from robustness, the basic advantage of this method is the relative simplicity of the sensory (hardware) structure required to obtain a non-trivial description of the physical input in terms of perceived real objects. The burden of the description is not on the hardware but on the processing (i.e. CCA) required to identify classes and objects.

Because of its emphasis on software, it is reasonable to expect that this type of perceptual structure appears in evolutionary higher biological systems, such as color vision. On ther other hand, non-trivial descriptions can be obtained also for relatively simple MDP, e.g. in 2-D perceptual spaces. Thus, we might speculate that most species use some variation of this type of MDP to form 'concepts' of real objects out of their sensory inputs. Because of this, robots designed with similar MDP might detect *naturally* occurring patterns as yet 'invisible' to our species.

Aknowledgements

We are grateful to J. Wang, M. Barth, and S. Parthasarathy for many stimulating conversations. This research has been carried out under NSF grant #0814285.

References

[1]See e.g. D.H.Ballard and C.M.Brown, "Computer Vision", Prentice-Hall, 1982.
[2]A.Rosenfeld and A.C.Kak, "Digital Picture Processing", Academic Press, 2nd Edition,1982.
[3] J.Wang and G.Beni, "Connectivity Analysis on Multi-dimensional Multi-valued Images", IEEE International Conference on Robotics and Automation (April, 1987)
[4] J.Wang and G.Beni, "Connectivity Analysis on Multiple Valued Images", Robotics Research Transactions, August 18-21, 1986, Scottsdale, Arizona. MS86-765.
[5] G. Wyszecki and W. S. Stiles, *Color Science Concepts and Methods, Quantitative Data and Formulae*, Wiley (New York, 1981).
[6] W. D. Wright, *Researches on Normal and Defective Color Vision,* Mosby (St. Louis, Missouri, 1947).
[7] See e.g. J. D. Mollon and L. T. Sharpe (Ed.), *Color Vision, Physiology and Psychophysics,* Academic Press (New York, 1983).
[8] W. K. Pratt, *Digital Image Processing,* Wiley (New York, 1978).
[9] J.R. Kender, "Saturation, Hue, and Normalized Color; Calculation, Digitization Effects, and Use", Technical Report, Dept. of Computer Science, Carnegie-Mellon University, 1976.
[10] W. N. Dember, *The Psychology of Perception,* Wiley (New York, 1960).
[11] Y. Ohta, *Knowledge-based Interpretation of Outdoor Natural Color Scenes.* Boston: Pitman Advanced Publishing Program, 1985.
[12] B. Noble, *Applied Linear Algebra*, Prentice-Hall (Englewood Cliffs, N.J., 1969).
[13] M. Barth, S. Parthasarathy, J. Wang, E. Hu, S. Hackwood, and G. Beni, "A color vision system for applications to microelectronics:application to oxide thickness measurements", IEEE International Conference on Robotics and Automation, San Francisco, April 1986 , p. 1242.
[14] S. Parthasarathy, D. Wolfe, E. Hu, S. Hackwood, and G. Beni, "Color Vision for Microelectronics Applications", SPIE Intelligent Robots and Computer Vision, Cambridge, MA (October 1986)
[15] S. Parthasarathy, D. Wolfe, E. Hu, S. Hackwood, and G. Beni, "A color vision system for thickness determination", IEEE International Conference on Robotics and Automation, (April1987)
[16] AT&T Truevision Image Capture Board (ICB), Electronic Photography and Imaging Center, Indianapolis, Indiana.
[17] Javelin JE-3010 solid state color camera, Javelin Electronics, Inc., Torrance, California.
[18] C. H. Graham (Ed.) *Vision and Visual Perception*, Wiley (New York, 1965).
[19] E. L. Hall, *Computer Image Processing and Recognition*, Academic Press (New York, 1979).

APPENDIX

AN ITERATIVE AND INTERACTIVE SIMULATION METHOD TO RECONSTRUCT UNKNOWN INPUTS CONTRIBUTING TO KNOWN OUTPUTS OF NEURONAL SYSTEMS

Manuel Hulliger

Brain Research Institute
University of Zürich
August-Forel-Strasse 1
CH-8029 Zürich
Switzerland

Abstract

A simulation technique, combining chronic recordings in freely moving cats with acute experiments on a nerve muscle preparation, has been developed to estimate (unknown) fusimotor activity profiles which permit a reconstruction of chronically recorded muscle spindle afferent (target) responses to movements.

Stimulation patterns are iteratively generated and successively tested for their ability to simulate a chronic target response during reproduction of the original movement. Semi-automatic computer algorithms, incorporating the error between a simulated and the target response into the current stimulation pattern, are combined with interactive features which, for instance, enable the user to redraw manually critical segments of a given stimulation profile.

The procedure tends to converge rapidly, and the solutions are unique, since target responses (first produced with known inputs) could be reconstructed by virtually identical, iterated, profiles. The method opens up the possibility of investigating complex transient adjustements of fusimotor drive, e.g. during motor learning.

Introduction

The discharge of muscle spindle afferents, which are classically regarded as stretch receptors, is determined by a complex and highly non-linear interaction of two kinds of inputs: variations of host muscle length and of fusimotor drive (for review see Hulliger, 1984). Depending on the relative strength of static and dynamic fusimotor drive, the sensitivity of spindle afferents (and hence the information content of proprioceptive feedback) can vary over a wide range (see Hulliger, this vol-

NATO ASI Series. Vol. F43
Sensors and Sensory Systems
for Advanced Robots
Edited by P. Dario
© Springer-Verlag Berlin Heidelberg 1988

ume). Therefore, the knowledge of type (static and/or dynamic) and of time course of activation fusimotor efferents is a pre-requisite to the understanding of the function of the feedback from spindle afferents during various forms of voluntary movement.

Whilst recordings of host muscle length and spindle afferent firing during movement have been achieved in unrestrained animals and in man, technical obstacles have so far ruled out direct recordings from fully identified fusimotor neurones. There are a number of reasons for these difficulties. One is that fusimotor efferents mostly are small γ-motoneurones and therefore difficult to record from (as is the case for other small nerve cells too). In addition, in an intact preparation the - indispensible - type-identification of the efferents (γ_S vs γ_D) poses a problem which can hardly be solved with present techniques, since surgical procedures to simplify the task can not be used nearly as effectively as in reduced preparations (for further details see Prochazka & Hulliger, 1983).

A simulation approach, which combines chronic afferent recordings in freely moving cats with acute experiments on a nerve muscle preparation (in anaesthetized cats), was therefore adopted to search for γ_S and/or γ_D stimulation profiles which, in the acute simulation experiment, permitted a reconstruction of spindle afferent responses recorded in the the alert animals (Hulliger & Prochazka, 1983). Whilst this method has made it possible to test experimentally a number of theories of fusimotor function it was, by the same token, restricted to **explicit** concepts of fusimotor action (Hulliger et al., 1986).

Description of the method

An extended version of this simulation technique, which is **independent** of specific hypotheses, has now been developed. The method is based on an iterative strategy to generate successive profiles of fusimotor activation, which in turn are tested for their ability to reproduce a chronic target response when, at the same time, the original movement is also reproduced. To this end semi-automatic computer algorithms (to minimize mismatch signals) are combined with interactive features which enable the operator to use his insight to steer the procedure.

In most respects the experimental procedures are identical with those of the first-generation technique (for details see Hulliger & Prochazka, 1983). Briefly, in the **chronic experiments** the discharge of carefully identified spindle afferents from cat hindlimb muscles is recorded from dorsal root ganglia (using chronically implanted wire electrodes and telemetry) whilst the animals are fully alert and perform a variety of movements. Also the length variations of the afferent's host muscle are recorded with suitably implanted length gauges (for details see Prochazka, 1984).

In the _acute simulation_ experiments the length changes of the original movements are reproduced identically by imposing them on the soleus muscle of the cat's hindlimb, using an electromagnetic puller which in turn is driven by a specially designed multi-channel hybrid signal generator which stores digitized length records and various stimulation functions (below; Frei et al., 1981). The responses of spindle Ia afferents to such reproduced movements are recorded from dorsal root filaments, and one or several functionally single static (γ_S) and/or dynamic (γ_D) fusimotor efferents acting on the spindle afferent under study are activated in ventral root filaments, using electrical stimulation (Goodwin et al., 1975, for further detail). The time course of efferent stimulation is determined by analogue signals which are first read from the hybrid signal generator and which are then converted into rate-modulated pulse trains, using a voltage controlled oscillator.

The _aim of_ the _simulations_ is to find fusimotor stimulation profiles which lead to spindle afferent responses to simultaneously reproduced movements that match chronic target responses. The _new feature_ of the second-generation technique is that this can now be achieved without preconceived notion of the successful stimulation pattern. The user is allowed to make a start with any profile of his choice, which is then modified in successive iteration cycles, until a target reproducing efferent driving function is arrived at.

Fig. 1 gives a schematic illustration of a single _iteration cycle_. The spindle receptor, shown as a combined summer and multiplier unit (elliptic shape to symbolize highly non-linear properties), receives two kinds of inputs. Input A are the length variations of the reproduced movement, which are invariant within an iteration sequence. Input X_n is the fusimotor driving function, which is modified in successive (n) iteration cycles. These two inputs combine to determine a simulated response, which is stored on-line by a laboratory computer and which is displayed as a cycle histogram. Simple subtraction (comparator in Fig. 1, centre right) of the target from the simulated response gives an error function, which reflects the degree of mismatch between the two responses. The scaled and phase-advanced error signal (Fig. 1: gain, shift) is then incorporated into the input X_n signal (using linear summation; Fig 1, bottom left) to give the driving function for the next iteration cycle (input X_{n+1}).

Whilst the principle of operation (feedback of error) is simple enough, the _details of implementation_ are not. The neuronal responses are first stored as digital cycle histograms. This simply reflects the discrete nature of nerve impulse trains (all-or-none action potentials). For computational simplicity the digital histograms are then converted into analogue signals, using linear _interpolation_. In addition, prior to comparison, simulated and target response are _low-pass filtered_, using a moving window average filter, in order to remove high-frequency (largely random) variability of discharge. Additional low-pass filtering (5th order Butterworth filter) can be

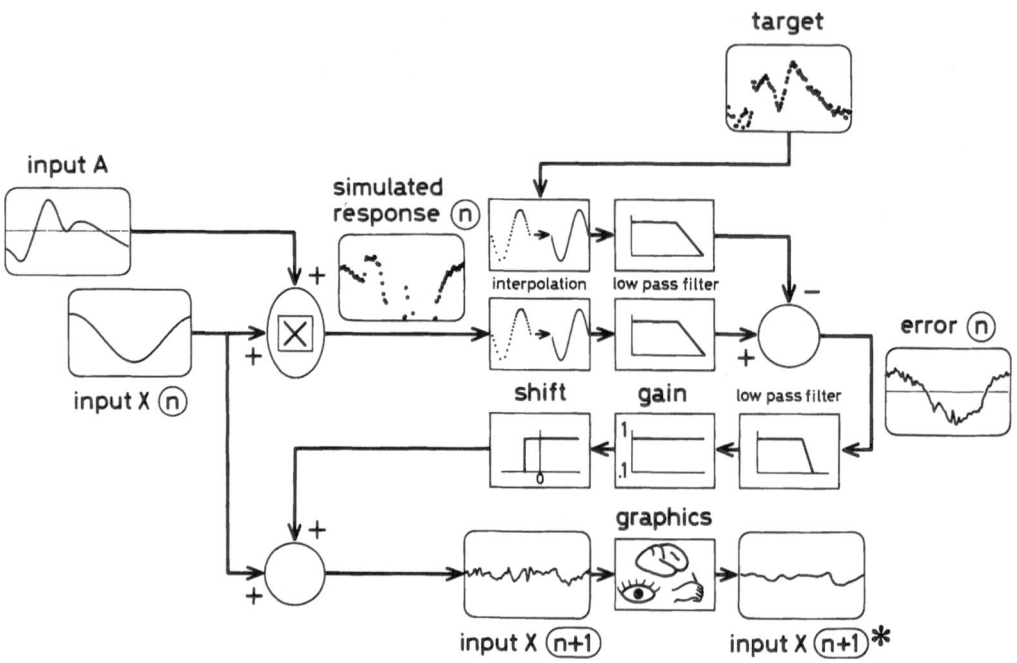

Figure 1. Schematic illustration of the nth iteration cycle. Two inputs, A (length) and X_n (γ drive), combine to elicit the nth simulated response of the spindle afferent under investigation. Subtraction of the target response gives the mismatch signal, $error_n$. Suitably scaled and phase-advanced, this error is summed with input X_n to provide input X_{n+1}, i.e. the fusimotor driving function of the next iteration cycle. For further details see text.

performed, if necessary, on the error and efferent driving signals (illustrated, but not executed, in Fig. 1). The justification of such multiple filtering lies in the well known low-pass filter properties of the efferent actuation system (intrafusal muscle fibres; see Hulliger, 1984). Since frequency components beyond 5-10 Hz give rise to only minimal fluctuations of afferent discharge (Zangger et al., 1985), high-frequency components of the driving signal (input X) tend to leave simulated responses virtuallly unaffected. Thus, unless removed, they are perpetuated and, in fact, easily boosted in successive iteration cycles.

The <u>shift operation</u> performed on the error signal is essential to avoid instability. Fusimotor action is fraught with actuation delays of the order of 30-60 ms, which arise from neuronal conduction and, mainly, muscular contraction delays. Unless compensated for, these rapidly cause increasingly large oscillations in successive simulations. In practice the activation delays of individual γ-Ia combinations are first estimated independently and then used for the error summation operation.

Iterative reconstruction of fusimotor input

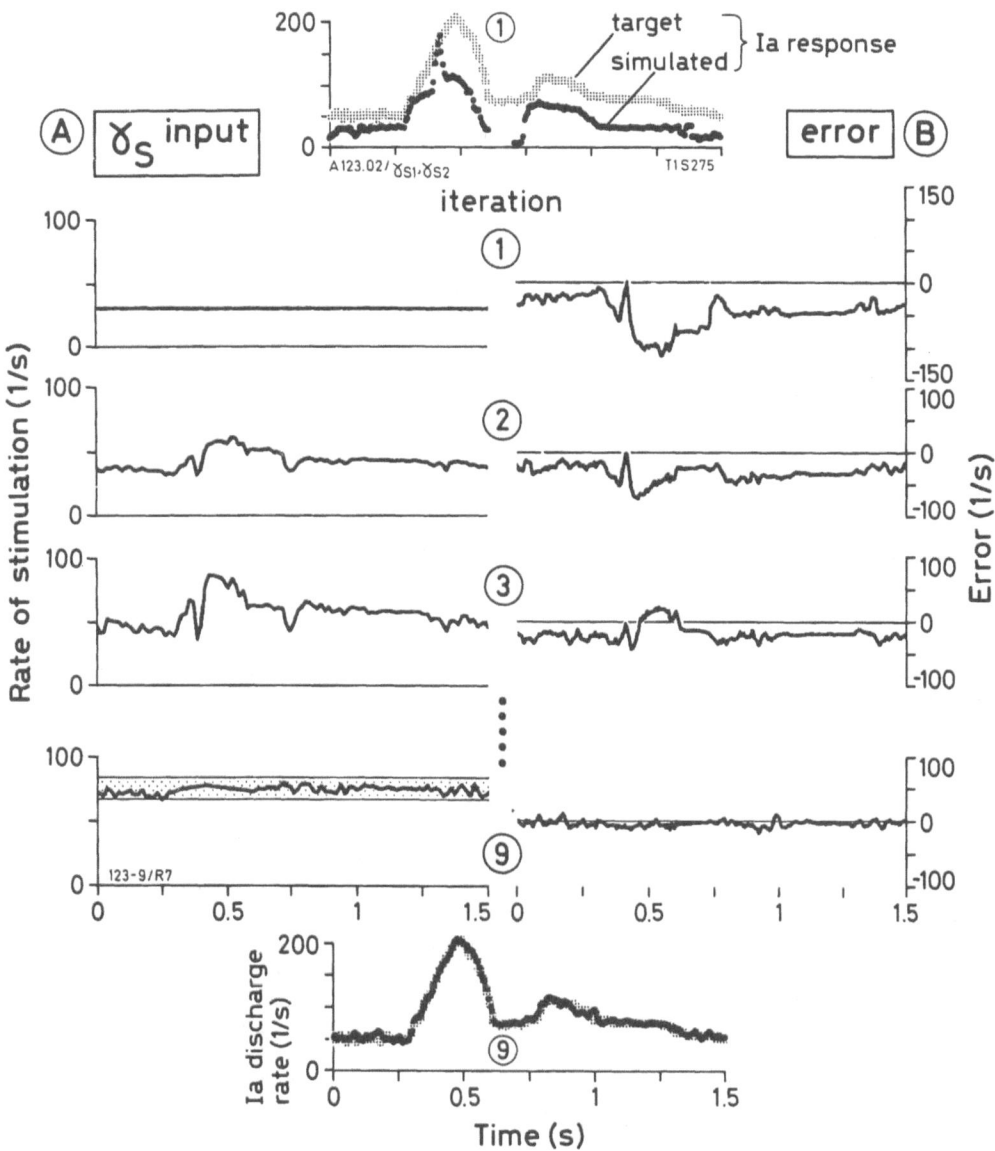

Figure 2. Iteratively generated fusimotor activation functions for reconstruction of Ia target response. Display of driving functions (A) and mismatch signals (B) for the first three and the last cycle of an iteration sequence. The target response is shown as grey raster profile in the top and bottom panels. It was generated in the same acute experiment during concomitant muscle stretch (same time course as input A, Fig. 1) and tonic stimulation of two γ_S axons at 75/s (rastered band in A.9). Simulated responses (first and last) are superimposed and shown as black dots. Note rapid decline of error, as the input profile approaches and reaches (in 9) its target band.

The <u>scaling of error</u> prior to summation (Fig. 1: gain) is equally important, since individual fusimotor efferents vary widely as regards their excitatory power (see Hulliger, 1984; review). Therefore suitable gain values have to be chosen for each γ-Ia combination, the choice often being by trial and error and by experience gathered in the course of an experiment.

The iterative, error-feedback based, generation of input profiles is further refined by important <u>interactive features</u>. The user can optimize performance by choosing suitable operation parameters for histogram resolution, filter frequencies and scaling and phase shift of the error function. In addition he is given the choice of using insight and intuition and to redraw, with a digitizing tablet (Fig. 1: graphics), critical segments of the input profile which is displayed on a graphics terminal.

Evaluation

The evaluation of the iterative simulation method so far has led to three main conclusions.

<u>First</u>, γ-activation profiles can indeed be found which give surprisingly detailed <u>reconstructions</u> of chronically recorded muscle spindle afferent target responses. In particular, segments of chronic recording which - with the first-generation simulation technique - were best matched by simple tonic levels of γ-drive (e.g. tonic $γ_S$ action for Ia responses during routine stepping; see Hulliger, this volume), could be reconstructed iteratively with driving profiles containing a dominant tonic component. Examples of successful simulations of target profiles are shown in Figures 2 and 3 (two separate iteration sequences). In the top and bottom panels the first and last simulated responses (black dots), respectively, of each sequence are shown against the background of the target response (grey raster profile; identical in Figs. 2 and 3). Whilst the simulated and target responses differed strikingly in the first iteration cycle (top panels; note also large errors in Figs. 2.B.1 and 3.B.1), they were nearly perfectly matched in the last iteration of each sequence (bottom panels) and, accordingly, the errors were very small and centered around the zero line (Figs. 2.B.9, 3.B.7). Note that the target responses of Figs. 2 and 3 were not from chronic recordings (see below, uniqueness).

<u>Second</u>, the procedure tends to <u>converge</u>, in that strikingly different initial profiles after only a few (3-5) iterations lead to practically identical, target reproducing, final products. In Fig. 2 the start-up profile (3.A.1) was a tonic level of $γ_S$ drive, whereas in Fig. 3 the initial profile was the envelope of the idealized smoothed e.m.g. during normal stepping (3.A.1). In spite of the clear difference between these two, the final driving functions were essentially identical (cf.

Iterative reconstruction of fusimotor input

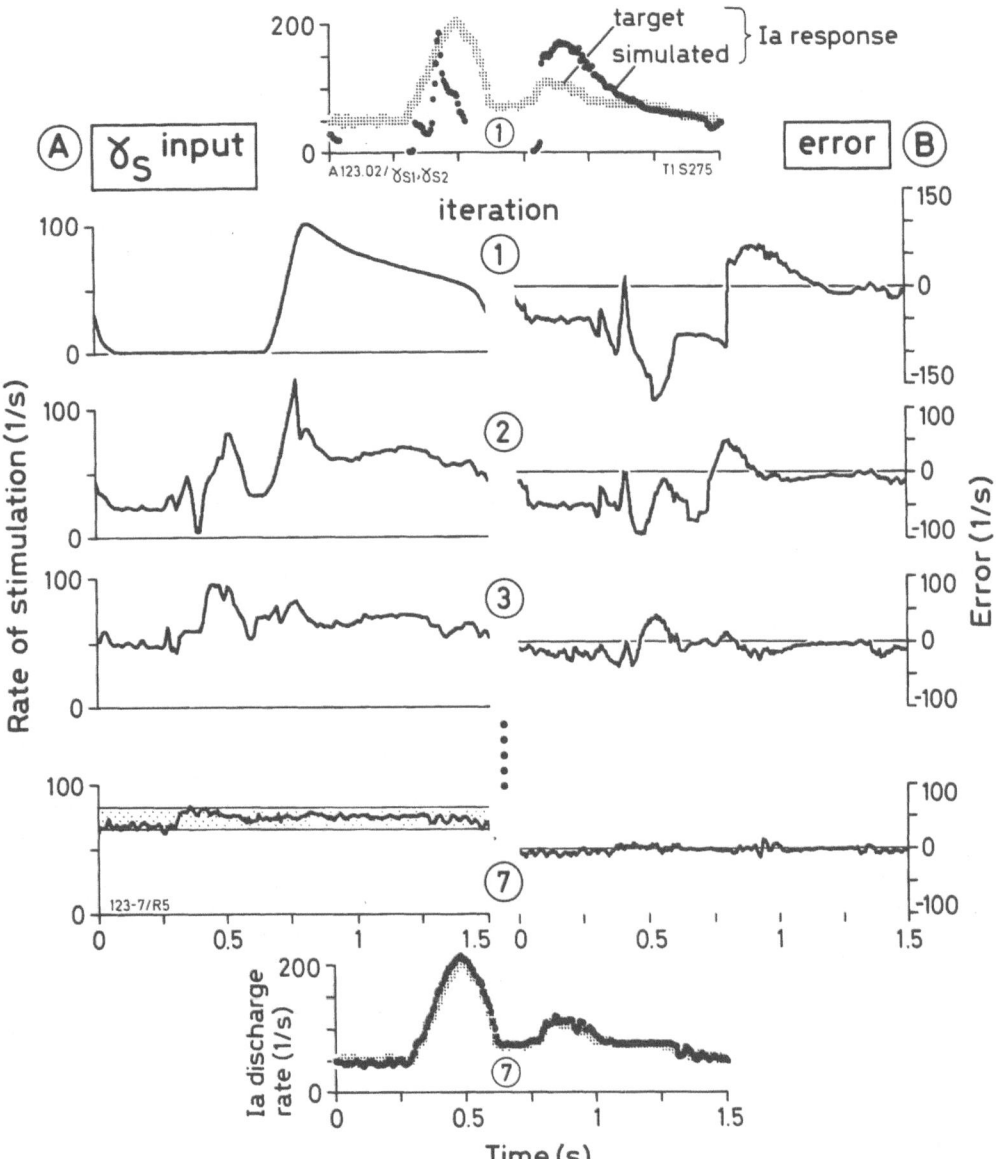

Figure 3. Illustration of input (A) and error (B) functions as generated in a second iteration sequence for reconstruction of the same Ia target response as in Fig. 2. Note, however, that the start-up profile (A.1, smoothed e.m.g. envelope) was different from that in Fig. 2.A.1 (tonic drive). In spite of this the final stimulation profile (A.7) was very similar to that of Fig. 2.A.9. Moreover, it also fell into the target band, which is centered around the level of tonic γ_S stimulation (at 75/s) which was initially used to generate the afferent target response. For further detail see text and legend to Fig. 2.

Figs. 2.A.9 and 3.A.7) apart from high-frequency fluctuations which, in the particular data illustrated, were not removed by filtering (see Method).

Third, for a given class of γ-input the iterative solutions are unique, since Ia target responses, which (in the same acute experiment) were first produced with defined stimulation patterns, could then be reconstructed by virtually identical, iteratively generated, profiles. In other words, known fusimotor inputs could be satisfactorily recovered. In Figs. 2 and 3 the target (raster profile in top and bottom panels) was the response of a spindle Ia afferent to a simulated stepping movement (same as shown in Fig. 1, input A) which was recorded during concomitant tonic stimulation of two γ_S fibres at 75/s. This tonic input level is illustrated by the rastered horizontal band in Figs. 2.A.9 and 3.A.7, the width of the band marking the ± 10% deviation from the 75/s line (i.e. the 'input target'). It can be seen that the iteratively generated input profiles, which gave convincing afferent target matching (bottom panels), also satisfactorily fell into the target band of fusimotor input.

Pending issues. It remains to be tested whether such uniqueness of solutions also extends to different types of fusimotor inputs, the question being whether targets generated e.g. by known patterns of γ_D drive can only be reconstructed by γ_D-, but not by γ_S-, iterations, and vice versa. At first sight such uniqueness of type would appear to be too optimistic an expectation, since both types of efferents exert qualitatively similar excitatory effects. However, there are also reasons to treat this, within limits, as a realistic possibility. Dynamic action can very strongly sensitize Ia afferents to dynamic stretch, often evoking peak discharge rates in excess of 500 impulses/s. Yet even powerful static action is rarely capable of evoking such pronounced afferent firing. Conversely, whilst γ_S action easily is effective enough to offset the effects of muscle shortening (which, on its own, would cause Ia silence), γ_D action is not. Thus powerful fusimotor action of a given type tends to induce features of afferent response, which may prove sufficiently discriminative to also convey uniqueness of type to the method of iterative analysis.

Limitations. The method is limited by the absence of unique solutions for targets with periods of afferent silence, since fusimotor drive then may cover the entire range of activation rates from zero up to the threshold of excitation of the afferent at issue. Preliminary evidence indicates that this limitation is less serious for static than for dynamic action.

In spite of this limitation, as long as afferents fire, the iterative reconstruction method provides a promising new tool to analyse complex transients of underlying fusimotor drive. The evidence described elsewhere (Hulliger, this volume) suggests that such transient processes may be of particular relevance for motor performance during periods of adjustment to unfamiliar conditions, conceivably also for motor learning.

Acknowledgements: This work was supported by grants from the Swiss National Science Foundation (3.157.81, 3.071.84), the Jubiläumsstiftung and Forschungsstiftung of the University of Zürich, the Dr. Eric Slack-Gyr Foundation, and the European Science Foundation (ETP).

References

Frei JB, Hulliger M, Lengacher D (1981) A programmable wide-range analogue signal generator based on digital memories for use in physiological experiments. J Physiol(Lond) 318: 2-3P

Goodwin GM, Hulliger M, Matthews PBC (1975) The effects of fusimotor stimulation during small amplitude stretching on the frequency response of the primary ending of the mammalian muscle spindle. J Physiol(Lond) 253:175-206

Hulliger M (1984) The mammalian muscle spindle and its central control. Rev Physiol Biochem Pharmacol 101:1-110

Hulliger M (this volume) Proprioceptive feedback for sensorymotor control

Hulliger M, Prochazka A (1983) A new simulation method to deduce fusimotor activity from afferent discharge recorded in freely moving cats. J Neurosci Methods 8:197-204

Hulliger M, Prochazka A, Zangger P (1986) Fusimotor activity in freely moving cats. Tests of concepts derived from reduced preparations. In: Grillner S, Stein PSG, Forssberg H, Herman RM, Stuart DG, Wallén P (eds) Neurobiology of vertebrate locomotion. Macmillan, London (in press)

Prochazka A (1984) Chronic techniques for studying neurophysiology of movement in cats. In: Lemon R (ed) Methods of neuronal recording in conscious animals. IBRO handbook series: methods in the neurosciences 4. Wiley, Chichester, pp 113-128

Prochazka A, Hulliger M (1983) Muscle afferent function and its significance for motor control mechanisms during voluntary movements in cat, monkey, and man. In: Desmedt JE (ed) Motor control mechanisms in health and disease. Adv Neurol 39:93-132

Zangger P, Hulliger M, Prochazka A (1985) Regular vs irregular fusimotor firing: the likely effects on Ia discharge during normal movement. In: Boyd IA, Gladden MH (eds) The Muscle spindle. Macmillan, London, pp 371-375

NON-CONVENTIONAL SMART COLOUR SENSING IN ROBOTICS

R.F. Wolffenbuttel

Delft University of Technology, Department of Electrical
Engineering, Laboratory for Electronic Instrumentation,
Mekelweg 4, 2628 CD Delft (The Netherlands)

ABSTRACT

A simple solid-state colour sensor suitable for robotic
applications is presented. The operating principle is based on
the strong dependence of the absorption of incident optical
radiation in silicon on the applied wavelength. The operation
of the sensor involves an adjustable confinement of the depth
in silicon, in which penetrated photons can contribute to the
photocurrent by using a controllable width of the depleted
region. The sensor is integrated in a bipolar process and the
average wavelength is measured independently of the optical
intensity by utilizing a compensation method.

INTRODUCTION

In solid state colour image sensors, a colour picture element
is, in principle, composed of three photodetectors, each with
a special red, blue or green dye deposited in a convenient
pattern on top of the sensor surface to perform the colour
filtering [1]. This construction enables the intensity of the
corresponding primary colour to be determined. From the
information provided by the three detectors it is possible to
determine the unique position of the colour in the CIE colour
triangle. Large colour imaging sensors have already been
realised using such techniques [2] and the fabrication and
alignment of even larger colour filter arrays do not seem to
imply a fundamental limitation to a further increase in the
sensor density [3].

In robotics research an increasing interest is perceptible in
colour imaging sensors. Image analysis has, so far, been
performed using black and white cameras, but naturally more
information can be obtained from a scene by using colour
cameras. The purpose for equipping an industrial robot with
such a colour vision system is to enable the distinction
between identically shaped objects of different colour.

The direct implementation of this objective using a colour
imager will increase the load of the available processing power
as the information content and thus the density of the
generated data triples compared to conventional vision because
of the three primary colours. Despite the introduction of fast
dedicated image processors the limitation of the processor

NATO ASI Series, Vol. F43
Sensors and Sensory Systems
for Advanced Robots
Edited by P. Dario
© Springer-Verlag Berlin Heidelberg 1988

capacity remains one of the strongest impediments in real-time
image analysis for conventional vision in industrial robot
applications. These problems will, obviously, increase in
computer-analysis of colour images.

Another approach involves a conventional vision system combined
with a simple one-point colour sensor [4]. In such a set-up the
amount of generated data will increase only very little while
it remains feasible to select an object based on the colour.
The exploration of the scene will result in the labelling of
several suitable objects based on insufficient sensor information supplied by the conventional vision system. The selection
can be completed by a closer investigation of these objects
using the simple and small colour sensor, which can easily be
mounted in the robot gripper.

Therefore colour vision is also possible using the simple
integrated colour sensor and reveals the additional advantage
of having to perform only minor modifications in an existing
robot configuration. This approach also fits in a trend, which
will lead to an increasing diversity of the environmental
explorative sensors. This trend includes the research on
tactile imaging sensors and proximity sensors and is intended
to persue an improved overall robot performance.

USING SILICON FOR COLOUR SENSING.

The characteristic course of the intrinsic absorption coefficient in silicon at different wavelengths of incident radiation
is shown in figure 1. Due to the indirect bandgap in silicon,
incident photons having an energy in excess of 3.5 eV (equivalent to a wavelength smaller than 350 nm) will allow a direct
transition of electrons from valence band to conduction band.

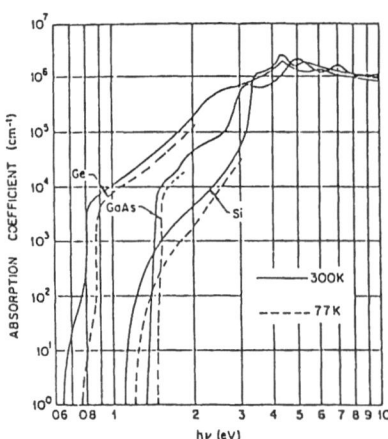

Figure 1, Absorption coefficient in silicon.

In the case of incident radiation with an energy in between
the bandgap energy of 1.12 eV and the energy sufficient to
allow a direct transition, an indirect transition could occur
provided energy and momentum are preserved. Incident photons
are not able to provide a momentum, so the probability of such
a transition taking place depends on lattice vibrations.
Therefore, the absorption increases at an increasing energy,
since less change in momentum will be required to generate an
electron-hole pair at larger energies, resulting in a larger
chance of an indirect transition occuring [5].

This wavelength-dependence causes a very shallow absorption of
blue light and enables red light to penetrate deep into the
silicon. The generated charge-carriers and thus the photocurrent
in a photodiode is proportional to the number of absorbed
photons and thus depends on the surface area. If a mechanism is
implemented that collects only the generated charge carriers in
a silicon layer down from the surface to a defined boundary
than the perceived photocurrent also depends on the wavelength
of the incident light. As the small-wavelength components in
the spectrum are shallow absorbed, all the incident blue light
is already absorbed at very thin layers.

Therefore at illumination with light having predominant small-
wavelength components the perceived photocurrent will remain
almost constant at an increasing width of this collecting
layer. However, at illumination with long-wavelength (red)
light the detected current will inrease.

In a practical photodiode the space charge region, which is
actually the desired collecting layer, is controlled using the
reverse biasing voltage across the junction. Figure 2 shows a
simulated (solid lines) and measured (dashed lines) response at
increasing reverse voltage for three different monochromatic
colours and clearly reveals the quicker saturation at low
wavelengths.

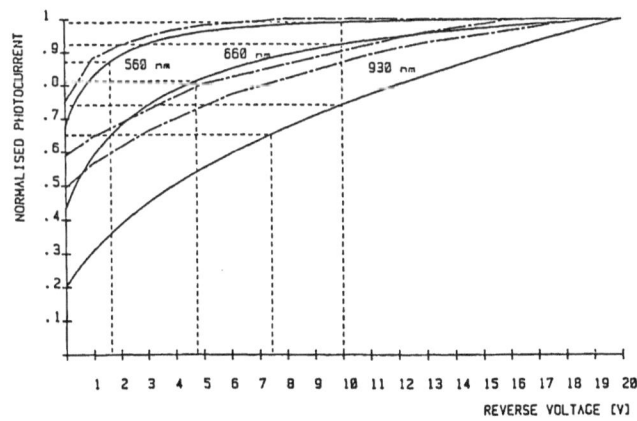

Figure 2, Photocurrent vs. reverse voltage at a 560 nm, 660nm.
         and 930nm. wavelength.

The detected photocurrent at a certain reverse voltage is therefore determined by the intensity and the wavelength of the incident light, providing that measures are taken to prevent charge carriers generated beyond the depleted region from diffusing to the depleted region and to contribute to the detected photocurrent.

In such a practical p+ n photodiode the large absorption coefficient at low wavelengths and the finite width of the p+ upper layer gives rise to a short wavelength limit. The vast majority of photons will be absorbed before penetrating into the junction and are not able to diffuse to the junction because of the relatively small lifetime of the minority charge carriers in this usually heavily doped upper layer.

The sensor consists, in principle, of two adjacent photodiodes with different areas. The photocurrent at a certain luminous intensity is proportional to the area of the photodiode and, as shown above, also depends on the width of the depleted region. It is possible to eliminate the effect of the intensity by imposing a current equality condition on both photodiodes in spite of the difference in sensitive area. In that case, the required reverse voltage across the larger photodiode is uniquely determined by the reverse reference voltage across the smaller diode, the difference in sensitive area and the colour of the incident radiation.

The imbalance in photocurrent, caused by the photodiode sensitive area inequality, can be compensated by a decrease in the larger photodiode reverse voltage, which narrows the space charge region of the diode and thus approaches a current balance. If at a constant width of the smaller diode a current balance is obtained, the width of the depleted region in the larger diode is established according to the current equality condition mentioned above.

Since the penetration depth and thus the required voltage to obtain a balance depends on the wavelength of the incident light, the sensor will reveal a response determined by the optical power distribution throughout the spectrum. It is necessary to elaborate the current balancing equation numeric- ally because of the complex equation describing the absorption coefficient. The results of a simulation, which involves the simultaneous illumination of the sensor with two different monochromatic light sources of variable relative intensity, are shown in figure 3 and indicates an almost linear relation between the reverse voltage ratio and the "centre of gravity" of the incident radiation within a limited part of the spectrum ranging from about 400 nm to 1000 nm. For the sake of simplici- ty, this principle will therefore be referred to as a method suitable for determining the "average colour" of the incident light. As shown above, it is possible to construct a simple solid-state colour sensor based on this principle, in which a compensation method, where the ratio between the drift currents in two controlled depletion layers are involved, is implemented.

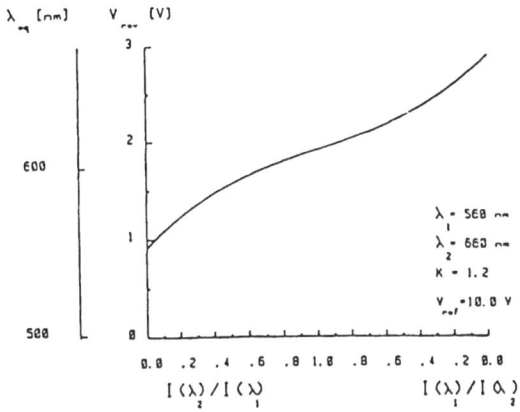

Figure 3, Simulation of the response to mixed colours using two
light sources with a variable relative intensity.

SEMICONDUCTOR COLOUR SENSOR.

The actual sensor consists of two adjacent and identical
photodiodes. One of the photodiodes is connected to a reverse
reference voltage and the other to an adjustable reverse
voltage. The two diodes are identical and therefore have the
same physical area. The effective sensitive area in one of the
diodes is implemented by using an electronically tuneable
attenuation k in the photocurrent supplied by this diode. This
approach reveals a higher flexibility in use compared to a
sensor having a fixed design-determined dimensioning. The
method used to apply this structure in silicon to colour
sensing is based on finding the appropriate reverse voltage
required across the compensating diode to compensate for the
loss in the effective sensitive area of the reference diode due
to the tuneable attenuation in its photocurrent.

The sensor is shown in plan and cross-section in figure 4 and
consists of two photodiodes with a drain and source on either
side of the diode. Thus the structure closely resembles a pair
of dual gate junction FETs. The photodiodes are formed by a
boron implantation layer and the n-epilayer with the junction
at a 0.5 um depth. The p-implantation layer is surrounded by a
p-diffusion guard ring to increase the breakdown voltage. Due
to the high doping of the p-top layer, the space charge region
will almost entirely be extended into the n-epilayer.

The source-substrate voltage removes charge carriers generated
in the bulk of the silicon and a drain-source voltage ensures
that only charge carriers generated in the depleted part of the
epilayer are collected. As mentioned before, the absorption
coefficient, and thus the penetration depth of photons in
silicon, depends on the wavelength of the radiation and therefore the current is also determined by the wavelength of the
incident light. The function i=f(intensity, wavelength) can be
solved by imposing the current balance in the sensor.

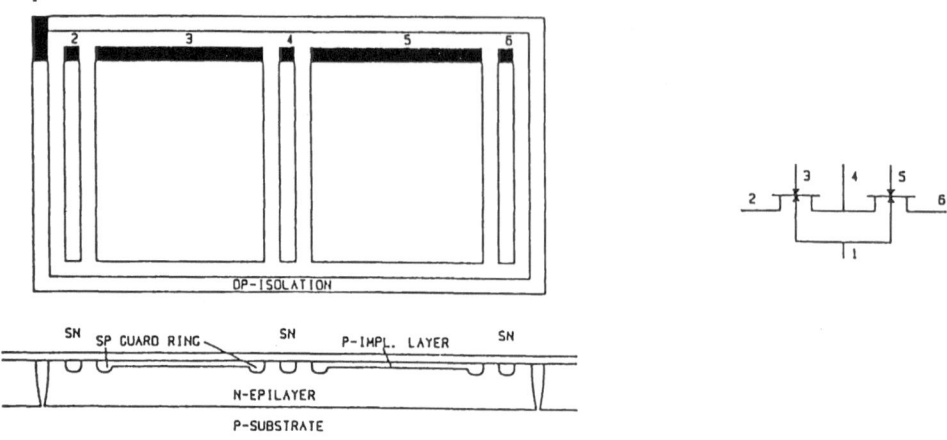

Figure 4, Cross-section and top view of the sensor structure.

The current balance is persued, in spite of the attenuation in
photocurrent of the reference photodiode, according to the
diagram depicted in figure 5. A controllable fraction k,
ranging from 0.5 to close to unity, of the photocurrent
provided by the reference photodiode is subtracted from the
compensating diode photocurrent. The remaining current is fed
to a controller, which generates the proper biasing voltage
across the compensating diode in order to reduce this current
difference, until, at stationary conditions, a balance is
obtained at a reverse voltage determined by the average
wavelength and the settings of the reference diode.

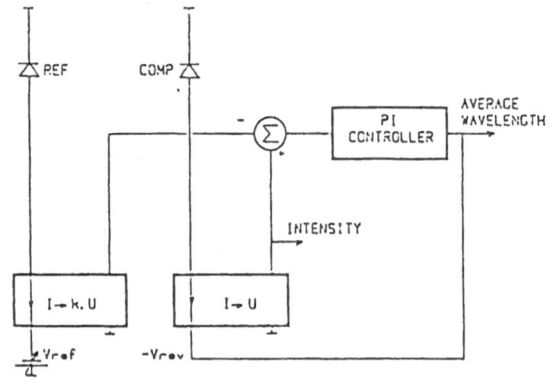

Figure 5, Functional diagram of the colour sensor.

A photomicrograph of the smart colour sensor, in which all
the functions are integrated is shown in figure 6 and consumes
a die area equal to 3000 x 1750 um. The photodiodes with dimensions 430x430 um can clearly be distinguished in the centre.

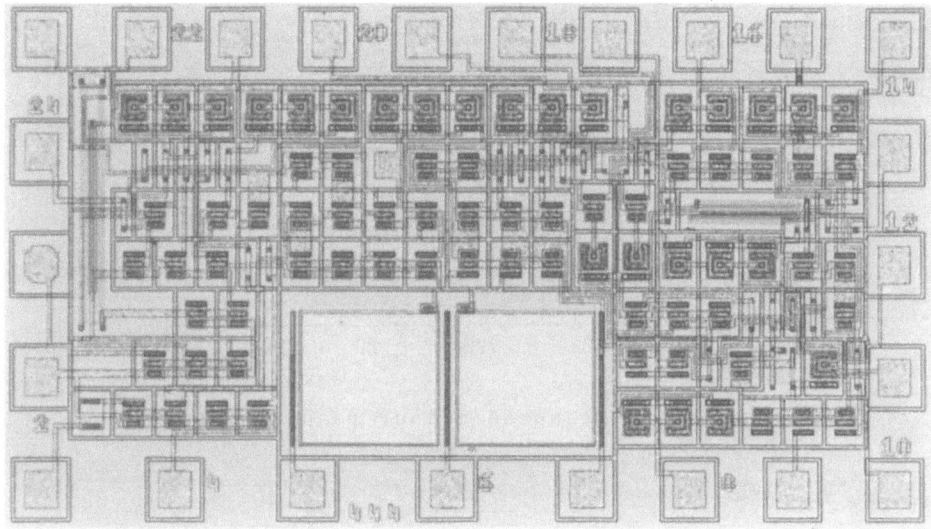

Figure 6, Photograph of the smart colour sensor.

Unlike the human eye, this sensor determines the average colour
of the incident light within the visual to near-infrared
spectrum ranging from 400 nm to 1000 nm. The observed response
is therefore not identical to that of the human eye and an
ambiguity is, in principle, possible.

EXPERIMENTAL RESULTS

The colour response of the above-described sensor can be
simulated by using a simple model, which calculates for each
wavelength the corresponding absorption coefficient in silicon
and the width of the depleted region necessary to obtain a
current balance in a specified sensor configuration. The
results of the simulations and measurements using an attenuation k equal to 0.9 and 0.6 are depicted in figure 7 and
figure 8, respectively. The curves show the required reverse
voltage across the compensating diode to obtain the current
balance vs. the wavelength at a reference voltage equal to 5 V.
The difference in lower wavelength limits clearly shows the
effect of the area ratio implemented in the variable attenuation. The results of experiments using the colour sensor are
depicted in the appropriate figures indicated by the squares.
The figures show that these experimental results are in
reasonable agreement with the theory down to a wavelength of
about 600 nm at an area ratio equal to 0.6 and down to 500 nm
when using an area ratio of 0.9. The calculated reverse voltage
that should be applied at a lower wavelength is not practical.

Figures 7 and 8 show that this sensor exhibits the particularly
convenient characteristic of having a geometric degree of
freedom left for the design of the sensor colour range. To a
certain extent this sensor reveals a 'low pass' behaviour that
can be affected by a proper choice of the attenuation and the
reference voltage.

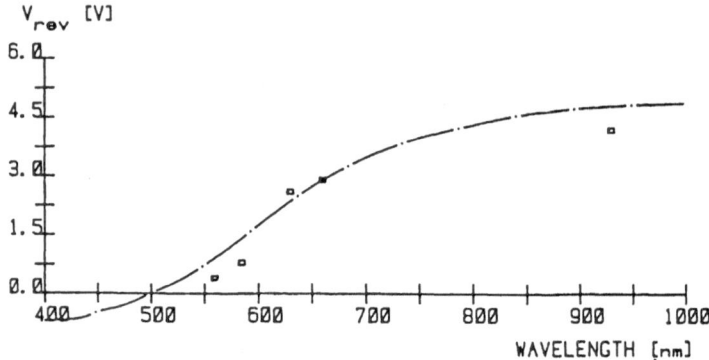

Figure 7, Simulated and measured colour response at an
attenuation of 0.9.

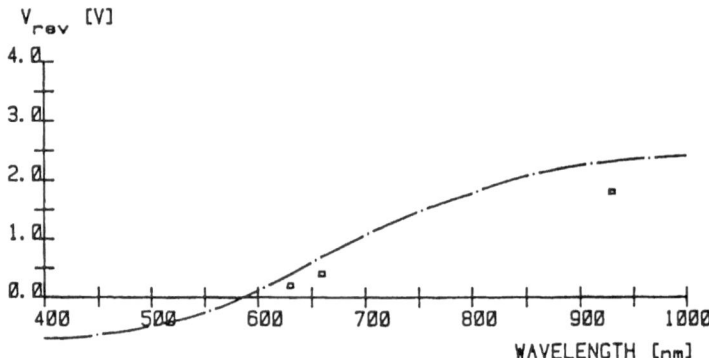

Figure 8, Simulated and measured colour response at an
attenuation of 0.6.

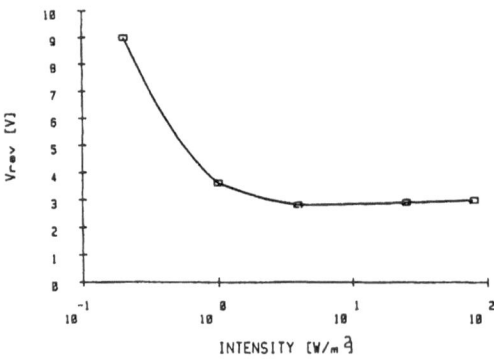

Figure 9, Colour response to a variable intensity of a fixed
wavelenghth (Vref=10).

The colour response should not depend on the intensity of the incident radiation. Such a demand can only be satisfied as long as the photocurrent is sufficiently large to determine the current balance. An experiment is performed in which the sensor is illuminated with an increasing intensity by an LED light source with a peak wavelength at 670 nm (red). At a very low intensity, the current balance will solely be established according to the dark current ratio. Since the photodiodes are of the same dimensions, the reverse voltage required to obtain the current balance necessarily equals the reference voltage. This reverse voltage is, naturally, independent of the source wavelength. The results are depicted in figure 9, which shows the uncertainty at an intensity of less than 1 W/m^2. This result is in accordance with the obvious fact that a colour can only be perceived if an illuminating source is present.

CONCLUSIONS

The response of the smart colour sensor to incident light with a wavelength ranging from 400 nm. to 1000 nm. is reproducible within a wide range of optical intensities. The shift of the minimum detectable wavelength to larger wavelengths at an increasing attenuation k can be used to solve the ambiguity, which could arise at the perception of multi-coulored objects. A blue-reddish object will, at an attenuation of the reference diode-photocurrent close to unity, give the same colour response as a uniform greenish object. A larger attenuation will emphasize the long-wavelength components of the incident spectrum and thus, in the case of multi-coloured objects, the detected average colour will shift to the longer wavelengths.

This sensor is therefore suitable for extracting unique colour features from the environment by determining the average colour for several values of k. In fact, it is possible to determine a kind of incremental spectrum of the incident radiation when increasing the attenuation from close to unity until 0.5 and simultaneously monitoring the colour output voltage.

The sampling of this spectrum for only a few values of k will, in many applications, provide sufficient information to distinguish between identically shaped objects of different colour, which makes this simple colour sensor very suitable for many applications in robotics.

REFERENCES

1 P.L. Dillon, D.M. Lewis and F.G. Kaspar, Color imaging system
  using a single CCD area array, IEEE Trans. Electr. Dev. ED-25
  No2(1978) pp. 102-107.

2 T.J. Tredwell, High-density solid-state image sensors, Proc.
  Solid-State Trans. 85, Philadelphia, Pennsylvania, USA, June
  1985, pp. 424-429.

3 P.L. Dillon, A.T. Brault, J.R. Horak, E. Garcia, T.W. Martin
  and W.A. Light, Fabrication and performance of color filter
  arrays for solid-state imagers, IEEE Trans. Electr. Dev. ED-25
  No2(1978) p. 97-101.

4 P.P.L. Regtien and R.F. Wolffenbuttel, A novel solid-state
  colour sensor suitable for robotic applications, Proc. 5th
  Int. Conf. on Robot Vision and Sensory Controls (RoViseC-5),
  Amsterdam, The Netherlands, October 1985, pp. 259-266.

5 H.R. Phillip and E.A. Taft, Optical constants of silicon in the
  region 1 to 10 eV, Phys. Rev., 120(1960) pp. 37-38.

6 R.A. Smith, Semiconductors, University Press, Cambridge,
  1st edn, 1959, pp. 189-211

Adaptable Accommodation in Assembly

C.J. Bland

Department of Mechanical and Manufacturing Systems Engineering

University of Wales Institute of Science and Technology

Cardiff, UK.

The Assembly Environment

Assembly traditionally is labour intensive. In comparison with component manufacture, mechanisation and automation are still in their infancy and are largely restricted to providing mechanical aids for the assembly worker.

In contrast with human assembly workers, automatic assembly machines are custom designed and cannot be easily modified to accommodate different products or design changes. This fact in conjunction with the very high captial cost of the machines means that they are only economically viable for the mass production environment. Also, assembly machines are very venerable to fluctuations in component quality and if not carefully controlled, machine jams develop leading to stopages in production.

NATO ASI Series, Vol. F43
Sensors and Sensory Systems
for Advanced Robots
Edited by P. Dario
© Springer-Verlag Berlin Heidelberg 1988

More recently, smaller, more flexible, assembly machines suitable for the batch production environment have been developed. These machines are increasingly being serviced by robots and are often integrated into manufacturing cells. Also, with the advent of assembly robots, flexibility has been further increased.

For all assembly machines to operate sucessfully, misalignment between components must be prevented or corrected. To prevent misalignments any jigs and fixtures have to be precision made for the components being assembled. This reduces flexibility, since new jigs and fixtures are required for a product change, and increases system cost. In addition, the components themselves must be maintained within tolerances which are suitable for the assembly machine. These tolerances may therefore be artificially high for the components function and supplementary quality control on the incoming components may be required.

Also, there are many industries which can not benefit from this technology because the components they assemble have clearances which are too small. For example, the assembly of hydraulic equipment often involves clearancess of 5 microns or less, whereas assembly robots only have repeatabilities of 0.02mm.

To help reduce the system tolerances and increase the application of these systems many institutions have developed devices which will accommodate for misalignment between components.

Passive Acommodation

Passive devices are mechanical structures which compensate for
misalignments during assembly by deforming. The number of occurrances
of component jams is not only reduced, but the insertion forces
necessary can be reduced hence reducing the risk of damage to the
components.

The best known device is the remote Centre Compliance, developed by
the Charles Stark Draper Laboratory (CSDL). It is manufactured under
licence and is commercially available in many countries.

However there are many other forms of passive compliance. One of these
is the inherent flexibility in machines causing them to deflect under
applied loads. One recent exploitation of this is the development of
the Selective Compliance Assembly Robot Arm (SCARA). This robot was
especially designed for extensive use in the assembly environment and
has passive compliance incorporated in its structure.

Although passive devices have been sucessfully used in a variety of
applications they have several limitations. Firstly, to reduce
assembly forces to a minimum, they need a low stiffness and hence a
low natural frequency and are susceptible to vibrations. Also, each
unit is designed for one particular peg length and can only be used
for cylindrical, chamfered components.

Active Accommodation

To reduce the limitations of passive devices, active devices have been developed in which the contact forces during assembly are measured by transducers and the signals used to control actuators which compensate for the misalignment.

One method of implementing active devices in robotic systems is to attach a sensor onto the robot wrist and use the signals produced to control the robot's movements. Six degree of freedom sensors are often used but are expensive and involve complex signal processing to decouple the axes before the signals can be used to control the robot arm. Therefore, some researchers have reduced the complexity of the sensor by using only three degree of freedom sensors. Other workers have tried to dispense altogether with any external sensing and instead measure the forces through the actuators.

The disadvantage of all these systems is that it is difficult to compensate for small misalignments because of the high inertia of the robot arm. Also, the clearance of the assembly components must be greater than the resolution of the robot system.

As outlined earlier, part misalignment is a problem not only for robotics, but also for classical assembly machines. Hence, the inclusion of a micro-manipulating device is an important part of the system.

The work at UWIST has concentrated on simple three degree of freedom sensing and two degree of freedom actuation. Particular emphasis being placed on the desire to obtain full information on the insertion process against the need to reduce the complexity of signal processing and hence increase the speed of insertion.

Pneumatic device

This has attracted attention from industrialists because of its simplistic construction and lack of software dependance, see Fig. 1. A lateral load applied to the central baffle [1] causes it to displace relative to the pneumatic back-pressure sensor [4] which is mounted in an intermediate assembly [2]. The change in the gap between the sensor and baffle cause a change in output pressure, which is fed through to the opposing bellows [5] placed between the intermediate assembly and the outer casing [3]. Variation in pressure in the bellows causes them to expand or contract and hence displace the central baffle to which the assembly is attached, in the same direction as the applied force.

The unit has sucessfully been used with a pneumatic robot having a repeatability of 0.83mm to assembly components with a 0.06mm clearance. It has also been used on a vertical station to assemble a gearpump shaft into a housing having a clearance of 0.03mm.

The limitation of this device is that it can only be used for cylindrical, chamfered assemblies. Also, at present it is too large for inclusion on a robot wrist. However, the unit is currently being redesigned to increases its sensitivity whilst reducing hysterisis, size, and weight.

Electric Device

One electric device developed consists of a three degree strain gauge sensor and two degrees of stepper motor actuation. The sensor outputs are processed in a micro-computer which then drives an X-Y table. Insertion is performed either directly by the robot or the assembly station. The combined system may be either completely attached to, for example a robot, or may if more convienient be separated into sensory and actuation components. Fig. 2 shows schematically one configuration of the system on an experimental assembly station.

The sensor consists of a flat plate in the shape of a cross. Onto each arm of the cross are mounted four strain gauges connected into a bridge. The output of each bridge is connected to an interface card where the appropriate signal conditioning and A→D convertion takes place. The four values are then available to be read by the micro-computer and used in the appropriate control strategy.

Routines have been developed for cylindrical chamfered components where the misalignment may be on or outside the chamfer. Selection of the appropriate routine is made by the computer upon measurement of the contact forces.

Chamfered assembly is accomplished by determining a force vector the direction of which is used to eliminate the misalignment. Non-chamfered assembly is achieved by implementing a spiral search pattern. Vectored non-chamfered assembly has been shown to be possible but requires ideal conditions and a carefully constructed sensor having a high sensitivity to bending moments. A new sensor is presently under evaluation and in early tests has reliably assembled non-chamfered components.

Conclusions

Smart sensors will be used in the assembly environment when they are simple, cheap and fast acting. Speed of operation is paramount when the desired assembly time is often under one second.

Misalignments occuring during the assembly of cylindrical components can be corrected using relatively simple devices whether or not initial contact is made on a chamfer. However, there is much more work to be carried out to compensate for misalignments on prismatic components.

Future work will consist of building up a data base from industrial assemblies, and using this to extract the essential assembly features. Strategies will then be written for each of these features individually and then will be combined to test the final result back on actual industrial assemblies.

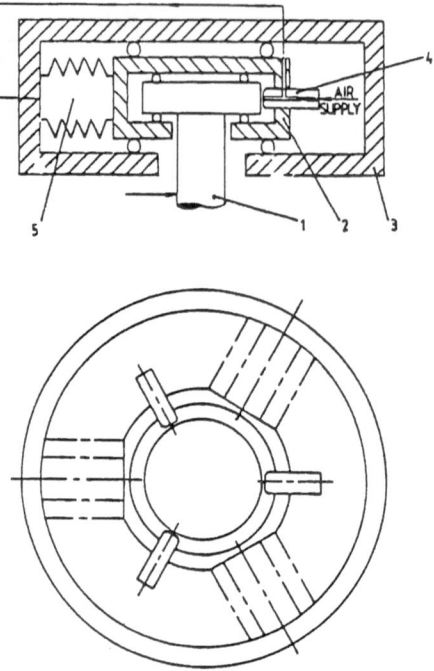

Fig1 Pneumatic Device

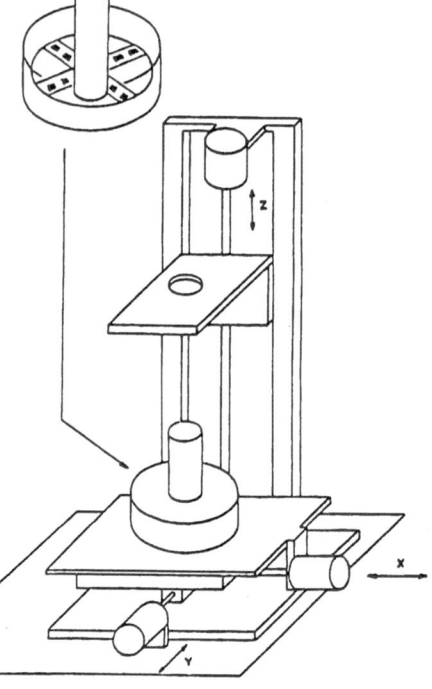

Fig2 Experimental Station

LIST OF PARTICIPANTS

Authors are indicated by an asterisk

* J. Albus
  National Bureau of Standards
  Building 220 - Room B124
  Gaithersburg, MD 20899, U.S.A.

  J.D.C. Allan
  Department of Engineering Science
  University of Oxford
  Parks Road
  OX1 3PJ Oxford
  U.K.

* R. Bajcsy
  Computer and Information
  Science Department
  University of Pennsylvania
  Philadelphia, PA 19104, U.S.A.

  A. Balestrino
  Istituto di Elettrotecnica
  Facoltà di Ingegneria
  Università di Pisa
  Via Diotisalvi, 2
  56100 Pisa, Italy

  A. Bejczy
  Robotics and Teleoperator Group
  Automated Systems Section
  Mail Sto198-330
  Jet Propulsion Laboratory
  4800 Oak Grove Dr.
  Pasadena, CA 91109, U.S.A.

* G. Beni
  Department of Electrical
  and Computer Engineering
  University of California
  Santa Barbara, CA.93106, U.S.A.

M. Bergamasco
Centro "E. Piaggio"
Facoltà di Ingegneria
Università di Pisa
Via Diotisalvi, 2
56100 Pisa, Italy

* P. Bergveld
Department of Electrical Engineering
Twente University of Technology
P.O. Box 217
7500 AE Enschede, The Netherlands

* L. Bjørnø
Industrial AcousticsLaboratory
Institute of Manufacturing Engineering
Technical University of Denmark
Building 352
DK-2800 Lyngby, Denmark

* C. Bland
Department of Mechanical and
Manufacturing Systems Engineering
University of Wales
Institute of Science and Technology
Bute Building-King Edward VII Avenue
Cardiff, U.K.

L. Bologni
Istituto di Progetti di Macchine
e Tecnologie Meccaniche
Facoltà di Ingegneria
Università di Bologna
Viale Risorgimento, 2
40136 Bologna, Italy

* M. Brady
Department of Engineering Science
University of Oxford
Parks Road
Oxford OX1 3PJ, U.K.

D.L. Brock
M.I.T.
A.I. Laboratory
545 Technology Square
02139 Cambridge, MA, U.S.A.

S. Caselli
Dipartimento di Elettronica,
Informatica e Sistemistica
Facoltà di Ingegneria
Viale Risorgimento, 2
40136 Bologna, Italy

* H. Clergeot
ENSET
61, Avenue du Président Wilson
94230 Cachan, France

* B. Culshaw
University of Strathclyde
Royal College Building
204 George Street
Glasgow G1 1XW, U.K.

* P. Dario
Scuola Superiore di Studi
Universitari e di Perfezionamento
"S. Anna"
Via Carducci, 40
56100 Pisa
and
Centro "E. Piaggio"
Facoltà di Ingegneria
Università di Pisa
Via Diotisalvi, 2
56100 - Pisa, Italy

I. De Lotto
Dipartimento di Informatica
e Sistemistica
Strada Nuova 106/C
27100 Pavia, Italy

Y. Denizhan
Department of Electrical and
Electronic Engineering
Bogazici University
P.K. 2, Bebek
Istanbul, Turkey

* D. De Rossi
Centro "E. Piaggio"
Facoltà di Ingegneria
Università di Pisa
Via Diotisalvi, 2
56100 Pisa, Italy

C. Domenici
Centro "E. Piaggio"
Facoltà di Ingegneria
Università di Pisa
Via Diotisalvi, 2
56100 Pisa, Italy

* G. Drunk
IPA
Nobelstraße 12
D-7000 Stuttgart 80, West Germany

A. Ersak
Department of Electrical and
Electronic Engineering
Middle East Technical University
Ankara, Turkey

* B. Espiau
IRISA
Campus Universitaire de Beaulieu
Avenue du Général Leclerc
35042 Rennes Cedex, France

A. S. Fiorillo
Centro "E. Piaggio"
Facoltà di Ingegneria
Università di Pisa
Via Diotisalvi, 2
56100 Pisa, Italy

V. Gerbig
SIEMENS AG
Forschungszentrum
D-8520 Erlangen, W. Germany

I. Gibson
Robotics Research Unit
Department of Electronic Eng.
Hull University
Cottingham Road, Hull
Nth Humberside, HU1 1HA, U.K.

R. Golini
IBM Italia S.p.A.
Via Ardeatine, 2491
00040 Santa Palomba (Roma), Italy

M. Gündüzalp
Dokuz Eylül University
Electrical and Electronic
Engineering Department
Bornova-Izmir, Turkey

* G. Hirzinger
DFVLR
Institut für Dynamik der Flugsysteme
Oberpfaffenhofen
D-8031 Weßling, W. Germany

* M. Hulliger
Institut für Hirnforschung
der Universität Zürich
August-Forel-Straße 1
CH-8029 Zürich, Switzerland

* **B.V. Jayawant**
The University of Sussex
School of Engineering and Applied Sciences
Falmer
Brighton, Sussex BN1 9QT, U. K.

* **W. Jüptner**
BIAS
59, Ermland Straße
2820 Bremen 71, W. Germany

O. M. Kaynak
Bogazici University
Department of Electrical and
Electronic Engineering
P.K.2 Bebek
Istanbul, Turkey

* **W. H. Ko**
Electronics Design Center
Case Western Reserve University
Bingham Bldg.
Cleveland, OH 44106, U.S.A.

A. Kurtoglu
Royal Institute of Technology
Department of Manufacturing System
S-100 44 Stockholm, Sweden

* **S. J. Lederman**
Department of Psychology
Queen's University
Kingston, Ontario K7L 3N6, Canada

* **J. F. Martin**
Science Center
Rockwell International Corporation
1049 Camino Dos Rios
P.O. Box 1085
Thousand Oaks, CA 91360, U.S.A.

P. Mataloni
Divisione COMB/CIVAL
ENEA CRE Trisala
Casella Postale 1
75025 - Policaro (MT), Italy

* S. Middelhoek
Delft University of Technology
Department of Electrical Engineering
P.O. Box 5031
2600 GA Delft, The Netherlands

* P. Morasso
Dipartimento di Informatica,
Sistemistica e Telematica
Università di Genova
Via Opera Pia, 11A
16145 Genova, Italy

G. Papadopoulos
Applied of Electronics Laboratory
School of Engineering
University of Patras
Patras, Greece

* P. Pelosi
Istituto di Industrie Agrarie
Università di Pisa
Via S. Michele degli Scalzi, 4
56100 Pisa, Italy

* K. C. Persaud
Department of Physiology and Biophysics
P.O. Box 551
Medical College of Virginia
Richmond, VA 23298, U.S.A.

* P.P.L. Regtien
Department of Electrical Engineering
Delft University of Technology
P.O. Box 5031
2600 GA Delft, The Netherlands

L.F. Requicha Ferreira
Departamento de Fisica
Universidade de Coimbra
3000 Coimbra, Portugal

A. Russell
Department of Electrical and
Computer Engineering
The University of Wollongong
P.O. Box 1144
Wollongong, N.S.W. 2500, Australia

* G. Sandini
Dipartimento di Informatica,
Sistemistica e Telematica
Università di Genova
Via Opera Pia, 11A
16145 Genova, Italy

M. Savini
Dipartimento di Informatica
e Sistemistica
Università di Pavia
Via Abbiategrasso, 209
27100 Pavia, Italy

L. Sciavicco
Dipartimento di Informatica
e Sistemistica
Università di Napoli
Via Claudio, 21
80125 Napoli, Italy

* M. Somalvico
Dipartimento di Elettronica
Politecnico di Milano
Piazza Leonardo da Vinci, 32
20133 Milano, Italy

S. Stansfield
Computer and Information
Science Department
University of Pennsylvania
Philadelphia, PA 19104, U.S.A.

J.A. Tenreiro Machado
Universidade do Porto
Faculdade de Engenharia
Departamento de Engenharia
Electrotécnica
Rua dos Bragas
4099 Porto, Portugal

* R.F. Wolffenbuttel
Delft University of Technology
Department of Electrical Engineering
P.O. Box 5031
2600 GA Delft, Netherlands

M. Wybrow
Department of Production Engineering and
Production Management
The University of Nottingham
Nottingham, U.K.

NATO ASI Series F

NATO ASI Series F

NATO ASI Series F